Microbial Remediation of Azo Dyes with Prokaryotes

Microbial Remediation of Azo Dyes with Prokaryotes

Edited by
Maulin P Shah

CRC Press
Taylor & Francis Group
Boca Raton London New York

CRC Press is an imprint of the
Taylor & Francis Group, an informa business

First edition published 2022
by CRC Press
6000 Broken Sound Parkway NW, Suite 300, Boca Raton, FL 33487-2742

and by CRC Press
2 Park Square, Milton Park, Abingdon, Oxon, OX14 4RN

ISBN: 9780367673451 (hbk)
ISBN: 9780367673468 (pbk)
ISBN: 9781003130932 (ebk)

DOI: 10.1201/9781003130932

Typeset in Times
by Newgen Publishing UK

Contents

Contributors

Komal Agrawal
Bioprocess and Bioenergy Laboratory,
Department of Microbiology, Central
University of Rajasthan, Bandarsindri,
Kishangarh, Ajmer Rajasthan, 305817, India

Moazam Ali
Department of Clinical Medicine and Surgery,
University of Agriculture Faisalabad, Pakistan

Ambreen Ashar
Department of Chemistry, University of
Agriculture, Faisalabad, 38000, Pakistan

Somok Banerjee
Department of Microbiology, Tripura
University (A Central University),
Suryamaninagar, Agartala, Tripura, India

Juliana Barden Schallemberger
Department of Sanitary and Environmental
Engineering, Federal University of
Santa Catarina, UFSC, 88040-970,
Florianópolis, Brazil

Tanay Bhagwat
Department of Chemical Engineering,
Northeastern University, Boston,
MA02115, USA

Sourish Bhattacharya
Process Design and Engineering Cell, Council
of Scientific and Industrial Research (CSIR),
Central Salt and Marine Chemicals Research
Institute, Bhavnagar, 364002, India

Zeeshan Ahmad Bhutta
Laboratory of Biochemistry and Immunology,
College of Veterinary Medicine, Chungbuk
National University, Cheongju, Chungbuk,
28644, Republic of Korea

Tanushree Chaterjee
Raipur Institute of Technology Raipur (C.G),
India 491102

Indranil Chattopadhyay
Department of Life Sciences, Central
University of Tamil Nadu, Thiruvarur,
Tamil Nadu, India

Ajay Kumar Chauhan
Department of Biological Science and
Bioengineering Indian Institute of Technology,
Roorkee, Uttarakhand 24667, India

Jyotsna Choubey
Chhattisgarh Swami Vivekanand Technical
University, Bhilai (C.G), India 491107
Raipur Institute of Technology Raipur (C.G),
India 491102

Jyoti Kant Choudhari
Chhattisgarh Swami Vivekanand Technical
University, Bhilai (C.G), India 491107

Bijan Choudhury
Department of Biological Science
and Bioengineering Indian Institute of
Technology, Roorkee,
Uttarakhand 24667, India

Praveen Dahiya
Amity Institute of Biotechnology,
Amity University Uttar Pradesh, Gautam
Buddha Nagar, Sector-125, Noida,
Uttar Pradesh, India

Sagar Daki
Department of Microbiology, Parul Institute
of Applied Sciences, Parul University,
Vadodara, Gujarat, India

Khac-Uan Do
School of Environmental Science and
Technology, Hanoi University of Science
and Technology, Hanoi, Vietnam

Sougata Ghosh
Department of Microbiology, School of
Science, RK. University, Rajkot 360020,
Gujarat, India
Department of Chemical Engineering,
Northeastern University, Boston,
MA02115, USA

Vandana Gupta
Department of Chemical Engineering,
National Institute of Technology Raipur,
Raipur, 492010, Chhattisgarh, India

Md. Milon Hossain
Department of Textile Engineering, Khulna
University of Engineering and Technology,
Khulna 9203, Bangladesh

Santosh Kumar Jha
Department of Bioengineering, Birla Institute
of Technology, Mesra, Ranchi-835215,
Jharkhand, India

Priyanka H. Jokhakar
Department of Biosciences, Veer Narmad
South Gujarat University, Surat (Gujarat)
395007, India

Rutu R. Kachhadiya
School of Science, P. P. Savani University,
Surat (Gujarat) 394125, India

Rishee K. Kalaria
ASPEE Shakilam Biotechnology Institute,
Navsari Agricultural University, Surat (Gujarat)
395007, India

Md. Rezaul Karim
Department of Textile Engineering, Port City
International University, Chittagong 4225,
Bangladesh

Ashutosh Kumar
Department of Microbiology,
Tripura University (A Central University),
Suryamaninagar, Agartala, Tripura,
India

Ishani Lahiri
Department of Bioengineering, Birla Institute
of Technology, Mesra, Ranchi-835215,
Jharkhand, India

Nelson Libardi
Department of Sanitary and Environmental
Engineering, Federal University of
Santa Catarina, UFSC, 88040-970,
Florianópolis, Brazil

Devendra Mohan
Department of Civil Engineering, IIT (BHU)
Varanasi, U.P., India

BM Muhilan
Department of Life Sciences, Central
University of Tamil Nadu, Thiruvarur, Tamil
Nadu, India

Huma Munir
Department of Medical Laboratory Technologist,
College of Allied Health Professionals, Faculty
of Medical Sciences, Government College
University Faisalabad, 38000, Pakistan

Iqra Muzammil
Department of Veterinary Medicine, University
of Veterinary and Animal Sciences, Lahore,
54000, Pakistan

Aditi Nag
Dr. B. Lal Institute of Biotechnology, Jaipur,
Rajasthan, India

Maria Eliza Nagel Hassemer
Department of Sanitary and Environmental
Engineering, Federal University of
Santa Catarina, UFSC, 88040-970,
Florianópolis, Brazil

Shubhangi Parmar
Department of Microbiology, Parul Institute of
Applied Sciences, Parul University, Vadodara,
Gujarat, India

Hiren K. Patel
School of Science, P. P. Savani University,
Surat (Gujarat) 394125, India

Swatilekha Pati
Department of Microbiology, Tripura
University (A Central University),
Suryamaninagar, Agartala, Tripura, India

Refat Pervin
Department of Textile Engineering, Khulna
University of Engineering and Technology,
Khulna 9203, Bangladesh

Luciana Porto de Souza Vandenberghe
Department of Bioprocess Engineering and
Biotechnology, Federal University of Paraná,
UFPR, 81531-980, Curitiba, Brazil

Yashsvi Raval
Department of Microbiology, Parul Institute of
Applied Sciences, Parul University, Vadodara,
Gujarat, India

Mayur R. Raviya
Reverse Osmosis Department, Central Salt and
Marine Chemicals Research Institute, CSIR,
Gijubhai Badheka Marg, Bhavnagar, 364002,
Gujarat, India

Rejane Helena Ribeiro da Costa
Department of Sanitary and Environmental
Engineering, Federal University of
Santa Catarina, UFSC, 88040-970,
Florianópolis, Brazil

Biju Prava Sahariah
Chhattisgarh Swami Vivekanand Technical
University, Bhilai (C.G), India 491107

Aveepsa Sengupta
Department of Microbiology, Tripura
University (A Central University),
Suryamaninagar, Agartala, Tripura, India

Sonali Sharma
Dr. B. Lal Institute of Biotechnology, Jaipur,
Rajasthan, India

Muhammad Shoaib
Institute of Microbiology, Faculty of Veterinary
Science, University of Agriculture Faisalabad,
38000, Pakistan

Key Laboratory of New Animal Drug Project,
Gansu Province, Key Laboratory of Veterinary
Pharmaceutical Development, Ministry of
Agriculture and Rural Affairs, Lanzhou
Institute of Husbandry

Pharmaceutical Sciences of Chinese Academy
of Agriculture Sciences, PR, China

Anupama Shrivastav
Department of Microbiology, Parul Institute of
Applied Sciences, Parul University, Vadodara,
Gujarat, India

Sushil Kumar Shukla
Department of Transport Science and
Technology, Central University of Jharkhand,
Brambe, Ranchi-835205, Jharkhand, India

Hare Ram Singh
Department of Bioengineering, Birla Institute
of Technology, Mesra, Ranchi-835215,
Jharkhand, India

R.S. Singh
Department of Chemical Engineering and
Technology, IIT (BHU) Varanasi, U.P., India

Carlos Ricardo Soccol
Department of Bioprocess Engineering and
Biotechnology, Federal University of Paraná,
UFPR, 81531-980, Curitiba, Brazil

Palla Mary Sulakshana
Raghu College of Pharmacy, Visakhapatnam,
Andhra Pradesh, India

Anandkumar J. Sweta Singh
Chhattisgarh Swami Vivekanand Technical
University, Bhilai, Chhattisgarh, India 491107
National Institute of Technology Raipur,
Raipur, Chhattisgarh, India 492010

Pushpa C. Tomar
Department of Biotechnology, Faculty of
Engineering and Technology, Manav Rachna
International Institute of Research and Studies,
Faridabad, Haryana, India

Pranjal Tripathi
Department of Chemical Engineering and
Technology, IIT (BHU) Varanasi, U.P., India

Divyesh K. Vasava
College of Agriculture, Junagadh Agricultural
University, Junagadh (Gujarat) 36200, India

Mukesh Kumar Verma
Chhattisgarh Swami Vivekanand Technical
University, Bhilai (C.G), India 491107
National Institute of Technology Raipur (C.G),
India 492010

Pradeep Verma
Bioprocess and Bioenergy Laboratory,
Department of Microbiology, Central
University of Rajasthan, Bandarsindri,
Kishangarh, Ajmer Rajasthan, 305817, India

Ngoc-Thuy Vu
School of Environmental Science and
Technology, Hanoi University of Science and
Technology, Hanoi, Vietnam

Thomas J. Webster
Department of Chemical Engineering,
Northeastern University, Boston,
MA02115, USA

Preface

Azo dyes represent the largest and most versatile class of synthetic dyes. Biodegradation is a cost-effective method for the removal of the residues from azo dyes before their release from dyeing product industries into wastewater. The effectiveness of this method of treatment strongly depends on the establishment of an effective degrading community and by the environmental conditions that support the growth and activity of the degradation organisms. Activated sludge is commonly used as a source of degradation organisms to initiate the process; however, the bioaugmentation of wastewater with efficient strains provides a much more reliable process. Therefore, the process manager could use bacterial strains that target specific chemicals and metabolites from dyes for complete mineralization. The most effective inoculants can degrade a wide concentration range of dyes, tolerate a range of environmental temperature conditions, pH and salinity, and persist at high population densities when competing with other microorganisms in microbial mixed cultures. Over the last decades, biological discoloration has been used for the transformation, degradation, or mineralization of azo dyes

This type of decolorization and degradation is environmentally friendly and is a cost-competitive alternative to chemical decomposition. Most azo dyes are not suitable to aerobic degradation by bacterial cells. However, there are a limited number of microorganisms that are known to reduce the cleavage of azo bonds under aerobic conditions. Compared with chemical and physical methods, there is increased interest in biological processes due to their cost effectiveness, lower sludge production, and environmental friendliness. Therefore, he tried to eradicate on eco-friendly and cost-effective dye-degrading organism in association with a microbial panel access.

In this book, various types of microorganisms that degrade azo dyes and their potential for bioaugmentation will be discussed.

Maulin P Shah
Editor

Editor's Biography

Maulin P Shah is an active researcher and scientific writer in his field with over 20 years of experience. His research interests include biological wastewater treatment, environmental microbiology, biodegradation, bioremediation, and phytoremediation of environmental pollutants from industrial wastewaters. He has published more than 250 research papers in national and international journals of repute on various aspects of microbial biodegradation and bioremediation of environmental pollutants. He is the editor of 70 books of international repute (Elsevier, Springer, CRC Press, RSC, De Gruyter, and Nova).

1 Azo Dye Degradation Using a Combination of Physicochemical and Biological Processes

Md. Rezaul Karim[1] and Md. Milon Hossain[2]*

[1]Department of Textile Engineering, Port City International University, Chittagong 4225, Bangladesh

[2]Department of Textile Engineering, Khulna University of Engineering and Technology, Khulna 9203, Bangladesh

*Corresponding author: milon.hossain@te.kuet.ac.bd

CONTENTS

DOI: 10.1201/9781003130932-1

1.1 INTRODUCTION

Textile effluents contain an enormous quantity of dyes and chemicals and pollute the water and surrounding environment if they are discharged without treatment. The World Health Organization reported that textile wet processing causes approximately 17%–20% of industrial water pollution [1]. Different groups of synthetic dyes have unique chemical structures, for example, azo, anthraquinone, phthalocyanine, oxazine, thiazine, xanthene, nitro, nitroso, triarylmethane, diarylmethane, indigoid, azine, methine, thiazole, indamine, indophenol, lactone, aminoketone, and hydroxyketone stilbenes. These groups are essential to produce colored textiles. Among these, the azo group containing dyes accounts for a 60%–70% share of textile wet processing due to their high colorfastness characteristics, wide color range, good brightness property, high intensity of color, and most importantly their cost-effectiveness. However, the excessive use of azo dyes is detrimental to the environment and has a toxic impact on living organisms.

Approximately 10%–15% of the unfixed dyes are released to the environment through effluents [2]. Among the lost dyes, 80% of the dyes are derivatives of azo groups, which are complex aromatic compounds with significant structural diversity and can be converted into an aromatic amine (an arylamine) anaerobically, which is very toxic, carcinogenic, mutagenic, and recalcitrant for the environment. In addition, these converted amines are toxic for water and increase the salinity in water reservoirs, total organic carbon (TOC), biological oxygen demand (BOD), chemical oxygen demand (COD), and the suspended solids in water by changing the pH and having an adverse effect to the aquatic ecosystem due to the low penetration of light into the water sources and the excessive consumption of oxygen. Toxic and carcinogenic amines are formed from the reductive cleavage of azo bonds and may cause diseases, such as skin irritation, risk of cancer, respiratory problems, renal damage, allergic reactions, gastritis, chromosomal aberrations, hypertension, and many other physiological and dermal disorders for humans and other organisms. For example, 4-phenylazoaniline and N-methyl- and N, N-dimethyl-4-phenylazoanilines are amino-substituted azo dyes known as mutagenic for humans. Some benzidine-based azo dyes can cause hepatocellular carcinoma, splenic sarcoma, and nuclear anomalies in mammalian bodies [3]. In addition to animals, azo dyes affect plants and aquatic systems, such as inhibiting the shoots and roots of plants, growth reduction, neurosensory damage, seed germination, plant growth, metabolic stress and death in fish, and productivity in plants.

Different physicochemical treatment techniques, such as membrane filtration, flocculation and coagulation, reverse osmosis, ion exchange, filtration, adsorption, electrolysis, activated carbon, fenton process, ozonation, and advanced oxidation process (AOPs) have emerged to treat wastewater that contains azo dyes [4]. In addition, biological treatments, such as aerobic and anaerobic treatments, phytoremediation, and microbes can convert the organic pollutants into nontoxic compounds. Physicochemical and biological processes have some limitations, and therefore, an integrated physicochemical–biological method that is more effective for the removal of toxicity, genotoxicity, mutagenicity, and carcinogenicity for plants and animals, which can remove azo dyes from wastewater is required. This chapter will discuss a combined physicochemical and biological treatment method for the degradation and decolorization of azo structures (–N=N–) in wastewater.

1.2 AZO CHEMISTRY AND CLASSIFICATION OF AZO DYES

1.2.1 Sources

In 1858, the German industrial chemist Johann Peter Griess discovered diazo compounds. Then, in 1863, Aniline Yellow (4-aminoazobenzene) was the first commercial azo dye to be produced in England [5]. The word azo comes from azote, which is the French name for nitrogen. Azo dyes have ≥1 azo group (–N=N–) and sulfonic SO_3^- groups in their aromatic structure, where ≥1 is an aromatic nucleus. The azo group, which usually establishes covalent bonds with textile substrates, can produce clear and strong colors for colored cotton, leather, cosmetics, and food.

1.2.2 Formation

The hues of the azo dyes are regulated by the azo bonds, chromophores, and auxochromes that are linked together. In addition, the chemical structure of azo dyes includes auxochrome, chromophore, and solubilizing groups (Figure 1.1).

Previously, azo dyes were produced via the treatment of arylamines in an acid solution. Therefore, part of the base was diazotized, and the residues were considered as the coupling components [5]. The production of azo dyes is based on the coupling of diazonium compounds with phenols, arylamines, pyrazolones, and naphthols, to provide hydroxy-azo or amino-azo compounds or their tautomeric analogs. Therefore, in azo dye structures, the azo groups are the chromophore groups, and the hydroxyl (OH) or amino groups are the auxochrome groups.

1.2.3 Classifications

Based on different chemical structures, there are a variety of azo dyes. A monoazo structure has one N=N double bond, a diazo has two, and triazo and polyazo dyes have three and more than three N=N double bonds, respectively (Figure 1.2). In addition, based on the reactive functional groups, azo dyes have been categorized into many subgroups (Figure 1.3).

1.2.3.1 Classification of Azo Dyes Based on Azo Bonds
1.2.3.2 Classification of Azo Dyes Based on Reactive Functional Groups

Based on the reactivity of the functional groups, the most reactive groups are dichlorotriazines, di- or tetrafluoropyrimidines, dichloroquinoxalines, and monofluorotriazines. Then, monochlorotriazines or vinyl sulfones are the medium reactive groups and low reactivity is aminochlorotriazines. Finally, the trichloropyrimidines are in the lowest reactive groups. These azo groups are linked with phenyl

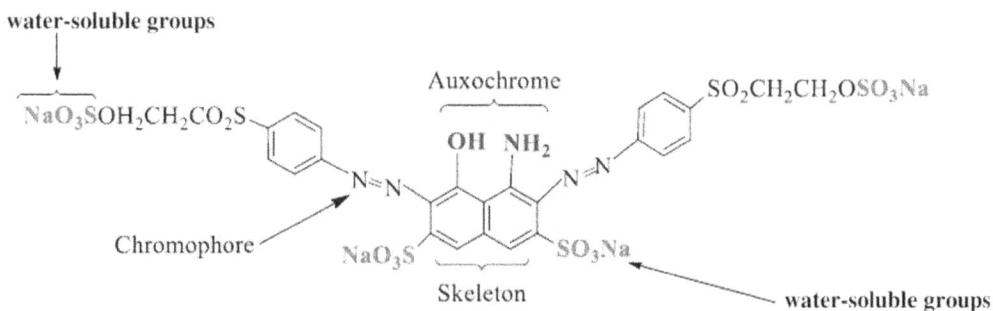

FIGURE 1.1 Azo dye structure [6].

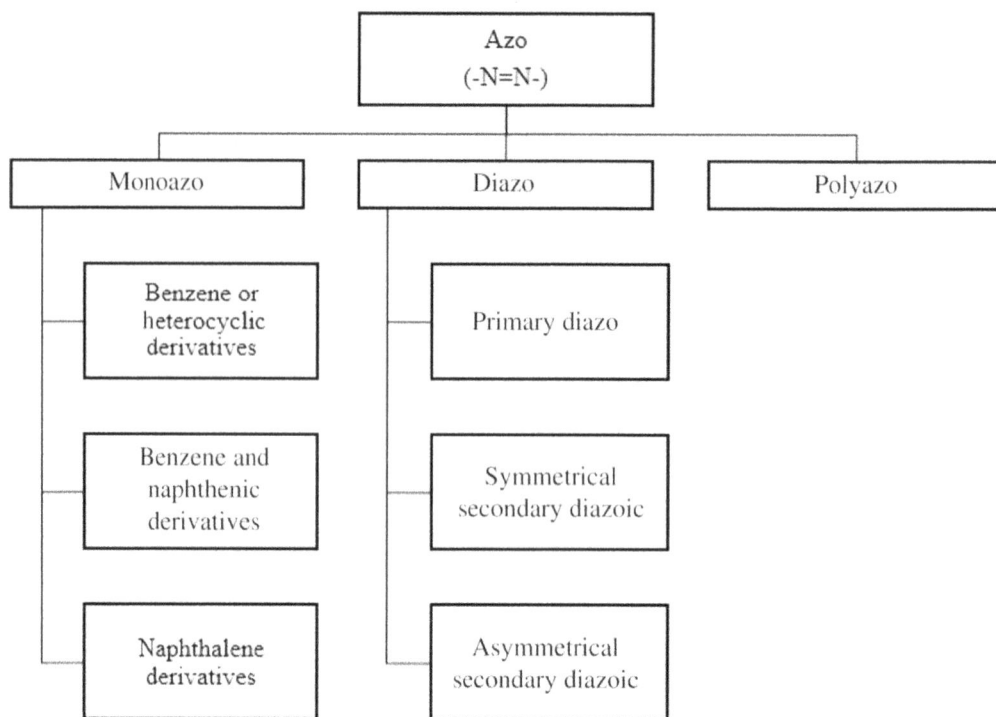

FIGURE 1.2	Different types of azo dyes according to azo bonds.

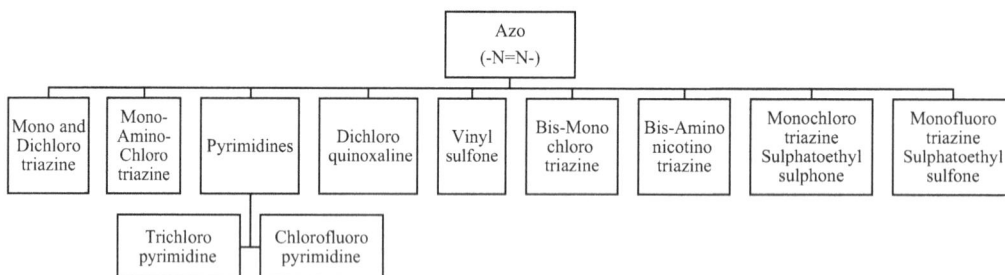

FIGURE 1.3	Different types of azo dyes based on reactive functional groups.

and naphthyl rings, which impart the color of the dye with different shades and intensities and are usually substituted with OH, methyl, triazine amine, chloro, nitro, and sulfonate groups [7].

1.3 ADVERSE EFFECTS OF AZO COMPOUNDS

The azo groups in azo dyes bond with an aromatic ring and via mineralization the dye molecules can be broken down to form aromatic amines. Aromatic amines are created by the reduction of an azo dye and are oxidized into reactive electrophilic species. These species bind to DNA via covalent bonds, which is an irreversible process. The different impacts of azo dyes will be discussed in the following sections.

1.3.1 Effects on Water

The exhausted effluents from textile industries have a variety of dyes and toxic chemicals that are exacerbated the pH, BOD, and COD in water resources, which create a disturbance and imbalance in the organic and inorganic chemical elements in the environment and cleave the biotic community in the water [8]. In addition, the wastewater that is mixed with azo dyes and released into the environment causes the pollution of natural water resources, because it creates problems, such as coloration, absorption of water, and the reflection of sunlight that falls on the water, which cannot penetrate the deep water. Therefore, the growth of bacteria, flora, shrubs, plants, and other living organisms deteriorates, and because of the mutagenicity and carcinogenicity of the dyes, it disturbs the ecological functions of the water. The mixture of these toxic substances in the water can easily penetrate our food chain and create sporadic fevers, renal damage, cramps, and other physiological disorders [8].

1.3.2 Effects on Health

1,4-diamino benzene or 1,4-phenylenediamine, which is an aromatic amine, and a major component of azo dyes contains toxic substances and is responsible for dermatological problems, such as skin irritation, contact dermatitis, chemosis, lacrimation, exophthalmos, and permanent blindness. In addition, this amine is known as para-phenylenediamine (PPD). Products that are manufactured from PPD can cause diseases, such as, rhabdomyolysis, edema in the face, neck, pharynx, tongue, and larynx, acute tubular necrosis, respiratory distress, vomiting gastritis, and hypertension [8]. Benzidine-based azo dyes have been banned, because they cause bladder and urinary problems in humans, for example, 1-amino-2-naphthol is produced from the reduction of Acid Orange 7, which is responsible for bladder tumors [9]. Malachite Green, a multiorgan toxin, decreases food intake, growth, fertility rates, and damages the liver, spleen, kidneys, and heart [8].

1.4 PHYSICOCHEMICAL PROCESSES TO REMOVE AZO DYES FROM WASTEWATER

The removal of synthetic dyes is difficult using separate conventional processes, such as physical, chemical, or biological treatments. Therefore, various physicochemical processes have been developed to treat wastewater that contains azo dyes as discussed in the following sections.

1.4.1 Membrane Filtration

Membrane filtration has been widely utilized for the treatment of secondary and tertiary wastewater because it can ensure a high degree of separation by filtering the particles that pass through the membrane. The general applications for membranes include the separation of fluids, dissolved, and suspended solids. The main feature of the membrane processes is to remove any harmful components that are suspended or dissolved in a colloidal form. The chemical formation of effluents and the required temperature for a particular procedure determines the type and porosity of the filters. First, during membrane filtration, clean and clear water is produced by the concentration of the dyestuffs and chemicals. Then, the dyestuff is removed, reused, or recovered [10].

A few types of pressure-driven techniques have been used for oily wastewater treatment. Among them, ultrafiltration (UF), nanofiltration (NF), and reverse osmosis are widely used (Figure 1.4).

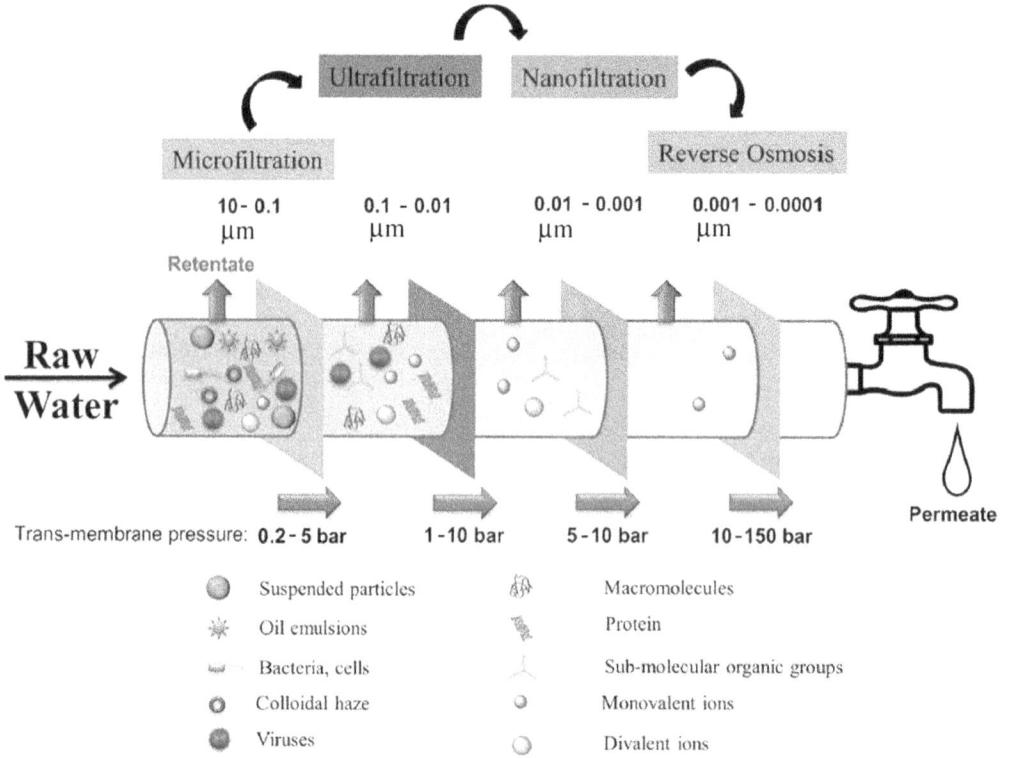

FIGURE 1.4 Pressure-driven membrane processes [11].

1.4.2 COAGULATION AND FLOCCULATION

1.4.2.1 Overview

Coagulation and flocculation are common wastewater treatment processes and by destabilizing and creating flocs, colloidal particles, finer solid suspensions, and soluble compounds can be removed. In addition, it can decolorize textile wastewaters and reduce the total load of suspended and organic pollutants in the effluents. It can remove approximately 70%–80% of the colorants and can significantly reduce the concentration of organic substances. Apart from anaerobic reduction, oxidation, and adsorption, it is the most effective procedure to minimize the aggregation of pollutants in effluents.

In general, coagulation and flocculation processes isolate the suspended solids in the water. Variations in the suspended solids largely depend on their origin, composition charge, size of particles, formation, and density. The relationship between these characteristics determines the selection of the coagulants when coagulation and flocculation are applied. The primary coagulants and coagulant aids are two types of coagulants. The electrical charges of the particles in the water that cause them to aggregate are neutralized by the primary coagulants. In addition, the coagulant aids add density to slow-settling flocs and strengthen the flocs to prevent damage during mixing and settling. To optimize the forces that stabilize the suspended substances and to allow particles to collide and floc growth, coagulation and flocculation occur consecutively [12].

The quality of the raw wastewater, pH, temperature of the solution, coagulant types and doses, and the intensity and duration of mixing determine the efficiency of the coagulation and flocculation method. Coagulation and flocculation that are followed by sedimentation, flotation, and filtration can remove the surfactants and dyes that have high molecular weights. However, during this process, sludge forms, which creates a problem coagulants are being used. In electrocoagulation systems, the efficiency of the electrode cell generally decreases over time because of the formation of an impenetrable layer of oxides on the cathode. Sometimes, this is required for high conductivity in the wastewater suspension. The sacrificial anode needs to be regularly replaced, to prevent dissolution in the wastewater due to oxidation. In a few cases, the use of electricity might increase the overall expenses [13].

1.4.3 ION EXCHANGE

To purify the water, separate the contaminants, and decontaminate solutions that contain ions, ion exchange methods have been applied for many years to treat water [14].

The interchange of different ions between two electrolytes or between an electrolytic solution and an ion is the main mechanism in this method. During ion exchange, the ambulant ions from an outermost solution are interchanged with electrostatic ions that are attached to the solid matrix that contains functional groups. A variety of ion exchange processes are available for wastewater treatment [13]. Natural ion exchangers involve natural inorganic ion exchangers and natural organic ion exchangers. Synthetic ion exchangers are produced by forming chemical substances that have good physical and chemical characteristics and can be either inorganic or organic, for instance, polystyrene divinylbenzene. Composite ion exchangers contain ≥ 1 ion exchangers that are incorporated with another compound, and this can be organic or inorganic [13].

Electrochemical ion exchange is a process of segregation that includes electrodialysis and traditional ion exchange, which integrates the advantages of ion exchange with an electric driving force [13]. During this integration, ion exchange is used to adsorb the ionic contents in the solution, and the electrical force is used to amplify the adsorption and elution reactions.

The waste from ion exchange is highly concentrated and the disposal of a used ion exchanger requires careful attention. The resins for ion exchange are dye-specific and their regeneration is a very expensive process.

1.4.4 ADSORPTION

Adsorption is an effective technique to remove dye particles from effluents due to its ability to thoroughly separate a variety of dyes from wastewater. In addition, it is easy to operate and involves reasonable costs and effectiveness. In this process, the molecules in the gaseous or liquid state concentrate on the surface without any association. Compounds that are in two phases, such as liquid phase or liquid–gas phase or gas–solid phase or solid–liquid phase can be aggregated in this process [15].

The efficiency of adsorption relies on the physical and chemical characteristics of the adsorbents and adsorbates. The adsorbate should be the substance that is adsorbed, and the adsorbent is the adsorbing substance. The adsorbent is a type of matter that has porosity with a high surface area and via intermolecular forces, it can adsorb material onto its surface. There are different types of adsorbents, such as round pellets, rods, moldings, and monoliths that have diameters between 0.5 and 10 mm and are hydrodynamic [16]. Activated carbon (powder/granular) has been extensively used as an adsorbent due to its large, expanded surface area, microporous structure, high adsorption capacity, and high degree of surface reactivity [10]. Preferences for the selection of adsorbents are nontoxicity, availability, renewability, recoverability in nature, low sludge generation, adsorption capacity, surface area, total cost, concentration, and type of microstructure.

Due to the low initial cost, simple design, ease of operation, and the use of harmless substances, adsorption is an effective segregation method. The important factors for adsorption are surface affinity, pH, surface area, chemical reactivity, and reduction in surface tension. In addition, dye adsorption is mainly dependent on the contact time, pH dose of adsorbent, and initial concentration of the dye. With an increase in contact time, the rate of dye removal increases.

1.4.5 AOPs

An AOP is designed to produce hydroxyl free (OH) radicals and uses them as a powerful oxidizing agent to break the particles that cannot be oxidized by traditional agents. Highly reactive groups, such as OH radicals that have an enduring oxidative potential are mainly responsible for the AOPs. A wide variety of organic contaminants are swiftly oxidized by OH radicals [10].

Because it is a recommended method to separate organic substances, in wastewater treatment plants AOPs are widely used to treat wastewater that contains surfactants, pesticides, colorants, waste from pharmaceuticals, and many other hazardous chemicals. The usual applications for AOPs cover various processes, such as Fenton oxidation, ozonation, photolysis with hydrogen peroxide (H_2O_2) and ozone (O_3), photochemical oxidation, electrochemical oxidation, corona process, titanium dioxide (TiO_2) photolysis, radiolysis, wet oxidation, and electronic beaming [17]. In AOPs, numerous reactive free radicals are produced and during decolorization this can exceed the number of conventional oxidizing agents. Total mineralization might not be achieved during this process; however, a large number of toxic substances can be produced. Many pH-dependent processes are negatively affected by the radical scavengers that decrease efficiency [18]. Different wastewater treatment processes that involve AOPs are discussed in the following sections.

1.4.5.1 Fenton Oxidation

Fenton's reagent is a catalyst that consists of H_2O_2 with ferrous iron; H_2O_2/Fe^{2+}, which participates in oxidation to oxidize the pollutants in effluents. This process produces hydroxyl ions (OH^-) under acidic conditions via the interaction between Fe^{2+} and H_2O_2. For instance, organic substances in wastewaters, such as trichloroethylene, and tetrachloroethylene can be broken down using Fenton's reagent. Fenton's reagent is widely for the oxidization of contaminants, such as aromatic amines, various dyestuffs, pesticides, surfactants, and insecticides rather than other AOPs because of functional characteristics [19].

The Fenton reaction involves two steps: (1) the reaction between H_2O_2 on the surface of Fe that is caused by the oxidation of the contaminants; and (2) the reaction between the contaminated H_2O_2 and Fe^{2+}, which liquefy on the surface of iron due to the transition to the liquid state [20]. Highly suspended and solid concentrated soluble and insoluble dyes are strongly decolorized by this process. Using this oxidation process, the COD in the water resources can be minimized by excluding the reactive dyes.

1.4.5.2 Ozonation

An eco-friendly, and excellent oxidant, O_3, is used to purify wastewater. In ozonation, the pollutants, colorants, surfactants, microorganisms, and other toxicants are removed, and it does not produce any hazardous chlorinated particles or compounds. O_3 is easily solubilized in water and can rapidly disintegrate. Free radicals are generally formed through this process, which can rapidly react with any type of organic matter, in particular, the molecules from the dyestuffs in the water. Some of the generated radicals are OH, HO_3, HO_4, and superoxide (O_2^-) [21].

1.4.5.3 Ultraviolet Treatments

Different types of ultraviolet (UV) processes, such as O_3/UV, H_2O_2/UV, and $O_3/H_2O_2/UV$ treatments are used to treat wastewater. In this process, UV light, which is an important portion of our visible light spectrum, can be generated from a medium-pressure mercury light source that is wrapped in a quartz sleeve. In general, the wavelength of UV light is from 200 to 280 nm. By applying two different radiations of UV light (150 W, $\lambda=254–578$ nm and 15 W, $\lambda=254$ nm) for 1–3 h, a notable reduction from 47% to 30% was observed in a microbial inhibitory action in exhausted textile effluents [22].

To decolorize the dyestuffs, O_3/UV treatment is more productive than UV oxidation by UV or the ozonation process. In the O_3/UV method, the particular substances break down through UV irradiation absorption and can be degraded by the reaction with OH radicals, which can be formed by the UV photons during O_3 activation. At pH 9, a large amount of COD can be separated using the O_3/UV process. At higher pH values, the H_2O_2/UV process has a rapid response to the scavenging effect of the carbonates. The highly concentrated H_2O_2 functions as a radical scavenger; however, a low concentration of H_2O_2 cannot generate enough OH radicals, which are taken up by the dye particles and produce a slow rate of oxidation. For discoloration, the H_2O_2/UV process is more effective in an acid medium [23]. In addition, via the photolysis of H_2O_2 under UV irradiation, two OH radicals are created that react with the organic pollutants. In contrast, H_2O_2 alone is less effective for oxidization during wastewater treatment. When H_2O_2 is added to the O_3/UV process, the O_3 decomposition accelerates, which results in higher OH⁻ generation. Integrated ozonation and H_2O_2/UV can increase the efficiency of wastewater by from 18% to 27% for COD removal. In the batch dyeing process for polyester and acetate fiber, 99% of COD can be eliminated from wastewater by applying the O_3/H_2O_2/UV integrated treatment for 90 min. In H_2O_2/UV-C treatment, the rate of TOC removal was 14% and the TOC removal rate was 17% for ozonation [23].

1.5 BIOLOGICAL PROCESSES TO REMOVE AZO DYES FROM WASTEWATER

Different biological treatment process has been developed to replace chemically intensive wastewater treatment process. Biological treatment processes are green and eco-friendly and can be performed under aerobic or anaerobic conditions or a combination of both [24].

1.5.1 AEROBIC TREATMENT

The process that includes the biological oxidation of the chemical components of wastewater in the presence of air or oxygen (O_2) is called aerobic treatment. Because of its low cost, this is the most widely used technique to remove BOD and chlorinated organic compounds [25]. To degrade organic wastes, the dissolved O_2 is used by microorganisms (aerobes) [26]. The rate of O_2 supply to the microorganisms should be fast to promote aerobic biochemical reactions, because of the O_2 feed limitation. For the aerobic treatment of wastewater, a contact stabilizer, sequencing batch reactor (SBR) system, microbubble aerator, or trickling filter are widely used [26,27].

1.5.1.1 Aerobic Granulation Technology

To generate granular sludge, biogranulation approaches are used and in the aerobic stage, 88% of the ammonia (NH_3) can be oxidized [28]. Short-time aerobic digestions (STAD) achieve improved flocculation in the sludge. An evaluation was carried out to determine the effect of cocoamidopropyl betaine (CAPB) on STAD. CAPB has a nonpolar linear hydrocarbon group that can generate micelles, which result in higher aqueous solubility. Two types of granular sludge can be generated by biogranulation, such as anaerobic and aerobic granular sludge (AGS). The aerobic granules are almost round in shape, compact, smooth and regular. High biomass retention and excellent settling capacity are found in aerobic granules. The aerobic granules are resistant to simultaneous COD, toxicity, and nitrogen and phosphate removal. Azo dyes and raw industrial wastewater can be treated by the SBR method. For different wastewaters, the maximum removal efficiency varies from 56% to 95%. According to a study, aerobic granular sludge developed in SBRs at temperatures from 30°C–50°C [29]. The efficiency of the dissolved organic carbon and COD removal was >95% in the reactor [26].

1.5.1.2 Biofilm Reactor

A biofilm is the community or clusters of microorganisms that are attached to a surface. The cells in the biofilm have higher capabilities of adjustment and survival in unfavorable conditions.

This is because of the matrix that acts as a barrier and conserves the cells within it from environmental extremes [26]. An extracellular polymeric substance (EPS) is significant for the growth of a biofilm due to its protective mechanism. The impact of the modifications on pH, concentration of toxic substances, and temperature can be minimized by the EPS. There are a variety of types of bioreactors: (1) integrated anaerobic–aerobic fluidized bed reactor; (2) rotating biological contactor; (3) anaerobic–aerobic fixed film bioreactor; (4) aerobic membrane bioreactor (MBR); and (5) moving-bed biofilm reactor (MBBR)

1.5.1.3 Activated Sludge Process (Microbubble Aerator)

A microbubble is a small bubble that has a diameter of 10–60 mm. The important characteristics of microbubble are: (1) fast dissolution rate; (2) long time in the liquid phase; and (3) a large gas–liquid interfacial area. Microbubbles can dissolve the O_2 in the water. Novel flotation, which is a microbubble aeration system, has been proposed to remove fine carbon particles that are suspended in the wastewater. The mechanical moving parts that have a large shear force act on the liquid. To remove the fine iron oxide particles that are suspended in wastewater, microbubble flotation technology has been developed [30].

1.5.2 ANAEROBIC TREATMENT

The fermentation process in which organic compounds are degraded is known as anaerobic treatment [31]. In the last few decades, anaerobic technology has improved and has different treatment processes, in particular, for industrial wastewater treatment due to its higher efficiency [32].

An anaerobic system is a complex and multistep process of degradation where organic materials are degraded into their basic constituents and produce methane gas in the absence of an electron acceptor, such as O_2. First, the anaerobic degradation is carried out with the help of hydrolytic enzymes. Then, the organics are fermented to produce an organic acid and hydrogen H_2 via fermentation by bacteria (acidogens). The organic acids are transformed into H_2 and acetate with the help of the bacteria. Then, methanogens use acetic acid and H_2. The steps of anaerobic digestion are given below (Figure 1.5) [32].

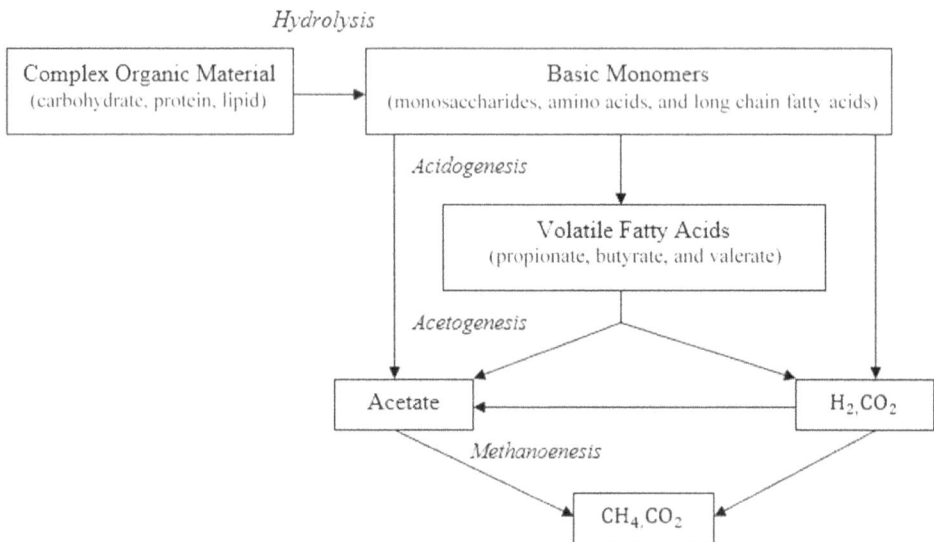

FIGURE 1.5 Anerobic mechanism in wastewater treatment.

1.5.3 DEGRADATION OF AZO DYES BY MICROBES

Bacterial degradation is usually faster than other degradation. The metabolism of azo dyes has been investigated by some researchers, who determined that under aerobic conditions azo dyes were not metabolized and mineralized and they degraded into an intermediate compound. However, it can be completely degraded in a combined aerobic and anaerobic degradation system. The azo bond in azo dyes generates aromatic amines under anaerobic conditions. Under aerobic conditions, it was only mineralized by nonspecific enzymes via ring cleavage. Therefore, combined anaerobic and aerobic treatment could be an effective method for degradation where bacterial strains show good growth in aerobic conditions; however, color removal was achieved in anoxic or anaerobic culture. For better degradation, a mixed bacterial culture could be used [33].

1.6 INTEGRATED PHYSICOCHEMICAL AND BIOLOGICAL METHODS TO REMOVE AZO DYES FROM WASTEWATER

Azo dyes are not susceptible to biodegradation under aerobic conditions where anaerobic treatment can operate; however, aromatic amines are produced, which will be mineralized under aerobic conditions. Therefore, a combination of aerobic and anaerobic treatments is required. Biological treatments are favorable, but their low removal efficiency restricts their applications. Physical methods only separate the toxic molecules from the dye effluents and fail to degrade the contaminants. Unlike physical methods, chemical methods can degrade dye molecules. However, physicochemical treatments have the disadvantages of generating a huge amount of sludge as a secondary pollutant. Therefore, it is necessary to develop hybrid technologies that integrate combined treatment techniques sequentially or simultaneously [34].

1.6.1 INTEGRATED (ELECTRO AND BIOOXIDATION) APPROACH FOR THE REMEDIATION OF INDUSTRIAL WASTEWATER THAT CONTAINS AZO DYES

Integrated electro and biooxidation is a highly effective process among the combined physico-chemical and biodegradation processes for the treatment of effluent that contains azo dyes. The ultimate degradation is achieved through biodegradation followed by removal of the toxic elements by exposure to sunlight. During electrooxidation, aromatic intermediates are produced by the reduction of azo bonds, which are further mineralized by bacterial consortia that convert them into organic acid and aldehydes. Various studies have shown that biological degradation is a slow process and there is the possibility of having toxic elements in the effluent due to the formation of toxic aromatic amines in the intermediate stage. In contrast, electrooxidation is an expensive but rapid process. The integrated electro- and biooxidation process could be a promising degradation method for azo dyes. In this method, the dye effluent is stored at 4°C before electrooxidation is performed for a certain time. Then, the effluent is transferred into an undivided polypropylene cell setting with a titanium, ruthenium, and iridium oxides (Ti/IrO_2–RuO_2–TiO_2) mesh anode and a Ti mesh cathode. Then, an electrolyte solution was added to the electrooxidation cell. By adjusting the anode potential and pH, the electrooxidation process is carried out for the removal of color. An investigation was carried out with different concentrations of NaCl from 1 to 5g/L and the pH was maintained from pH 2 to 8. From the investigation, the maximum efficiency of decolorization was obtained with 3g/L NaCl at pH 5. Electrolyte concentrations >5 g/L did not increase the efficiency rate; however, they increased the toxicity due to the generation of chlorine. The resulting chlorine inhibits the subsequent biodegradation. Then, the solution obtained from the electrochemical oxidation cell was exposed to sunlight and stored by adding a sufficient amount of water, which then evaporated. Then, the resulting dye solution was treated by naphthalene-degrading bacteria, such as *Ochrobactrum intermedium*, *Rhodococcus aetherivorans*, *Microbacterium barkeri*, *Bacillus subtilis*, and *Pseudomonas plecoglossicida*. By following this treatment, the toxic aromatic intermediates present in the dye solution were reduced [35,36].

1.6.2 DEGRADATION OF AZO DYES USING A COMBINED OZONATION AND BIOLOGICAL TREATMENT

Using a combination of ozonation and biological processes, the toxic elements that are produced by ozonation can be eliminated. The resulting toxic elements that are produced during ozonation are degraded during the biological process. In this process, ozonation acts as a pretreatment. In addition, >96% efficiency can be obtained by considering pH values from 3 to 11 at 500 mg/L dye [37].

1.7   CONCLUSIONS

The treatment of textile effluents is complicated because of the high variation in their composition. Textile effluents from dyeing and finishing processes can cause serious problems in the environment due to the presence of various carcinogenic and mutagenic dyes and chemicals. Among those hazardous dyes, the most prominent threat is posed by azo dyes due to their toxicity and slow degradation. Various methods have been employed to degrade the azo dyes contained in wastewaters, such as physicochemical and biological processes and AOPs. Although individual treatments can be effective in removing specific contaminants from wastewater, they are not effective at removing multiple contaminants. In addition, individual treatment processes have some limitations. For example, a biological process is cost-effective and eco-friendly; however, it produces toxic aromatic amines during the intermediate stage that increases certain toxicity and slows down the degradation. In contrast, a physiochemical process is quicker but has higher operating costs and produces a concentrated sludge. Therefore, an integrated method, such as a combination of a physiochemical and biological process could have a promising role in the degradation of azo dyes contained in wastewater. The improper selection of methods might mean that the process is unsuccessful and expensive. Therefore, the combined process should be selected according to the types of contamination and the characteristics of the wastewater.

REFERENCES

1. Adedayo O, Javadpour S, Taylor C, Anderson WA, Moo-Young M. Decolourization and detoxification of methyl red by aerobic bacteria from a wastewater treatment plant. World J of Microb Biot. 2004; 20:545–50.
2. Baban A, Yediler A, Lienert D, Kemerderea N, Kettrup A. Ozonation of high strength segregated effluents from a woollen textile dyeing and finishing plant. Dyes Pigments. 2003; 58(2):93–8.
3. McCann J, Ames BN. Detection of carcinogens as mutagens in the Salmonella/microsome test. Assay of 300 chemicals: discussion. Proc. Natl. Acad. Sci. USA. 1975; 73:950–4.
4. Hossain MM, Mahmud MI, Parvez MS, Cho HM. Impact of current density, operating time and pH of textile wastewater treatment by electrocoagulation process. Environ Eng Res. 2013; 18(3):157–61.
5. Allen RLM. Colour Chemistry, Springer US. 1971. Available from: doi: 10.1007/978-1-4615-6663-2.
6. Benkhaya S et al. Classifications, properties, recent synthesis, and applications of azo dyes. Heliyon. 2020; 6: e03271.
7. Bell J, Plumb JJ, Buckley CA et al. Treatment and decolorization of dyes in an anaerobic baffled reactor. J Environ Eng. 2000; 126(11):1026–32.
8. Sudha M, Saranya A, Selvakumar G, Sivakumar N. Microbial degradation of azo dyes: a review. Int J Curr Microbiol Appl Sci. 2014; 3(2):670–90.
9. Bonser GM, Bradshaw L, Clayson DB. A further study on the carcinogenic properties of ortho-hydroxyamines and related compounds by bladder implantation in the mouse. Br J Cancer. 1956; 10:539–46.

10. Shah M. Effective treatment systems for azo dye degradation: A joint venture between physico-chemical & microbiological process. Int J Environ Bioremed Biodegradation. 2014; 2: (5):231–42.
11. Liao Y, Loh C-H, Tian M, Wang R, Fane AG. Progress in electrospun polymeric nanofibrous membranes for water treatment: Fabrication, modification and applications. Prog Polym Sci. 2018; 77: 69–94.
12. Venkat S, Mane PV Babu V. Evaluation of performance of Coagulation/Flocculation method for the removal of dyes from aqueous solutions. Paper presented at International Conference on Current Trends in Technology, Ahmedabad, India (2011, December 8–10).
13. Sharma S, Saxena R, Gaur G. Study of removal techniques for azo dyes by biosorption: a review. IOSR J. Appl. Chem. 2014; 7:6–21.
14. Dyer A, Hudson HJ, Williams PA. Ion Exchange Processes: Advances and Applications. Cambridge: Royal Society of Chemistry;1993.
15. Mrowetz M, Pirola C, Selli, E. Degradation of organic water pollutants through sonophotocatalysis in the presence of TiO2. Ultrason Sonochem. 2003; 10(247):247–54.
16. Ouslimani N, Boureghda MZM. Removal of directs dyes from wastewater by cotton fiber waste. Int J Waste Res. 2018; 8(2):1–8.
17. Verma M, Ghaly, AE. Treatment of Remazol Brilliant Blue dye effluent by advanced photo oxidation process in TiO_2/UV and H_2O_2/UV reactors. Am J Eng App Sci. 2008; 1(3):230–40.
18. Robinson T, McMullan G, Marchant R, Nigam P. Remediation of dyes in textile effluent: A critical review on current treatment technologies with a proposed alternative. Bio Res Technol. 2011; 77 (3): 247–55.
19. Barbusiński K. The modified Fenton process for decolorization of dye wastewater. Pol J Environ Stud. 2005; 14 (3):281–5.
20. Ozdemir C, Oden MK. Decolorization of azo dyes by modified Fenton process. Selcuk University Journal of Engineering, Science and Technology. 2011; 26(2):33–7.
21. García-Morales MA, Roa-Morales G, Barrera-Díaz C, Miranda VM, Hernández PB, Silva TBP. Integrated advanced oxidation process (ozonation) and electrocoagulation treatments for dye removal in denim effluents. Int. J. Electro Chem Sci. 2013; 8: 8752–63.
22. StanisŁaw L, Monika G. Optimization of oxidants dose for combined chemical and biological treatment of textile wastewater. Water Res. 1999; 33:2511–6.
23. Arslan IA, Isil AB. Advanced oxidation of raw and biotreated textile industry wastewater with O_3, H_2O_2, /UV-C and their sequential application. J Chem Technol Biot. 2001; 76:53–60.
24. Anijiofor SC, Mohd Jamil NA, Jabbar S, Sakyat S, Gomes C. Aerobic and anaerobic sewage biodegradable processes: The gap analysis. Int. J Res. Env. Sci. 2017; 3(3):9–19.
25. Kumara Swamy N, Singh P, Sarethy IP. Aerobic and anaerobic treatment of paper industry wastewater. Res. Environ. Life Sci. 2011; 4(4):141–8.
26. Mondal T, Jana A, Kundu D. Aerobic wastewater treatment technologies: A mini review. Int J Env Tech Sci. 2017; 4:135–40.
27. Almuktar SAAAN, Scholz M, Al-Isawi RHK, Sani A. Recycling of domestic wastewater treated by vertical-flow wetlands for irrigating chillies and sweet peppers. Agric Water Manag. 2015; 149:1–22.
28. Zupancˇicˇ GD, Uranjek-Zˇevart N, Roš M. Full-scale anaerobic co-digestion of organic waste and municipal sludge. Biomass Bioenerg. 2008; 32:162–7.
29. Halim MH, Anuar AN, Jamal NSA, Azmi SI, Ujang Z, Bob MM. Influence of high temperature on the performance of aerobic granular sludge in biological treatment of wastewater. J Environ Manag. 2016; 184:271–80.
30. Terasaka K, Aoki S, Kobayashi D. Recovery of fine iron oxide particulates from waste water using microbubble flotation. In: Proceedings of the 19th International Symposium on Transport Phenomena; 17–20 August 2008 Reykjavik, Iceland 2008. 25.
31. van Lier JB, Mahmoud N, Zeeman G. Anaerobic wastewater treatment. IWA Publishing. 2008;415–57.
32. Ersahin ME, Ozgun H, Dereli RC, Ozturk I. Anaerobic treatment of industrial effluents: An Overview of Applications [Internet]. 2011: Available from: doi: 10.5772/16032.
33. Sudha M, Saranya A, Selvakumar G, Sivakumar N. Microbial degradation of azo dyes: A review. Int J Curr Microbiol App Sci 2014; 3(2): 670–90.

34. Shah M. Effective treatment systems for azo dye Ddgradation: A joint venture between physico-chemical & microbiological process. Int J Environ Biorem Biodeg. 2014; 2(5):231–42.

35. Aravind P, Selvaraja H, Ferro S, Sundarama M. An integrated (electro- and bio-oxidation) approach for remediation of industrial wastewater containing azo-dyes: Understanding the degradation mechanism and toxicity assessment. J Hazar Mater. 2016; 318:203–15.

36. Senthilkumar S, Ahmed Basha C, Perumalsamy M, Prabhu HJ. Electrochemical oxidation and aerobic biodegradation with isolated bacterial. Electro chimica. Acta. 2012; 77:171–8.

37. de Arruda Guelli Ulson de Souza SM, Santos Bonilla KA, Ulson de Souza AA. Removal of COD and color from hydrolyzed textile azo dye by combined ozonation and biological treatment. J Hazard Mater. 2010; 179:35–42.

2 An Insight into the Past, Present, and Future of Azo Dyes

Komal Agrawal[1] and Pradeep Verma[1]*

[1]Bioprocess and Bioenergy Laboratory, Department of Microbiology, Central University of Rajasthan, Bandarsindri, Kishangarh, Ajmer Rajasthan, 305817, India

*Corresponding author: pradeepverma@curaj.ac.in

CONTENTS

2.1 INTRODUCTION

The azo group in a compound is represented as -N=N- where R and R' (R-N=N-R') represent either the aryl or the alkyl compounds. According to the International Union of Pure and Applied Chemistry (IUPAC), azo compounds are the derivatives of diazene (HN=NH) where both H are replaced by azobenzene or diphenyldiazene (IUPAC 2009). The azo dyes have played an integral part in the chemical industry and compose two-thirds of the synthetic dyes and are the most abundantly used dyes and >50% of the dyes synthesized globally are azo dyes (Mock & Freeman 2007; Puvaneswari et al., 2006). It has been assessed that approximately 3,000 azo dyes were used previously in various industries, for example, pharmaceuticals and paper (Meyer, 1981). Azo dyes are used in various sectors, such as foods, drugs, and cosmetics (Moosvi et al., 2007; Puvaneswari et al., 2006). Azo dyes are not easily removed following the conventional treatment of wastewater, because they are light stable, resistant to microbial degradation, and do not fade after washing. Because the dyes used in these industries are not utilized completely, approximately 10% of the dyestuffs are released into the environment (Hildenbrand et al., 1999; Puvaneswari et al., 2006). In addition, azo dyes are toxic and can cause mutagenicity, genotoxicity, and carcinogenicity in, animals and humans (de Lima et al., 2007).

DOI: 10.1201/9781003130932-2

This chapter deal with the timeline of various approaches that have been used for the removal and degradation of azo dyes from the 1980s till 2010s. The various mechanism involved in the removal of the azo dyes, such as physical, chemical, and biological treatment methods will be discussed in detail. In addition, the challenges and prospects that are associated with the removal of azo dye will be discussed in this chapter.

2.2   TIMELINE OF VARIOUS APPROACHES USED IN THE TREATMENT OF AZO DYES

2.2.1   1980s

In the 1980s, the studies on the removal of azo dyes focused on various aspects, such as biological treatment, ozonolysis, rat models. In addition, experimental and theoretical studies were carried out. The diversity or the area explored varied. Previous literature focused on the harmful effects of the dye (Lamas et al., 1986; Akao & Kuroda, 1981) (Table 2.1).

2.2.2   1990s

In the 1990s the focus of dye removal further expanded to the use of bioreactors, a consortium of microorganisms, and algae. The research into dye removal gained attention, because the harmful effects from the discharge of the dye gained increasing attention (Table 2.2).

TABLE 2.1
Various Approaches Used for the Removal of Azo Dyes in the 1980s

Serial Number	Azo Dyes	Treatment	References
1	Amaranth, sunset yellow, new coccine, and tartrazine	Anaerobic reduction by human fecal flora and flora from rat gut	Watabe et al. (1980)
2	Vermelho Reanil P8B	*Neurospora crassa* strain 74A	Corso et al. (1981)
3	Acid Yellow 23, Acid Orange 7, Acid Orange 8, Acid Red 88, Acid Black 1	Ozonation	Matsui et al. (1984)
4	Direct Black 38	Semicontinuous culture system inoculated with freshly voided human feces and used to investigate the metabolism of the BZ-based dye DB38 over a protracted time	Manning et al. (1985)
5	Acid Orange 12, Orange 20, and Red 88	*Pseudomonas cepacia* 13NA	Ogawa et al. (1986)
6	Procion Scarlet MX-G	Experimental adsorption was studied using *N. crassa* and the adsorption isotherms were evaluated using Freundlich and Langmuir equation	Corso et al. (1987)
7	Direct Yellow 4, Direct Yellow 50, Direct Orange 102, Direct Violet 9, Direct Red 24, and Direct Yellow 12	Ozonolysis	Gould and Groff (1987)

TABLE 2.2
Various Approaches Used for the Removal of Azo Dyes in the 1990s

Sl. Number	Azo Dyes	Treatment	References
1	Acid Red 1 and Acid Yellow 23	AOPs	Shu et al. (1994)
2	Reactive Black 5	AOPs	Ince and Gönenç (1997)
3	Red RBN	*P. mirabilis*	Chen et al. (1999)
4	Acid Orange 8, Acid Orange 10, and Acid Red 14	Aerobic biodegradation in biofilms	Jiang and Bishop (1994)
5	Reactive Red 22, Reactive Violet 2, and Reactive Yellow 2	*P. luteola*	Hu (1998)
6	Acid Black 1, Acid Red 1, Acid Red 14, Acid Red 18, Acid Orange 10, Acid Yellow 17, Acid Yellow 23, and Direct Yellow 4	Ozonation and photooxidation process	Shu and Huang (1995)
7	Mordant Orange 1	Continuous upflow anaerobic sludge blanket reactors	Donlon et al. (1997)
8	Orange II	Photoassisted Fenton Degradation	Fernandez et al. (1999)
9	Orange II and Amino black	Cloning of DNA from a *Rhodococcus* strain that could remove sulfonated azo dyes	Heiss et al. (1992)
10	Mordant Yellow 3	6-Aminonaphthalene-2-Sulfonate-Degrading Bacterial Consortium	Haug et al. (1991)
11	Disperse Red 1	Photoinduced poling (PIP)	Blanchard and Mitchell (1993)
12	Disperse Blue 79	Chemical and sediment mediated reduction	Weber and Adams (1995)
13	Azo dyes and heavy metal ions	algae (*U. lactuca L.*)	Jianwei et al. (1997)

2.2.3   2000s

In the 2000s, the approaches for the removal of azo dye removals improved in various sectors due to the integration of new techniques. With time, new and improved removal techniques were explored. In the 1990s, the advanced oxidation process (AOP), a consortium of microorganisms, photoinduced poling (PIP), sedimentation, cloning, and reactors were used. In the 2000s the approaches extended to immobilization, optimization studies using statistical tools, and composites. In addition to the physical and chemical methods, biological and plant-based removal of azo dyes started to gain attention. In the 2000s, the disadvantages associated with the physical and chemical treatments gained increased attention from the scientific community and researchers explored new areas for biological treatment methods (Table 2.3)

2.2.4   2010s

In the 2010s, the study of the biological removal of azo dye gained attention and the best concept for the waste was developed. In this concept, various wastes, such as hen feathers and lignocellulosic biomass were used for the removal of the waste. In addition, bioinformatic studies gained impetuous in the research associated with dye removal (Table 2.4).

TABLE 2.3
Various Approaches Used for the Treatment of Azo Dyes in the 2000s

Sl. Number	Azo Dyes	Treatment	References
1	Acid Red B	Magnetic powder MnO–Fe_2O_3 composite	Wu et al. (2005)
2	Acid Red 18	Photocatalytic membrane reactor	Mozia et al. (2005)
3	Acid Red B	Adsorption and catalytic combustion using magnetic $CuFe_2O_4$ powder	Wu et al. (2004)
4	Acid Red 18	Two-hybrid membrane systems employing a photodegradation process	Mozia et al. (2006)
5	Acid Red 14	Fenton, UV/H_2O_2, UV/H_2O_2/Fe(II), UV/H_2O_2/Fe(III), and UV/H_2O_2/Fe(III)/oxalate processes	Daneshvar and Khataee (2006)
6	Congo Red and Direct Blue I	Use of polycations	Dragan and Dinu (2008)
7	Direct Green 99	Photocatalysis with ultrafiltration or membrane distillation	Mozia and Morawski (2009)
8	Acid Orange 7	Adsorption over two waste materials, for instance, bottom ash and de-oiled soya	Gupta et al. (2006)
9	Direct Brown	Coal-based sorbents were performed using kinetic and mechanistic study	Mohan et al. (2002a)
10	4BS	Microbial consortium consisting of a white-rot fungus and a *Pseudomonas* consisted of azo dye Direct Fast Scarlet 4BS as the carbon and energy source	He et al. (2004)
	Chicago Sky Blue 6B	*Moringa oleifera* seed extract coagulation	Beltrán-Heredia and Sánchez (2008)
11	Acid red 14	Optimization electrocoagulation batch process with response surface methodology (RSM)	Aleboyeh et al. (2008)
12	Metanil Yellow	Batch and bulk removal and recovery of dye by adsorption over waste materials	Mittal et al. (2008)
13	Acid Orange 7	Upflow constructed wetlands	Ong et al. (2009)
14	Reactive Black 5, Trapaeolin 000, Methyl Orange, and Direct Violet 51	Calix[n]arene derivatives from aqueous solution into the organic phase to explore the potential use of calixarenes as low-cost efficient extractants for wastewater dye removal	Gungor et al. (2008)
	Reactive Red 2	AC as redox mediator	Van Der Zee et al. (2003)
	Reactive Yellow 22	*Spirogyra sp.*	Mohan et al. (2002b)
	Procion Red MX-5B	Photocatalytic oxidation	So et al. (2002)
	Red RBN	A microbial consortium was immobilized by a phosphorylated polyvinyl alcohol gel	Chen et al. (2003)

TABLE 2.4
Various Approaches Used for the Removal of Azo Dyes in the 2010s

Sl. Number	Azo dyes	Treatment	References
1	Direct Congo Red, Reactive Green HE4BD, and Golden Yellow MR	Multiwalled carbon nanotubes	Mishra et al. (2010)
2	Reactive Blue 5	Fabrication of nanospinel $ZnCr_2O_4$ using sol–gel method	Yazdanbakhsh et al. (2010)
3	Biodegrade a mixture RBRX-3B, DB-6 and DB-19	Mixed microbial culture in batch shake flask	Krishnan et al. (2017)
4	Acid Orange 7	Upflow membraneless bioelectrochemical system integrated with biocontact oxidation reactor	Pan et al. (2017)
5	Reactive Black 5	$ZnO–Fe_3O_4$ Nanocomposite	Farrokhi et al. (2014)
6	Remazol Black B	Heterogeneous Fenton/photo Fenton type processes using a Fe-exchanged zeolite of Y-type	Blanco et al. (2014)
7	Acid Orange 7	A highly graphitized and heteroatom doped porous carbon was prepared from fish waste	Liu et al. (2016)
8	Methyl Orange	A constructed wetland for pollutants removal and influence of iron scrap and sulfate-reducing bacterial enrichment in *Canna indica* planted constructed wetlands microcosms	Yadav et al. (2012)
9	Congo Red	The use of lignocellulosic jute fiber as a bioadsorbent in batch and column studies	Roy et al. (2013)
10	Reactive Orange dye	Ultrasonic/Fe_3O_4/H_2O_2	Jaafarzadeh et al. (2018)
11	Methyl Orange and Amaranth	Spherical ZnO nanoparticles	Zafar et al. (2019)
	Congo Red	RSM with central composite design was used to analyze the effects of main factors and their interaction for dye removal	Azharul et al. (2010)
12	Reactive Green 19 and Reactive Violet 5	Spent tea leaves as a potential low-cost adsorbent	Zuorro et al. (2013)
13	Remazol Brilliant Violet	Sonochemical method was employed in the synthesis of nickel aluminum layered double hydroxides (NiAl-LDH)	Pahalagedara et al. (2014)
14	Amido Black 10B	Hen feathers as adsorbent	Mittal et al. (2013)
15	Drimaren Red CL-5B	*Aeromonas hydrophila* MTCC 1739 and *Lysinibacillus sphaericus* MTCC 9523. Exploring docking and aerobic–microaerophilic biodegradation	Srinivasan and Sadasivam (2018)

2.3 METHODS EMPLOYED FOR THE REMOVAL OF DYES

Due to the accumulation and extensive use of dyes, various treatment methods have been employed for the removal of the dyes. The techniques in the following sections have been used for the removal of the dyes.

2.3.1 PHYSICAL METHODS

Various physical methods used for the removal of the azo dyes include adsorption, ion exchange, and membrane filtration (Figure 2.1). The major drawback of the physical treatment methods is that they do not break or degrade the dye. They transfer the dye molecules from one phase to another. Therefore, they only are used when dyes are present in low concentrations and can be ineffective if used at higher dye concentrations (Labanda et al., 2009; Chatterjee et al., 2009a, 2009b; Ahmad & Puasa, 2007; Robinson et al., 2001). In adsorption, the dye is transferred between the interface of two immiscible phases that are in contact, for instance, granulated carbon has proved to be an economically feasible approach for the removal of dyes. Ion exchange methods have been used for the removal of cations and anions from dyes. In this method, a bed of resins is prepared, and wastewater is passed through. In the bed, the wastewater is exchanged for sodium or hydrogen ions in the resin (Vijayaraghavan et al., 2013). Finally, in membrane filtration, reverse osmosis and electrodialysis are used. In dyeing, electrolytes play an important role in the exhaustion of the dye. Therefore, reverse osmosis has been effectively used for the removal of the salts, total dissolved solids, ions, and larger species. In addition, the reject can be reused; however, it has disadvantages, such as membrane clogging and a high cost. In electrodialysis, an electric potential is applied across water to remove the dissolved salts that are ionic. This process results in the movement of the ions (cations and anions) to the electrodes via permeable (cationic and anionic) membranes. However, in the electrodialysis membrane, fouling is a major drawback and to avoid this issue when operating the process, suspended solids, turbidity, trace organics, and colloids must be removed before electrodialysis (Vijayaraghavan et al., 2013).

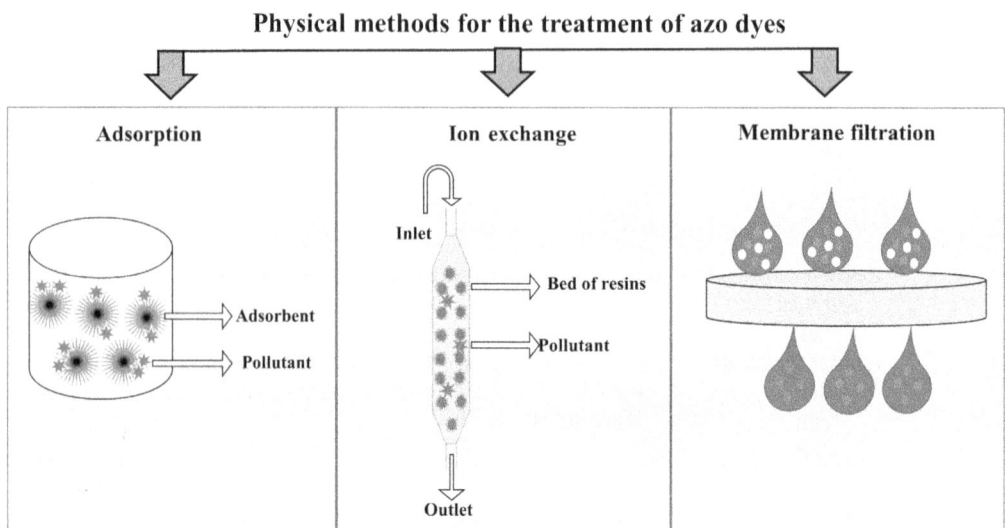

FIGURE 2.1 Various physical treatment methods used for the removal of azo dyes.

2.3.2 CHEMICAL METHODS

In the chemical methods, various approaches, such as chemical oxidation (Türgay et al., 2011), electrochemical degradation (Song et al., 2010), and ozonation (Ulson et al., 2010) have been used to remove dyes. In Zhu et al. (2020) chemical oxidation was used to remove azo dye (Procion Red MX-5B). In this study, fluorescence excitation-emission matrix spectroscopy was performed to understand the decay of Procion Red MX-5B using a hydroxyl radical that was generated from catalytic iron [Fe (III)] on hydrogen peroxide (H_2O_2). The changes in the chemical groups were observed during degradation. However, in Song et al. (2010), the electrochemical degradation of azo dye C.I. Reactive Red 195 was studied in an aqueous solution on a titanium, tin, lead (Ti/SnO_2–Sb/PbO_2) electrode. With an increase in the concentration of dye the mineralization efficiency decreased. In addition, the current density had a positive and negative impact on dye mineralization. A further increase in the concertation of sodium chloride (NaCl) increased dye removal and decreased the mineralization efficiency. Ozonation of the hydrolyzed azo dye Reactive Yellow 84 was studied in Koch et al. (2002) in ultrapure water in a laboratory-scale cylindric batch reactor. After 60 and 90 min, complete decolourization was observed at 18.5 and 9.1 mg/L ozone (O_3) concentration. Therefore, these chemical methods of dye removal have been effectively used for the removal of dyes. However, the major limitations of chemical treatment are the high cost and the sludge generated at the end of the process is toxic and requires further chemical treatment for its conversion into less toxic products. This increases the disposal related issues (Vijayaraghavan et al., 2013).

2.3.3 BIOLOGICAL METHODS

Because the physical and chemical methods suffer limitations, the treatment methods employed for the removal of azo dyes changed from physical–chemical to biological treatment methods. In biological methods, bacteria, fungi, and enzymatic equipment were used for the removal of azo dye and will be discussed in the following sections.

2.3.3.1 Approaches for Microbial Biodegradation

The microbial biodegradation of the azo dyes occurs via two approaches, for example, biosorption and enzymatic degradation and they are discussed as follows (Figure 2.2).

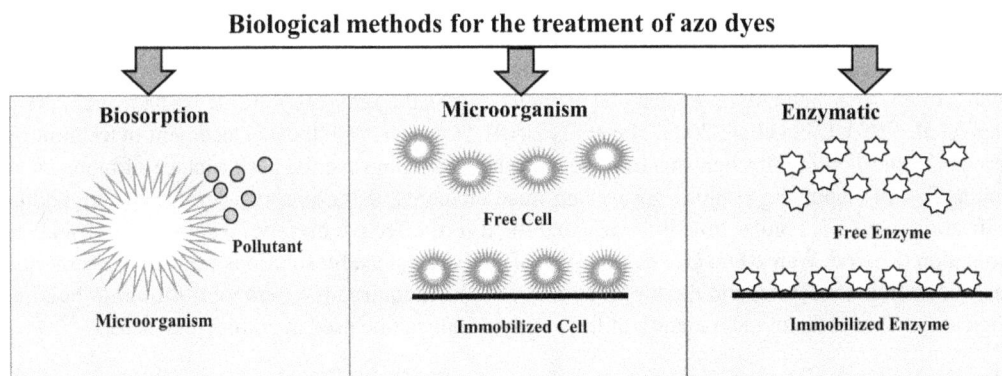

Biological methods for the treatment of azo dyes

FIGURE 2.2 Various biological treatment methods employed for the removal of azo dyes.

2.3.3.1.1   Biosorption and Enzymatic Degradation

The biosorption of dyes has been reported via the utilization of biomass from various microorganisms, such as bacteria, fungi, yeast, and algae (Bhatnagar & Sillanpää, 2010). The major factor that affects the biosorption potential of the microorganism is the presence of the heteropolysaccharide and the lipid components of the cell wall. These components affect biosorption due to the presence of the functional group and charged ions (Vitor & Corso, 2008; Srinivasan & Viraraghavan, 2010; Charumathi & Das, 2012; Aksu & Dönmez, 2003). The treatment of biomass by autoclaving ruptures the cells in the biomass and increase the adsorption and acid treatment might either increase or decrease the efficiency of the binding sites (Lim et al., 2010; Fu & Viraraghavan, 2001; Ambrósio et al., 2012; Fu & Viraraghavan, 2000; Vijayaraghavan & Yun, 2007). The major advantages of biosorption are that the dead cells do not require nutrients, and therefore, can be stored for an extended time and can be used after regeneration using solvents (Fu & Viraraghavan, 2001).

In enzymatic degradation, reductases have been reported for the degradation of azo dyes (Majeau et al., 2010; Misal et al., 2011). The anaerobic microbial degradation of azo dyes occurs with the help of anaerobic azoreductase (McMullan et al., 2001). In the literature, the activity of azoreductase has been associated with various genes; therefore, helping to determine that more than one reductase might be involved in the removal of azo dyes by various microorganisms and the condition of the culture (Ramalho et al., 2002). The various azoreductases identified for the removal of the azo dyes have been reported in bacteria, algae, and yeast. The various azoreductases includes FMN-dependent reductases, FMN-independent reductases, NADH-dependent reductases, NADPH-dependent reductases, and NADH–DCIP reductases (Solís et al., 2012). However, the oxidative removal of azo dyes has been reported by peroxidases and phenoloxidases, such as manganese peroxidase, lignin peroxidase, laccase, and tyrosinase (Verma & Madamwar, 2002, 2005; Harazono et al., 2003; Ferreira-Leitão et al., 2007).

2.3.3.2   Role of Various Microorganisms in the Biodegradation of Azo Dyes

Various microorganisms have been reported in the biodegradation of azo dyes, such as fungi, bacteria, algae, and yeast. The role of these microorganisms in the removal of the dyes is discussed in the following section.

2.3.3.2.1   Fungi

Fungi, either living or dead, have been reported for the decolourization and biodegradation of azo dyes. Fungal systems have been effectively used for the removal of azo dyes and the various mechanisms include biosorption and enzymatic degradation. Combined with free cells it has been effectively used in bioreactors for the removal and biodegradation of azo dyes. The various bioreactors, such as membrane (Kim et al., 2004), immobilized (Singh et al., 2010), and continuous bioreactor (Libra et al., 2003) have been used; however, a commercial-scale application has not been achieved. The major problems with a continuous bioreactor are clogging, bacterial contamination (fungi grows slowly), and the loss of enzymes and mediators that play a crucial role in the removal of the dyes (Kim et al., 2004; Singh et al., 2010; Libra et al., 2003; Hai et al., 2012). However, with the advancement in technology, various troubleshooting mechanisms have been evaluated to improve the performance of bioreactors, such as low pH, applying osmosis for certain trace elements, using nitrogen-limited media, adding fresh sterile media at regular time intervals, and the use of effective bactericidal agents to avoid contamination (Zhou & Wen 2009; Gao et al., 2006). These are not useful solutions for the long-term utilization of column reactors, and therefore, more research is required (i.e., immobilization) to build an efficient system for the biodegradation of the azo dyes that can be used at commercial scales.

2.3.3.2.1   Yeast

Yeasts have been reported for the removal of dyes via biosorption because they have high potential to accumulate dyes (e.g., *Galactomyces geotrichum*, *Saccharomyces cerevisiae*, and *Trichosporon beigelii*) (Jadhav et al., 2008), fast growth, and high potential to survive under unfavorable conditions (Ertuğrul & Dönmez, 2008; Fairhead & Thöny-Meyer, 2012; Martorell et al., 2012). Based on Meehan et al. (2000), *Kluyveromyces marxianus* IMB3 was reported for the decolorization of Remazol Black B. The yeast strain *T. beigelii* NCIM-3326 effectively removed various azo dyes, such as Navy Blue HER, Red HE7B, Golden Yellow 4BD, Green HE4BD, and Orange HE2R (Saratale et al., 2008). Other yeast strains that have been reported for the removal of the dyes include *Candida zeylanoides* (Ramalho et al., 2002), *Candida albicans* (Vitor & Corso, 2008), and *Candida tropicalis* (Charumathi & Das, 2010). Yeast removes dyes using adsorption, or an enzyme, or both (Safarik et al., 2007). Various enzymes that are involved in the removal of the dyes by yeast include manganese peroxidase, tyrosinase, oxidase, NADH–DCIP reductase, and reductase (Waghmode et al., 2011). Therefore, yeast can potentially be explored for the removal of various azo dyes.

2.3.3.2.2   Filamentous Fungi

Filamentous fungi (e.g., *Phanerochaete, Trametes, Bjerkandera, Aspergillus, Pleurotus,* and *Phlebia*) have been used for the removal of the azo dyes because they allow the total mineralization of the dye at low-cost via adsorption or enzymatic biodegradation (Asgher et al., 2008). At low pH, the dye removal is enhanced due to the electrostatic attraction between the charged dye molecules and the cell surface of the microorganism. However, the high temperature reduces the dye removal potential, which is probably due to the deactivation of the adsorbent surface of the active site (Iqbal & Saeed, 2007). The inactive mycelia of certain filamentous fungi, such as *Cunninghamella elegans, Trametes versicolor,* and *Fusarium solani* (Ambrosio et al., 2012; Erden et al., 2011; Abedin, 2008) have better potential than activated carbon or amberlite (Ambrósio et al., 2012; Erden et al., 2011).

2.3.3.2.3   Fungal Consortia for the Removal of Azo Dyes

Because axenic cultures have been used for the removal of the azo dyes, consortia of cultures have been reported for the efficient removal of the dyes; however, the reported data is limited. In Pan et al. (2009), two fungi *Penicillium sp.* A1 and *Fusarium sp.* A19 had better dye removal efficiency than a pure culture and in Yang et al. (2011a, 2011b) a consortium that consisted of *Trametes sp.* SQ01 and *Chaetomium sp.* R01 was used to remove triphenylmethane dyes.

2.3.3.2.4   Bacteria

The removal of azo dyes by bacteria occurs under aerobic, anaerobic, and facultative anaerobic conditions. The removal of the dye by bacteria is favored because the technique is cheap, fast, eco-friendly, and generates minimum sludge (Verma & Madamwar, 2003). For pure cultures, the first reports were published in the 1970s with strains of *Bacillus subtilis, Aeromonas hydrophila,* and *Bacillus cereus* (Wuhrmann et al., 1980). Later, strains including *Proteus mirabilis, Pseudomonas luteola,* and *Pseudomonas sp.* were reported by Chen et al. (1999), Hu (1998), and Tuttolomondo et al. (2014). The advantage of using a pure culture is that the data becomes reproducible, can be verified experimentally easily, the detailed mechanism of dye removal can be elucidated using biochemistry and molecular biology methods. In addition, this type of detailed study can help to regulate the enzyme systems associated with the microbes for its efficient utilization in dye removal (Telke et al., 2008). However, in some examples, mixed microbial consortia have better dye removal potential than axenic cultures (Joshi et al., 2008). In these cases, a microorganism might metabolize the dye to its cometabolite, which can be reutilized by the other microorganisms in the mixed culture; therefore, further biodegrading the compound (Saratale et al., 2011).

2.4   CONCLUSION: PRESENT TRENDS, CHALLENGES, AND PROSPECTS

According to the trends, for the persistent in the removal of azo dyes the concept "best out of waste" has been gaining impetuous where biological treatment methods, immobilization, nanoparticles, nanocomposite (Agrawal and Verma 2021), and bioreactor studies have been explored during the last decade. In addition, bioinformatic studies have gained the attention of the scientific community and have enabled a better understanding of the dye removal system. However, the main challenge associated with biological treatment systems for azo dyes is commercialization. Therefore, more rigorous and detailed studies are required for their implementation at wastewater treatment plants. In addition, in physical treatment, the molecules are not degraded but are transferred from one phase to another and for chemical treatment methods the process is expensive and the disposal of the sludge is a major problem. Therefore, a biological treatment system is more effective because, following optimization studies it could be economically feasible, the end product generated following treatment by an axenic, mixed microbial consortium biodegrades the products into less or nontoxic end products. In addition, because sludge is not generated during biological treatment, the disposal related issues that are faced from chemical treatment are not a drawback and it is economically and environmentally benign.

REFERENCES

Abedin, R. M. (2008). Decolorization and biodegradation of crystal violet and malachite green by *Fusarium solani* (Martius) Saccardo. A comparative study on biosorption of dyes by the dead fungal biomass. *American-Eurasian Journal of Botany, 12*:17–31.

Agrawal, K. & Verma, P. (2021). The Interest in Nanotechnology: A Step Towards Bioremediation. In *Removal of Emerging Contaminants Through Microbial Processes*, Springer, Singapore, 265–282.

Ahmad, A. L. & Puasa, S. W. (2007). Reactive dyes decolourization from an aqueous solution by combined coagulation/micellar-enhanced ultrafiltration process. *Chemical Engineering Journal, 132*:257–265.

Akao, M. & Kuroda, K. (1981). Analysis of loss of nuclear RNA in azo dye-induced hepatoma by DNA-RNA competitive hybridization. *Cancer Research, 41*:735–740.

Aksu, Z. & Dönmez, G. (2003) A comparative study on the biosorption characteristics of some yeasts for Remazol Blue reactive dye. *Chemosphere*, 50:1075–1083.

Aleboyeh, A., Daneshvar, N. & Kasiri, M. B. (2008). Optimization of CI Acid Red 14 azo dye removal by electrocoagulation batch process with response surface methodology. *Chemical Engineering and Processing: Process Intensification, 47*:827–832.

Ambrósio, S. T., Vilar Júnior, J. C., Da Silva, C. A. A., Okada, K., Nascimento, A. E., Longo, R.L. & Campos-Takaki, G. M. (2012). A biosorption isotherm model for the removal of reactive azo dyes by inactivated mycelia of *Cunninghamella elegans* UCP542. *Molecules, 17*:452–462.

Asgher, M., Batool, S., Bhatti, H. N., Noreen, R., Rahman, S. U. & Asad, M. J. (2008). Laccase mediated decolorization of vat dyes by Coriolus versicolor IBL-04. *International Biodeterioration & Biodegradation, 62*:465–470.

Azharul, I. M., Nikoloutsou, Z., Sakkas, V., Papatheodorou, M. & Albanis, T. (2010). Statistical optimisation by combination of response surface methodology and desirability function for removal of azo dye from aqueous solution. *International Journal of Environmental and Analytical Chemistry, 90*:497–509.

Beltrán-Heredia, J. & Sánchez Martín, J. (2008). Azo dye removal by *Moringa oleifera* seed extract coagulation. *Coloration Technology, 124*:310–317.

Bhatnagar, A. & Sillanpää, M. (2010). Utilization of agro-industrial and municipal waste materials as potential adsorbents for water treatment—a review. *Chemical engineering Journal, 157*:277–296.

Blanchard, P. M. & Mitchell, G. R. (1993). A comparison of photoinduced poling and thermal poling of azo-dye-doped polymer films for second order nonlinear optical applications. *Applied Physics Letters, 63*:2038–2040.

Blanco, M., Martinez, A., Marcaide, A., Aranzabe, E. & Aranzabe, A. (2014). Heterogeneous Fenton catalyst for the efficient removal of azo dyes in water. *American Journal of Analytical Chemistry, 5*:490.

Charumathi, D. & Das, N. (2010). Bioaccumulation of synthetic dyes by *Candida tropicalis* growing in sugarcane bagasse extract medium. *Advances in* Biological Research, *4*:233–240.

Charumathi, D. & Das, N. (2012). Packed bed column studies for the removal of synthetic dyes from textile wastewater using immobilised dead C. tropicalis. *Desalination*, *285*:22–30.

Chatterjee, S., Lee, D. S., Lee, M. W. & Woo, S. H. (2009a). Congo red adsorption from aqueous solutions by using chitosan hydrogel beads impregnated with nonionic or anionic surfactant. *Bioresource Technology, 100*:3862–3868.

Chatterjee, S., Lee, M. W. & Woo, S. H. (2009b). Influence of impregnation of chitosan beads with cetyl trimethyl ammonium bromide on their structure and adsorption of congo red from aqueous solutions. *Chemical Engineering Journal, 155*:254–259.

Chen, K. C., Huang, W. T., Wu, J. Y. & Houng, J. Y. (1999). Microbial decolorization of azo dyes by Proteus mirabilis. *Journal of Industrial Microbiology and Biotechnology, 23*:686–690.

Chen, K. C., Wu, J. Y., Huang, C. C., Liang, Y. M. & Hwang, S. C. J. (2003). Decolorization of azo dye using PVA-immobilized microorganisms. *Journal of Biotechnology, 101*:241–252.

Corso, C. R., De Angelis, D. F., De Oliveira, J. E. & Kiyan, C. (1981). Interaction between the diazo dye "Vermelho Reanil" P8B, and Neurospora crassa strain 74A. *European Journal of Applied Microbiology and Biotechnology, 13*:64–66.

Corso, C. R., Marcanti, I. & Yamaoka, E. M. (1987). Applicability of the equations of Freundlich and Langmuir to the adsorption of the azo dye procion scarlet on paramorphic colonies of *Neurospora crassa*. *Brazilian Journal of Medical and Biological Research, 20*:623–626.

Daneshvar, N. & Khataee, A. R. (2006). Removal of azo dye CI acid red 14 from contaminated water using Fenton, UV/H2O2, UV/H2O2/Fe (II), UV/H2O2/Fe (III) and UV/H2O2/Fe (III)/oxalate processes: a comparative study. *Journal of Environmental Science and Health Part A, 41*:315–328.

de Lima, R. O. A., Bazo, A. P., Salvadori, D. M. F., Rech, C. M., de Palma Oliveira, D. & de Aragão Umbuzeiro, G. (2007). Mutagenic and carcinogenic potential of a textile azo dye processing plant effluent that impacts a drinking water source. *Mutation Research/Genetic Toxicology and Environmental Mutagenesis, 626*:53–60.

Donlon, B., Razo-Flores, E., Luijten, M., Swarts, H., Lettinga, G. & Field, J. (1997). Detoxification and partial mineralization of the azo dye mordant orange 1 in a continuous upflow anaerobic sludge-blanket reactor. *Applied Microbiology and Biotechnology, 47*:83–90.

Dragan, E. S. & Dinu, I. A. (2008). Removal of azo dyes from aqueous solution by coagulation/flocculation with strong polycations. *Research Journal of Chemistry and Environment, 12*:5.

Erden, E., Kaymaz, Y. & Pazarlioglu, N. K. (2011). Biosorption kinetics of a direct azo dye Sirius Blue K-CFN by *Trametes versicolor*. *Electronic Journal of Biotechnology, 14*:3–3.

Ertuğrul, S., Bakır, M. & Dönmez, G. (2008). Treatment of dye-rich wastewater by an immobilized thermo-philic cyanobacterial strain: *Phormidium* sp. *Ecological Engineering, 32*:244–248.

Fairhead, M. & Thöny-Meyer, L. (2012). Bacterial tyrosinases: old enzymes with new relevance to biotech-nology. *New Biotechnology, 29*:183–191.

Farrokhi, M., Hosseini, S. C., Yang, J. K. & Shirzad-Siboni, M. (2014). Application of ZnO–Fe3O4 nanocomposite on the removal of azo dye from aqueous solutions: kinetics and equilibrium studies. *Water, Air, & Soil Pollution, 225*:2113.

Fernandez, J., Bandara, J., Lopez, A., Buffat, P. & Kiwi, J. (1999). Photoassisted Fenton degradation of nonbiodegradable azo dye (Orange II) in Fe-free solutions mediated by cation transfer membranes. *Langmuir, 15*:185–192.

Ferreira-Leitão, V. S., de Carvalho, M. E. A. & Bon, E. P. (2007). Lignin peroxidase efficiency for methylene blue decolouration: comparison to reported methods. *Dyes and Pigments, 74*:230–236.

Freeman, H. S., & Mock, G. N. (2007). Dye application, manufacture of dye intermediates and dyes. In *Kent and Riegel's Handbook of Industrial Chemistry and Biotechnology* (pp. 499–590) Boston, MA: Springer.

Fu, Y., & Viraraghavan, T. (2000). Removal of a dye from an aqueous solution by the fungus Aspergillus niger. *Water Quality Research Journal, 35*:95–112.

Fu, Y., & Viraraghavan, T. (2001). Fungal decolorization of dye wastewaters: a review. *Bioresource Technology, 79*:251–262.

Gao, D. W., Wen, X. H., & Qian, Y. (2006). Decolorization of reactive brilliant red K-2BP by white rot fungus under sterile and non-sterile conditions. *Journal of Environmental Sciences, 18*:428–432.

Gould, J.P., & Groff, K. A. (1987). The kinetics of ozonolysis of synthetic dyes. *Ozone Science and Engineering, 9*:153–164

Gungor, O., Yilmaz, A., Memon, S., & Yilmaz, M. (2008). Evaluation of the performance of calix [n] arene derivatives as liquid phase extraction material for the removal of azo dyes. *Journal of Hazardous Materials, 158*:202–207.

Gupta, V. K., Mittal, A., Gajbe, V., & Mittal, J. (2006). Removal and recovery of the hazardous azo dye acid orange 7 through adsorption over waste materials: bottom ash and de-oiled soya. *Industrial & Engineering Chemistry Research, 45*:1446–1453.

Hai, F. I., Yamamoto, K., Nakajima, F., & Fukushi, K. (2012). Application of a GAC-coated hollow fiber module to couple enzymatic degradation of dye on membrane to whole cell biodegradation within a membrane bioreactor. *Journal of Membrane Science, 389*:67–75.

Harazono, K., Watanabe, Y., & Nakamura, K. (2003). Decolorization of azo dye by the white-rot basidiomycete *Phanerochaete sordida* and by its manganese peroxidase. *Journal of Bioscience and Bioengineering, 95*:455–459.

Haug, W., Schmidt, A., Nörtemann, B., Hempel, D. C., Stolz, A. & Knackmuss, H. J. (1991). Mineralization of the sulfonated azo dye Mordant Yellow 3 by a 6-aminonaphthalene-2-sulfonate-degrading bacterial consortium. *Applied and Environmental Microbiology, 57*:3144–3149.

He, F., Hu, W. & Li, Y. (2004). Biodegradation mechanisms and kinetics of azo dye 4BS by a microbial consortium. *Chemosphere, 57*:293–301.

Heiss, G. S., Gowan, B. & Dabbs, E. R., 1992. Cloning of DNA from a *Rhodococcus* strain conferring the ability to decolorize sulfonated azo dyes. *FEMS Microbiology Letters, 99*:221–226.

Hildenbrand, S., Schmahl, F.W., Wodarz, R., Kimmel, R. & Dartsch, P. C. (1999). Azo dyes and carcinogenic aromatic amines in cell cultures. *International Archives of Occupational and Environmental Health, 72*:M052–M056.

Hu, T. L. (1998). Degradation of azo dye RP2B by *Pseudomonas luteola*. *Water Science and Technology, 38*:299–306.

Ince, N. H., & Gönenç, D. T. (1997). Treatability of a textile azo dye by UV/H_2O_2. *Environmental Technology, 18*:179–185.

Iqbal, M., & Saeed, A. (2007). Biosorption of reactive dye by loofa sponge-immobilized fungal biomass of *Phanerochaete chrysosporium*. *Process Biochemistry, 42*:1160–1164.

IUPAC. (1997). Compendium of Chemical Terminology (2nd ed.). Online corrected version (2009) "azo compounds".

Jaafarzadeh, N., Takdastan, A., Jorfi, S., Ghanbari, F., Ahmadi, M., & Barzegar, G. (2018). The performance study on ultrasonic/Fe_3O_4/H_2O_2 for degradation of azo dye and real textile wastewater treatment. *Journal of Molecular Liquids, 256*:462–470.

Jadhav, U. U., Dawkar, V. V., Ghodake, G. S., & Govindwar, S. P. (2008). Biodegradation of Direct Red 5B, a textile dye by newly isolated Comamonas sp. UVS. *Journal of Hazardous Materials, 158*:507–516.

Jiang, H., & Bishop, P. L. (1994). Aerobic biodegradation of azo dyes in biofilms. *Water Science and Technology, 29*:525.

Jianwei, M., Xingdong, Y., Gulan, Z., & Yingfen, Z. (1997). Study on the removal mechanism of azo dyes and heavy metal ions in water with algae (*U. lactuca L.*). *Zhongguo Huanjing Kexue, 17*:241–243.

Joshi, T., Iyengar, L., Singh, K., &Garg, S. (2008). Isolation, identification and application of novel bacterial consortium TJ-1 for the decolourization of structurally different azo dyes. *Bioresource Technology, 99*:7115–7121.

Kim, T. H., Lee, Y., Yang, J., Lee, B., Park, C. & Kim, S. (2004). Decolorization of dye solutions by a membrane bioreactor (MBR) using white-rot fungi. *Desalination, 168*:287–293.

Kock, M., Yediler, A., Lienert, D. & Kettrup, A. (2002). *Ozonation of hydrolyzed azo dye reactive yellow 84 (CI)*. *Chemosphere, 46*(1):109–113.

Krishnan, J., Kishore, A. A., Suresh, A., Madhumeetha, B. & Prakash, D. G. (2017). Effect of pH, inoculum dose and initial dye concentration on the removal of azo dye mixture under aerobic conditions. *International Biodeterioration & Biodegradation, 119*:16–27.

Labanda, J., Sabaté, J., & Llorens, J. (2009). Modeling of the dynamic adsorption of an anionic dye through ion-exchange membrane adsorber. *Journal of Membrane Science, 340*:234–240.

Lamas, E., Schweighoffer, F., & Kahn, A. (1986). Early modifications of gene expression induced in liver by azo-dye diet. *FEBS Letters, 206*:229–232.

Libra, J. A., Borchert, M., & Banit, S. (2003). Competition strategies for the decolorization of a textile-reactive dye with the white-rot fungi *Trametes versicolor* under non-sterile conditions. *Biotechnology and Bioengineering, 82*:736–744.

Lim, S. L., Chu, W. L., & Phang, S.M. (2010). Use of *Chlorella vulgaris* for bioremediation of textile wastewater. *Bioresource Technology, 101*:7314–7322.

Liu, Z., Zhang, F., Liu, T., Peng, N., & Gai, C. (2016). Removal of azo dye by a highly graphitized and heteroatom doped carbon derived from fish waste: adsorption equilibrium and kinetics. *Journal of Environmental Management, 182*:446–454.

Majeau, J. A., Brar, S. K., & Tyagi, R. D. (2010). Laccases for removal of recalcitrant and emerging pollutants. *Bioresource Technology, 101*:2331–2350.

Manning, B. W., Cerniglia, C. E., & Federle, T. W. (1985). Metabolism of the benzidine-based azo dye Direct Black 38 by human intestinal microbiota. *Applied and Environmental Microbiology, 50*:10–15.

Martorell, M. M., Pajot, H. F., & de Figueroa, L.I. (2012). Dye-decolourizing yeasts isolated from Las Yungas rainforest. Dye assimilation and removal used as selection criteria. *International Biodeterioration & Biodegradation, 66*:25–32.

Matsui, M., Kimura, T., Nambu, T., Shibata, K., & Takase, Y. (1984). Reaction of Water–soluble Dyes with Ozone. *Journal of the Society of Dyers and Colourists, 100*:125–127.

McMullan, G., Meehan, C., Conneely, A., Kirby, N., Robinson, T., Nigam, P., … Smyth, W.F., 2001. Microbial decolourisation and degradation of textile dyes. *Applied Microbiology and Biotechnology, 56*:81–87.

Meehan, C., Banat, I. M., McMullan, G., Nigam, P., Smyth, F., & Marchant, R. (2000). Decolorization of Remazol Black-B using a thermotolerant yeast, *Kluyveromyces marxianus* IMB3. *Environment International, 26*:75–79.

Meyer, U. (1981). Biodegradation of synthetic organic colorants. In FEMS Symposium (*12*: pp. 371–385). London: Academic Press.

Misal, S. A., Lingojwar, D. P., Shinde, R. M., & Gawai, K. R. (2011). Purification and characterization of azoreductase from alkaliphilic strain *Bacillus badius*. *Process Biochemistry, 46*:1264–1269.

Mishra, A. K., Arockiadoss, T., & Ramaprabhu, S. (2010). Study of removal of azo dye by functionalized multi-walled carbon nanotubes. *Chemical Engineering Journal, 162*:1026–1034.

Mittal, A., Gupta, V. K., Malviya, A., & Mittal, J. (2008). Process development for the batch and bulk removal and recovery of a hazardous, water-soluble azo dye (Metanil Yellow) by adsorption over waste materials (Bottom Ash and De-Oiled Soya). *Journal of Hazardous Materials, 151*:821–832.

Mittal, A., Thakur, V., & Gajbe, V. (2013). Adsorptive removal of toxic azo dye Amido Black 10B by hen feather. *Environmental Science and Pollution Research, 20*:260–269.

Mohan, S. V., Rao, N. C., & Karthikeyan, J. (2002a). Adsorptive removal of direct azo dye from aqueous phase onto coal based sorbents: a kinetic and mechanistic study. *Journal of Hazardous Materials, 90*:189–204.

Mohan, S. V., Rao, N. C., Prasad, K. K., & Karthikeyan, J. (2002b). Treatment of simulated Reactive Yellow 22 (Azo) dye effluents using Spirogyra species. *Waste Management, 22*:575–582.

Moosvi, S., Kher, X., & Madamwar, D. (2007). Isolation, characterization and decolorization of textile dyes by a mixed bacterial consortium JW-2. *Dyes and Pigments, 74*:723–729.

Mozia, S., & Morawski, A.W. (2009). Integration of photocatalysis with ultrafiltration or membrane distillation for removal of azo dye direct green 99 from water. *Journal of Advanced Oxidation Technologies, 12*:111–121.

Mozia, S., Tomaszewska, M., & Morawski, A.W. (2005). A new photocatalytic membrane reactor (PMR) for removal of azo-dye Acid Red 18 from water. *Applied Catalysis B: Environmental, 59*:131–137.

Mozia, S., Tomaszewska, M., & Morawski, A.W. (2006). Removal of azo-dye Acid Red 18 in two hybrid membrane systems employing a photodegradation process. *Desalination, 198*:183–190.

Ogawa, T., Yatome, C., Idaka, E., & Kamiya, H. (1986). Biodegradation of azo acid dyes by continuous cultivation of *Pseudomonas cepacia* 13NA. *Journal of the Society of Dyers and Colourists, 102*:12–14.

Ong, S. A., Uchiyama, K., Inadama, D., & Yamagiwa, K. (2009). Simultaneous removal of color, organic compounds and nutrients in azo dye-containing wastewater using up-flow constructed wetland. *Journal of Hazardous Materials, 165*:696–703.

Pahalagedara, M. N., Samaraweera, M., Dharmarathna, S., Kuo, C. H., Pahalagedara, L. R., Gascón, J.A., & Suib, S. L. (2014). Removal of azo dyes: intercalation into sonochemically synthesized NiAl layered double hydroxide. *The Journal of Physical Chemistry C, 118*:17801–17809.

Pan, R., Cao, L., & Zhang, R. (2009). Combined effects of Cu, Cd, Pb, and Zn on the growth and uptake of consortium of Cu-resistant *Penicillium* sp. A1 and Cd-resistant *Fusarium* sp. A19. *Journal of Hazardous Materials, 171*:761–766.

Pan, Y., Wang, Y., Zhou, A., Wang, A., Wu, Z., Lv, L., … Zhu, T. (2017). Removal of azo dye in an up-flow membrane-less bioelectrochemical system integrated with bio-contact oxidation reactor. *Chemical Engineering Journal, 326*:454–461.

Puvaneswari, N., Muthukrishnan, J., & Gunasekaran, P. (2006). Toxicity assessment and microbial degradation of azo dyes. *Indian Journal of Experimental Biology, 44*:618–626.

Ramalho, P. A., Scholze, H., Cardoso, M. H., Ramalho, M. T., & Oliveira-Campos, A. M. (2002). Improved conditions for the aerobic reductive decolourisation of azo dyes by *Candida zeylanoides*. *Enzyme and Microbial Technology, 31*:848–854.

Robinson, T., McMullan, G., Marchant, R., & Nigam, P. (2001). Remediation of dyes in textile effluent: a critical review on current treatment technologies with a proposed alternative. *Bioresource Technology, 77*:247–255.

Roy, A., Chakraborty, S., Kundu, S. P., Adhikari, B., & Majumder, S. B. (2013). Lignocellulosic jute fiber as a bioadsorbent for the removal of azo dye from its aqueous solution: Batch and column studies. *Journal of Applied Polymer Science, 129*:15–27.

Safarik, I., Rego, L. F. T., Borovska, M., Mosiniewicz-Szablewska, E., Weyda, F., & Safarikova, M. (2007). New magnetically responsive yeast-based biosorbent for the efficient removal of water-soluble dyes. *Enzyme and Microbial Technology, 40*:1551–1556.

Saratale, G. D., Chen, S. D., Lo, Y. C., Saratale, R. G., & Chang, J. S. (2008). Outlook of biohydrogen production from lignocellulosic feedstock using dark fermentation–a review. *Journal of Scientific & Industrial Research, 67*:962–979.

Saratale, G. D., Chien, I. J., & Chang, J. S. (2011). *Enzymatic pretreatment of cellulosic wastes for anaerobic treatment and bioenergy production. Environmental Anaerobic Technology Applications and New Developments*. London: London Imperial College Press.

Shu, H. Y. & Huang, C. R. (1995). Degradation of commercial azo dyes in water using ozonation and UV enhanced ozonation process. *Chemosphere, 31*:3813–3825.

Shu, H. Y., Huang, C. R., & Chang, M. C. (1994). Decolorization of mono-azo dyes in wastewater by advanced oxidation process: a case study of acid red 1 and acid yellow 23. *Chemosphere, 29*:2597–2607.

Singh, S., Pakshirajan, K., & Daverey, A. (2010). Enhanced decolourization of Direct Red-80 dye by the white rot fungus *Phanerochaete chrysosporium* employing sequential design of experiments. *Biodegradation, 21*:501–511.

So, C. M., Cheng, M. Y., Yu, J. C., & Wong, P. K. (2002). Degradation of azo dye Procion Red MX-5B by photocatalytic oxidation. *Chemosphere, 46*:905–912.

Solís, M., Solís, A., Pérez, H. I., Manjarrez, N., & Flores, M. (2012). Microbial decolouration of azo dyes: a review. *Process Biochemistry, 47*:1723–1748.

Song, S., Fan, J., He, Z., Zhan, L., Liu, Z., Chen, J., & Xu, X. (2010). Electrochemical degradation of azo dye CI Reactive Red 195 by anodic oxidation on $Ti/SnO_2–Sb/PbO_2$ electrodes. *Electrochimica Acta, 55*:3606–3613.

Srinivasan, A., & Viraraghavan, T. (2010). Decolorization of dye wastewaters by biosorbents: a review. *Journal of Environmental Management, 91*:1915–1929.

Srinivasan, S., & Sadasivam, S. K. (2018). Exploring docking and aerobic-microaerophilic biodegradation of textile azo dye by bacterial systems. *Journal of Water Process Engineering, 22*:180–191.

Telke, A., Kalyani, D., Jadhav, J., & Govindwar, S. (2008). Kinetics and mechanism of reactive red 141 degradation by a bacterial isolate *Rhizobium radiobacter* MTCC 8161. *Acta Chimica Slovenica, 55*:320–329.

Türgay, O., Ersöz, G., Atalay, S., Forss, J., & Welander, U. (2011). The treatment of azo dyes found in textile industry wastewater by anaerobic biological method and chemical oxidation. *Separation and Purification Technology, 79*:26–33.

Tuttolomondo, M. V., Alvarez, G. S., Desimone, M. F., & Diaz, L. E. (2014). Removal of azo dyes from water by sol–gel immobilized *Pseudomonas* sp. *Journal of Environmental Chemical Engineering, 2*:131–136.

Ulson, S. M. D. A. G., Bonilla, K. A. S., & de Souza, A. A. U. (2010). Removal of COD and color from hydrolyzed textile azo dye by combined ozonation and biological treatment. *Journal of Hazardous Materials, 179*:35–42.

Van Der Zee, F. P., Bisschops, I. A., Lettinga, G., & Field, J. A. (2003). Activated carbon as an electron acceptor and redox mediator during the anaerobic biotransformation of azo dyes. *Environmental Science & Technology, 37*:402–408.

Verma, P., & Madamwar, D. (2002). Comparative study on transformation of azo dyes by different white rot fungi. *Indian Journal of Biotechnology, 1*:393–396.

Verma, P., & Madamwar, D. (2003). Decolourization of synthetic dyes by a newly isolated strain of *Serratia marcescens*. *World Journal of Microbiology and Biotechnology, 19*:615–618.

Verma, P., & Madamwar, D., (2005). Decolorization of azo dyes using Basidiomycete strain PV 002. *World Journal of Microbiology and Biotechnology*, 21:481–485.

Vijayaraghavan, J., Basha, S. S., & Jegan, J. (2013). A review on efficacious methods to decolorize reactive azo dye. *Journal of Urban and Environmental Engineering, 7*:30–47.

Vijayaraghavan, K., & Yun, Y. S. (2007). Utilization of fermentation waste (Corynebacterium glutamicum) for biosorption of Reactive Black 5 from aqueous solution. *Journal of Hazardous Materials, 141*:45–52.

Vitor, V., & Corso, C. R. (2008). Decolorization of textile dye by *Candida albicans* isolated from industrial effluents. *Journal of Industrial Microbiology & Biotechnology, 35*:1353–1357.

Waghmode, T. R., Kurade, M. B., & Govindwar, S. P. (2011). Time dependent degradation of mixture of structurally different azo and non azo dyes by using *Galactomyces geotrichum* MTCC 1360. *International Biodeterioration & Biodegradation, 65*:479–486.

Watabe, T., Ozawa, N., Kobayashi, F., & Kurata, H. (1980). Reduction of sulphonated water-soluble azo dyes by micro-organisms from human faeces. *Food and Cosmetics Toxicology, 18*:349–352.

Weber, E. J., & Adams, R. L. (1995). Chemical-and sediment-mediated reduction of the azo dye disperse blue 79. *Environmental Science & Technology, 29*:1163–1170.

Wu, R., Qu, J., & Chen, Y. (2005). Magnetic powder MnO–Fe$_2$O$_3$ composite—a novel material for the removal of azo-dye from water. *Water Research, 39*:630–638.

Wu, R., Qu, J., He, H., & Yu, Y. (2004). Removal of azo-dye Acid Red B (ARB) by adsorption and catalytic combustion using magnetic CuFe$_2$O$_4$ powder. *Applied Catalysis B: Environmental, 48*:49–56.

Wuhrmann, K., Mechsner, K. L., & Kappeler, T. H. (1980). Investigation on rate—Determining factors in the microbial reduction of azo dyes. *European Journal of Applied Microbiology and Biotechnology, 9*:325–338.

Yadav, A. K., Jena, S., Acharya, B. C., & Mishra, B. K. (2012). Removal of azo dye in innovative constructed wetlands: Influence of iron scrap and sulfate reducing bacterial enrichment. *Ecological Engineering, 49*:53–58.

Yang, Q., Angly, F. E., Wang, Z., & Zhang, H. (2011a). Wastewater treatment systems harbor specific and diverse yeast communities. *Biochemical Engineering Journal, 58*:168–176.

Yang, X., Wang, J., Zhao, X., Wang, Q., & Xue, R. (2011b). Increasing manganese peroxidase production and biodecolorization of triphenylmethane dyes by novel fungal consortium. *Bioresource Technology, 102*:10535–10541.

Yazdanbakhsh, M., Khosravi, I., Goharshadi, E. K,. & Youssefi, A. (2010). Fabrication of nanospinel ZnCr2O4 using sol–gel method and its application on removal of azo dye from aqueous solution. *Journal of Hazardous Materials, 184*:684–689.

Zafar, M. N., Dar, Q., Nawaz, F., Zafar, M. N., Iqbal, M., & Nazar, M. F., (2019). Effective adsorptive removal of azo dyes over spherical ZnO nanoparticles. *Journal of Materials Research and Technology, 8*:713–725.

Zhou, C., & Wen, X. H. (2009). Degradation of acid blue 45 in a white-rot fungi reactor operated under non-sterile conditions. *Huan Jing Ke Xue, 30*:1797.

Zhu, G., Fang, H., Xiao, Y., & Hursthouse, A. S. (2020). The application of fluorescence spectroscopy for the investigation of dye degradation by chemical oxidation. *Journal of Fluorescence, 30*:1271–1279.

Zuorro, A., Lavecchia, R., Medici, F., & Piga, L. (2013). Spent tea leaves as a potential low-cost adsorbent for the removal of azo dyes from wastewater. *Chemical Engineering, 32*:19–24.

3 Microbial Remediation of Azo Dyes Using Bacterial Approaches

Refat Pervin[1] and Md. Milon Hossain[1]*

[1]Department of Textile Engineering, Khulna University of Engineering and Technology, Khulna 9203, Bangladesh

*Corresponding author: milon.hossain@te.kuet.ac.bd

CONTENTS

DOI: 10.1201/9781003130932-3

3.1   INTRODUCTION

In textile manufacturing, dyes are indispensable to enhance the aesthetic of garments. Every day, many types of dyes are used and most of them are synthetic dyes. Approximately 60%–70% of artificial dyes are some form of azo dyes, and these dangerous and toxic waste products cause an extreme level of water pollution in the water bodies close to the industries [1]. Azo dyes are cheaper and are used the most, because of the variable color range, different shades, and they have good physical performance, such as colorfastness to rubbing, perspiration, saliva, and washing. Due to their low cost and high efficiency, azo dyes are versatile and are used in the pharmaceutical, cosmetics, plastics and rubber, leather, paper, and food industries.

Azo dyes are xenobiotic compounds that contain ≥1 (-N=N-) bond along with aromatic benzene and naphthalene cyclic rings that contain different types of substitute chromophore groups, which creates a variety of azo dyes of varied color and shade. During dyeing, all azo molecules cannot bond with the textiles. The residual dye molecules and auxiliaries enter the water and produce a large amount of wastewater. This wastewater is quantified by extreme fluctuations in various parameters, such as dye molecules, chemicals, colors, biological oxygen demand, chemical oxygen demand, high salinity, and pH [2]. It alters the pH of the water, reduces dissolved oxygen concentration, prevents sunlight penetration into deep water, and it breaks down the ecological balance if it is not treated before discharge into the water sources. Azo dyes are carcinogenic, mutagenic, and toxic compounds; therefore, contaminated wastewater adversely affects water quality, soil health, biotic lives, and environmental harmony.

Various methods, such as physicochemical and biological have been used to treat wastewater effluents. Physicochemical methods are used more for industrial wastewater treatment; however, these processes are expensive, not bio-friendly, and produce more sludge that is toxic and recalcitrant. These circumstances have influenced the search for an alternative way to treat azo dyes that are contaminated in wastewater so that they could be detoxicated completely. Because of the abundance of microorganisms and their high effectiveness, researchers have been working on different microbial treatment methods, such as bacterial, fungal, algae, yeast, and bacterial–fungal consortia for the bioremediation of azo dyes for the last 20 years. The bacterial approach has been studied extensively for the bioremediation of azo dyes, because bacteria proliferate under different conditions, such as anaerobic, aerobic, and facultative. In addition, they remain active in adverse environmental conditions, such as high salinity, high temperature, and different pH [3]. This chapter will describe azo dye decolorization and degradation using different mechanisms under anaerobic, anoxic, and aerobic conditions, and in a redox mediator system that uses immobilized bacteria cells, bacterial enzymes, and bacterial consortia.

3.2   BENEFITS OF WASTEWATER TREATMENT BY BACTERIA

Physical and chemical treatment processes that are based on coagulation–flocculation, adsorptions, filtration (ultrafiltration and nanofiltration) and reverse osmosis, electrochemical oxidation, advanced oxygen processes (photochemical and photocatalytic), photolysis, and ozonation have been widely used to detoxify and degrade aromatic azo compounds contained in effluents. However, these processes are not economically feasible, cannot eliminate recalcitrant and resistant azo dyes and their other organic substrates because of the resistance, colorfastness, and durability of the azo dyes to degradation [4]. Numerous biotechnological and microbial approaches have attracted attention to tackle the pollution caused by azo dyes in an environmentally acceptable way, with the application of bacteria in addition to combined physicochemical treatments. The application of microbial and enzymatic treatment methods for the complete mineralization of azo compounds in effluents has different benefits, such as eco-friendly, cost-effective, lower sludge generation, production of nontoxic end products, and they consume less water compared with physicochemical approaches [5]. In addition, the bacterial approach to azo dyes degradation involves less energy and is a less chemical-intensive method.

3.3 COLOR REMOVAL MECHANISMS USING A BACTERIAL APPROACH

There are a significant number of azo dye decolorization and degradation mechanisms by bacteria that involve intracellular enzymes or nonspecified extracellular reduction [6]. Under anaerobic conditions, there are two methods for azo dye decolorization. The azo bonds can be broken down or induced by transporting electrons to the azo dye, which is produced during the catabolic process that generates ATP [6]. In general, the mineralization of aromatic azo compounds by microorganisms involves the reductive breakdown of the azo bond (-N=N-) with the help of azo reductase enzymes under anaerobic conditions, and this includes the transfer of four electrons (reduced equivalents). Then, two steps involve the azo bond. In each step, two electrons are transferred to the azo dyes, which acts as a terminal electron acceptor. Therefore, decolorization occurs and a colorless solution is produced [7]. It forms a low redox potential (≤ 50 mv) under anaerobic conditions, which causes suitable decolorization of the azo dyes. Another mechanism produces inorganic substrates as the end products by reductive action that cleaves the azo bond via catabolic reactions. Under a nonoxidation system, many bacteria can decolorize azo dyes and their auxiliaries. However, this operation is considered the nonspecific breakdown of azo bonds, which often results in toxic and carcinogenic products [8]. Therefore, an anaerobic–aerobic bioremediation process has been proposed that can break down the waste products produced under anoxic conditions in the aerobic process [9].

In aerobic systems, many bacteria can decolorize azo dyes, and the auxiliaries and the mono- and dioxygenase enzymes can catalyze with oxygen (O_2) into the aromatic cyclic ring of organic compounds for ring fission [10]. In addition, azo reductase has been associated with aerobic systems that could use nicotinamide adenine dinucleotide phosphate hydrogen (NAD(P)H) and nicotinamide adenine dinucleotide-hydrogen (reduced) (NADH) as cofactors and reductively cleaved carboxylated substrates and sulfonated structural analogs [11]. In addition, electrons move through the electron transfer chain to the terminal electron receiver (azo compounds), which is reduced, and therefore, decolorization is achieved and the flavin nucleotide is reoxidized [12]. If the reduction occurs intracellularly, the molecules must disperse through the cell membrane. Sulfonated compounds have been incorporated into the reduction and transfer of molecules through the cell membrane [13]. However, the actual transport systems remain unknown and they could transfer other sulfonate compounds, in particular, p-toluenesulfonate, taurine, or alkanesulfonates [14]. Redox mediators are employed to enhance the breakdown of azo bonds. The mediators accelerate the reactions by decreasing the activation energy of the overall reaction and the organic particles are oxidized or reduced alternatively. Therefore, under different redox reactions, they are considered to be electron carriers.

3.4 BACTERIAL DEGRADATION OF AZO DYES

3.4.1 Immobilized Whole Bacteria

Dye degradation by immobilized bacterial cells is another attractive process. The immobilization of these biocatalysts have higher functional stability than free systems. In general, immobilization accelerates the performance of the enzymes that are less exposed to copper (Cu) halides and the dyeing additives than when free enzymes are used [15]. Immobilization of bacteria cells can be carried out by four different methods, such as microencapsulation, covalent binding or attachment, matrix entrapment, and adsorption [13]. During the attachment process, cells adhere to the external layer of an inert substance or other microorganisms. In contrast, cells become trapped in the perforated porous or fibrous substances during entrapment. Bacterial cells can survive a very high level of changes in pH or exposure to higher concentrations of dyes. The degradation of dyes by the immobilization of bacterial cells is a favorable system because of its easy operation under sterile conditions. It prevents cell washouts and allows a greater cell density, which has to be maintained in the continuous reactor. Carriers, such as sintered glass, nylon web, polyurethane foam, activated carbon, pinewood, and porous polystyrene have been used for the immobilization of bacterial cells. A study described the degradation of different azo compounds where immobilized bacterial cultures *of Morganella sp., Pseudomonas sp.,* and *Enterobacter sp.,* on kaolin, powdered activated carbon, and bentonite were used [16]. Various functioning states, such as cell bead number, diameters of the beads, bed extension, density, primal dye concentration, and hydraulic retention time for the decolorization of dyes have been optimized [17]. In addition, bacterial entrapment under natural or synthetic substances could be applied to the degradation of azo dyes on a large-scale (reactor scale), and it generates anaerobic conditions that promote dye degradation by the enzymes in bacteria cells. The immobilization of bacterial cells can accelerate the breakdown of aromatic azo dyes that are contained in industrial waste [18].

3.4.2 Bacterial Enzymes

3.4.2.1 Reductive Enzymes

For the bioremediation of azo dyes using a bacterial approach, reductive enzymes are very effective for azo dye degradation. Azoreductase enzymes induce reductive reactions and break down the azo bonds to produce colorless cyclic amine products [19]. According to their mechanism, these enzymes are classified as flavin-dependent azo reductases [20] and flavin independent azo reductase [21]. Based on flavin dependency, azoreductases are categorized according to their cofactors, NADH, or NADP (H) [20], or both NADH and NADP (H) [22]. Therefore, these coenzymes act as electron donors. Azo reductases collected from bacteria represent the special groups of minuscule enzymes similar to other reductases. The purification and characterization of azoreductase enzymes from various bacteria, such as *Staphylococcus aureus, Xenophilus azovorans* KF46F, *Escherichia coli, Enterococcus faecalis, Bacillus sp.,* and *Rhodobacter sphaeroides* have been studied [21]. In addition, NADH–DCIP reductases are marker enzymes for bacterial and fungi mixed functional oxidase systems. These take part in detoxifying xenobiotic azo compounds [23]. In addition, riboflavin reductase enzymes participate in the degradation of xenobiotic azo dyes.

3.4.2.2 Oxidative Enzymes

Oxidative enzymes, such as laccase tyrosinase and lignin peroxidase, are used for bacterial decolorization and biodegradation of aromatic azo compounds combined with the reductive enzymes and are responsible for the complete mineralization of azo compounds. Laccases are known as phenolic oxidase and are specified by Cu holding enzymes that can oxidize different cyclic azo and other inorganic compounds combined with the reductive process [24]. The structure of this enzyme delivers four adjacent Cu atoms, and they are categorized into three classes: Cu I, Cu II, and Cu III that have different characteristics. The molecular weight of laccase is 60–390 kDa [25]. Due to the indirect breaking down of azo bonds, the decolorization of some azo dyes is performed by this enzyme through a highly nonspecific free radical mechanism that avoids the formation of toxic aromatic amines [25]. Tyrosinase is a monophenol monooxygenase and is involved in phenolic oxidization. Tyrosinase is used to remove phenol where molecular O_2 is the oxidation agent instead of hydrogen peroxide (H_2O_2) and decreases the potential cost of use [26]. Lignin peroxidase is a glycoprotein type of oxidative enzyme, which has been used in the complete mineralization of a wide variety of toxic cyclic compounds, such as three and four-ring polyaromatic hydrocarbons (PAHs), polychlorinated biphenyls, and commercial synthetic dyes [27]. The oxidation mechanism of lignin peroxidase involves the oxidization of phenolic groups by the generation of radicals at the carbon that holds the azo bonds. Water molecules attack the phenolic carbon; therefore, it can break down the molecule that produces phenyldiazene. Then, phenyldiazene can be oxidized using a single electron reaction that generates nitrogen [28].

3.4.3 BACTERIAL CONSORTIA

Under various conditions, it has been recorded that mixed cultures are more efficient than pure cultures for dye breakdown and mineralization, because of the synergistic assimilation processes in bacteria [29,30]. Some isolated strains focus on attacking dye compounds and molecules at various stages of the treatment process, which results in waste that can be carcinogenic and harmful. Further down the process, another strain of bacteria could process the molecules into useful sources of carbon dioxide, ammonia, and water. Bioremediation is the most suitable method to ensure that the effluent is not toxic and can be discharged to the water bodies. Mixed cultures are usually resistant to changes in pH, temperature, and nutrients compared with pure cultures. In addition, they do not require sterile conditions [31], and therefore, mixed cultures are more suitable for bioreactors.

3.5 FACTORS THAT AFFECT BACTERIAL DECOLORIZATION AND DEGRADATION

3.5.1 O_2 AND AGITATION

To maintain the efficiency of dye degradation, aeration and agitation should be avoided, which increase the O_2 concentration in the solution [30]. Azo dyes have strong resistance to reduction due to bacteria at high concentrations of O_2. This was caused by the direct inhibition of the azoreductase enzyme or a reduced concentration of O_2. The wastewater from textile industries is treated sequentially by anaerobic and aerophilic processes. Under anaerobic conditions, bacteria degrade the dyes into colorless toxic (carcinogenic) aromatic amines. However, under aerophilic conditions, some of the toxic amines are easily metabolized. Under anaerobic conditions, most aromatic amines generated from azo dyes decolorization are resistant to biodegradation. Therefore, although anaerobic biodegradation is preferred over aerobic biodegradation; however, sequential degradation from both is required to metabolize the carcinogenic aromatic components [32].

3.5.2 BIOAVAILABILITY

Different physicochemical characteristics of dye molecules and their intermediates can individually affect bioavailability in aqueous state or solution containing microbes or enzymes and could penetrate the cellular membrane that is to be metabolized. Hydrophilicity and hydrophobicity are crucial and affect the azo dyes when they are exposed to living organisms. Sulfonated azo compounds that are water-soluble dyes are polar molecules [33]. In addition, due to their larger molecular size, they cannot penetrate the membrane barrier. A biogenic approach, such as enzymes, or redox mediators, or both are responsible for the reduction of aromatic azo bonding and the degradation of these compounds. Sudan azo dyes are hydrophobic and fat-soluble and can pass through the cell membrane's steric barrier. In addition, they can be degraded and absorbed across the membrane. However, the bioavailability in the fluid state is low. Liposomes are active when inducing a higher decolorization of Acid Orange 7 via anaerobic biomass [33]. A selective microbial group can improve the bioavailability of insoluble dyes, and this can be carried out by Shewanella strain J18 143, which is characterized by its ability to break down large pigment molecule aggregates to disperse the dye and generate separate pigment particles [34].

3.5.3 Carbon and Nitrogen

Biodegradation of the azo dyes requires additional nitrogen and carbon sources. Complex organic compounds, such as peptone, yeast extract, and a mixture that contains carbohydrates are required for the decolorization of azo dyes. During decolorization, reduced equivalents from various carbon sources result in transporting the organic sources to the dyes. However, the additional carbon sources do not effectively enhance dye degradation, which is probably due to the tendency of the cells to assimilate the excess sources by over-utilizing the dye particles and carbon sources for biodegradation [35]. Currently, glucose is the most readily usable carbon source for most types of microorganisms. In addition to glucose, inexpensive sources of carbon, such as fructose, molasses, and starch have been utilized in dye degradation [36]. Using organic nitrogen sources, such as yeast, peptone, beef extract, and urea can reproduce NADH, which acts as an electron donor during azo dye degradation by microorganisms, which results in effective decolorization. Lignocellulosic agricultural wastes could be a potentially cost-effective and economically viable supplement for decolorization, as reported in the literature [35].

3.5.4 Temperature

Temperature changes change significantly affect microorganisms assisted biodegradation. For every 10°C increase in temperature, the reaction rate increases twice; however, this loses linearity outside of the ideal temperature [30]. The increase in the degradation activity of the microorganisms is due to a reduced growth and reproduction rate combined with the deactivation of enzymes that results in dye degradation. Different types of microorganisms have different ideal temperatures for growth and degradation; however, but most grow between 25° and 35°C [30]. The denaturation of enzymes causes the loss of viable cells and the dye decolorization rate reduces at higher temperatures.

3.5.5 pH

Textile wastewater has different pH values that depend on the processing steps. Usually, at an acidic pH (3.0–5.0) fungi and yeasts cause maximum decolorization and biodegradation. For bacteria, the maximum biodegradation occurs under neutral and basic pH conditions [37]. Under acidic or alkaline conditions, the dye decolorization efficiency of bacteria is lower. pH affects the dissolving and conveyance of dye particles through the plasma membrane combined with the growth of the bacteria and the substrate-limiting stage for dye decolorization [37]. Therefore, microorganisms that can degrade dyes across a wide range of pHs are required for improved performance.

3.5.6 Dye Concentration

The concentration of azo dye is proportional to the time required for the decolorization of the dye in the wastewater system [38]. The decolorization rate decreases with an increase in the concentration of the dye. This decreased rate is due to the toxic properties of the dyes that affect the multiplying of the microorganisms at increased dye concentrations. In contrast, the reduction rate of the dye does not depend on the dye concentration [39].

3.5.7 Dye Structure

The type and structure of azo based dyes, combined with the changing chemical properties and structures, such as isomerism and diverse functional groups, are responsible for the decolorization and degradation of the dyes. Of note, dyes of high molecular weight are more resistant to degradation than low molecular weight dyes. Dyes with branched and crosslinked polymer chains are resistant to degradation. For example, monoazo dyes have a faster breakdown rate than diazo and triazo dyes. Azo based dyes that contain nitro, sulfo, methyl, or methoxy are more resistant toward degradation than the ones that contain hydroxyl and amino groups [35]. Rapid single electron transfer chemical reactions force the dyes to be reduced into anionic forms, which results in a second slower electron transfer chemical reaction that creates a much more stable dianion. Therefore, dianion formation might be hampered in the presence of high electron density groups, which results in reduced or no decolorization. The main rate-limiting factor to break the sulfonate group dyes using bacteria is the permeation of the dye through the bacterial cell membrane. Of note, a significant impact on the dye reduction rate was due to the presence of electron-dense hydrogen bonds near the azo bond that resulted in enhanced azo–hydrazone tautomerism of the hydroxy azo compound [35].

3.5.8 ELECTRON DONORS

Electron donors are the prerequisite for the remediation of the azo dyes. Theoretically, azo bonds require four reducing equivalents, which indicates that a lower quantity of electron donors are required. However, the demand for other reactions for the electron donors leads to an increase in the quantity of the required reducing equivalents. Azo dyes combined with other organic components of textile wastewater do not contain the necessary electron donors [35]. Therefore, there are not enough components to support the growth of anaerobic bacteria. Therefore, the electron donors, such as formate ions, acetate, and glucose act as supplements to induce the reductive cleavage of azo bonds. In addition, the high concentration of electron donors kinetically benefits the reduction of azo dye. Therefore, it is very important to determine the physiological electron donors for each bioremediation cycle, because the donors trigger the chemical reactions in azo dye reduction and induce the enzyme clusters that are responsible for the reduction process. In some cases, certain electron donors obstruct electron transportation due to competition for electrons from the electron donors [35].

3.5.9 REDOX MEDIATORS

Redox mediators are a crucial factor for the bioremediation of cyclic azo compounds. In these mediators, reduced compounds are transferred from the principal electron sources to the final electron receiver (azo dyes). In general, they act as a rate reducing step in the dye reduction reactions under anaerobic conditions [40]. The decolorization of dye compounds is related to the electron density of the azo bonding region, and the color removal rate could be enhanced by ensuring a low level of electron density in the cyclic azo bonds [41].

3.5.10 HIGH SALINITY

Residues of azo dyes usually contain 15%–20% of salt that originates from the processing industries. Therefore, halotolerant bacterial strains that can degrade azo dyes might help in the biotreatments used for the remediation of wastewater that contains salt and azo dyes from the textile industries [42]. Initially, a salt concentration of 3–5 g/L causes an increase in the dye decolorization rate; however, further increasing the salt concentration decreases the decolorization. This is due to the variations in enzyme activity and molecular transportation across the plasma membrane of the microorganisms. The membrane properties are affected by the elevated osmotic pressure that is caused by the increase in salt concentration. The decolorization rate of the dye decreases due to the decrease in transportation. Some nonmarine bacterial strains might grow at a lower concentration of salt. However, a higher concentration could be lethal for bacterial growth [41].

3.6 DECOLORIZATION OF AZO DYES BY BACTERIA UNDER DIFFERENT CONDITIONS AND SYSTEMS

3.6.1 ANAEROBIC CONDITIONS

Anaerobic methanogenesis involves various trophic species of bacteria, which includes acetogenic, acidogenic, and methanogenic bacteria [43]. Dye decolorization depends on the primary electron donor types under methanogenic conditions. Volatile fatty acids and acetate are very poor sources to act as electron sources donors, H_2, CO_2, glucose, formate, and ethanol act as electron sources or donors for dye reduction reactions [44]. A lot of studies have been carried out to determine the role of different types of bacteria in decolorization under anaerobic conditions. Currently, the reduction is carried out by the co-metabolic mechanism and the reduced compounds produced during degradation can be transferred by chemical means. In the anaerobic consortium systems, fermentative bacteria produce reduced compounds during the degradation of pure substrates that can be chemically transferred to the azo dyes. In an anaerobic consortia system, fermentative bacteria form reducing equivalents and consume methanogens to form methane. However, instead of methanogenesis, some methanogens could be used to reduce analogous compounds during azo dye reduction. Therefore, methanogenic archaea and fermentative bacteria might have a crucial role in reducing azo dyes. Several studies have been performed under anaerobic conditions. Congo red and textile wastewater were investigated at different stages under anaerobic conditions [45]. Using the molecular method, the bacteria populations could be characterized in anaerobic baffled reactors. The members of Proteobacteria and sulfate-reducing bacteria were most prominent in the mixture of dye wastewater. In addition, Methanosaeta strains and *Methanomethylovorans hollandica* have been used for the treatment of industrial dyes contained in wastewater [46]. The mechanism of anaerobic azo dye reduction is presented in Figure 3.1.

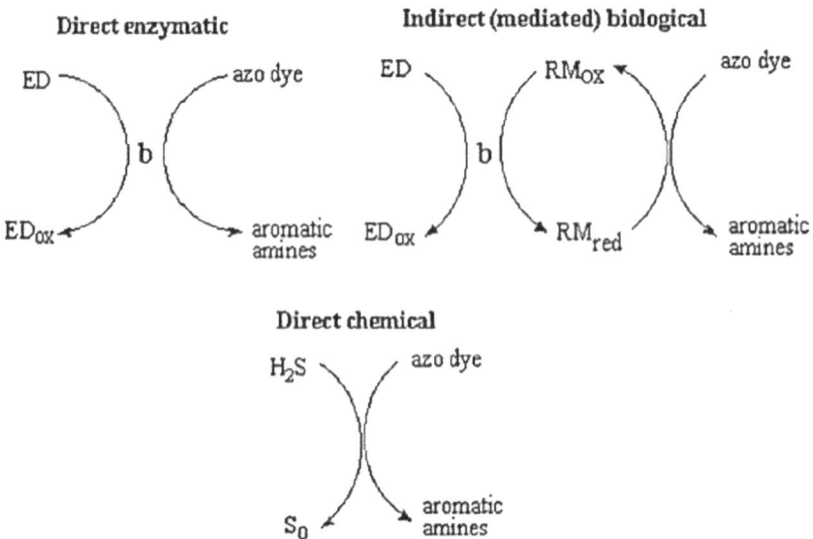

FIGURE 3.1   Schematic representation of different mechanisms of anaerobic azo dye reduction.

RM = Redox mediator, ED = Electron donor, b = bacteria (enzymes) [47].

3.6.2 ANOXIC CONDITIONS

Anoxic conditions are free from O_2, and a microaerophilic medium requires a very low level of dissolved O_2 [48]. Various types of azo dyes, which have been decolorized under anoxic conditions, have been reported [49]. Although several strains can grow under aerobic conditions, decolorization can only be achieved aerobically. Pure bacterial strains, such as *Pseudomonas luteola, Aeromonas hydrophila, Bacillus subtilis, Pseudomonas sp.,* and *Proteus mirabilis* decolorized azo dyes under anoxic conditions [19]. Complex sources, such as peptone, yeast, carbohydrates, or a mixture of complex organic compounds are generally required to decolorize azo dyes by mixed and pure cultures [49]. In anaerobic decolorization, glucose is the most used feedstock under methanogenic conditions. Glucose enhances the decolorizing rate of Mordant Yellow 3 by *Sphingomonas xenophaga* BN6, although a significant reduction in decolorization was shown by *P. leuteola, Aeromonas sp.,* and a few other mixed strains [50]. Glucose negatively affects anoxic decolorization due to the decreased pH.

3.6.3 AEROBIC CONDITIONS

Aerobic decolorization of aromatic azo compounds is carried out by different bacterial strains. However, different bacterial strains cannot use azo compounds as the sole source of organic carbon sources for growth [51]. Under aerobic conditions, and in the presence of glucose, *P. aeruginosa* decolorized industrial, leather, and textile dyes, such as Navitan Fast blue S5R. In addition, this microorganism can degrade other types of cyclic azo compounds [52]. There are very few bacterial strains that utilize aromatic azo dyes as the carbon source for growth and break down the azo bonds using a reductive approach and utilize amines substrates as the sole source for growth and energy. Bacterial groups, such as *X. azovorans* KF46, and *Pigmentiphaga kullae* K24, which can grow aerobically on Carboxy Orange I and II, respectively are used for the decolorization of azo dyes [53]. Four types of bacteria have been separated that use Methyl Red as a carbon source. Two groups (*P. nitroreducens* and *Vibrio logei*) have been identified and the amine products were not detected in the culture medium, which indicated degradation [54]. In addition, three bacterial species, *Pseudomonas, Arthrobacter,* and *Rhizobium* can decolorize Acid Orange 7 [55].

3.6.4 REDOX MEDIATOR SYSTEM

Redox mediators increase the transport of electrons from the primary electron sources to the final electron receiver. Quinone based compounds, such as 2-Hydroxy-1, 4-Naphthoquinone (Lawsone), Anthraquinone-2-Sulfonate (AQS), Anthraquinone-2, 6-Disulfonate (AQDS), and flavin based substrates, such as riboflavin, flavin adenine dinucleotide, and flavin adenine mononucleotide are the most common redox mediators [56]At the lower activation energy of the reaction, these quinone compounds are highly effective [57]. Therefore, the activation energy must be half of the two ultimate reactions that are the reductive breakdown of cyclic azo compounds and the oxidation of the primary electron sources. Reductive decolorization is carried out by two phases in the presence of redox mediators. The first phase is a nonspecific enzymatic reduction, and the second phase is re-oxidation by aromatic azo compounds [58]. The reductive decolorization accelerates the transfer of the reduced compounds to the final electron receiver. They decrease the steric hindrance and reduce the activation energy to facilitate the reactions. Therefore, redox mediators enhance the reaction speed by ≥1 order of magnitude [59]. The proposed mechanism of a redox mediator is shown in Figure 3.2.

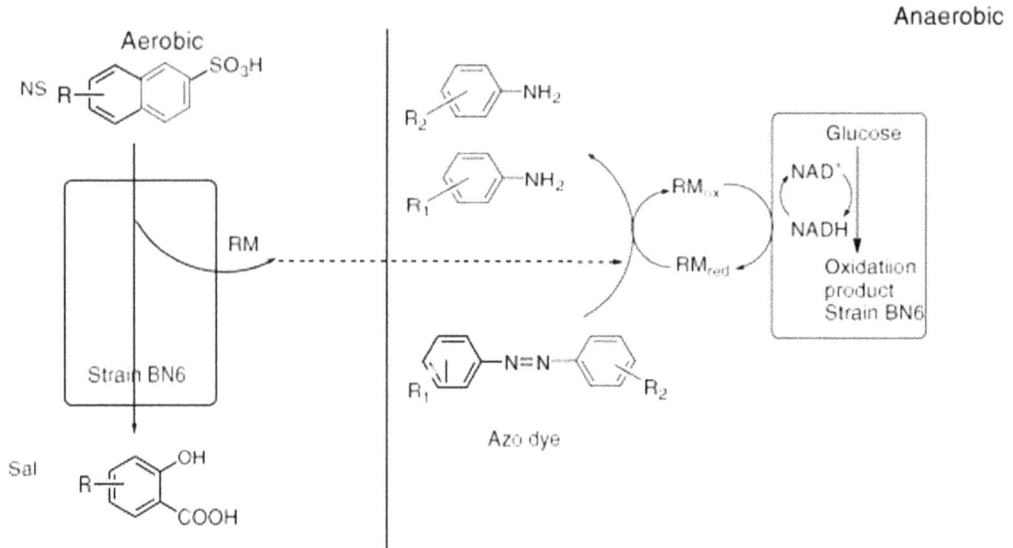

FIGURE 3.2 Proposed mechanism for the redox mediator-dependent reduction of azo dyes by strain *Sphingomonus xenophaga* BN6 [58].

3.7 FUTURE ASPECTS OF BACTERIAL REMEDIATION

The microbial remediation of azo dyes has been studied by many researchers mainly in the laboratory. However, microbial degradation is not cost-effective or efficient enough to be used commercially. Therefore, further research is required to make the process commercially viable. Genetically engineered and highly efficient microorganisms could have a huge impact on remediation and require better understanding. In addition, by applying molecular biology the pathways of dye degradation could be determined, which could help to develop more efficient genetically engineered bacterial strains. In addition, the study of the immobilization of enzymes could help to make the system more efficient. A microbial fuel cell could be developed to reduce the use of industrial electricity and lower operating costs. Multidisciplinary studies should be conducted that focus on lower water use during processing, reducing the waste generated and developing and producing dyes that can easily be biodegraded.

3.8 CONCLUSION

In this chapter, enzymes were studied, and the characteristics of the metabolites produced during different treatment processes were discussed. The xenobiotics, which are toxic waste products of azo dyes, cause an extreme level of pollution in the nearby water bodies. Microbial remediation could be a viable solution to improve the process. The method is environmentally friendly and economically inexpensive, and therefore, could be feasible in the long term. In addition, it does not produce a large volume of harmful sludge. Moreover, bacteria have the advantages of rapid growth and high hydraulic retention time, which makes them effective when treating the complex structures in organic wastes. Although many large-scale studies have been carried out into the bioremediation of azo dyes that used pure and mixed bacterial strains in the laboratory, more industrial trials are required to optimize the process. The reductive cleavage of azo bonds results in the formation of toxic aromatic amines under anaerobic conditions. Therefore, it is essential to develop an efficient bacterial strain that can degrade aromatic amines and assess the level of mineralization. Anaerobic and aerobic methods and redox mediators have different efficiencies when treating wastewater that contains different azo compounds. Further research is required into bacterial bioremediation, which could develop a feasible solution to treat azo dyes that contaminate wastewater that is produced by the textile industries.

REFERENCES

1. Ambrósio ST, Campos-Takaki GM. Decolorization of reactive azo dyes by *Cunninghamella elegans* UCP 542 under cometabolic conditions. Bioresour. Technol. 2004; 91:69–5.
2. Dos Santos AB, Bisschops IAE, Cervantes FJ. In: Cervantes FJ, Van Haandel AC, Pavlostathis SG, editors. Closing process water cycles and product recovery in textile industry: perspective for biological treatment. Advanced Biological Treatment Processes for Industrial Wastewaters. London: International Water Association London; 2006. pp. 298–320.
3. Solis M, Solis A, Perez HI, Manjarrez N, Flores M. Microbial decolorization of azo dyes: a review. Process Biochem. 2012; 47:1723–48.
4. Anjaneyulu YN, Chary S, Raj SSD. Decolourization of industrial effluents – Available methods and emerging technologies – A review. Rev. Environ. Sci Biotechnol. 2005; 4:245.
5. Banat IM, Nigam P, Singh D, Merchant R. Microbial decolorization of textile-dye-containing effluents: A review. Bioresour. Technol. 1996; 58:217.
6. Double M, Kumar A. Biotreatment of Industrial Effluents. UK: Elsevier Butterworth-Heimann; 2005.
7. Chang JS, Kuo TS. Kinetics of bacterial decolorization of azo dye with *Escherichia coli* NO_3. Bioresour Technol. 2000; 75:107.
8. Mc Mullan G, Meehan C, Conneely A, Kirby N, Robinson T, Nigam P et al. Microbial decolourisation & degradation of textile dyes. Appl Microbiol Biotechnol. 2001; 56:81–7.
9. Rajagura P, Kalaiselvi K, Palanivel M, Subburam V. Biodegradation of azo dyes in a sequential anaerobic-aerobic system. Appl Microbiol Biotechnol. 2000; 54:268–73.
10. Sarayu K. Sandhya S. Aerobic biodegradation pathway for Remazol Orange by *Pseudomonas aeruginosa*. Appl. Biochem. Biotechnol. 2010; 160:1241–53.
11. Nachiyar CV, Rajkumar GS. Purification and characterization of an oxygen insensitive azoreductase from *Pseudomonas aeruginosa*. Enzyme Microb. Technol. 2005; 36:503–9.
12. Robinson T, McMullan G, Marchant R, Nigam P. Remediation of dyes in textile effluent: A critical review on current treatment technologies with a proposed alternative. Bioresour Technol. 2001; 77:247–55.
13. Monsan P. Les methodes immobilisation enzymes. In: Durand G, Monsan P, editors. Les enzymes, productions utilizations industrielles, Paris: Gauthier-Villards; 1982. pp. 81–118.
14. Blümel S, Knachmuss H-J, Stolz A. Molecular cloning and characterization of the gene coding for the aerobic azoreductase from *Xenophilus azovorans* KF46F. Appl Environ Microbiol. 2002; 68:3948–55.
15. Rogalski J, Jozwik E, Hatakka A, Leonowick A. Immobilization of laccase from *Phlebia radiata* on controlled porosity glass. J Mol Catal A Enzym. 1995; 95:99–108.
16. Barragan BE, Costa C, Marquez MC. Biodegradation of azo dyes by bacteria inoculated on solid media. Dyes Pigments. 2007; 75:73–81.
17. Puvaneswari N, Muthukrishnan J, Gunasekaran P. Toxicity assessment and microbial degradation of azo dyes. Indian J Exp Biol. 2006; 44:618–26.
18. Sheth N, Dave S. Enhanced biodegradation of Reactive Violet 5R manufacturing wastewater using down flow fixed film bioreactor. Bioresour Technol. 2010; 101:8627–31.
19. Chang JS, Chou C, Lin Y, Ho J, Hu TL. Kinetic characteristics of bacterial azo-dye decolorization by *Pseudomonas luteola*. Water Res. 2001; 35:2841–50.
20. Chen H, Hopper SL, Cerniglia CE. Biochemical and molecular characterization of an azoreductase from *Staphylococcus aureus*, a tetrameric NADPH-dependent flavoprotein. Microbiology. 2005; 151:1433.
21. Blümel S, Stolz A. Cloning and characterization of the gene coding for the aerobic azoreductase from *Pigmentiphaga kullae* K24. Appl. Microbiol. Biotechnol. 2003; 62:186.
22. Wang CJ, Hagemeier C, Rahman N, Lowe E, Noble M, Coughtrie M, et al. Molecular cloning, characterisation and ligand-bound structure of an azoreductase from *Pseudomonas aeruginosa*. J. Mol. Biol. 2007; 373:1213–28.

23. Bhosale S, Saratale G, Govindwar S. Mixed function oxidase in *Cunninghamella blakesleeana* (NCIM-687). J. Basic Microbiol. 2006; 46:444.

24. Majcherczyk A, Johannes C, Hüttermann A. Oxidation of polycyclic aromatic hydrocarbons (PAH) by Laccase of *Trametes versicolor*. Enzyme Microb. Technol. 1998; 22:335.

25. Kalme S, Jadhav S, Jadhav M, Govindwar S. Textile dye degrading Laccase from *Pseudomonas desmolyticum* NCIM 2112. Enzyme Microb. Technol. 2009; 44:65.

26. Wu FC, Tseng RL, Juang RS. Enhanced abilities of highly swollen chitosan beads for color removal and tyrosinase immobilization. J. Hazard. Mater. 2001; 81:167.

27. Dawkar VV, Jadhav UU, Ghodake GS, Govindwar SP. Effect of inducers on the decolorization and biodegradation of textile azo dye Navy Blue 2GL by *Bacillus sp.* VUS. Biodegradation. 2009; 20:777.

28. Chivukula M, Renganathan V. Phenolic azo dye oxidation by Laccase from *Pyricularia oryzae*. Appl. Environ. Microbiol. 1995; 61:4374.

29. Sudha M, Saranya A, Selvakumar G, Sivakumar N. Microbial degradation of Azo dyes: A review. Int. J. Curr. Microbiol. Applied Sci. 2014; 3:670–90.

30. Ali H. Biodegradation of synthetic dyes-a review. Water Air Soil Pollut. 2010; 213:251–73.

31. Ramalho PA, Cardoso MH, Cavaco-Paulo A, Ramalho MT. Characterization of azo reduction activity in a novel ascomycete yeast strain. Applied Environ. Microbiol. 2004; 70: 2279–88.

32. Rai HS, Bhattacharyya MS, Singh J, Bansal TK, Vats P, Banerjee UC. Removal of dyes from the effluent of textile and dyestuff manufacturing industry: A review of emerging techniques with reference to biological treatment. Crit Rev Environ Sci Technol. 2005; 35:219–38. doi:10.1080/10643380590917932

33. Carvalho MC, Pereira C, Goncalves IC, Pinheiro HM, Santos AR, Lopes A, et al. Assessment of the biodegradability of a monosulfonated azo dye and aromatic amines. Int Biodeterior Biodegradation. 2008; 62(2):96–103.

34. Pearce C, Guthrie JT, Lloyd JR. Reduction of pigment dispersions by *Shewanella* strain J18 143. Dyes Pigm. 2008; 76(3):696–705.

35. Saratale RG, Saratale GD, Chang JS, Govindwar SP. Bacterial decolourization and degradation of azo dyes: a review. J Taiwan Inst Chem Eng. 2011; 42 :138–57. doi:10.1016/j.jtice.2010.06.006

36. Kaushik P, Malik A. Fungal dye decolourization: Recent advances and future potential. Environ Int. 2009; 35:127–41. doi:10.1016/j.envint.2008.05.010

37. Agrawal S, Tipre D, Patel B, Dave S. Optimization of triazo Acid Black 210 dye degradation by *Providencia sp.* SRS82 and elucidation of degradation pathway. Proc Biochem. 2014; 49:110–19. doi:10.1016/j.procbio.2013.10.006

38. Sheth NT, Dave SR. Optimisation for enhanced decolourization and degradation of reactive Red BS C.I. 111 by *Pseudomonas aeruginosa* NGKCTS. Biodegradation. 2009; 20:827–36. doi:10. 1007/s10532-009-9270-2

39. Dave SR, Patel TL, Tipre DR. Bacterial degradation of azo dye containing wastes. In: Singh SN, editor. Microbial degradation of synthetic dyes in waste water. Cham, Switzerland: Springer. pp. 57–83. doi:10.1007/978-3-319-10942-8_3

40. Van der Zee FP, Bouwman RHM, Strik DPBTB, Lettinga G, Field JA. Application of redox mediators to accelerate the transformation of reactive azo dyes in anaerobic bioreactors. Biotechnol. Bioeng. 2001; 756:691.

41. Walker R, Ryan AJ. Some molecular parameters influencing rate of reduction of azo compounds by intestinal microflora. Xenobiotica. 1971; 1:483.

42. Khalid A, Arshad M, Crowley DE. Decolorization of azo dyes by Shewanella sp. under saline conditions. Applied Microbiol. Biotechnol. 2008; 79:1053–9.

43. Kasper HF, Wuhrmann K. Kinetic parameters and relative turnover of some important catabolic reactions in digesting sludge. Appl Environ Microbiol. 1978; 36:1–7.

44. Pearce CI, Christie R, Boothman C, Von Canstein H, Guthrie JT, Lloyd JR. Reactive azo dye reduction by *Shewanella* Strain J18 143. Biotechnol Bioeng. 2006; 95:692–703.

45. Firmino PIM, da Silva MER, Cervantes FJ, dos Santos AB. Colour removal of dyes from synthetic and real textile wastewaters in one- and two-stage anaerobic systems. Biores Technol. 2010; 101:7773–9.

46. Plumb JJ, Bell J, Stuckey DC. Microbial populations associated with treatment of an industrial dye effluent in an anaerobic baffled reactor. Appl Environ Microbiol. 2001; 67:3226–35.

47. Pandey A, Singh P, Iyengar L. Bacterial decolorization and degradation of azo dyes. Intl Biodeter Biodegrad. 2007; 59(2):73–84.

48. Moosvi S, Keharia H, Madamawar D. Decolorization of textile dye reactive violet by a newly isolated bacterial consortium RVM 11.1. World J Microbiol Biotechnol. 2005; 21:667–72.
49. Khehra MS, Saini HS, Sharma DK, Chadha BS, Chimni SS. Decolorization of various azo dyes by bacterial consortia. Dyes Pigment. 2005; 67:55–61.
50. Chen KC, Wu JY, Liou DJ, Hwang SJ. Decolorization of textile dyes by newly isolated bacterial strains. J Biotechnol. 2003; 101:57–68.
51. Kodam KM, Soojhawon I, Lokhande PD, Gawai KR. Microbial decolorization of reactive azo dyes under aerobic conditions. World J Microbiol Biotechnol. 2005; 21:367–70.
52. Nachiyar CV, Rajkumar GS. Degradation of tannery and textile dye, Navitan Fast Blue S5R by *Pseudomonas aeruginosa*. World J Microbiol Biotechnol. 2003; 19:609–14.
53. Kulla HG, Klausener F, Mayer U, Ludeke B, Leisinger T. Interference of aromatic sulfo groups in the microbial degradation of the azo dyes Orange I and Orange II. Arch Microbiol. 1983; 135:1–7.
54. Adedayo O, Javadpour S, Taylor C, Anderson WA, Moo-Young M. Decolourization and detoxification of methyl red by aerobic bacteria from a wastewater treatment plant. World J Microbiol Biotechnol. 2004; 20:545–50.
55. Ruiz-Arias A, Juárez-Ramírez C, Cobos-Vasconcelos DD, Ruiz-Ordaz N, Salmerón-Alcocer A, Ahuatzi-Chacón D, et al. Aerobic biodegradation of a sulfonated phenylazonaphthol dye by a bacterial community immobilized in a multistage packed-bed BAC reactor. Appl Biochem Biotechnol. 2010; 162(6):1689–1707.
56. Dos Santos AB. Reductive decolourisation of dyes by thermophilic anaerobic granular sludge Sub-department of Environmental Technology. Wageningen: Wageningen University. 2005. p. 176.
57. Van der Zee FP, Cervantes FJ. Impact and application of electron shuttles on the redox (Bio) transformation of contaminants: a review. Biotechnol Adv. 2009; 27:256–77.
58. Keck A, Klein J, Kudlich M, Stolz A, Knackmuss HJ, Mattes R. Reduction of azo dyes by redox mediators originating in the naphthalenesulfonic acid degradation pathway of *Sphingomonas sp.* strain BN6. Appl Environ Microbiol. 1997; 63:3684–90.
59. Costa MC, Mota S, Nascimento RF, Dos Santos AB. Anthraquinone-2,6-disulfonate (AQDS) as a catalyst to enhance the reductive decolourisation of the azo dyes Reactive Red 2 and Congo Red under anaerobic conditions. Biores Technol. 2010; 101(1):105–10.

4 Molecular Approaches for the Microbial Remediation of Azo Dyes Using a Bacterial Approach

Aditi Nag[1]* and Sonali Sharma[1]
[1]Dr. B. Lal Institute of Biotechnology, Jaipur, Rajasthan, India
*Corresponding author: aditinag.bibt@gmail.com

CONTENTS

4.1 INTRODUCTION TO MICROBIAL BIOREMEDIATION

Industrial revolution benefits mankind and increases the economic growth of nations; however, it comes with a cost that is environmental pollution, which results in the pollution of water, air, and soil. Soil and water have been polluted due to mining processes, industrial waste production processes, the disposal or accidental spillage of chemicals, and many other manufacturing processes. Gulf war oil spills (1991), and the Deepwater Horizon oil spill, Mexico, 2010 was one of the largest oil spills in history (Dozniek et al., 2016). Chemicals that have been released into the environment have an impact on fauna and flora in the affected area and disturb the biodiversity of the area. Humans and the environment are affected by these toxins globally. Due to the discharge of these toxins into the water, it increases the biological oxygen demand (BOD), chemical oxygen demand (COD), total dissolved solids, and heavy metals, such as chromium (Cr), lead (Pb), cadmium (Cd), and nickle (Ni) due to which this wastewater becomes unsuitable for drinking and irrigation (Shah, 2019). Therefore, cleaning up these environmental pollutants is a major concern today.

Microbial remediation or bioremediation is the use of microorganisms for the treatment of these polluted and contaminated sites and helps to decrease their harmful effect on the environment. These microorganisms can use these pollutants as their food sources and in their metabolic reactions; they break these compounds into simpler forms. Therefore, microorganisms are helping to maintain the natural environment and to prevent further pollution. Bioremediation based methods make use of the naturally occurring catabolic diversity to degrade and transform these compounds, such as radionuclides, metals, organic dyes, pharmaceutical substances, polyaromatic hydrocarbons, and polychlorinated biphenyls (Muneer et al., 2005). Microbial remediation is an environmentally friendly, low cost, and sustainable approach compared with the conventional methods that are used for the treatment of pollutants.

The microorganisms that are mainly used are bacteria (aerobic and anaerobic) and fungi, these indigenous microorganisms can degrade pollutants via various enzymatic processes (Yap et al., 2019). All of the metabolic reactions are catalyzed by enzymes, for example, oxidoreductases, hydrolases, lyases, transferases, isomerases, and ligases. Aerobic processes require oxygen (O_2), for the degradation, and carbon dioxide (CO_2), water, and salts are mainly the end products. Anaerobic processes do not require O_2, and methane, hydrogen, sulfides, elemental sulfur, and dinitrogen are mainly the end products obtained.

Microbial remediation can either be carried out by culturing the microorganisms in high numbers and then introducing them into the contaminated soil site or by creating the ideal growth conditions for the growth of the microorganisms at the contaminated soil site, which provides direct contact between the contaminants and the indigenous microorganisms. Some of the bacteria that are used include *Acinethobacter sp., Actinobacter sp., Acaligenes sp, Arthrobacter sp., Bacillins sp., Berijerinckia sp., Flavobacterium sp., Methylosinus sp., Mycrobacterium sp., Mycococcus sp., Nitrosomonas sp., Nocardia sp., Penicillium sp., Phanerochaete sp., Pseudomonas sp., Rhizoctomia sp., Serratio sp., Trametes sp.,* and *Xanthofacter sp.*

Microorganisms can use different processes, such as binding, volatilization, oxidation, and immobilization for the pollutants. Among these techniques, the most common is the oxidation of these organic contaminants to nontoxic forms. O_2 is the most common electron acceptor for microbial respiration and the agent for the aerobic degradation of a wide range of organic pollutants, which range from arenes, such as benzene to xenobiotics (pesticides) (Muneer et al., 2005). Different biological and environmental factors can affect the bioremediating capacity of the microbes at the contaminated site, such as availability of the contaminants to the microbial population, ability of the microbes to degrade the contaminants, and environmental factors, such as type of soil, temperature, pH, presence of O_2 or any other electron acceptor in the contaminated site, and nutrients. For better results from bioremediation, an environment that supports the growth of the microbes at the contaminated site is important. This can be achieved by the manipulation of the environmental parameters; therefore, microbial growth is enhanced (Abatenh et al., 2017).

4.1.1 Advantages of Microbial Remediation

The advantages of microbial remediation include.

1. The complete breakdown of organic pollutants into non-harmful compounds is possible. Complex structures, which are resistant to degradation can be broken down into simpler forms by microbes that are nontoxic to the environment.
2. Minimum requirement of equipment. For microbial remediation, only microorganisms are required, which are easy to isolate from the contamination site or other sites and they can efficiently degrade the contaminants.
3. Depending on the type of contaminated site *in situ* or *ex situ* methods can be used. Sometimes, an *in situ* approach is not suitable for the contaminated site, because it can affect the

surrounding area by generating hazardous material. In addition, it can come in contact with the wildlife; therefore, it is necessary to remove the contaminated site and an *ex situ* method should be used.

4. Low cost, natural-process, eco-friendly. Microbial remediation does not harm the environment by generating secondary pollutants compared with the physical and chemical processes used for degradation, and its operational costs are low.
5. It does not use any dangerous chemicals that can be ecotoxic.

4.1.2 Disadvantages of Microbial Remediation

The disadvantages of microbial remediation include.

1. It should be controlled; otherwise, other types of toxic byproducts can be produced. Sometimes, metabolic reactions generate compounds that are more harmful than the original compound; therefore, the structure and chemical nature should be known before treatment with a particular microbe.
2. The environmental conditions can be a limiting factor. The presence or absence of O_2 can be a limiting factor for microbes that use aerobic or anaerobic methods for degradation. In addition, temperature, humidity, and soil pH can affect the growth of microbes.
3. The treatment time is longer. Sometimes, microbes require time to adapt to the contaminated site and grow under these conditions, which can be a limiting factor.
4. Not all compounds are susceptible to biodegradation; these include some heavy metals, pesticides, DDT, radioactive elements, and some hydrocarbons.
5. It is difficult to extrapolate from bench to field-scale operations, the microbes grow easily under laboratory conditions when all of the nutrients are available for their growth. However, under environmental conditions, they need to acclimatize to the contaminated environment and sometimes they fail to do this, which is a limiting factor for bioremediation.

Different types of bioremediation are available for remediation. They mainly include *in situ* bioremediation; in which bioremediation occurs at the contaminated site. It is further classified into natural attenuation, biostimulation, and bioaugmentation. Natural attenuation or bioattenuation are concerned with the degradation ability of the indigenous microorganism and it is a slower process. It is carried out within biological processes, such as aerobic and anaerobic biodegradation, plant and animal uptake, physical phenomena, such as advection, dispersion, dilution, diffusion, volatilization, sorption or desorption, and chemical reactions, such as ion exchange, complexation, and abiotic transformation. Bioaugmentation is used to increase the efficiency of bioremediation. It is used when the indigenous microorganisms cannot break down the contaminants. Then, specific degraders are introduced at the contamination site for efficient degradation. In addition, microbes can be collected from the site and cultured separately and then genetically modified before they are returned to the site. Biostimulation involves the physical and chemical parameters of the soil being modified by the addition of nutrients, such as biogas slurry, manure, rice straw, and spent mushroom compost or by adding electron acceptors, such as phosphorus (P), nitrogen (N), O_2 and carbon (C).

The other method is an *ex situ* bioremediation technique in which bioremediation is carried out at a specially prepared location; the contaminated area is excavated and moved to the processing location. It is a more efficient technique that controls the physicochemical parameters of the soil that results in a decrease in the remediation time. It mainly includes constructed wetlands, landfarming, biopiles, and composting. Constructed wetlands are used for liquid sources, such as wastewater from domestic use, agricultural, and industrial sources. Here, bioremediation and phytoremediation (remediation by use of plants) techniques are used. In landfarming, a thin layer of the excavated contaminated soil is prepared on the ground surface and aerobic microbial activity is enhanced using aeration and the addition of minerals and other nutrients. Composting is used for the treatment of agricultural and municipal solid waste under thermophilic and aerobic conditions. Biopiles are an advanced process for composting, it is more expensive; however, it is more efficient. It is used for petroleum-contaminated soil. Air is supplied to the biopile system with the help of pumps that either forces air into the pile under positive pressure or draws air through the pile under negative pressure (Dzionek et al., 2016; Abatenh et al., 2017).

4.2   DYES

Dyes are used to provide color to substances by attaching themselves to the surfaces of the substrates by different interactions, such as covalent bonds or by making complexes with metals, or physical adsorption (Chequer et al., 2013). Before the mid-nineteenth century, natural dyes were used for coloring, which were obtained from animals or vegetables. At the beginning of the twentieth century, natural dyes were almost completely replaced by synthetic dyes. The first synthetic dye was mauveine, which was made from aniline by W. H. Perkin (father of the dye industry) in 1856 when he was trying to synthesize quinine to treat malaria (Chengalroyen & Dabbs, 2012). Currently, a large number of synthetic dyes are used in the textile industries, printing, food and drink, leather, pharmaceutical, and cosmetic industries. More than 10,000 different dyes are available for this, and the annual production of these dyes is >700,000 t. Due to the discharge of these industrial wastes, water bodies are affected, and water becomes polluted (Hao et al., 2018). Dyes can be classified as acid, basic, direct, fluorescent, reactive, sulfurous, and vat dyes based on their application and chemical structures (Bruna & Maria, 2013). Among all the available dyes, azo dyes are mainly used in the textile industries and are used to add permanent color to the fabrics and account for >50% of the synthetic dyes. The annual production of azo dye is estimated to be 420,000 t, of which 15% is lost to effluent manufacturing (Zhuang et al., 2020). Azo dyes are easy to synthesize and have great fixative properties and are available in different colors, which is why they are used so much in textile industries.

All dye molecules absorb electromagnetic radiation, but of different wavelengths. Dyes contain chromophores (delocalized electron systems with conjugated double bonds), and auxochromes (electron-withdrawing or electron-donating substituents) groups. The chromophore groups add color to the dyes, such as $-C=C-$, $-N=N-$, $-C\equiv N-$, and auxochromes (functional groups) intensify the color of the chromophore and are required for fixation to the fibers, such as $-NH_2$, $-OH$, $-COOH$, and $-SO_3H$ (Benkhaya et al., 2020).

4.2.1   STRUCTURE AND PROPERTIES OF AZO DYES

Azo dyes contain ≥1 R1-N=N-R2 (azo or diamine) bonds in their structure in which R1 and R2 can be either aryl or alkyl groups (Dong et al., 2019). The azo group attaches with two sp2 hybridized C atoms. Mainly, these carbons are part of aromatic systems; however, this is not always necessary (i.e., diethyldiazene is an aliphatic azo compound). The aromatic side groups help to stabilize the $-N=N-$ group by making it part of an extended delocalized system. Due to the chromophore group that is present in the azo dyes they absorb light of a specific intensity (visible range) and give a bright color to the azo dyes. Azo dyes are classified based on the number of azo bonds they have, for example, monoazo (e.g., Acid Orange 52, Reactive Yellow 201, Disperse Blue 399), diazo (e.g., Reactive Brown 1, Brown 2, Acid Black 1, Amido Black), triazo (Direct Blue 78, Direct Black 19), and polyazo (>3 azo bonds, i.e., Direct Red 80) (Shah, 2014; Chengalroyen et al., 2013). Many of the major brown and black dyes that are applied to leather are polyazo compounds. The aromatic rings present in the azo dyes usually have chloride, hydroxyl (OH), sulfate, or nitro groups, which increase the solubility of the dye in water.

The planar -N=N- bond shows stereoisomerism, because of that cis and trans forms can be observed in their structures. The trans form, which is preferred, can be converted into the unstable cis form, by exposure to ultraviolet (UV) radiation or strong light; however, this is a reversible change and can be reversed when stored under dark conditions, which is a phenomenon known as photochromism. The trans form is stable because of the formation of a hydrogen bond between the hydrogen atom of OH or amino groups and the N atom of azo groups. This partial trans to cis conversion (color change due to light) is prevented in the dyed fibers in which dye molecules are held by adsorption.

In addition, azo dyes display tautomerism, in which the hydrogen atom (proton) from one side (OH group) is removed and becomes attached to the other side (keto group). It is more common when there is an –OH group present on the ortho or para position to the azo group. These forms can be identified by their characteristic absorption spectra. However, all azo dyes do not show tautomerism (Allen, 1971).

4.2.2 Synthesis of Azo Dyes

Azo dyes are simply synthesized via diazotization and coupling reactions with phenols, naphthols, arylamines, pyrazolones, or any other compounds that result in hydroxyazo or aminoazo compounds. To obtain the desired color, properties, and other characteristics of that particular compound, different methods can be used. In the hydroxyazo or aminoazo compound, the azo group is the chromophore, and the OH or amino group is the auxochrome group. To synthesize an azo dye, two main compounds are required: (1) a diazonium salt; and (2) a coupling component. The synthesis is mainly composed of two stages: (1) diazonium salt formation occurs by the treatment of a primary aromatic amine with sodium nitrite at low temperatures; and (2) the diazonium salt reacts with the coupling component (e.g., phenols, naphthols, and arylamines) to form a stable azo dye structure (Figure 4.1).

Most of the diazonium salts are unstable and explosive, which is why they are prepared under acidic conditions and at low temperatures (0°C) and they are used immediately in the reaction. The diazonium salt reacts with the electron-rich coupling component and acts as an electrophile. This reaction is an electrophilic aromatic substitution mechanism (Benkhaya et al., 2020).

4.2.3 Dyeing Process

Textile materials can be dyed using different processes that include batch, continuous, and semi-continuous, the type of process used depends on the type of dye, fabric, and fiber, and quality of

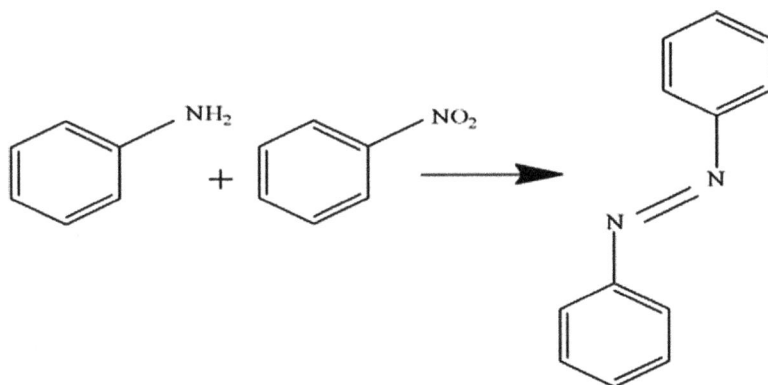

FIGURE 4.1   Synthesis of azo dye.

the final product. The most common method used is a batch process. Due to the cost-effectiveness, ease, and great structural diversity of azo dyes they are widely used in textile, pharmaceutical, and cosmetic industries. Consumers look at the design and the color of the fabric. In addition, the quality of the product, such as color fixation after washing and reaction to light is assessed. To demonstrate these properties the dyes must be tightly bound to the fiber and have to be resistant to fading. The dyeing process involves three steps: (1) preparation; (2) dyeing; and (3) finishing.

The first step is a preparation step, where the fabrics are treated with an alkaline solution, enzymes, or detergents to remove all the impurities. To remove the natural color of a fabric bleaching agents, such as hydrogen peroxide (H_2O_2) and compounds that contain chlorine are used. In the second step, which is the dyeing step, dyes are used in an aqueous solution and the fabric is treated with them. In addition, with the dyes, different chemicals are used to obtain uniform color depth and the color fastness properties of the fabric, such as surfactants, acids, alkali/bases, electrolytes, carriers, leveling agents, promoting agents, chelating agents, emulsifying oils, and softening agents. First, the dye is diffused into the liquid and then it is adsorbed onto the outer surface of the fibers, followed by adsorption onto the inner surface of the fibers. In the final step, the finishing step, the fabric is treated with chemical compounds to improve the quality and the fastness properties, for example, waterproofing, softening, antistatic protection, soil resistance, stain release, and protection against microbes/fungus.

The dye fixation depends on the type of fabric used. Two main types of fibers are used; natural and synthetic. Natural fibers are obtained from the environment, plants, or animals (e.g., wool, cotton, flax, silk, and jute) and synthetic fibers are organic polymers and are mainly derived from petroleum sources (e.g., polyester, polyamide, rayon, acetate, and acrylic) (Chequer et al., 2013).

4.3 TOXIC AND MUTAGENIC EFFECTS OF AZO DYES

Azo dyes are a major threat to the environment and are toxic due to their recalcitrant nature, solubility in water, and photolytic stability. The wastewater that is released from textile industries during manufacturing has high COD, BOD, salinity, pH, dissolved O_2, and micropollutants. Because of this, the wastewater is difficult to treat, and these properties interfere in the biotransformation and inhibit the growth of indigenous microorganisms that are present at that contaminated site (Zhuang et al., 2020). These azo dyes and their degradation products can harm the fauna and flora and can cause health effects in humans, such as allergies, heart defects, jaundice, tumors, skin irritation, and cancers. Because of the low O_2 and photosynthetic ability of water that is polluted by dyes, the phytoplankton and zooplankton are affected, and it harms the aquatic habitat (Sridhar et al., 2019). Through mineralization, azo dyes can be broken down and can produce aromatic amines that are used as intermediates in their synthesis and are carcinogenic to human health. The first carcinogenic aromatic amine that caused cancer was observed in 1995 by Ludwig Rehn, who reported the effect of aniline on workers that were involved in the dyeing industry, which caused bladder cancer (Carreon et al., 2004). Brüschweiler (2017) conducted a study, in which 470 azo dyes were checked for the generation of aromatic amines that were toxic to health. They found 40 mutagenic aromatic amines by conducting an Ames test on the azo dyes. The metabolic activation of the amino group generates the intermediate, hydroxylamine that can damage DNA. Para-phenylene diamine, known as 1,4-diamino benzene, is an aromatic amine that is a major component of azo dyes and cause skin irritation, contact dermatitis, chemosis, lacrimation, exophthalmos, permanent blindness, rhabdomyolysis, acute tubular necrosis supervene, vomiting gastritis, hypertension, and vertigo (Sudha et al., 2014).

A 58-year-old male was diagnosed with basal cell carcinoma and the probable reason was their continuous exposure to a colorant, which was used to color fishing bait. The colorant was identified as the azo pigment Solvent Red 8 using chromatographic-mass spectrometry and infrared spectroscopic studies (Engel et al., 2008). Disperse Red 1 is an azo dye, which created DNA adducts that are mutagenic in humans, Disperse Orange 1 can damage DNA and has cytotoxic effects. Congo Red (CR) can be metabolized into benzidine, which is a carcinogen in humans (Asses et al., 2018). Sudan I dye can cause neoplastic liver nodules in rats and this azo dye can enzymatically be transformed by the intestinal microflora (Bruno et al., 2019). This dye can be metabolized to form benzene diazonium ions, which reacts with DNA and forms a stable 8-(phenylazo) guanine adduct (Chen, 2006). Methyl Orange (MO) was analyzed and causes genetic variability and results in mutations in rhizobium obtained from root nodules of *Arachis hypogaea* (Anand et al., 2012). In addition, it can form dimethyl-4-phenylenediamine (DMPD), which is mutagenic and remains untreated in culture (Ayed et al., 2011). Tartrazine and carmoisine azo dyes were analyzed for their toxicity in albino male rats and these dyes change the levels of hepatic and renal parameters and cause oxidative stress by forming free radicals. Methyl Yellow (p-Dimethylaminobenzene), another azo dye, causes liver tumors and chromosomal abnormalities when fed to rats. Similarly, many other azo dyes, such as Disperse Blue 291, Sudan II, Sudan IV, and Reactive Red 120 harm DNA, chromosome, and proteins; therefore, they are cytotoxic, genotoxic, and mutagenic (Bruna & Maria, 2013). In addition, azo dyes inhibit the enzyme tyrosinase, an oxidase that controls the synthesis of melanin, which inhibits melanin synthesis and causes hypopigmentation (Lade et al., 2015).

Azo dyes can be activated into carcinogenic compounds by three basic mechanisms:

1. Azo bond can be reduced to form aromatic amines by cleavage.
2. Azo dyes that have free aromatic amines can be oxidized.
3. Azo bonds can be directly oxidized to form highly reactive electrophilic diazonium salts (Chen, 2006).

Azo dyes can be toxic, nontoxic, or potentially toxic. Toxic dyes do not need any enzymatic cleavage to be hazardous in the environment; the parent dye molecule is toxic because of its low molecular weight and lipophilicity; it can diffuse into the cell membrane. Nontoxic dyes might form potentially toxic dyes due to interactions with biotic or abiotic components. Potential toxic dyes have multiple sulfonic groups attached to them and because of their bulk and high polarity, they are unable to pass through the cell membrane. However, due to the degradation of these dyes, hydrophobic metabolites are generated that can easily enter the cell membrane. When nontoxic food dyes are extensively consumed, they can be metabolically activated by the intestinal bacteria. Intestinal bacteria remove the sulfo and nitro groups to increase their lipophilicity, which makes them available for metabolic activation. Sunset Yellow food dye can be degraded to form 5-amino-6-hydroxy-naphthalene2-sulfonic acid and 4-amino-benzenesulfonic acid, which are genotoxic. Extrinsic and intrinsic factors can affect the toxicity potential of the dyes. O_2 availability (aerobic or anaerobic), light (photolysis and photooxygenation), moisture, temperature, and intrinsic factors, such as microbial interactions with dyes and enzymatic action on the dyes can influence the ecotoxicity of the dyes (Deepak, Vandana, & Shyam, 2016). All of these studies demonstrated the potentially toxic nature of azo dyes and that they can harm human health. In addition, they can harm the environment, which is why it is important to deal with wastewater and to find efficient methods that can treat these dyes contained in the textile industry effluents.

4.4   METHODS FOR AZO DYE DEGRADATION

Remediation of wastewater is required to protect the environment from its toxic and hazardous effects. Different physical, chemical, and biological methods are available for the treatment of wastewater from the textile industry. Based on the type of dye and the contaminants that are present in the wastewater, cost of operation, environmental fate, and the efficiency of the process, any method can be used. Sometimes a single process is not sufficient for the removal of the contaminant; therefore, a combination of these could be used for efficient degradation. These techniques include membrane filtration, nanofiltration, flocculation, precipitation, ozonation, adsorption by activated C, electrolysis, advanced oxidation, and chemical reduction for physicochemical processes and bacterial, fungal, or enzymatic degradation for biological degradation.

4.4.1 Physical and Chemical Methods

Adsorption is one of the common methods that is used for degradation. Activated C and silica gel or natural materials, such as peat and agricultural lignocellulosic residues are effective adsorbents that can be used. Membrane filtration works via reverse osmosis, in which a semi-permeable membrane is used for effluent treatment. In coagulation, coagulants, such as lime, aluminum, magnesium, and iron salts are used, which become linked to the pollutant and the resultant flocks are precipitated and removed by flotation or settling to obtain sludge. Ion exchange uses an ion exchange resin for the treatment of anionic and cationic dyes. Different photochemical methods, such as, UV/H_2O_2 and UV/titanium dioxide (UV/TiO_2) can be used, which are based on UV radiation and free radical formation (Shah, MP. 2019). UV light decomposes H_2O_2, which is an oxidizing agent and forms OH radicals, which helps to degrade the dyes. Fenton treatment ($H_2O_2/FeSo_4$) can be used for the treatment, ferrous ion is catalyzed by H_2O_2 and forms OH radicals, which oxidize organic matter. Another good oxidizing agent, ozone (O_3) and electrolysis can be used for the chemical treatment of wastewater. Chemical oxidation works by cleaving the aromatic ring of the dye. The advanced oxidation process is one of the chemical processes that are based on the formation of highly reactive OH radicals, which are highly oxidative, and can treat organic waste (Kuhad et al., 2004). Photocatalytic degradation was used to degrade dyes, in which transition metal [Cr, copper (Cu), iron (Fe), manganese (Mn), zinc (Zn), or silver (Ag)] doped TiO_2 was used as the photocatalyst. The reactant is transferred onto the catalyst and adsorbed onto its surface, where the reaction occurs and then the final product is desorbed, it is based on a photoredox process (Rauf et al., 2011). Photocatalysis of Cr was carried out by UV at 254 nm and this was increased when an H_2O_2/UV and $S_2O^{2-}_8/UV$ process was used (Djebbar et al., 2009). Protein nanofibrils have been used for the removal of azo dyes by coagulation, they disperse easily in the aqueous dye solution and showed better results than charcoal (Morshedi et al., 2013). TiO_2 nanoparticles entrapped in biopolymer calcium (Ca) alginate was used to treat dye solutions by photocatalysis, which showed promising results for degradation by nanoparticles for future applications (Harikumar et al., 2013). Studies have used a combination of chemical and biological processes for better results. Sequential ZnO/photocatalysis and biological treatment used a microbial consortium of *Galactomyces geotrichum* and *Bervibacillus laterosporus* for the complete decolorization of Methyl Red (MR) within 4 h, and decreased COD and the formation of noncytotoxic products (Waghmode et al., 2019). Similarly, electrochemical oxidation that used ruthenium–iridium (RuO_2-IrO_2) coated Ti electrode followed by biological treatment using *Pseudomonas stutzeri* MN1 and *Acinetobacter baumannii* MN3 was used for CR decolorization and showed efficient degradation within 7.2 min. All these methods have some disadvantages, such as high operational cost and some produce secondary pollutants.

4.4.2 Biological Treatment

Biological treatment is a more reliable approach for the treatment of the dyes than physical and chemical approaches. Different types of bacteria, fungus, and algae are used to degrade dyes, which have shown promising results. Indigenous microorganisms have enzymes that help to degrade the dyes; bacteria achieve this aerobically or anaerobically. During anaerobic degradation of the dyes, reductive cleavage of the azo bond occurs, which results in the formation of aromatic amines. During the anaerobic process, the azo dye acts as an electron acceptor that is derived from the electron transport chain carriers. The disadvantage of this process is that these aromatic amines cannot be mineralized further. Various bacteria use aerobic degradation, by the reductive cleavage of dyes.

These bacteria do not use the azo dye as a carbon (C) source. They use other C and energy sources for degradation. The presence of O_2, which acts as an electron acceptor, might inhibit the effective degradation of the azo dyes, because it can interfere with the electron transfer from NADH to the azo bond. In addition, some studies used a combination of both techniques for the efficient degradation of aromatic amines via aerobic processes, which were generated during the anaerobic process. Microbes are acclimatized into the dyeing environment and use enzymatic biotransformation, which helps in the biodegradation of recalcitrant compounds. The efficiency of the process depends on the type of dye and the adaptability of the microbe to use it. New and efficient remediating strains can be identified by bioremediation setup and the molecules responsible for the degradation can be identified and used in similar contexts.

Fungi can degrade azo dyes by producing intracellular or extracellular enzymes, such as lignin peroxidase, Mn peroxidase, and laccase that can carry out various reactions. The most studied fungi are white-rot fungi, *Phanerochaete chrysosporium* and *Trametes versicolor, Bjerkandera adusta, Pleurotus*, and *Aspergillus* species have been used (Shah, 2014). *Poria sp., Ganoderma sp.,* and *Trametes sp.* of lignin-degrading white-rot fungi have been used for the decolorization of CR, Rhodamine 6G, and Malachite green with complete decolorization (Selvam et al., 2012). *Candida oleophila, Candida zeylanoides, Saccharomyces cerevisiae, Debaryomyces polymorphus, Galactomyces geotrichum* and *Trichosporon beigelii* are examples of yeasts that use azoreductase and other enzymes to degrade dyes. *S. cerevisiae* MTCC 463 showed complete degradation of MR under static anoxic conditions within 16 min in distilled water at room temperature (Jadhav et al., 2007).

Various studies showed potential bacterial species, such as *Bacillus cereus, Bacillus subtilis, Proteus mirabilis, Pseudomonas luteola,* and *Aeromonas hydrophila* use various oxidoreductase enzymes that cleave the azo bond by inserting O_2 into the aromatic ring under aerobic conditions (Jamee & Siddiqui, 2019). Pure culture or consortia of these bacteria can be used. In mixed culture, the advantage is the synergistic activity of the microorganisms. Individual microbes can attack dye structures differently and the degradation products can be further decomposed by other strains that are present in the culture. A bacterial consortium of *Providencia rettgeri* strain HSL1 and *Pseudomonas sp.* SUK1 was used under aerobic, microaerophilic, sequential microaerophilic/aerobic and aerobic/microaerophilic processes for the degradation of Reactive Black 5, Reactive Orange 16, Disperse Red 78, and Direct Red 81. Under all these conditions, the dyes were decolorized completely. The most effective method was the sequential microaerophilic/aerobic process in which the formation of aromatic amines was not detected (Lade et al., 2015).

4.5 FACTORS THAT AFFECT THE DEGRADATION OF AZO DYES

Dye structure plays an essential role in the degradation of the dyes; dyes with high molecular weight have a correspondingly lower rate of degradation than lower molecular weight dyes. The concentration of the dye in the culture affects the growth of the microbes in the media. Bacterial growth becomes slower at increased concentrations of dyes. The composition of the media affects microbial growth, the presence of C and N sources is an essential factor, because some bacteria do not use the azo dye as a C source; therefore, yeast extract or peptone can be used for better results. The requirement for different C and N sources can be different for individual isolates. CR decolorization was highest when yeast extract was used, apart from other C sources, such as glucose, lactose, maltose, and sucrose by *Kocuria rosea* (Parshetti et al., 2010). Consortia from *Bacillus sp.* showed better decolorization of CR dye in the presence of sucrose and sodium nitrate as the C and N sources, respectively (Kumari & Rajoriya, 2019). *Bacillus megaterium (MTCC 8371)* was used for degradation and cultured in Kirk's media and the highest decolorization was observed when glucose was used as the C source and potassium hydrogen phosphate (K_2HPO_4) as the N source (Lekha et al., 2017). Studies showed that in the presence of an additional N source, decolorization by *A. hydrophila* was enhanced but was inhibited when glucose was the C source. Similarly, a mixed bacterial culture showed enhanced degradation when starch was the C source.

Temperature and pH conditions should be optimized for better degradation, the optimal pH should be between 6.0 and 10.0 (Saratale et al., 2011). The pH tolerance of the bacteria that is used is important, because the textile effluents contain high salts and are alkaline and these alkaline conditions can be a limiting factor; therefore, it is preferable to use alkaliphilic and halophilic bacteria. These extremophilic microorganisms can easily adapt to these conditions, a halotolerant and alkaliphilic bacteria *Nesterenkonia lacusekhoensis* EMLA3 was used for the degradation of MR in alkaline conditions, and it showed 97% degradation and the optimum pH range was 8.0–11.5 and the temperature range was 30°C–35°C (Bhattacharya et al., 2017). A bacterial consortium was analyzed at different pH concentrations to observe the optimum pH, and at pH 9.5, the decolorization was maximum, and the optimum temperature was 36°C for Direct Blue 151 and Direct Red 31 (Lalnunhlimi & Krishnaswamy, 2016). *Vibrio logei* showed the highest percent decolorization of MR at pH 6–7 and 30°C–35°C (Adedayo et al., 2004).

4.6 MOLECULES INVOLVED IN DYE DEGRADATION

Enzymes, such as azoreductase, laccase, lignin peroxidase, Mn peroxidase, polyphenol oxidase can degrade azo dyes. Various bacteria, fungi, yeast, and animals have these enzymes that assist degradation by intracellular or extracellular secretion.

Azoreductase can cleave azo bonds by reductive cleavage, which results in the formation of aromatic amines, and it requires electron donor molecules that are reducing agents, such as NADH, NADPH, and $FADH_2$. Bacteria, for example, *Xenophilus azovorans* KF46F, *Pigmentiphaga kullae* K24, *Enterococcus faecalis*, *Staphylococcus aureus*, *Escherichia coli*, *Bacillus sp.* OY1-2, and *Rhodobacter sphaeroides* contain azoreductase. There are two types of azoreductase, flavin-dependent and flavin independent. There are three types of flavin-dependent azoreductase: azoreductase that uses NADH; azoreductase that uses NADPH; and azoreductase that uses both as a coenzyme. Dyes that contain sulfo groups are high molecular weight, which means that they cannot enter bacterial cells and bacterial membranes are impermeable for the flavin-containing cofactors, which is why they cannot be transferred from the cytoplasm to the extracellular environment. Therefore, for the degradation of these dyes, a specific transport system is required, and these transport systems have to be localized on the outer cell membrane, where they are in contact with the azo dye or with the redox mediator, which acts as an electron shuttle between the dye and them. A dye without the sulfo group can be degraded by intracellular azoreductase, because they diffuse through the cell membrane (Joshni & Kalidas, 2011). A study showed that *azoR1* was responsible for the production of azoreductase. A mutant of *B. subtilis* ORB7106, in which negative control of *yodB* was lost; therefore, azoreductase was produced constitutively, which was more efficient at decolorization than its parent wild type, *B. subtilis* JH642, and therefore, azoreductase is mainly responsible for the decolorization of MR (Montira & Sukallaya, 2012).

Laccases are the multicopper oxidase enzymes that degrade azo dyes via a free radical mechanism that result in the formation of phenolic compounds instead of toxic aromatic amines and do not require any cofactors. In the presence of O_2 as an electron acceptor, they catalyze the oxidation of phenolic azo dyes, due to the oxidation of the phenolic ring, phenoxy radicals are formed, which are oxidized to form carbonium ions. Then, nucleophilic attack by water occurs on the C of the phenolic ring that has azo bonds and produces 4-sulfophenyldiazene and a benzoquinone (Singh et al., 2015; Sarkar et al., 2017).

Peroxidases are hemoproteins, which work in the presence of H_2O_2. They are efficient in degrading aromatic amines and phenolic compounds and treating textile effluent. Lignin peroxidase is a classic peroxidase that catalyzes the oxidation of the lignin chain and forms reactive radicals. Its molecular weight is between 38 and 47 kDa. Studies found increased activity during dye decolorization by bacterial strains. In addition, Mn peroxidases have a similar reaction mechanism to lignin peroxidase, which is via oxidation by H_2O_2. It oxidizes Mn^{2+} to Mn^{3+}, which oxidizes many phenolic compounds (Joshni & Kalidas, 2011).

Polyphenol oxidase, which is a tetramer, has four molecules of Cu per molecule and can bind with two aromatic compounds and O_2. They catalyze the o-hydroxylation of monophenols to o-diphenol and then to o-quinones. They are oxidoreductase enzymes that can remove aromatic pollutants from contaminated sites. They can oxidize the phenolic ring of tyrosine amino acid to form o-quinone, which is why they are known as tyrosinases. Studies have shown their presence in bacteria, such as *Streptomyces glaucescens, Streptomyces antibioticus, Bacillus licheniformis, Bacillus natto,* and *Bacillus sphaericus* (Telke et al., 2014).

A comparative study was carried out between three dye degrading enzymes laccase, azoreductase, and peroxidase, laccase had more basic amino acids and azoreductase and peroxidase had more acidic amino acids. In addition, azoreductase was more stable than the other enzymes. All three are hydrophilic and nonpolar and because of the presence of a high percentage of alpha-helix regions, azoreductase and peroxidase were more thermostable than laccase (Sarkar et al., 2019).

In a study carried out by our research group, six isolates were isolated from textile dye wastewater bioremediation and the proteins responsible for the degradation of the dyes, CR, MR, and MO were identified. The six isolates (RB1, RB2, RB3, RB4, RB6, and B1-B3) were cultured in the media that contained the dyes for 24 and 48 h. After 24 and 48 h of incubation, the culture that contained the dye and the nutrient broth culture, which served as the control, were analyzed to observe the proteins produced under different conditions, using SDS-PAGE analysis. Then, the bands were compared with the control and bands that were specific to particular dyes were not present in the control. The specific bands that were observed could be responsible for the decolorization of the dyes in the media. In Table 4.1 the dyes were labeled as MR, CR, and MO, and NB stands for nutrient broth culture that was the control for every dye-containing culture that contained dye. Table 4.1 lists the protein band sizes that were analyzed after 24 h incubation.

Table 4.2 lists the protein profiling results after 48 h culture of the isolates in their respective dye and controls where the isolates were labeled RB1, RB2, RB3, RB4, RB6, and B1-B3.

From the differential protein profiling, for isolate RB1, an 11 kDa protein band was observed for MR at 24 and 48 h. For isolate RB2, an 11 kDa band was observed in MR at 48 h. For isolate RB3, a band at <11 kDa and an 11 kDa band were observed at 24 and 48 h. respectively. For RB4, a band at <75 kDa were observed in MO and CR and a band at <11 kDa for MR at 24 h. In addition, RB4 showed a band at approximately 11 kDa band for CR at 48 h, which was not present in the NB protein sample. For RB6a band at <11 kDa was observed at 24 and 48 h in MR and a band at 11 kDa was observed for CR at 48 h. For B1-B3, 75 kDa, <75 kDa, <48 kDa, 35 kDa, and 11 kDa protein bands were observed in MO at 24 h. In addition, B1-B3 showed <75 kDa, <48 kDa, 35 kDa,

TABLE 4.1
Protein Band Size for Different Isolates for Their Dyes and Control at 24 h

Isolates	NB	MO	CR	MR
RB1	25 kDa	25 kDa	25 kDa	11 kDa
RB2	25 kDa	25 kDa	25 kDa	–
RB3	25 kDa	25 kDa	25 kDa	<11 kDa
RB4	–	<75 kDa	<75 kDa	<11 kDa
RB6	35 kDa	35 kDa	35 kDa	<11 kDa
	25 kDa	25 kDa	25 kDa	
B1-B3	–	75 kDa, <75 kDa, <48 kDa, 35 kDa 11 kDa	<75 kDa, <48 kDa, 35 kDa, 11 kDa	<75 kDa, approx. 35 kDa, approx. 11 kDa

TABLE 4.2
Protein Band Size for Different Isolates for Their Dyes and Control at 48 h

Isolates	NB	MO	CR	MR
RB1	–	–	20 kDa	11 kDa
RB2	–	–	–	11 kDa
RB3	100 kDa, 135 kDa 63 kDa, <25 kDa	–	–	11 kDa
RB4	–	–	approx. 11 kDa	
RB6	–	–	11 kDa	< 11 kDa
B1-B3	–	–	approx. 11 kDa	–

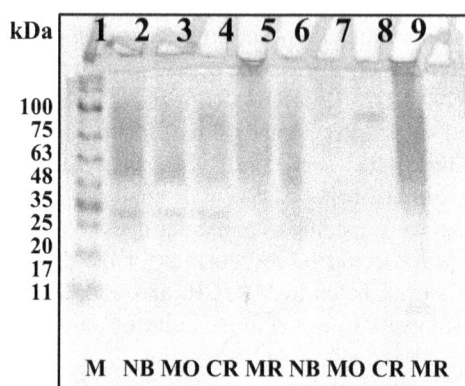

FIGURE 4.2   Protein profiling of RB3 (2–5) and RB4 (wells 6–9) at 24 h.

and 11 kDa protein bands in CR at 24 h and a protein band at approximately 11 kDa at 48 h for CR, which were not present in the NB protein sample. B1-B3 showed protein bands at <75 kDa, and at approximately 35 kDa and 11 kDa in MR at 24 h. The protein bands mentioned previously were not present in the control, which showed that these proteins were uniquely synthesized by the respective isolates in the presence of their respective dyes. Other protein bands that were observed under normal conditions showed that those isolates did not need to synthesize any proteins under those particular conditions.

In Figures 4.2 and 4.3 the dyes are represented as MR, CR, and MO and NB is the control. Figure 4.2 shows the protein band pattern for RB3 (well numbers 2–5) and RB4 (well numbers 6–9). Figure 4.3 shows the protein band patterns for RB6 (well numbers 2–5) and B1-B3 (well numbers 6–9), obtained by SDS-PAGE analysis after 24 h.

4.7   ANALYSIS OF BIODEGRADED PRODUCTS

To understand the mechanisms of biodegradation, techniques such as mass spectrometry, Fourier–transform infrared spectroscopy (FTIR) spectroscopy, UV–visible spectroscopy (UV–Vis), and gas chromatography–mass spectroscopy (GS–MS) have been used in various studies to iden-tify the products formed during biotransformation. GC–MS analysis of MO biodegradation by *Pseudomonas sp.* SUK1 was performed, and the end product was identified as 1, 4-benzenediamine, N, N-dimethyl. This degradation pathway showed that MO was first asymmetrically cleaved by

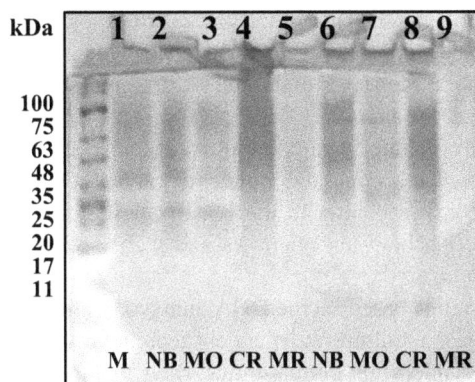

FIGURE 4.3   Protein profiling of RB6 (2–5) and B1-B3 (wells 6–9) at 24 h.

lignin peroxidase into an intermediate product. Then, the azo bond of this intermediate was cleaved by reductase enzyme and the final product was formed (Kalyani et al., 2009). Another study showed that the MO azo bond was symmetrically cleaved by an azoreductase from *Kocuria rosea* MTCC 1532, then it formed 4-amino sulfonic acid and N, N-dimethyl p-phenylenediamine (Parshetti et al., 2010; Ramesh et al., 2018). Degradation of MR by *Bacillus sp.* UN2 was analyzed by liquid chromatography–mass spectroscopy (LC-MS) analysis and the final degraded products were identified as N, N'-dimethyl-p-phenylenediamine and 2-aminobenzoic acid. The enzymes responsible were identified as azoreductase, laccase, and NADH–DCIP reductase, because of an increase in their activity (Zhao et al., 2014). In addition, degradation can be confirmed using UV–Vis and FTIR spectroscopy. The peak observed in FTIR in the control for CR and MR disappeared in the decolorized sample and new peaks were observed in the sample. The disappearance of the peak at 1599.75 cm^{-1} for N=N stretching vibrations, showed the reductive cleavage of azo bonds (Lekha et al., 2017). *Pseudomonas sp.* SU-EBT was used for the decolorization of CR, and FTIR analysis was performed on the samples after decolorization. The peak at 1,587 cm^{-1} for N=N stretching vibrations in the control CR was absent in the decolorized sample, which showed that the azo bond was cleaved during decolorization. The process for the biodegradation of CR was determined by GC–MS analysis, which showed that in the first step CR was subjected to asymmetric cleavage by oxidoreductase, which formed p-dihydroxy biphenyl and suggested a diazine intermediate, which was further converted into 8-amino naphthol 3-sulfonic acid and then, due to laccase activity, the final product 3-hydroperoxy 8-nitrosonaphthol was formed (Telke et al., 2009). FTIR analysis of decolorized samples of Reactive Violet 5 by *Paracoccus sp.* GSM2, showed peaks at 1651.91 and 3252.48 cm^{-1}, which were for primary amine and secondary amides, respectively that formed during the degradation of Reactive Violet 5, and the breakdown of the azo bond was confirmed by the absence of the 1549.25 cm^{-1} peaks in the decolorized sample (Bheemaraddi et al., 2014). Reactive Black- B, which is a multisulfonated azo dye, was decolorized with the help of *Morganella sp.* HK-1, which was analyzed by FTIR analysis, and the degradation pathway was analyzed by GC–MS and the degradation products were identified as, disodium 3,4,6-Triamino-5-hydroxynaphthalene-2, 7-disulfonate, 4-aminophenyl sulfonyl ethyl hydrogen sulfate, naphthalene-1-ol, aniline, and benzene via azoreductase activity (Jadhav et al., 2014).

4.8   MOLECULAR APPROACHES FOR BIOREMEDIATION

Traditional methods for bioremediation have some limitations; therefore, advanced methods, such as synthetic biology, molecular methods, genomics, metatranscriptomics, metaproteomics, metabolomics, protein engineering, proteomics, metabolic engineering, and whole-transcriptome profiling could be used to enhance it. DNA shuffling could be used to introduce mutations in the DNA sequence, which could synthesize proteins with new functions. Whole-transcriptome profiling could be used to analyze the relative amount of proteins in the transcriptome and it could be used to predict changes in the proteins (Wood, 2008). The 16S rRNA gene sequence analysis enabled the understanding of the microbial community at the species level and helped to differentiate the closely related species, this could help to identify some novel bacterial species for efficient degradation. All these approaches help with the understanding of microbial diversity, microbial metabolism, and the interaction between microbes and their environment, which is necessary to understand bioremediation.

The metagenomic approach is based on the DNA analysis of microbes that are isolated directly from the environment and it is a culture-independent technique that helps to analyze the genetic makeup and structure of the microbial communities, their metabolism, evolution, functions, and their response to environmental changes. In addition, this is to identify the interactions between microbes and how they efficiently degrade contaminants, which is why it is beneficial to use consortia rather than single microbes. Metagenomic databases have information on genes, which could be used to develop novel strains for the degradation of particular contaminants. There are two types of metagenomic approaches: sequence-based, and function-based. Sequence analysis provides a base to predict the function. Genes can be identified, organisms from different communities can be compared, and the metabolic pathways can be understood. Functional analysis helps us to understand the function of a particular gene. In this process, DNA is isolated and cloned and then expressed in the host, which is then screened for enzyme activity. For a more detailed analysis of gene activity, expression metatranscriptomics and metaproteomics approaches are used today. Metatanscriptomics is used to identify the gene function in a given environment. Protein profiling studies by SDS-PAGE analysis can be used to characterize the microbes and the environment involved in bioremediation. Proteomics can be used to identify how microbes adapt to the environments, such as thermophilic and halophilic conditions. Proteins that are synthesized by these microbes under these conditions are of great importance, because they help them to survive under these conditions (Muneer et al., 2005). A thermophilic bacterium was used to degrade Direct black G, and metagenomic and metaproteomic analysis was carried out to understand its degradation mechanism. Genetic information from the bacteria was analyzed using shotgun sequencing and to observe the mRNA expression pattern quantitative real-time polymerase chain reaction (qRT-PCR) was carried out. The mRNA expression profiling showed higher levels of some genes, which were responsible for degradation, for example, genes that encoded proteins, such as FAD/NAD(P)-binding protein, catechol-2-3-dioxygenase, and 4-oxalocrotonate tautomerase, had higher expression levels. From these results, these genes are important for the degradation of dyes. Metaproteome analysis was carried out to confirm the relationship with metagenomic analysis and it was highly related (An et al., 2020). Another study used a proteomics approach to understand the genes and enzymes responsible for the degradation of Reactive Blue 160 by *Proteus hauseri* ZMd44. The major protein bands were obtained using SDS-PAGE analysis, then LC-ESI-MS/MS analysis was used, and they were identified as, Zn-dependent metalloprotease, phosphopyruvate hydratase, glycine/D-amino oxidase and outer membrane protein (porin). By using its closely related species, *Proteus penneri* ATCC 35198, degenerated primers were designed and the genes for laccase and porin were cloned. This study showed that the important proteins for decolorization were laccase, porin, and NADH dehydrogenase (Ng et al., 2012). 16S rRna sequencing of residue DNA was carried out to identify the functional genes and the distribution of the azo dye degrading bacteria in estuaries and coastal environments. The azoreductase gene (*azoR*) and naphthalene degrading gene were more abundant than the laccase genes. Genes related to azo dye and naphthalene degradation were found, for example, azoreductase (*azoR*), laccase, *nahAb*, tyrosinase, glutamine synthetase, cytochrome c peroxidase, and NADH peroxidase. The azo dye degrading genes were highly abundant in coastal and estuarine sediments. The distribution of these genes was influenced by the

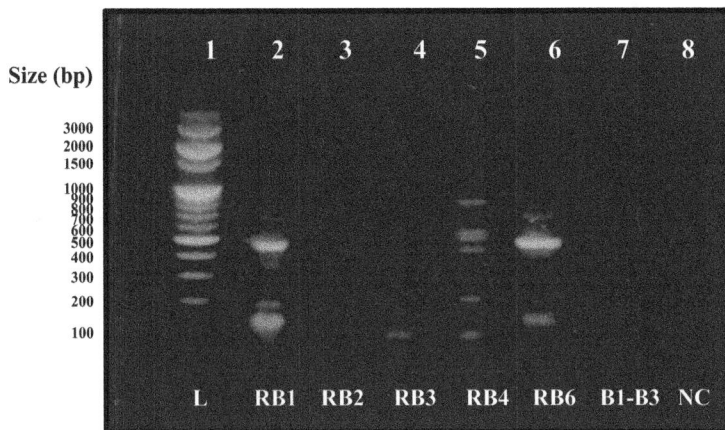

FIGURE 4.4   RAPD profile of microorganisms.

environment, and high concentrations of heavy metals and high salinity negatively affected their abundance (Zhuang et al., 2020). To compare and quantify the similarity between microbial communities the random amplified polymorphic DNA (RAPD) technique can be used. It is used to differentiate between closely related strains of bacteria. In our study, the molecular genetic diversity of the isolates RB1, RB2, RB3, RB4, RB6, and B1-B3 were determined. DNA was isolated using a chloroform:isoamyl alcohol (CI) method and the primers were used to amplify the DNA. PCR and agarose gel electrophoresis were used, and RB1 (well number 2) and RB6 (well number 6) had the most similar band patterns, which suggested that these bacterial isolates were more similar and belonged to the same group (bacilli), because the primer that was used for both of these isolates was for bacilli. Figure 4.4 shows the RAPD results for these isolates. The ladder was labeled as L (well number 1), RB1 (well number 2), RB2 (well number 3), RB3 (well number 4), RB4 (well number 5), RB6 (well number 6), B1-B3 (well number 7), and the negative control was labeled as NC (well number 8).

4.9   THE FUTURE OF DYE DEGRADATION

Methods that can decolorize and degrade dyes more easily and in less time are of great use to enhance degradation. Genetic engineering could be used to manufacture novel strains, with specific genes that are responsible for individual contaminants and given higher decomposition rates. Bioaugmentation could achieve this, in which dye degrading genes are introduced into the genome and are then expressed, which results in the desired protein formation. Studies showed that the cloning and expression of azoreductase genes from bacteria, such as *Clostridium perfringens*, *Bacillus sp.*, *P. luteola*, and *E. coli* were achieved successfully (Esther et al., 2004). By using integrated data and multiomics approaches (e.g., metagenomics, transcriptomics, proteomics, and metabolomic data) more detailed mechanism for bioremediation could be understood, which could help to improve the process. In addition, it could help to generate standard protocols (Muneer et al, 2005). The treated water could be used for irrigation and other nondrinking applications, for example, industrial, cleaning, domestic, and toilet flushing.

REFERENCES

Abatenh, E., Gizaw, B., Tsegaye, Z., & Wassie, M. (2017). The role of microorganisms in bioremediation: A review. Retrieved from www.peertechz.com/articles/OJEB-2-107.php

Adedayo, O., Javadpour, S., & Taylor, C. (2004). Decolourization and detoxification of methyl red by aerobic bacteria from a wastewater treatment [lant. *World J Microb Biot, 20*, 545–550. doi.org/10.1023/B:WIBI.0000043150.37318.5f

Ahlström, L.-H., Eskilsson, C., & Björklund, E. (2005). Determination of banned azo dyes in consumer goods. *TrAC, 2*, 49–56. doi:10.1016/j.trac.2004.09.004.

Allen, R., L., M. (1971). The chemistry of azo dyes. In *Colour Chemistry. Studies in Modern Chemistry.* Boston, MA: Springer. doi.org/10.1007/978-1-4615-6663-2_3

An, X., Chen, Y., Chen, G., Feng, L., & Zhang, Q. (2020). Integrated metagenomic and metaproteomic analyses reveal potential degradation mechanism of azo dye-Direct Black G by thermophilic microflora. *Ecotoxic Environ Safe, 196*, 110557. doi.org/10.1016/j.ecoenv.2020.110557

Anand, T., Kalaiselvan, A., Gokulakrishnan, K., Chandramohan, S., & Elambarathi, P. (2012). Analysis of genetic variation by Methyl Orange in Rhizobium using RAPD–PCR. *Int J Pharm Tech Res, 4*(3), 1101–1109.

Asses, N., Ayed, L., Hkiri, N., & Hamdi, M. (2018). Congo red decolorization and detoxification by *Aspergillus niger*: Removal mechanisms and dye degradation pathway. *BioMed Res Int, 2018*, 3049686. doi.org/10.1155/2018/3049686

Ayed, L., Mahdhi, A., Cheref, A., & Bakhrouf, A. (2011). Decolorization and degradation of azo dye Methyl Red by an isolated *Sphingomonas paucimobilis*: Biotoxicity and metabolites characterization. *Desalination, 274*, 272–277. doi:10.1016/j.desal.2011.02.024.

Benkhaya, S., M'rabet, S., & El Harfi, A. (2020). Classifications, properties, recent synthesis and applications of azo dyes. *Heliyon, 6*(1), e03271. doi.org/10.1016/j.heliyon.2020.e03271

Bhattacharya, A., Goyal, N., & Gupta, A. (2017). Degradation of azo dye methyl red by alkaliphilic, halotolerant *Nesterenkonia lacusekhoensis* EMLA3: application in alkaline and salt-rich dyeing effluent treatment. *Extremophiles, 21*(3), 479–490. doi.org/10.1007/s00792-017-0918-2

Bheemaraddi, M. C., Patil, S., Shivannavar, C. T., & Gaddad, S. M. (2014). Isolation and characterization of *Paracoccus sp*. GSM2 capable of degrading textile azo dye Reactive Violet 5. *The Scientific World Journal, 2014*, 410704. doi.org/10.1155/2014/410704

Brüschweiler, B. J., & Merlot, C. (2017). Azo dyes in clothing textiles can be cleaved into a series of mutagenic aromatic amines which are not regulated yet. *Regul Toxicol Pharm, 88*, 214–226. doi.org/10.1016/j.yrtph.2017.06.012

Carreón, T., Hein, M. J., Hanley, K. W., Viet, S. M., & Ruder, A. M. (2014). Bladder cancer incidence among workers exposed to o-toluidine, aniline and nitrobenzene at a rubber chemical manufacturing plant. *Occup Environ Med, 71*(3), 175–182.

Chacko, J., & Kalidass, S., (2011). Enzymatic degradation of azo dyes: A review. *Int J Environ Sci, 1*, 1250–1260.

Chen H. (2006). Recent advances in azo dye degrading enzyme research. *Curr Protein Pept Sci, 7*(2), 101–111. doi.org/10.2174/138920306776359786

Chengalroyen, M. D., & Dabbs, E. R. (2013). The microbial degradation of azo dyes: minireview. *World J Microb Miot, 29*(3), 389–399. doi.org/10.1007/s11274-012-1198-8

Chequer, F., Olivera, G., A., R., Ferraz, E., Cardoso, J., Zanoni, M., & Olivera, D., P. (2013). Textile dyes: Dyeing process and environmental impact. *Eco-friendly Textile Dyeing and Finishing, 6*, 151–176. doi:10.5772/53659.

Dong, H., Guo, T., Zhang, W., Ying, H., Wang, P., Wang, Y., & Chen, Y. (2019). Biochemical characterization of a novel azoreductase from *Streptomyces sp*. Application in eco-friendly decolorization of azo dye wastewater. *Int J Biol Macromol, 140*, 1037–1046. doi.org/10.1016/j.ijbiomac.2019.08.196

Dzionek, A., Wojcieszyńska, D., & Guzik, U. (2016). Natural carriers in bioremediation: A review. *Electron J Biotechn, 23*. doi:10.1016/j.ejbt.2016.07.003.

Engel, E., Ulrich, H., Vasold, R., König, B., Landthaler, M., Süttinger, R., & Bäumler, W. (2008). Azo pigments and a basal cell carcinoma at the thumb. *Dermatology, 216*(1), 76–80. doi.org/10.1159/000109363

Forgacs, E., Cserháti, T., & Oros, G. (2004). Removal of synthetic dyes from wastewaters: A review. *Environ Int, 30*, 953–971. doi:10.1016/j.envint.2004.02.001.

Hao, D. C., Song, S. M., Cheng, Y., Qin, Z. Q., Ge, G. B., An, B. L., & Xiao, P. G. (2018). Functional and transcriptomic characterization of a dye-decolorizing fungus from Taxus Rhizosphere. *Pol J Microbiol, 67*(4), 417–430. doi.org/10.21307/pjm-2018-050

Harikumar, P. S., Joseph, L., & Arikal, D. (2013). Photocatalytic degradation of textile dyes by hydrogel supported titanium dioxide nanoparticles. *J Environ Eng Ecol Sci, 2*(2). doi:10.7243/2050-1323-2-2.

Jadhav, J. P., Parshetti, G. K., Kalme, S. D., & Govindwar, S. P. (2007). Decolourization of azo dye methyl red by *Saccharomyces cerevisiae* MTCC 463. *Chemosphere, 68*(2), 394–400. doi.org/10.1016/j.chemosphere.2006.12.087

Jamee, R., & Siddique, R. (2019). Biodegradation of synthetic dyes of textile effluent by microorganisms: An environmentally and economically sustainable approach. *Eur J Micro Immunol, 9*(4), 114–118. doi.org/10.1556/1886.2019.00018

Kalyani, D. C., Telke, A. A., Govindwar, S. P., & Jadhav, J. P. (2009). Biodegradation and detoxification of reactive textile dye by isolated *Pseudomonas sp.* SUK1. *Water Environm Res, 81*(3), 298–307. doi:10.2175/106143008x357147

Kamel, D., Sihem, A., Halima, & C., Sehili, T. (2009). Decolorization process of an Azoïque Dye (Congo Red) by photochemical methods in homogeneous medium. *Desalination, 247.* 412–422. doi:10.1016/j.desal.2009.02.052.

Kuhad, R., Sood, N., Tripathi, K., Singh, A., & Ward, O. P. (2004). Developments in microbial methods for the treatment of dye effluents. *Adv Appl Microbiol, 56,* 185–213. doi:10.1016/S0065-2164(04)56006-9.

Kumari, S., & Rajoriya, P. (2019). Textile industrial effluent treatment by azo dye decolorizing bacterial consortium. *Int J Curr Microbiol App Sci, 8*(11): 884–907. doi.org/10.20546/ijcmas.2019.811.105

Lade, H., Kadam, A., Paul, D., & Govindwar, S. (2015). Biodegradation and detoxification of textile azo dyes by bacterial consortium under sequential microaerophilic/aerobic processes. *EXCLI Journal, 14,* 158–174. doi:org/10.17179/excli2014-642

Lalnunhlimi, S., & Krishnaswamy, V. (2016). Decolorization of azo dyes (Direct Blue 151 and Direct Red 31) by moderately alkaliphilic bacterial consortium. *Braz J Microbiol, 47*(1), 39–46. doi.org/10.1016/j.bjm.2015.11.013

Leelakriangsak, M., & Borisut, S. (2012). Characterization of the decolorizing activity of azo dyes by *Bacillus subtilis* azoreductase AzoR1. *SJST, 34,* 509–516.

Lekha, R., Begum, M., & Ragunathan, R. (2017). Biodegradation of azo dyes using *Bacillus megaterium* and its phytotoxicity study. *IOSR-JESTFT, 11,* 12–20. doi:10.9790/2402-1107011220.

Lellis, B., Fávaro-Polonio, C., Pamphile, J., & Polonio, J. (2019). Effects of textile dyes on health and the environment and bioremediation potential of living organisms. *Biotechnol Res Innovation, 3,* 10. doi:1016/j.biori.2019.09.001.

Malla, M. A., Dubey, A., Yadav, S., Kumar, A., Hashem, A., & Abd Allah, E. F. (2018). Understanding and designing the strategies for the microbe-mediated remediation of environmental contaminants using omics approaches. *Front Microbiol, 9,* 1132. doi.org/10.3389/fmicb.2018.01132

Masarbo, R., Ismailsab, M., Monisha, T. R., Nayak, A., & Karegoudar, T. B. (2018). Enhanced decolorization of sulfonated azo dye methyl orange by single and mixed bacterial strains AK1, AK2 and VKY1. *Bioremediat J, 22,*1–11. doi:10.1080/10889868.2018.1516612.

Mohan, C. (2012). Analysis of genetic variation by Methyl Orange in *Rhizobium* using RAPD–PCR. *Internat J PharmTech Res, 4,* 1101–1109.

Morshedi, D., Mohammadi, Z., Akbar Boojar, M. M., & Aliakbari, F. (2013). Using protein nanofibrils to remove azo dyes from aqueous solution by the coagulation process. *Colloids Surf B Biointerfaces, 112,* 245–254. doi.org/10.1016/j.colsurfb.2013.08.004

Muneer, M., Qamar, M., Saquib, M., & Bahnemann, D. W. (2005). Heterogeneous photocatalysed reaction of three selected pesticide derivatives, propham, propachlor and tebuthiuron in aqueous suspensions of titanium dioxide. *Chemosphere, 61*(4), 457–468.

Ng, I-S., Zheng, X., Chen, B-Y., Chi, X., Lu, Y., & Chang, C-S. (2013). Proteomics approach to decipher novel genes and enzymes characterization of a bioelectricity-generating and dye-decolorizing bacterium *Proteus hauseri* ZMd44. *Biotechnol Bioproc Eng, 18.* doi:10.1007/s12257-012-0340-7.

Parshetti, G. K., Telke, A. A., Kalyani, D. C., & Govindwar, S. P. (2010). Decolorization and detoxification of sulfonated azo dye methyl orange by *Kocuria rosea* MTCC 1532. *J Hazard Mat, 176*(1–3), 503–509. doi.org/10.1016/j.jhazmat.2009.11.058

Pathak, H., Soni, D., & Chauhan, K. (2014). Evaluation of in vitro efficacy for decolorization and degradation of commercial azo dye RB-B by *Morganella sp.* HK-1 isolated from dye contaminated industrial landfill. *Chemosphere, 105,* 126–132. doi:10.1016/j.chemosphere.2014.01.004

Rauf, M., Meetani, M., & Hisaindee, S. (2011). An overview on the photocatalytic degradation of azo dyes in the presence of TiO$_2$ doped with selective transition metals. *Desalination, 276,* 13. doi:10.1016/j.desal.2011.03.071

Rawat, D., Mishra, V., & Sharma, R. S. (2016). Detoxification of azo dyes in the context of environmental processes. *Chemosphere, 155,* 591–605. doi.org/10.1016/j.chemosphere.2016.04.068

Saratale, R., Saratale, G., Chang, Jo-S., & Govindwar, S. (2011). Bacterial Decolorization and Degradation of Azo Dyes: A Review. *J Taiwan Inst Chem E, 42,* 138–157. doi:10.1016/j.jtice.2010.06.006.

Sari, I. P., & Simarani, K. (2019). Comparative static and shaking culture of metabolite derived from methyl red degradation by *Lysinibacillus fusiformis* strain W1B6. *Royal Soc Open Sci, 6*(7), 190152. doi.org/10.1098/rsos.190152

Sarkar, S., Banerjee, A., Halder, U., Biswas, R., & Bandopadhyay, R. (2017). Degradation of Synthetic Azo Dyes of Textile Industry: a Sustainable Approach Using Microbial Enzymes. *Water Conserv Sci Eng, 2,* 121–131 doi.org/10.1007/s41101-017-0031-5

Sarkar, S., Banerjee, A., Chakraborty, N., Soren, K., Chakraborty, P., & Bandopadhyay, R. (2019). Structural-functional analyses of textile dye degrading azoreductase, laccase and peroxidase: a comparative in silico study. *Electron J Biotechn, 43,* 48–54.doi: 10.1016/j.ejbt.2019.12.004.

Sathishkumar, K., Al Salhi, M. S., Sanganyado, E., Devanesan, S., Arulprakash, A., & Rajasekar, A. (2019). Sequential electrochemical oxidation and bio-treatment of the azo dye congo red and textile effluent. *J Photoch Photobio B, 200,* 111655. doi.org/10.1016/j.jphotobiol.2019.111655

Selvam K., Arungandhi K., Rajenderan G., Yamuna M. (2012). Biodegradation of azo dyes and textile industry effluent by newly isolated white-rot fungi. *Sci Rep, 1,* 564. doi:10.4172/scientificreports.564

Shah, K. (2014). Biodegradation of azo dye compounds. *Biochem Biotechnol, 1,* 5–3.

Shah, M. P. (2019). Bioremediation of azo dyes. Retrieved from: www.sciencedirect.com/science/article/pii/B9780128168097000063.

Singh, R., Singh, P., & Singh, R. (2015). Enzymatic decolorization and degradation of azo dyes: A review. *Int Biodeterior Biodegr, 104.* doi:10.1016/j.ibiod.2015.04.027.

Sridharan, R., Krishnaswamy, V. G., Archana, K. M., Rajagopal, R., Kumar, D. T., & Doss, C. G. P. (2021). Integrated approach on azo dyes degradation using laccase enzyme and CuI nanoparticle. *SN Appl Sci, 3*(3), 1–12.

Sudha, M., Arul, S., Gopal, S., & Natesan, S. (2014). Microbial degradation of Azo Dyes: A review. *Int J Curr Microbiol App Sci, 3,* 670–690.

Telke, A. A., Joshi, S. M., Jadhav, S. U., Tamboli, D. P., & Govindwar, S. P. (2010). Decolorization and detoxification of Congo red and textile industry effluent by an isolated bacterium *Pseudomonas sp.* SU-EBT. *Biodegradation, 21*(2), 283–296. doi.org/10.1007/s10532-009-9300-0

Telke, A., Kadam, A., & Govindwar, S. (2014). Bacterial enzymes and their role in decolorization of azo dyes. doi.10.1007/978-3-319-10942-8_7.

Ventura-Camargo, B. & Marin-Morales, M. (2013). Azo Dyes: Characterization and toxicity: A review. *Textiles and Light Industrial Science and Technology. 2,* 85–103.

Waghmode, T. R., Kurade, M. B., Sapkal, R. T., Bhosale, C. H., Jeon, B. H., & Govindwar, S. P. (2019). Sequential photocatalysis and biological treatment for the enhanced degradation of the persistent azo dye methyl red. *J Haz Mat 371,* 115–122. doi.org/10.1016/j.jhazmat.2019.03.004

Wood, T. (2008). Molecular approaches in bioremediation. *Curr Opin Biotechn, 19,* 572–578. 10.1016/j.copbio.2008.10.003.

Yap, C. (2019). Cleaning up of contaminated soils by using microbial remediation: A review and challenges to the weaknesses. *Am J Biomed Sci Res, 2.* doi:10.34297/AJBSR.2019.02.000589.

Zhao, M., Sun, P. F., Du, L. N., Wang, G., Jia, X. M., & Zhao, Y. H. (2014). Biodegradation of methyl red by *Bacillus sp.* strain UN2: decolorization capacity, metabolites characterization, and enzyme analysis. *Environ Sci Pollut R Int, 21*(9), 6136–6145. doi.org/10.1007/s11356-014-2579-3

Zhuang, M., Sanganyado, E., Xu, L., Zhu, J., Li, P., & Liu, W. (2020). High throughput sediment DNA sequencing reveals azo dye degrading bacteria inhabit nearshore sediments. *Microorganisms, 8*(2), 233. doi.org/10.3390/microorganisms8020233

Zhuang, M., Sanganyado, E., Zhang, X., Xu, L., Zhu, J., Liu, W., & Song, H. (2020). Azo dye degrading bacteria tolerant to extreme conditions inhabit nearshore ecosystems: Optimization and degradation pathways. *J Environ Manage, 261,* 110222. doi.org/10.1016/j.jenvman.2020.110222

5 Microbes and Microbial Enzymes in the Bioremediation of Environmental Pollutants

Somok Banerjee[1], Swatilekha Pati[1], Palla Mary Sulakshana[2], Aveepsa Sengupta[1], and Ashutosh Kumar[1,]*

[1]Department of Microbiology, Tripura University (A Central University), Suryamaninagar, Agartala, Tripura, India

[2]Raghu College of Pharmacy, Visakhapatnam, Andhra Pradesh, India

*Corresponding author: AK: ashu.mtb@gmail.com, ashutoshkumar@ tripurauniv.in

CONTENTS

5.1 ENVIRONMENTAL POLLUTANTS

Pollutants are particles that are released into the ecosystem and have serious effects on inhabitants. The major part of pollutants comes from anthropogenic sources. Wastewater from industrial sources contains toxic chemicals, which includes heavy metals, dyes, and acids, and adversely affects the health of workers and residents. Polycyclic aromatic hydrocarbons (PAHs) are highly toxic and persistent pollutants that can disturb both the ecosystem and the physiological processes in plants, animals, and microbes when discharged into the ecosystem. Moreover, these pollutants also cause other health problems such as rashes, nausea, impaired immune response, and many more. Recent development, in industrialization has resulted in an increased amount of metal pollution globally.

DOI: 10.1201/9781003130932-5

The majority of metallic species are highly toxic, are a danger to human health, and harm the ecosystem. Continuous exposure to heavy metal pollutants causes serious health diseases through interactions with macromolecules. It leads to immune, neurological, and circulatory disorders and can cause cancers, mutagenicity, and teratogenicity.

One of the primary pollutants that are present in the waste effluent from textile industries is the dyes since most of the locations do not have efficient systems to treat the water before discharging it into the ecosystem. The untreated water constitutes the majority of the total waste that is discharged from the industry. Synthetic dyes are difficult to break down in the water and others release toxic substances when they degrade. In general, these dyes are mostly toxic and can cause several types of cancers. In addition, synthetic dyes affect the photosynthesis of aquatic flora due to low light penetration in rivers, which significantly affects the aquatic biota and can increase the biological and chemical oxygen demand. In addition, the dyeing process, especially with azo-type textile dyes, harms microbial communities in the soil and plant growth and germination.

Recently, nuclear activities, rapid industrial development, and natural biogeochemical cycle disruptions have affected large areas of land or ecosystems worldwide with radionuclidesPlants uptake the radionuclides present in the soil and distribute them throughout the environment, affecting living beings. The radiation generated by radioactive wastes causes severe health hazards, such as neurological disorders, infertility, birth defects, and cancers.

Technological advancement has left the entire ecosystem exposed to a variety of chemicals, which include pesticides and fungicides. The increase in population places large demands on the supply of food, which enhances the use of pesticides to increase the production of crops, fruits, and vegetables to meet these requirements. There are different types of pesticides: fungicide, rodenticides, nematicides, herbicides, molluscicides, acaricides, and herbicides, which are categorized based on their action against pests. Though these chemicals eliminate pests from the crops and the livestock besides improving yields, their distribution or accumulation throughout the trophic level has serious effects on the ecosystem and the living beings. When these chemicals are applied on the crops, some fractions remain in the area and some part gets into the ecosystem as pollutants through agricultural runoff. The invasion of pesticides into the body might be through dermal, inhalation, or the oral route. Moreover, few of the pesticides and insecticides remain in the ecosystem for a long time which makes them more hazardous. Organochlorine pesticides are a group of hydrocarbon pesticides, minute levels of which can cause serious symptoms, such as anxiety, gastrointestinal problems, headaches, insensibility, and irritability in individuals. Organophosphate pesticides (OPPs) are even more toxic to human health. In children, there is higher pesticide absorption due to their immature metabolic processes, which leads to high toxicity. Several reports have suggested that these chemicals are toxic and sometimes cause lymphoma, brain tumors etc. in humans.

5.2 MICROBES INVOLVED IN BIOREMEDIATION

Microorganisms are widely distributed across the biosphere because of their ability to grow under various environmental conditions. Using bioremediation, the nutritional versatility of these microorganisms is exploited to break down environmental contaminants that are present in the soil, water, and sediments. Several microorganisms, such as bacteria, yeasts, fungi, archaea, and algae are biologically active. They can convert or modifying the harmful pollutants into less toxic or nontoxic compounds. These microbes naturally reside in contaminated areas and a few have been isolated from different parts of the environment and artificially inoculated into the bioremediation site. Because various types of contaminants might be present at the bioremediation site, various types of microbes are required for successful bioremediation. Some microbes can degrade petroleum hydrocarbons and then use them as an energy source, and other groups of microorganisms utilize oil as their nutrient source and then produce surface-active complexes that emulsify oil in water, and therefore, help the removal process. In addition, numerous microbes have natural capacities for the biosorption of heavy metal ions and help in the treatment of heavy metal toxicity. The efficiency or effectiveness of bioremediation depends on the potential of the microbes to interact with the environment and its pollutants. The expected growth and activity of the microbes are affected by several environmental factors, for example, pH, temperature, and soil structure, and the physicochemical characteristics of the pollutants, which includes concentration, type, solubility, chemical structure, and toxicity. In addition to the extremophiles, genetic engineering has provided a method to deal with these problems. Genetically engineered microorganisms possess mechanisms to tolerate highly adverse environmental conditions and have a high growth rate compared with other microorganisms and are economical.

Bacteria are a diverse group of microorganisms that play an essential role in bioremediation. With the help of their universal toxins, bacteria could probably degrade pollutants under favorable environmental conditions *Deinococcus radiodurans* is an extremophile (radiation-resistant) that has been genetically modified to bioremediate heavy metals and solvents. For example, a genetically engineered *Deinococcus sp.* helped to break down mercuric ions and toluene from radioactive wastes that were present in the environment (Brim et al., 2000). *Bacillus cereus,* which is a gram-positive soil bacterium, helped with the bioremediation of chromium (Cr) in contaminated sites. Two Cr tolerant strains, *B. cereus* D and *B. cereus* 332 have a high Cr removal rate in both the planktonic and immobilized forms (Li et al., 2020). *Alkanivorax borkumensis* is a marine bacterium that utilizes hydrocarbons that are present in fuels and generates carbon dioxide (Biello et al., 2010). *Bacillus sp.* and *Staphylococcus sp.* degrade endosulfan, *Sporosarcina ginsengisoli* degrades arsenic (As), *Kocuria flava* degrades copper (CU, *Rhizobium meliloti* degrades Dibenzothiophene, *Acinetobacter sp. and Ralstonia sp.* degrade aromatic hydrocarbons (Mohamed et al., 2011; Achal et al., 2012; Achal et al., 2011; Frassinetti et al., 1998; Simarro et al., 2013).

One of the most hazardous heavy metal pollutants in the environment is Hg. Hg pollution might be caused by natural events (e.g., volcanos, wood fires, and ocean volatilizations) and man-made events (e.g., mining, paper pulp and cosmetic industries*)*. Various studies have shown that many microorganisms that are present in the environment can degrade Hg by involving their Hg-resistant genes. The Hg-resistant operon consists of the *merA, merB, merT, merP*, and *merF*, which are responsible for Hg resistance (Osborn et al., 1997). Proteins encoded by the mer operon convert elemental Hg to inorganic Hg mediated by *merB*. The product of *merB* is organomercurial lyase (Osborn et al., 1997; Dash et al., 2012).

As is another harmful environmental pollutant. Many industrial effluents contain As, which is a by-products. The accumulation of As is hazardous in plants and animals. The bioremediation of As by bacteria depends on *arsM*, which mediates the conversion of As into volatile methylated As (Liu et al., 2011). Two nongenetically engineered highly potent As-removing bacterial strains are *Marinomonas communis* (Takeuchi et al., 2007) and *Lactobacillus acidophilus* (Singh et al., 2010). Genetically engineered organisms with *arsM* can accumulate As and convert As into volatile arsenic (Rahman et al., 2014).

Hydrocarbon pollution is mainly caused by petroleum and its derivatives. Hydrocarbons are categorized in aliphatic hydrocarbons (e.g., alkane, alkene, and alkyne), aromatic hydrocarbons (e.g., toluene and naphthalene), and heterocyclic compounds (e.g., carbazole and xanthenes). One of the most prevalent genetically engineered hydrocarbon degraders is *Pseudomonas putida*, which is a superbug and a marine bacterium. Engineered plasmids consist of various hydrocarbon-degrading genes, such as the camphor degrading and naphthalene degrading genes. Hydrocarbon degradation is a process where microorganisms use complex hydrocarbons as their metabolites and produce the required energy and simpler organic substances that have no harmful impact on the environment. Other hydrocarbon-degrading species include *Bacillus sp.*, *Pseudomonas sp.*, *Micrococcus sp.*, and *Syntrophus sp.*

Fungal species are superior to bacteria in metabolic versatility and environmental flexibility. Some of them can oxidize a wide range of chemical compounds. They can adapt well in extreme environments, such as in low moisture and areas with a high level of pollutants. In addition, fungi are more effective at degrading PAHs than bacteria. They can degrade high molecular weight PAHs, and bacteria mostly degrade smaller molecules. Moreover, fungi can function in conditions where PAHs accumulate, such as in a nonaqueous environment and under low oxygen (O_2) concentrations. In addition, they can remediate metals and pesticides. *Phanerochaete chrysosporium* was the first white-rot fungal species discovered and could degrade key organic pollutants. *Penicillium chrysogenum* is a halotolerant fungus that degrades high hydroquinone concentrations under hypersaline conditions; therefore, making it an efficient candidate for the remediation of habitats that are contaminated with this aromatic compound (Pereira et al., 2014). *Aspergillus niger* is a haploid filamentous fungus that is often used in the bioremediation of petroleum hydrocarbon contaminated sites, because of its high tolerance of crude oil and its ability to utilize this as a nutrient source (Al-Jawhari, 2014). Other notable fungi, such as *Candida viswanathii*, degrade phenanthrene and benzopyrene, *Ganoderma lucidum* degrades As, *Penicillium canescens* degrades Cr, and *Myrothecium roridum* IM 6482 degrades industrial dyes (Hesham et al., 2012; Loukidou et al., 2003; Say et al., 2003; Jasinska et al., 2012). Other hydrocarbon-degrading fungi include *Mucor sp., Alternaria sp., Aspergillus sp., Cephalosporium sp., Penicillium sp., Cladosporium sp., Fusarium sp., Geotrichum sp., Saccharomyces sp., Gliocladium sp., Paecilomyces sp., Talaromyces sp., Pleurotus sp., Rhizopus sp., Polyporus sp., Candida sp., Rhodotorula sp.,* and *Torulopsis sp.* Fungal enzymes involved in hydrocarbon degradation include laccase, lignin peroxidase, and Mn-dependent peroxidase. Polyhydroxyalkanoids are bacteria-derived polyesters that have short and medium-length carbon chains. Many fungi have PHA depolymerizing activity (Sang et al., 2002). Fungi, such as *Aspergillus sp.,* and *Zalerion maritimum* can break down polyethylene using their degradative enzymes (Paço et al., 2017). Dyes from the textile and paper industry can be remediated with the help of various fungi, such as *Cerrena unicolor, A. niger* and *Penicillium janthinellum.*

Extreme halophilic archaea can degrade pollutants, such as hydrocarbons under harsh environmental conditions. They use hydrocarbons as their energy source and can withstand several antibiotics, such as penicillin, streptomycin, and many more. Algae are used to remove a variety of environmental pollutants, which include organic pollutants and heavy metals, via phycoremediation. Algae eliminate organic pollutants and heavy metals using two techniques: biosorption and bioaccumulation. Some of the agal species, such as *Ulva lactuca,* degrade chloramphenicol, *Selenastrum capricornutum* degrades atrazine, *Cyclindrotheca sp.* and *Euglena gracilis* degrade Dichlorodiphenyltrichloroethane, *Ascophyllum nodosum* absorbs lead (Pb), Cu, zinc (Zn) and Cd, *Spirogyra sp.* and *Cladophora sp.* absorb Pb and Cu, *Acrosiphonia coalita* degrades Trinitrotoluene (Leston et al., 2013; Friesen-Pankratz et al., 2003; Semple et al., 1999; Kobayashi et al., 1982; Romera et al., 2007; Lee et al., 2011; Cruz-Uribe et al., 2007).

5.3 EXTREMOPHILES IN BIOREMEDIATION

Recently, the technological and industrial growth in various sectors has led to an increase in the complexity and toxicity of wastes. Microbes from extreme environments provide a biocatalytic system that is more effective under extreme conditions compared with other typical bioconversions. They use various strategies to tackle the adverse effects of the environment. Microorganisms are distributed across various temperature niches. Based on the optimum temperatures for growth, the microorganisms are classified into different groups, such as psychrotolerants that grow optimally at temperatures from 5°C to 20°C. The optimum growth temperature of psychrophiles is <15°C and it does not grow >20°C. Some examples of psychrophilic bacteria are *Chryseobacterium greenlandensis* and *Arthrobacter sp.* Moderate thermophiles grow at temperatures from 45°C to 70°C. One example is *Methanothermobacter tenebrarum,* which requires a temperature between 45°C and 80°C for growth. Extreme thermophiles optimally grow at a temperature from 70°C to 80°C. An example of an extreme thermophile is *Thermus aquaticus.* The microorganisms that belong to hyperthermophiles include *Geogemma barossii* and *Methanopyrus kandleri* and they can grow ≥80°C. To counter extreme temperatures, extremophiles have adapted to their environment in numerous ways. For instance, psychrophiles possess high levels of unsaturated fatty acids in their plasma membranes to maintain fluidity at low temperatures (Chintalapati et al. 2004; Russell 2008). In addition, the cold-adapted proteins are structurally flexible and catalytically effective at low temperatures due to several protein modifications, which includes decreased disulfide bonds, hydrogen (H) bonds and salt bridges, increased solvent interactions, and specific amino acid sequence modifications. Heat-stable proteins in the thermophilic microbes adapt by an increased number of salt bridges and H bonds and increased hydrophobic interactions. In bioremediation, the psychrophiles and psychrotolerants are used in sludge treatment and the removal of hydrocarbon contamination, and the thermophilic microbes are used in the bioremediation of industrial effluents and deep subsurface sites at high temperatures. Cold-adapted bacteria, such as *Pseudomonas sp.* ST41 could degrade alkanes in soil with marine gas oil at 4°C (Stallwood et al., 2005).

Microbes under extreme pH environments are grouped into acidophile and alkaliphile, which is based on the optimum pH required for growth. Acidophiles are microbes that thrive in acidic environments at a pH <5. A few examples of acidophilic microbes are *Sulfolobus sp., Acidiphilium sp., Bacillus sp.,* and *Acidithiobacillus sp.* Arulazhagan et al. (2017) showed the capability of *Stenotrophomonas maltophilia*s AJH1 in degrading PAH in an acidic mineral salt medium at pH 2. Alkaliphilic microorganisms adapt to a high pH environment by altering the physiological reactions of the cell, mainly with the help of proton and sodium pumps. A few examples of the alkaliphilic microbes are *Clostridium sp., Cladosporium sp.,* and *Alkalibacter sp.* In bioremediation, the acidophiles are used for the bioleaching of tanks and ore pipes, and the alkaliphilic microorganisms are used in the bioconversion of starch, lignocellulosic feedstock, fats, and proteins (Peeples, 2014).

The microorganisms that grow in high salt environments are known as halophiles. *Natrialba sp.* C21 exhibited good potential in the biodegradation of naphthalene and pyrene under high salinity conditions (Khemili-Talbi et al., 2015). The halophilic microbes accumulate molar concentrations of potassium chloride, which helps their enzymatic pathways to adapt to be functionally active in such high salt stress. The second strategy is to synthesize osmolytes, such as ectoin, betaine and glycine which excludes salt rapidly from the cytosol (Oren et al., 2008). The halophilic microbes and their enzymes are known to be involved in bioremediation in brine and hydrocarbon-rich environments (Sei and Fathepure, 2009; Al-Mailem et al., 2010, 2012, 2013; Najera-Fernandez et al., 2012; Erdogmus et al., 2013; Harada et al., 2013).

There are only a few microorganisms that can survive under extreme radioactive environments, such as weapon proliferation sites, radioactive ore deposits, and the energy industries. *D. radiodurans* is one such radiation-tolerant microbe. According to a study, this bacterium uses several antioxidants that confer radiation resistance, which includes DNA repair. In addition to being used for the bioremediation of toluene and Hg at radioactive sites, they are utilized in the industry (Brim et al., 2006). Many microorganisms, which include archaea and yeasts, can tolerate high metal and hydrocarbon content under high-pressure environments, such as deep oceanic and lithospheric sites. Many extremophile microorganisms, which include bacteria and archaea are microbial sources for metal biodegradation. For instance, thermophilic microbes, such as *Methanothermobacter thermoautotrophicus, Thermus scotoductus,* and *Pyrobaculumis landicum* can degrade Cr (Cr^{6+}) and *Methanothermobacter sp.* can degrade Cr^{6+} internally and externally by utilizing Cr(VI), which is the final electron acceptor (Singh et al., 2015). According to Tomova et al. (2014), psychrotolerant and halotolerant bacteria (*Bacillus sp.* and *Stenotrophomonas sp.*) are involved in the treatment of wastewater contaminated with heavy metals. Microbes are involved in bioremediating textile wastewater that has high salinity, temperature, and alkalinity with the help of several enzymes, which include laccases and azoreductases. Microbes, such as a thermophilic fungus *Thermomucorindicae seudatica* can adsorb a high concentration of azo and anthracenedione dyes at 55°C (Taha et al., 2014) and cyanobacteria remediate textile dyes from industrial effluents at an alkaline pH of 10.27 (Bela et al., 2015).

5.4 GENETICALLY ENGINEERED MICROBES IN BIOREMEDIATION

In general, a variety of specialized microbial communities aid the population responsible for degradation; however, they do not possess the appropriate catabolic pathways required for the removal of various novel elements that constitute the pollutants or contaminants. One way of countering this situation is through bioaugmentation. This involves the introduction of exogenous or genetically engineered microbes that contain the necessary genes to augment the indigenous population. The current knowledge of molecular biology has developed novel strains of microbes that have beneficial properties, which includes constructing novel pathways, modifying the specificity of catabolic enzymes and affinity toward the substrate, and cellular localization and its expression to make bioremediation more efficient. Some of the genetically modified microbes used in bioremediation include *Ralstonia eutropha* CH34 that is used in Cd, *D. radiodurans* is used in Hg, *P. fluorescens* HK44 is used for PAHs, *P.putida* S12 used for naphthalene, toluene, and biphenyl, and *E.coli* SE5000 is used in Ni bioremediation (Valls et al., 2000; Brim et al., 2000; Sayler et al., 2000; Fujita et al., 1995; Deng et al., 2005).

Bioremediation of aromatic contaminants from soil and water using engineered bacteria has been a focus for a long time. It has been shown that engineered recombinant strains can be used to completely degrade chloroaromatics. In this technique, an overall pathway that can mineralize a particular compound was studied by combining several pathways from different bacteria, which were transferred to a single recombinant host through conjugation. For instance, the transformation of *Comamonas testosteroni* VP44 with plasmids bearing the *ohb* operon from *P.aeruginosa* and the *fcb* operon from *Arthrobacter globiformis* provided the capability to degrade monochlorobiphenyls (Hrywna et al., 1999). Another example includes the development of a novel pathway that is responsible for degrading nitrobenzene and 4-chloronitrobenzene by transforming the CNB-1 cloned genes in a *Comamonas sp.* strain into *E. coli* (Wu et al., 2006). Heavy metals and mixed wastes that contain radionuclides and organic pollutants are sometimes toxic to most bacteria; therefore, the development of bacteria that can perform efficient bioremediation is required. For example, *D.radiodurans* was engineered with the toluene dioxygenase genes from *P. putida* F1 for the efficient oxidation of toluene, 3,4-dichloro-1-butene, chlorobenzene, and trichloroethylene under extreme irradiation environmental conditions (Lange et al., 1998). Renninger et al. (2004) showed that an engineered strain of *P. aeruginosa* in which the genes responsible for the production and degradation of polyphosphate were cloned together under the *tac-lac* promoter, effectively degraded uranium via the uranyl group. On degradation, a significant amount of phosphate was obtained, combined with the uranyl groups that were precipitated at the plasma membrane.

To augment the previous efforts on the transfer of genes to increase efficiency, in vitro site-directed mutagenesis or recombinant DNA technology were used to enhance enzyme expression or to produce alternative transcriptional promoters. Various molecular techniques, such as gene shuffling and oligonucleotide-directed mutagenesis have been used that reduced substrate specificity and improved the degradation rates of multiple oxygenases. Another study showed that the modification of two amino acids in the active site of the cytochrome P450 enzyme from *P. putida* through site-specific mutagenesis produced variants with an enhanced capability required to oxidize hexane and 3-methylpentane (Stevenson et al., 1998). Genetic engineering of *Pseudomonas sp.* KKS102 resulted in enhanced polychlorinated biphenyl (PCB) and biphenyl degrading capability in which the *bph* promoter was replaced with a variety of constitutive promoters using various techniques, such as recombinant DNA construction, homologous recombination, and polymerase chain reaction (Ohtsubo et al., 2003). Another study reported the incorporation of a gene, chlorobenzene dioxygenase (*CDO*) from *P. putida* P51 into *E. coli* strain under the control of the Palk promoter,

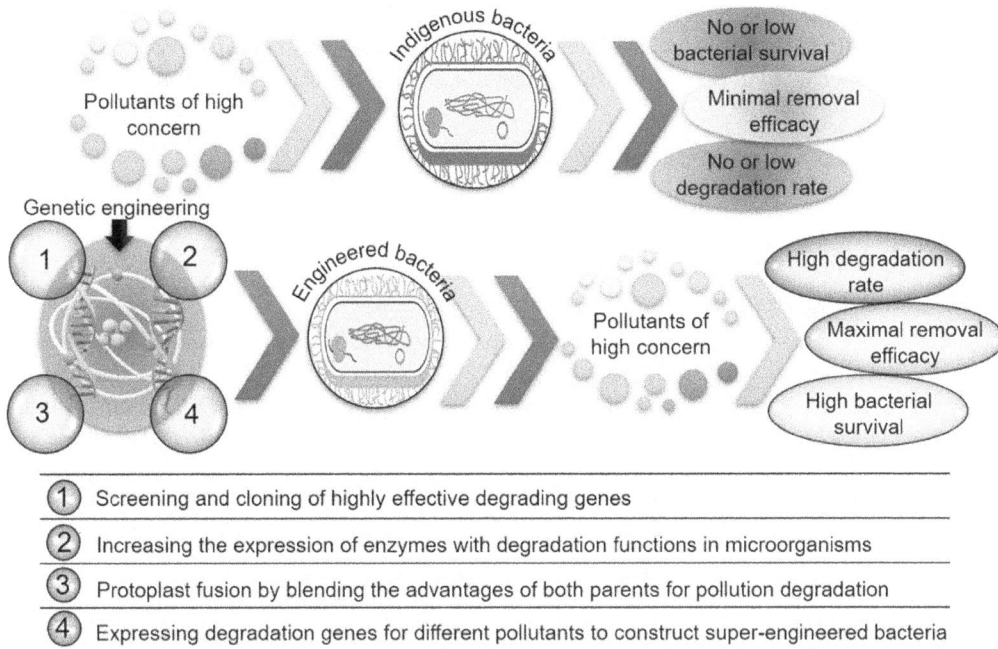

FIGURE 5.1 Strategies to construct engineered microbes for efficient bioremediation. (Figure adapted, with permission, from Liu et al. (2019) © 2019 Elsevier B.V.)

which improved the rate of production of chlorobenzene dioxygenase by three times compared with an *E.coli* strain, where *CDO* was regulated by the *lac* promoter. The engineered strain had enhanced capability in catalyzing the *cis*-dihydroxylation of benzonitrile and other aromatics, and therefore, could be used in biosynthesis and bioremediation (Yildirim et al., 2005). Various strategies for the construction of genetically engineered microbes (GEMs) are shown for efficient bioremediation (Figure 5.1). Despite having such significant advantages of using GEMs for bioremediation, their field applications are limited. The application of GEMs could alter the ecological framework, which could be due to the persistence of undesired genes in the environment, and the risk of unwanted gene transfer to indigenous species. The plasmids, combined with recombinant genes and selective markers could continue to spread to other bacteria and the environment due to their physical and chemical stability.

5.5 MICROBIAL ENZYMES INVOLVED IN BIOREMEDIATION

Enzymes might be proteins or glycoproteins that act as biocatalysts and facilitate the conversion of substrates into products. Many microbe specific proteins are involved in degrading contaminants from the environment. Chen et al. (2014) reported the degradation of acetochlor by an oxygenase enzyme in *Sphongomonas sp.* DC-6. *Comamonas acidovorans* esterase helps in the degradation of polyurethane (Masaki et al., 2005). Kundys et al. (2017) reported the biodegradation of aliphatic polyesters and fatty acid methyl esters by lipase enzymes from *C. antarctica*. *Agrobacterium radiobacter* phosphotriesterases help in the removal of insecticides from contaminated sites (Riya & Jagatpati, 2012). *Ideonella sakaiensis* polyesterase is used in the biodegradation of plastic wastes, such as poly(ethylene terephthalate) and mono(2-hydroxyethyl) terephthalic acid (Yoshida et al., 2016; Austin et al., 2018). The degradation of plastic waste, such as poly(ethylene terephthalate) involves *Clostridium botulinum* esterase (Biundo et al., 2016). Bioremediation mainly depends on a broad range of enzymes. The intracellular and extracellular enzymes from bacteria, fungi, and algae interact with toxic pollutants and degrade or convert them into nontoxic products. Some of the enzymes produced by the microbes for bioremediation are discussed in the following sections.

5.5.1 MICROBIAL OXIDOREDUCTASES

Bacteria and fungi produce oxidoreductase for the detoxification of organic compounds via oxidative coupling.

5.5.1.1 Microbial Oxygenases

These enzymes are from the oxidoreductase group of enzymes. The oxygenases play an essential role in the bioremediation of halogenated organic compounds by breaking the aromatic ring or raising reactivity. They carry out oxidation processes using FAD/NADH/NADPH as cosubstrates. In bioremediation, these enzymes are used as biocatalysts due to their ability to degrade several substrates, such as alkenes, steroids, and fatty acids via biotransformation, biodegradation, denitrification, and dehalogenation.

5.5.1.2 Microbial Laccases

Certain fungi and bacteria produce laccases (which are a group of multicopper oxidases), facilitate the oxidation of an extensive range of reduced phenolics and aromatic substrates. Laccases have various isozyme forms that are encoded by separate genes, and the genes are expressed differently based on the nature of the inducer.

5.5.1.3 Peroxidases

These enzymes oxidize lignin and other phenolic groups in the presence of a mediator and utilize hydrogen peroxide (H_2O_2). Among the peroxidases, lignin and manganese peroxidases are important in bioremediation because of their capability to degrade toxic compounds. Lignin peroxidize has a crucial role in degrading lignin that is present in plant cell walls. During lignin degradation, H_2O_2 acts as a cosubstrate, and veratryl alcohol is the mediator. Manganese peroxidase, which is produced by a fungus, can oxidize numerous phenolic compounds. Microbial versatile peroxidases are used in the bioremediation of recalcitrant pollutants due to their broad substrate specificity and capability of oxidizing substrates, such as manganese peroxidase (MnP), lignin peroxidase (LiP), and horseradish peroxidases in the absence of manganese compared with its counterparts (Karigar et al., 2011).

5.5.2 MICROBIAL HYDROLYTIC ENZYMES

Various microbes, mainly bacteria, produce hydrolytic enzymes that are used in bioremediation. These enzymes cleave the chemical bonds in the toxic contaminants; therefore, attempting to reduce their toxicity. They are widely used in the biodegradation of oil spills, organophosphates, and carbamate insecticides. In addition, their availability, absence of stereoselectivity, and capability of tolerating water-miscible events make them advantageous in bioremediation. Many hydrolytic enzymes, which include including lipases, amylases, proteases, and cellulases are used in other diverse fields, such as the food industry, biomass degradation, biomedical, and chemical industries (Karigar et al., 2011).

5.5.2.1 Lipases

Several microbes, including bacteria and actinomycetes, produce lipases that catalyze the hydrolysis of fatty acids and triacylglycerols. One of the main components of oils and fats are triglycerides, and lipases can convert triglycerides into glycerol. Other than hydrolysis, these enzymes are involved in esterification, alcoholysis, and aminolysis. A study reported that microbes, such as *P. putida*,

Acinetobacter sp., Rhodococcus sp., and *Mycobacterium sp.,* etc. could degrade petroleum hydrocarbons in cold environments by secreting microbial lipases (Margesin et al., 2003).

5.5.2.2 Microbial Cellulases

Some microbes, such as bacteria and fungi produce a group of inducible enzymes when growing on cellulosic materials, which are collectively known as microbial cellulases that are responsible for the degradation of cellulose to glucose. The cellulases consist of three groups of enzymes: (1) endoglucanase attacks the internal O-glycosidic bonds of cellulose and create free glucan chain ends; (2) exoglucanase further degrades the cellulose chain to release β-cellobiose; and (3) β-glucosidase hydrolyzes the cellobiose to release glucose units. In addition to bioremediation, microbial cellulases are used in a wide range of industrial fields, such as paper, textiles, and bioethanol (Karigar et al., 2011).

5.5.2.3 Microbial Proteases

These are a group of ubiquitous enzymes that hydrolyze amide bonds in amino acids in an aqueous environment (Karigar et al., 2011). In general, proteases are classified into two groups; exopeptidases cleave the peptide bond at the amino or carboxy-terminal end of the substrate, and endopeptidases cleave the peptide bond at a distance from the terminal ends of the substrate.

5.6 CONCLUSION

Pollution has a direct impact on human and animal health; therefore, affecting overall diversity. For a long time, science and technology have been employed when the environment is polluted. However, with scientific and technological advancements, environmental pollution could be reduced or eliminated under some conditions. Bioremediation is a potential alternative to the traditional methods used to clean up pollution by enhancing natural biodegrading processes. In addition, bioremediation is a promising upcoming technique in biotechnology, because it uses plants, microorganisms, and their enzymes to eradicate pollutants from the sites in a fast and eco-friendly manner. As discussed in this chapter, because of their diversified nature and stress tolerating mechanisms, various microorganisms can withstand several harsh environmental conditions. In addition, their ability and applications could be enhanced by implementing more novel techniques that could finally offer better potential for pollution removal. Based on these novel approaches and further focused research, new, clean, and environmentally accepted technologies could be developed that are commercially feasible.

REFERENCES

Achal V., Pan X., Fu Q., & Zhang D. (2012). Biomineralization based remediation of As (III) contaminated soil by *Sporosarcina ginsengisoli*. *J Haz Mat*, 201–202, 178–184.

Achal V., Pan X., & Zhang D. (2011). Remediation of copper-contaminated soil by *Kocuriaflava* CR1, based on microbially induced calcite precipitation. *Ecol Eng*, 37, (10), 1601–1605.

Al-Jawhari, I. F. H. (2014). Ability of some soil fungi in biodegradation of petroleum hydrocarbon. *J Appl Environ Microbiol*, 2, 46–52.

Al-Mailem, D.M., Eliyas, M., Radwan, S.S., 2012. Enhanced haloarchaeal oil removal in hypersaline environments via organic nitrogen fertilization and illumination. Extremophiles. 16 (5), 751–758.

Al-Mailem, D.M., Eliyas, M., Radwan, S.S., 2013. Oil-bioremediation potential of two hydrocarbonoclastic, diazotrophic marinobacter strains from hypersaline areas along the Arabian gulf coasts. Extremophiles. 17 (3), 463–470.

Al-Mailem, D.M., Sorkhoh, N.A., Al-Awadhi, H., Eliyas, M., Radwan, S.S., 2010. Biodegradation of crude oil and pure hydrocarbons by extreme halophilic archaea from hypersaline coasts of the Arabian gulf. Extremophiles. 14 (3), 321–328.

Arulazhagan, P., Al-Shekri, K., Huda, Q., Godon, J. J., Basahi, J. M., & Jeyakumar, D. (2017). Biodegradation of polycyclic aromatic hydrocarbons by an acidophilic *Stenotrophomonas maltophilia* strain AJH1 isolated from a mineral mining site in Saudi Arabia. *Extremophiles, 21*, 163–174. doi.org/10.1007/s00792-016-0892-0

Austin, H. P., Allen, M. D., Donohoe, B. S., Rorrer, N. A., Kearns, F. L., Silveira, R. L., … Beckham, G. T. (2018). Characterization and engineering of a plastic-degrading aromatic polyesterase. *P Natl Acad Sci USA, 115*(19), E4350–E4357

Bela, R. B., & Malliga, P. (2015). Treatment of textile dye effluent using marine cyanobacterium *Lyngbya sp.* with different agrowastes and its effect on the growth of cyanobacterium. *J. Environ. Biol*, 36, 623–626.

Biello, D. (2010). Slick solution: How microbes will clean up the deep water horizon oil spill. *Sci Am* (n.d.), 25.

Biundo, A., Ribitsch, D., Steinkellner, G., Gruber, K., & Guebitz, G. M. (2017). Polyester hydrolysis is enhanced by a truncated esterase: Less is more. *Biotechnol J, 12*(8). 10.1002/biot.201600450

Brim, H., McFarlan, S. C., & Fredrickson, J. K. (2000). Engineering *Deinococcus radiodurans* for metal remediation in radioactive mixed waste environments. *Nat Biotechnol, 18*, 85–90.

Brim, H., Osborne, J. P., Kostandarithes, H. M., Fredrickson, J. K., Wackett, L. P., & Daly, M. J. (2006). *Deinococcus radiodurans* engineered for complete toluene degradation facilitates cr(VI) reduction. *Microbiology-SGM, 152*, 2469–2477.

Chen, Q, Wang, C. H., Deng, S. K., Wu, Y. D., Li, Y., Yao, L., & Li, S. P. (2014). Novel three-component Rieske non-heme iron oxygenase system catalyzing the 419 N-dealkylation of chloroacetanilide herbicides in *Sphingomonads* DC-6 and DC-2. *Appl Environ Microbiol, 420*, 5078–5085.

Chintalapati S, Kiran MD, Shivaji S (2004) Role of membrane lipid fatty acids in cold adaptation. Cell Mol Biol 50:631–642.

Cruz-Uribe, O., Cheney, D. P., & Rorrer, G. L. (2007). Comparison of TNT removal from seawater by three marine microalgae. *Chemosphere, 67*, 1469–1476.

Dash, H. R., & Das, S. (2012). Bioremediation of mercury and importance of bacterial *mer* genes. *Int Biodeter. Biodegrad, 75*, 207–213.

Deng, X., Li, Q. B., Lu, Y. H. (2005). Genetic engineering of *Escherichia coli* SE5000 and its potential for Ni2? bioremediation. *Process Biochem, 40*, 425–430.

Erdogmus, S.F., Mutlu, B., Korcan, S.E., Guven, K., Konuk, M., 2013. Aromatic hydrocarbon degradation by halophilic archaea isolated from Camalti Saltern, Turkey. Water Air Soil Pollut. 224 (3), 1449.

Frassinetti, S., Setti, L., Corti, A., Farrinelli, P., Montevecchi, P., & Vallini, G. (1998). Biodegradation of dibenzothiophene by a nodulating isolate of *Rhizobium meliloti*. *Can J Microbiol, 44*(3), 289–297.

Friesen-Pankratz, B., Doebel, C., Farenhorst, A., & Goldsborough, L.G. (2003). Influence of algae (*Selenastrum capricornutum*) on the aqueous persistence of atrazine and lindane: implications for managing constructed wetlands for pesticide removal. *J Environ Sci Heal B, 38*, 147–155.

Fujita, M., Ike, M., Hioki, J. I. (1995). Trichloroethylene degradation by genetically-engineered bacteria carrying coned phenol catabolic genes. *J Ferment Bioeng, 79*, 100–106.

Harada, R.M., Yoza, B.A., Masutani, S.M., Li, Q.X., 2013. Diversity of archaea communities within contaminated sand samples from Johnston atoll. Bioremed. J. 17 (3), 182–189.

Hesham, A., Khan, S., Tao, Y., Li, D., & Zhang, Y. (2012). Biodegradation of high molecular weight PAHs using isolated yeast mixtures: application of metagenomic methods for community structure analyses. *Environ Sci Pollut Res Int, 19*, 3568–3578.

Hrywna, Y., Tsoi, T. V., Maltseva, O. V., Quensen, J. F. (III)., & Tiedje, J. M. (1999). Construction and characterization of two recombinant bacteria that grow on ortho-and para-substituted chlorobiphenyls. *Appl Environ Microbiol, 65*, 2163–2169.

Jasinska, A., Rozalska, S., Bernat, P., Paraszkiewicz, K., & Dlugonski J. (2012). Malachite green decolorization by non-basidiomycete filamentous fungi of P*enicillium pinophilum* and *Myrothecium roridum*. *Int Biodeterior Biodegrad, 73*, 33–40.

Karigar, C. S., & Rao, S. S. (2011). Role of microbial enzymes in the bioremediation of pollutants: A Review. doi:10.4061/2011/805187

Khemili-Talbi, S., Kebbouche-Gana, S., Akmoussi-Toumi, S., Angar, Y., & Gana, M.L. (2015). Isolation of an extremely halophilic *Arhaeon natrialba sp.* C21 able to degrade aromatic compounds and to produce stable biosurfactant at high salinity. *Extremophiles, 19*, 1109–1120. doi.org/10.1007/s00792-015-0783-9

Kobayashi, H., & Rittman B. E. (1982). Microbial removal of hazardous organic compounds. *Environ Sci Technol, 16,* 170A–183A.

Kundys, A., Bialecka-Florjanczyk, E., Fabiszewska, A., & Malajowicz J. (2017). *Candida antarctica* lipase B as catalyst for cyclic esters synthesis, their polymerization and degradation of aliphatic polyesters. *J Poly Environ, 26,* 396–407.

Lange, C. C., Wackett, L. P., Minton, K. W., & Daly, M. J. (1998). Engineering a recombinant *Deinococcus radiodurans* for organopollutant degradation in radioactive mixed waste environments. *Nat Biotechnol, 16,* 929–933.

Lee, Y. C., & Chang, S. P. (2011).The biosorption of heavy metals from aqueous solution by Spirogyra and Cladophora filamentous macroalgae. *Bioresource Technol, 102*(9), 5297–5304.

Leston, S., Nunes, M., Viegas, I., Ramos, F., & Pardal, M. A. (2013). The effects of chloramphenicol on Ulva lactuca. *Chemosphere, 91,* 552–557.

Li, M. H., Gao, X. Y., Li, C., Yang C. L., Liu J, Rui W., … Pang, X. (2020). Isolation and identification of chromium reducing *Bacillus cereus* species from Chromium-contaminated soil for the biological detoxification of chromium. Int J Environ Res Public Health. 2020 Mar 23;17(6):2118. doi: 10.3390/ijerph17062118.

Liu, L., Bilal, M., Duan, X., & Iqbal, H. M. (2019). Mitigation of environmental pollution by genetically engineered bacteria: Current challenges and future perspectives. *Sci Total Environ, 667,* 444–454.

Liu, S., Zhang, F., Chen, J., & Sun, G. (2011). Arsenic removal from contaminated soil via biovolatilization by genetically engineered bacteria under laboratory conditions. *J Environ Sci, 23,* 1544–1550.

Loukidou, M. X., Matis, K. A., & Zouboulis, A. I. (2003). Removal of As (V) from wastewaters by chemically modified fungal biomass. *Water Res, 37*(18), 4544–4552.

Margesin, R., Labbé, D., Schinner, F., Greer, C. W., & Whyte, L. G. (2003). Characterization of hydrocarbon-degrading microbial populations in contaminated and pristine Alpine soils. *Appl Environ Microbiol, 69*(6), 3085–92. doi: 10.1128/aem.69.6.3085-392.2003

Masaki, K., Kamini, N. R., Ikeda, H., & Iefuji, H. (2005). Cutinase-like enzyme from the yeast *Cryptococcus sp.* strain S-2 hydrolyzes polylactic acid and other biodegradable plastics. *Appl Environ Microb, 71*(11), 7548–7550.

Mohamed, A. T., El Hussein, A. A., El Siddig, M. A., & Osman, A. G. (2011). Degradation of oxyfluorfen herbicide by Soil microorganisms: Biodegradation of herbicides. *Biotechnol, 10,* 274–279.

Najera-Fernandez, C., Zafrilla, B., Jose Bonete, M., Maria Martinez-Espinosa, R., 2012. Role of the denitrifying haloarchaea in the treatment of nitrite-brines. Int. Microbiol. 15 (3), 111–119.

Ohtsubo, Y., Shimura, M., Delawary, M., Kimbara, K., Takagi, M., Kudo, T., … Nagata, Y. (2003). Novel approach to the improvement of biphenyl and polychlorinated biphenyl degradation activity: Promoter implantation by homologous recombination. *Appl Environ Microbiol, 69,* 146–153.

Osborn, A. M., Bruce, K. D., Strike, P., & Ritchie, D.A. (1997). Distribution, diversity and evolution of the bacterial mercury resistance (*mer*) operon. *FEMS Microbiol Rev, 19,* 239–262.

Oren, A. (2008). Microbial life at high salt concentrations: phylogenetic and metabolic diversity. *Saline systems, 4*(1), 1–13.

Paço, A., Duarte, K., da Costa, J. P., Santos, P. S., Pereira, R., Pereira, M. E. (2017). Biodegradation of polyethylene microplastics by the marine fungus *Zalerion maritimum. Sci Total Environ, 586,* 10–15.

Peeples, T.L. (2014). Bioremediation using Extremophiles. doi.org/10.1016/B978-0-12-800021-2.00010-8

Pereira, P., Enguita, F. J., Ferreira, J., & Leitao, A. L. (2014). DNA damaged induced by hydroquinone can be prevented by fungal detoxification. *Toxicol Rep, 1,* 1096–1105.

Rahman, S., Ki-Hyun, K., Saha, S.K., Swaraz, A. M., & Paul, D. K. (2014). Review of remediation techniques for arsenic (As) contamination: a novel approach utilizing bio-organisms. *J Environ Manag, 134,* 175–185.

Reineke, W. (1998). Development of hybrid strains for the mineralization of chloroaromatics by patchwork assembly. *Annu Rev Microbiol, 52,* 287–331.

Renninger, N., Knopp, R., Nitsche, H., Clark, D. S., & Keasling J. D. (2004). Uranyl precipitation by *Pseudomonas aeruginosa* via controlled polyphosphate metabolism. *Appl Environ Microbiol, 70,* 7404–7412.

Riya, P., & Jagatpati, T. (2012). Biodegradation and bioremediation of pesticides in soil: its objectives, classification of pesticides, factors and recent developments. *World J Sci Technol, 2,* 36–41.

Romera, E., González, F., Ballester, A., Blázques, M. I., & Munoz J. A. (2007). Comparative study of biosorption of heavy metals using different types of algae. *Bioresource Technol, 98*(17), 3344–3353.

Russell NJ (2008) Membrane components and cold sensing. In: Psychrophiles: from biodiversity to biotechnology. Springer, Berlin, pp 177–190.

Sang, B-I., Hori, K., Tanji, Y., & Unno, H. (2002). Fungal contribution to in situ biodegradation of poly(3-hydroxybutyrate-co-3-hydroxyvalerate) film in soil. *Appl Microbiol Biotechnol, 58,* 241–247.

Say, R., Yimaz N., & Denizli A. (2003).Removal of heavy metal ions using the fungus *Penicillium canescens*. *Adsorpt Sci Technol, 21*(7), 643–650.

Sayler, G. S., & Ripp, S. (2000). Field applications of genetically engineered microorganisms for bioremediation processes. Current opinion in biotechnology, 11(3), 286–289.

Sei, A., & Fathepure, B. Z. (2009). Biodegradation of BTEX at high salinity by an enrichment culture from hypersaline sediments of rozel point at great salt lake. *J Appl Microbiol, 107*(6), 2001–2008.

Semple, K. T., Cain, R. B., & Schmidt, S. (1999). Biodegradation of aromatic compounds by micro-algae. *FEMS Microbial Lett, 170,* 291–300.

Simarro, R., Gonzalez, N., Bautista, L. F., & Molina, M. C. (2013). Assessment of the efficiency of in situ bioremediation techniques in a creosote polluted soil: change in bacterial community. *J Hazard Mater,* 262: 158–167.

Singh, A. L., & Sarma, P. N. (2010). Removal of arsenic(III) from waste water using *Lactobacillus acidophilus*. *Biorem J, 14,* 92–97.

Singh, R., Dong, H., Liu, D., Zhao, L., Marts, A. R., Farquhar, E., Tierney, D. L., … Briggs, B.R. (2015). Reduction of hexavalent chromium by the thermophilic methanogen *Methanothermobacter thermautotrophicus. Geochim Cosmochim Acta, 148,* 442–456. doi.org/10.1016/j.gca.2014.10.012

Stallwood, B., Shears, J., Williams, P.A., & Hughes, K.A. (2005). Low temperature bioremediation of oil-contaminated soil using biostimulation and bioaugmentation with a Pseudomonas sp. from maritime Antarctica. *J Appl Microbiol, 99*(4), 794–802. doi.org/10.1111/j.1365-2672.2005.02678.x

Stevenson, J. A., Bearpark, J. K., & Wong, L. L. (1998). Engineering molecular recognition in alkane oxidation catalyzed by cytochrome P450cam. *New J Chem, 22,* 551–552.

Taha, M., Adetutu, E. M., Shahsavari, E., Smith, A. T., & Ball, A. S. (2014). Azo and anthraquinone dye mixture decolourization at elevated temperature and concentration by a newly isolated thermophilic fungus, *Thermomucorindicae seudaticae. J Environ Chem Eng, 2,* 415–423. doi.org/10.1016/j.jece.2014.01.015

Takeuchi, M., Kawahata, H., Gupta, L.P., Kita, N., Morishita, Y., Yoshiro, O., & Komai, T. (2007) Arsenic resistance and removal by marine and non-marine bacteria. *J Biotechnol, 127,* 434–442.

Tomova, I., Stoilova-Disheva, M., & Vasileva-Tonkova, E. (2014). Characterization of heavy metals resistant heterotrophic bacteria from soils in the Windmill Islands region, Wilkes Land, East Antarctica. *Polish Polar Res, 35*(4), 593–607. doi.org/10. 2478/popore-2014-0028

Valls, M., Atrian, S., de Lorenzo, V. (2000). Engineering a mouse metallothionein on the cell surface of *Ralstonia eutropha* CH34 for immobilization of heavy metals in soil. *Nat Biotechnol, 18,* 661–665.

Wu, J-F., Jiang, C., Wang, B., Liu, Z., & Liu, S. (2006). Novel partial reductive pathway for 4-chloronitrobenzene and nitrobenzene degradation in *Comamonas sp.* strain CNB-1. *Appl Environ Microbiol, 72,* 1759–1765.

Yildirim, S., Franko, T. T., Wohlgemuth, R., Kohler, H. P. E., Witholt, B., & Schmid, A. (2005). Recombinant chlorobenzene dioxygenase from Pseudomonas sp. P51: a biocatalyst for regioselective oxidation of aromatic nitriles. *Advanced Synthesis & Catalysis, 347*(7–8), 1060–1072.

Yoshida, S., Hiraga, K., Takehana, T., Taniguchi, I., Yamaji, H., Maeda, Y., Toyohara, K., … Oda, K. (2016). A bacterium that degrades and assimilates poly(ethylene terephthalate). *Science, 351*(6278), 1196–1199.

6 Microbial Communities for the Removal of Ammonium from Wastewater in an Activated Sludge System Combined with Low-Cost Biochar
A Review

Ngoc-Thuy Vu[1] and Khac-Uan Do[1]*

[1]School of Environmental Science and Technology, Hanoi University of Science and Technology, Hanoi, Vietnam

*Corresponding author: uan.dokhac@hust.edu.vn

CONTENTS

6.1 INTRODUCTION

Removing ammonium (NH_4) from wastewater is required to prevent eutrophication in lakes and rivers. As shown in Figure 6.1, NH_4 could be removed in several ways, such as adsorption and biological methods. Among them, the NH_4 in most wastewaters could be removed effectively by biological technology, which is known as biological nutrient removal.

Organic compounds, nitrogen (N), and phosphate, could be removed simultaneously and effectively in anaerobic, anoxic, and oxic systems (Jena et al., 2020). During operation, several working conditions must be optimized to achieve high performance. In addition, sequential batch reactors (SBRs) could be used to remove nutrients by adjusting the operational conditions for each phase.

In this biological process, NH_4 was removed using two steps, including nitrification and denitrification. In the nitrification step, NH_4 is oxidized to NO_2 [by ammonia oxidizing bacteria (AOBs)] and NO_3 [by the nitrite oxidizing bacteria NOBs)]. In the subsequent denitrification step, NO_3 is

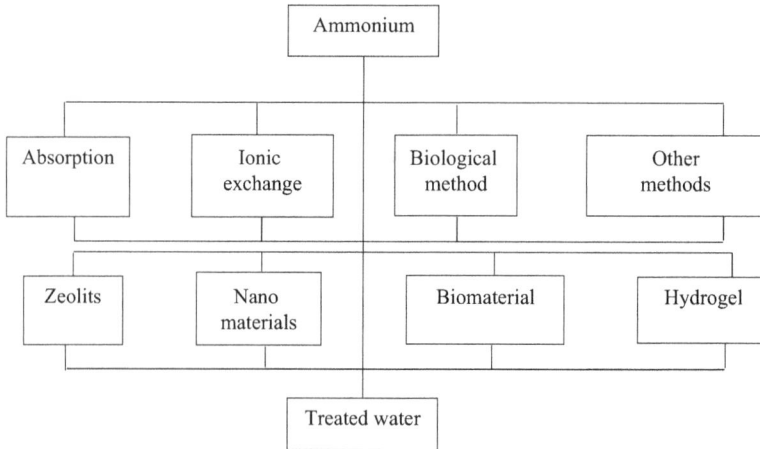

FIGURE 6.1 NH_4 removal by different methods.

reduced to nitrogen gas (N_2) by several reactions that are based on denitrifying microorganisms (de Sousa Rollemberg et al., 2018). During nitrification, the NH_4 oxidation rate was slower than the NO_2 oxidation rate. Therefore, it normally existed at low concentrations in the system.

Of note, if NO_2 oxidation is well controlled, N removal can be conducted by partial nitrification. If a system contains a large number of bacteria treating nutrient and organic (e.g., *Betaproteobacteria, Deltaproteobacteria, Flavobacteria,* and *Cytophagia*), the sludge concentration in the system would increase. Therefore, the system performance would be enhanced as well (Świątczak & Cydzik-Kwiatkowska, 2018a). Several N removing bacteria (e.g., *Nitrosomonas, Nitrospira,* and *Nitrobacter)* have been affected strongly by the chemical oxygen demand (COD) and TN ratio. If the COD/TN ratio was low, the number of *Nitrospira* and *Nitrobacter* would decrease, and *Nitrosomonas* would increase and be dominant in the system (de Sousa Rollemberg et al., 2018).

Granular sludge was developed >20 years ago (Franca et al., 2018). It was a satisfactory solution to upgrade the conventional aerobic systems when treating different wastewater types. Bacteria that can oxidize NH_4 were very low in the granular sludge flocs. However, the bacteria that could oxidize NO_2, such as *Nitrospira sp.* (which included the main genera *Tetrasphaera, Sphingopyxis, Dechloromonas, Flavobacterium,* and *Ohtaekwangia*) grew well and were prevalent (Świątczak & Cydzik-Kwiatkowska, 2018a).

Therefore, understanding the granular sludge structures could give a good basis to design and operate wastewater treatment plants where the activated sludges were upgraded by granular sludges.

6.2 CULTIVATION AND FORMATION OF GRANULAR SLUDGE

Modifications should be made when designing a wastewater treatment plant when conventional activated sludge is changed to granular sludge. Creating granularity is not easy. However, when it has formed, it could play an important role in treating nutrients. Of note, the cultivation time and operational conditions strongly affect the granule formation and composition. In some cases, the filamentous bacteria could be the backbone that helps to create the aerobic granular sludges, such as *Sphingomonadales* and *Xanthomonadales* (Świątczak & Cydzik-Kwiatkowska, 2018a). These bacteria could generate extracellular polymeric substances (EPS). EPS could play a good function to link and bind the granules tightly. In addition, the EPS could help to enhance the phosphorus (P) removal because P could accumulate in the granules.

Granular sludge is hard to form in a system, because it depended on many factors. During operation, some bacteria, for example, *Meganema sp.* and *Zooglea sp.*, from granular sludge could be washed out (Szabó et al., 2017a). They could be detected in the effluent from the treatment system. This could be because these bacteria mainly lived on the granular surface. Therefore, they were affected by the washing flow. However, other bacteria, for example, *Flavobacterium sp.* and *Bdellovibrio sp.* could almost exist in the granular sludge. They were hard to wash out and they were not detected in the effluent from the system. This could be because they mainly lived inside the granules. Therefore, they were protected from the washing flow (Szabó et al. 2017a).

Activated sludge can create granular sludge under suitable operational conditions. For example, continuous or intermittent feeding, high influent flowrate, low mixing, or low sedimentation time could be the correct conditions for sludge to aggregate (Wilén et al., 2018). Granular granules could exist in a stable structure. Their functionality could be achieved using a wide range of different seeding wastewaters. It could be formed by different sludge sources and under different operational conditions (Wilén et al., 2018). Therefore, the bacterial communities inside the granular sludge could develop differently.

Granular sludge is formed by self-granulated flocs that are based on a gel-like consistency. Wash out conditions in the SBR could cause the fast granulation of flocs. Bacteria in the granular sludge could be enriched by *Accumulibacter* and *Competibacter* (Weissbrodt et al., 2013). Granular sludge could be created within 2 weeks if the system was supplemented with *Zoogloea* colonies around the flocs. Zoogloeal biofilms formed slowly (Weissbrodt et al., 2013). In the first month of operation, when *Zoogloea* reached 37%–79%, they could reduce nutrient removal. Then, the nutrient removal could be improved, because the bacteria developed stably. Nitrification could be improved ≤80%–100%). In addition, N removal could be enhanced by 43%–83%. More important, dephosphatation improved but was unstable at approximately 75% (Weissbrodt et al., 2013). Nitrifiers, for example, *Accumulibacter* and *Competibacter* grew and concentrated in the core of the granular sludge. They created dense clusters of heterogeneous bioaggregates. In the biofilm, *Zoogloea* was present at approximately 5% (Weissbrodt et al., 2013). Different extracellular glycoconjugates have been found in granular sludges. They could play a role as a gel substance to create stable granules.

Granular sludge developed using an SBR system could contain high *Accumulibacter* communities (>56%) (Weissbrodt et al., 2013). They could create heterogeneous granules. Therefore, they could promote the anaerobic uptake of volatile fatty acids. Self-granulated flocs could be formed in an SBR under over-aerated conditions (Weissbrodt et al., 2013). In some systems, the granular sludge could be formed after a very long operational time, even after 400 days when the system treated real wastewater (Lin et al., 2010). Bacterial communities in granular sludge could be different from those in a conventional activated sludge. *Proteobacteria, Sphingobacteria,* and *Flavobacteria* have been found in granular sludge in a system that treated real wastewater (Zhou et al., 2014a). Figure 6.2 shows an example of granular sludge in a laboratory-scale SBR with the addition of biochar.

The seed sludge that is added to the reactor contains several bacteria, for example, *Arcobacter, Aeromonas, Flavobacterium,* and *Acinetobacter* (Ye et al., 2011). They grow slowly and become dominant in the granular sludge. During operation, some of the NH$_4$ oxidizing archaea could be washed out, and the NOB developed and were retained inside the granules.

6.3 CHARACTERISTICS OF GRANULAR SLUDGE

A granular sludge contains many different bacteria. The bacteria could depend on different mixing conditions and the hydraulic flow in the system. The hydraulic retention and sludge retention times could strongly affect the granular sludge formation and the bacterial communities. This could be determined by experiments with granular and suspended sludges. The experiments should be carried out long enough for granular sludge to form (Szabó et al., 2017a).

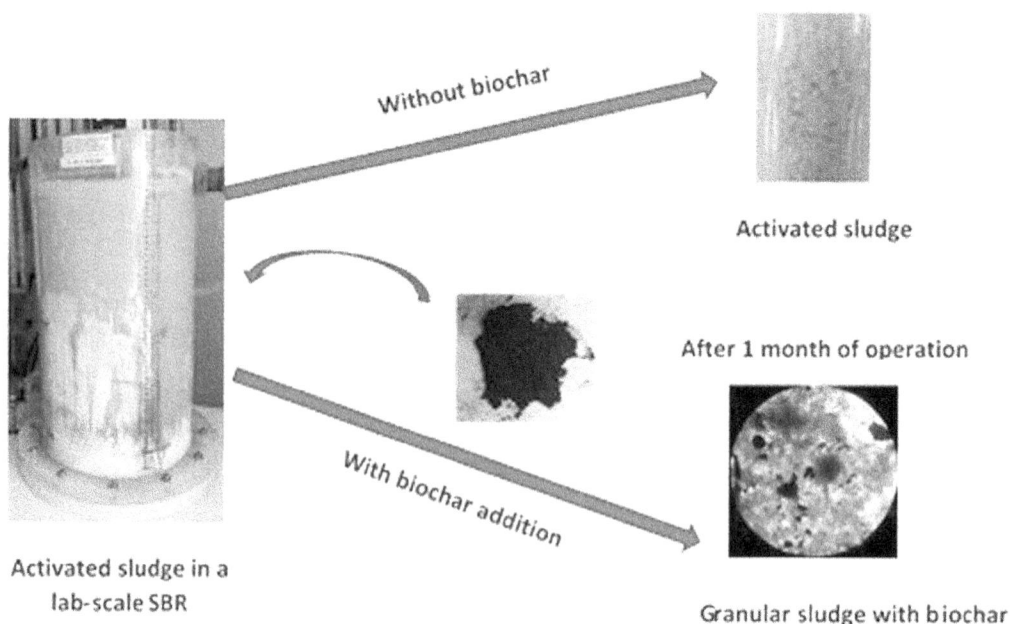

FIGURE 6.2   Granular sludge formation in laboratory-scale SBR with biochar.

Of note, after washing out of the granular sludge, *Meganema sp.* and *Zooglea sp.* grew rapidly in the system (Szabó et al., 2017a). However, it is not easy to identify the ratio of the bacteria that exist inside granular sludge and the number of bacteria that are washed out. Under these conditions, *Zoogloea, Accumulibacter,* and *Competibacter* were dominant in the granular sludge. Among them, *Competibacter* accounted for >37%. *Zoogloea* and *Accumulibacter* contributed to the formation of the granules, but their portion was lower than *Competibacter* in the granular sludge (Weissbrodt et al., 2013).

Granular sludge contains a bacterial density that is higher than in a normal activated sludge. This is because the granular sludge could have more ecological niches, which could provide more substrates for bacterial growth (Weissbrodt et al., 2013). Bacteria in granular and activated sludge could carry out the same function in removing nutrients (N and P) and organic compounds [(e.g., COD and biological oxygen demand (BOD)]; however, this depends on the configuration of the treatment systems. In addition, *Rhodanobacter* is part of the microbial diversity in granular sludge (Xu et al., 2017). It could contribute by increasing the protein portion of the EPS. The bacterial community decreased during the formation of granular sludge and they become stable in the formed granules.

In addition, granular sludge could contain a high number of bacterial communities, such as *Planctomycetes, Proteobacteria, Bacterioidetes,* and the archaea *Euryarchaeota* (Świątczak & Cydzik-Kwiatkowska, 2018b). Granular sludge tends to increase in size and its structure could change due to various bacteria that grow inside the granules. In addition, they could develop on the surface of granules. It appeared that bacteria were more abundant within the granules than at the surface of them (de Kreuk et al., 2005). In addition, bacterial communities that lived on the granular surface were easily washed out and moved into the suspended phase in the reactor, because of the shearing force from mixing (Gonzalez-Gil & Holliger, 2011). That is why most nitrifiers are attached to the granular surface. They could be retained in that area for >10 days. However, the denitrifiers and P-accumulating organisms mainly grew on the inner granules with a longer retention time (>15 days) (Gonzalez-Gil & Holliger, 2014).

TABLE 6.1
Granular Sludge that Contains Different Bacterial Species

Number	Group of Bacteria	Prevalent Bacterial Species	References
1	AOB	*Nitrosomonas europaea, Paracoccus aminovorans*	Coats et al. (2017)
2	NOB	*Nitrospira defluvii, Nitrobacter, Nitrospiraceae, Bradyrhizobiaceae,*	Lv et al. (2014)
3	Denitrifying glycogen-accumulating organisms	*Aquincola tertiaricarbonis*	Layer et al. (2019)
4	Denitrifying ordinary heterotrophic organisms	*Pseudomonas, Acidorovax, Rhizobiales, Zoogloea, Acinetobacter, Methylobacteriaceae*	Szabó et al. (2017b)
5	DPAO	*Competibacter, Tetrasphaera, Rhodoplanes elegans, Thauera, Propionivibrio aalborgensis*	de Sousa Rollemberg et al. (2018)

Bacterial communities, such as *Flavobacterium spp.* and *Bdellovibrio spp.* were hard to wash out from the granules. However, other bacteria, for example, *Meganema sp. Zoogloea sp.* were abundant in the suspended phase (Layer et al., 2019). *Accumulibacter* was abundant in a 200 μm outer layer of granules. *Competibacter* was more dominant in the granular core (He et al., 2016; Lemaire et al., 2008). Table 6.1 lists different bacteria in granular sludge.

The determination of the ammonia oxidizing bacteria (AOB) and NOB kinetics is required to determine the mechanism for N removal via biological processes. An important parameter is the maximum specific growth rate of the AOB. This factor depends on several operational conditions. It is strongly affected by sludge retention time. In a conventional activated sludge process (ASP), when the sludge retention time was from 8 to 20 d, the specific growth rates of the AOB were from 0.45/d and 0.72/d. Of note, at the same sludge retention time of 20 d, the specific growth rate of the AOB in the membrane bioreactor was approximately 0.49/d, which was slightly higher than in a conventional activated sludge (de Sousa Rollemberg et al., 2018). In addition, the endogenous decay coefficients of the NOB and AOBs should be determined to determine the number of dead cells. In the nitrification kinetic experiments, inhibition tests are required with different concentrations of allylthiourea (ATU). This test could help to determine the sensitivity of the NOB and AOBs under different conditions, which could be used to determine the maximum nitrification activity of the bacteria (de Sousa Rollemberg et al., 2018). In this test, granular sludge that contained NOB was cultivated for approximately 20 days. Therefore, the granular sludge diameter increased to approximately 0.8 mm. The granular sludge could contain approximately 60% *Nitrosomonas* and approximately 30% of *Nitrospira*. However, after converting the NH_4 to NO_2, the percent of *Nitrosomonas* reduced to 45% and *Nitrospira* increased to 40% (Vázquez-Padín et al., 2009).

6.4 APPLICATION OF GRANULAR SLUDGE IN NUTRIENT TREATMENT SYSTEMS

Wastewater treatment with granular sludge could give a high removal efficiency. Table 6.2 lists systems that use granular sludge for nutrient removal. Granular sludge could help to reduce the working volume of the treatment system due to the high sludge concentration. Granular sludge has been widely applied in aerobic systems. Of interest, granular sludge could be partially formed in an SBR. The bacteria in the granular sludge could enhance P and N uptake at low concentrations and under different conditions. In particular, several bacteria, such as *Betaproteobacteria,*

TABLE 6.2
Nutrient Removal in Typical Wastewater Systems

Number	Type of Reactor	Main Function	Sludge Type/Form	References
1	ASP	Organic removal, nitrification	Suspended, granular	Huang et al. (2014)
2	Anoxic oxic process	Organic removal, nitrification, denitrification	Suspended, granular	Franca et al. (2018)
3	Anaerobic anoxic oxic process	Organic removal, nitrification, denitrification, phosphorus removal	Suspended, granular	Świątczak and Cydzik-Kwiatkowska (2018c)
4	SBR	Organic removal; nutrient removal	Suspended, granular	Huang et al. (2014)
5	Membrane bioreactor	Organic removal, nitrification, completed solids separation	Suspended	Downing and Nerenberg (2008)
6	Trickling filter process	Organic removal, nutrient removal	Attached	Świątczak et al. (2019)

Alphaproteobacteria, Gammaproteobacteria, and *Bacteroidia* were detected in the SBRs (Jena et al., 2020). *Lactobacteriales* and *Enterobacteriales* have been found in the granular sludge in an SBR system (Jena et al., 2020). These bacteria could contribute to the removal of N and P via denitrification and dephosphatation. They could join in the tricarboxylic acid (TCA) cycle, glycolysis process, production of EPS, and storage of polyhydroxyalkanoate (PHA) (Jena et al., 2020). In these processes, the carbon (C) source could be a key factor that affects the TCA cycle and glycolysis process in SBRs.

SBRs have been used as a good solution to treat domestic and industrial wastewaters. To develop a granular sludge, an SBR could remove organic pollutants and nutrients effectively. Of interest, a study that used an SBR system was operated with a high sludge retention time and could remove NH_4, P, and COD consistently over 263 d (Huang et al., 2014). An SBR system was operated in several phases. The anoxic phase was only controlled by mixing. In this phase, denitrification occurred. However, in this phase, the organic C source could be reduced. Therefore, P removal was maintained at a high efficiency, which demonstrated the appearance of denitrifying polyphosphate accumulating organisms (DPAOs).

In addition, under these operational conditions, the AOBs developed. In particular, in addition to DPAOs, *Candidatus Accumulibacter phosphatis* was detected in the system (Huang et al., 2014). Nitrogen removal could be removed by a shortcut way. By reducing NO_2 but not by NO_3, N could be removed. This could be conducted in a membrane-aerated biofilm (MAB) (Downing & Nerenberg, 2008). The kinetics of this process should be studied in detail to control the factors that could affect NO_2 formation in MABs. The dissolved O_2 (DO) could be an important factor that impacts bacterial communities and the nitrification rate in the biofilm on the membrane surface (Downing & Nerenberg, 2008).

In addition, the sludge concentration suspended in the anoxic zone in the MAB reactor could conduct denitrification. Finally, total N removal was enhanced in the MAB (Downing & Nerenberg, 2008). Nitrifying bacteria in a MAB that used a hollow-fiber type membrane could develop and carry out nitrification well at NH_4 of 3 mg/L (Downing & Nerenberg, 2008). By using MABs, the

DO at the membrane surface could easily be controlled at 2 mg/L, which is suitable for nitrification (Downing & Nerenberg, 2008). To maintain the DO between 2.2 and 3.5 mg/L in the membrane biofilm, the NH$_4$ fluxes were controlled between 0.75 and 1.0 g/m^2/d (Downing & Nerenberg, 2008). Under these conditions, most of the NH$_4$ was oxidized into NO$_2$. If the membrane DO in the membrane biofilm increased to 5.5 mg/L, and the NH$_4$ flux increased to approximately 1.3 g/m^2/d, NO$_3$ increased and became the main product of nitrification (Downing & Nerenberg, 2008).

Under all operational conditions, the AOBs were relatively stable in the membrane biofilm. However, the NOBs reduced, which corresponded to the reduction in DO in the membrane biofilm. N removal in a reactor that works with granular sludge was reduced in the start-up period. This could be because the bacteria washed out of the granules. This would reduce the sludge retention time in the reactor (Edwards et al., 2013). Therefore, in these systems, the nutrient removal could become stable and efficient after 2–4 months of operation (Weissbrodt et al., 2014). The AOB and NOBs grew slowly. They were sensitive to the variations in sludge retention time (Gruber-Dorninger et al., 2015). The formation of granules could be faster at lower settling times. Various bacterial communities, which contain *Nitrosomonas* and *Nitrosospira* in the granular sludge, enhanced the start-up of the nitrification period. Nitrification could be stable after 10 d operation (Szabó et al., 2017a).

6.5 FACTORS THAT AFFECT GRANULAR SLUDGE FORMATION

Operational factors could have a significant impact on the bacterial communities in the granular sludge. A system that was operated at a low organic loading rate could treat COD, N, and P simultaneously and effectively (Kanimozhi & Vasudevan, 2014).

Bacteria in the granular sludge were affected by the organic loading rate (OLR) variation. Of interest, the bacteria in the granular sludge were much different at different OLRs. This could be because the substrate uptake rate of the bacteria for NO$_2$ oxidation was high. Therefore, the C source could be limited for NOB growth. The reduction in *Nitrosomonas* could be due to the limited substrate. Therefore, *Nitrobacter* could develop well under poor substrate conditions. The main adjustment should create suitable conditions for different bacterial communities to grow. Therefore, the dominant bacteria to remove the organic pollutants and nutrients could appear. This system could achieve >80% removal of organic compounds. Of interest, N and P could be removed ≤90% under stable and suitable operational conditions (Jena et al., 2020). For example, when the treatment system operated at different organic loading rates (i.e., from 0.9 to 3.7 kg/m^3/d), and the NH$_4$ loading rate was controlled at 0.2 kg NH$_4$-N/m/^{3}/d, the organic pollutant was removed well for both organic loading rates. However, N was only removed well at the high organic loading rate (Szabó et al., 2017b). Although the percent of bacteria changed, the size and structure of the granular sludge were approximately the same when NH$_4$ and NO$_2$ were added to the system.

With a limited C source, the wasted sludge would reduce significantly. In this situation, the removal of N and P would be affected, because the bacterial community reduced (Huang et al., 2014). The rate of organic loading, the ratio of food and microorganism, the ratio of COD/N, sludge retention, sedimentation, and hydraulic retention times were important operational conditions that affected the bacteria in the granular sludge. Granular sludge could become smaller and more compact at low COD/N values from 2 to 5. Under this condition, the granules contained slow-growing heterotrophs, for instance, nitrifiers. At higher COD/N ratios (>7.5), the granular sludge tended to be larger due to the growth of more heterotrophs (Kocaturk & Erguder, 2016; Wei et al., 2014). During operation, the recalcitrant organic compounds in the wastewater could have a strong effect on the development of the bacterial communities in the granular sludge (Świątczak & Cydzik-Kwiatkowska, 2018c).

In addition, denitrifiers prefer to use a suitable C source, for instance, poly-β-hydroxybutyrate, for efficient denitrification (Mosquera-Corral et al., 2005; Vázquez-Padín et al., 2010). In particular, *Zoogloea, Acidovorax,* and *Thiothrix* were abundant with propionate feeding. However, *Accumulibacter, Competibacter, Chloroflexi, Acidivorax, Thiothrix* were dominant with acetate feeding (Gonzalez-Gil & Holliger, 2011).

During the operation of an aerobic system, pH and temperature are factors that have a significant effect on nitrification and denitrification. At high temperatures, the conversion of ammonia into NO_3 could be achieved with a high reaction rate (Hellinga et al., 1999). The DO concentration is an important factor to simultaneously control nitrification and denitrification. DO should be controlled at a high enough level to provide good conditions for nitrification, which occur at the surface of the granules. In addition, the DO must be controlled low enough to maintain the anoxic conditions inside the granules for denitrification. More important, the granular sludge could become more stable at a low DO value (de Kreuk et al., 2005).

Nitrifiers can grow on the surface and in the core of granules (Coats et al., 2017). This meant that the O_2 and ammonia could be transported inside the granules to help the growth of nitrifiers. Granular sludge mainly developed due to the growth of bacteria inside the granules. In addition, the aggregation of small bacterial flocs could contribute to the formation of granules (Szabó et al., 2017a). If the granules could be a spherical structure, the O_2 and substrate concentration gradients would reduce from the surface to the center of the granules. Therefore, nitrifiers were normally dominant on the surface due to the high O_2 concentration. Denitrifiers and P-accumulating organisms (e.g., *Rhodocyclaceae, Flavobacteriaceae, Xanthomonadaceae, Rhodobacteraceae,* and *Microbacteriaceaea*) grew and dominated the core of granules (Borowik et al., 2020; Lv et al., 2014; Pohlner et al., 2019). The bacterial communities grew slowly at the granular surface. A lower shear rate in the reactor could lead to a low DO, which could accelerate the formation of compact aggregates of sludge flocs (Mosquera-Corral et al., 2005). The structure and dimensions of the reactor, for instance, based on the heigh/diameter (H/D) ratio, could affect the formation of granular sludge.

For a low H/D ratio, the aeration should be increased to improve the shearing stress to help granular sludge formation. When aeration is maintained at a high rate, there could be a higher hydraulic shear force, which enhances granular formation. More important, high aeration rates will prevent the growth of filamentous bacteria. This could be a good solution to maintain a stable granular sludge (Adav et al., 2008; Paulo et al., 2021; Zhou et al., 2014b).

Sludge retention time could be an important factor for nitrification and denitrification (Hellinga et al., 1999). A treatment system that was operated with a high sludge retention time could reduce excess sludge and could enhance N removal. In particular, some systems could be operated with very high sludge retention times (infinite). This means that no excess sludge would be discharged during operations, which could be a solution for sludge minimization. However, this method could strongly influence efficient nutrient removal, because the bacterial communities could be changed based on the sludge retention time (Huang et al., 2014).

Granular sludge could contain some types of predatory bacteria (Feng et al., 2017). They could grow on the surface of granules (i.e., stalked ciliates) or inside the granules (i.e., *Vorticella*-like protist) (Bratanis et al., 2020). Predatory bacteria, such as *Micavibrio*-like bacteria, could be a factor that affects the bacterial communities in the granules. They could predate *Nitrospira sp.*, which directly influences nitrification (Dolinšek et al., 2016; Liébana et al., 2019).

6.6 NH$_4$ ADSORBED BY GRANULAR SLUDGE

NH_4 could be effectively adsorbed by aerobic granular sludge, activated sludge, and anammox granules (Xie et al., 2021). The NH_4 concentrations in the influent and effluent should be monitored to determine the adsorption of NH_4 by the granules (Bassin et al., 2011). Granular sludge has shown high potential adsorption capacity for NH_4. Activated sludge could adsorb approximately 0.2 mg

NH$_4$-N/g VSS (Bassin et al., 2011). However, granular sludge could enhance the adsorption capacity of NH$_4$ up to 0.9 mg NH$_4$-N/g VSS (Bassin et al., 2011).

Temperatures and salt concentration could affect the NH$_4$ adsorption by the granular sludge. The NH$_4$ adsorption was reduced at high salt concentrations (Bassin et al., 2011). In addition to adsorption, ion exchange of NH$_4$ in the granular sludge could occur, which could help to improve nitrification and denitrification (Bassin et al., 2011). When treating synthetic wastewater, different bacterial communities, for example, *Proteobacteria, Firmicutes, Bacteroidetes, Chloroflexi,* and *Actinobacteria* could appear in the granular sludge (Martin Vincent et al., 2018). In addition, other bacteria (e.g., *β-proteobacteria, δ-Proteobacteria, Flavobacteria,* and *Cytophagia*) in the activated sludge could bind to the granules and contribute to the removal of COD, N, and P in domestic and industrial wastewater (Ambuchi et al., 2016; Świątczak et al., 2019).

The concentration of NO$_2$ could depend on the reactor type. N is at a low concentration in continuous treatment systems and a high concentration in intermittent treatment systems. This could be due to the variations in the substrate in the influent flow. *Nitrobacter* and *Nitrospira* are present in continuous and intermittent treatment systems. Of note, in the continuous system, *Nitrospira* accounted for approximately 59% of the total bacteria, and *Nitrobacter* was approximately 5%. However, in an intermittent system, *Nitrobacter* was approximately 64% and *Nitrospira* was approximately 3% (Kim & Kim, 2006). Specific NO$_2$ oxidation by NOBs could be determined by kinetic studies. In the same treatment system, *Nitrobacter* could achieve a NO$_2$ oxidation rate of 93.8 mg/g/h, and *Nitrospira* was approximately 10.5 mg/g/h. This meant that *Nitrospira* activity was nine times lower than Nitrobacter's activity when converting NO$_2$ to NO$_3$ (Kim & Kim, 2006). Therefore, NO$_2$ concentration was a key factor that affected the levels of *Nitrobacter* and *Nitrospira* in the treatment system. More important, NO$_2$ concentration and organic load could significantly impact the NO$_2$ oxidation rate of NOBs in the treatment system. Varing NO$_2$ concentrations could be used to test the competition between *Nitrospira* and *Nitrobacter* (Vázquez-Padín et al., 2009).

6.7 FUTURE PROSPECTS FOR GRANULAR SLUDGE WHEN REMOVING NUTRIENTS

During nitrification, *Nitrospira* and *Nitrobacter* play important roles when converting NH$_4$ to NO$_2$ and NO$_3$. However, it appears that *Nitrospira* could be better in the oxidation step of NO$_2$ to NO$_3$ than *Nitrobacter* at low NO$_2$ concentrations. However, *Nitrospira* grew slower than *Nitrobacter* (Kim & Kim, 2006). This could be determined by conducting a test with different NO$_2$ concentrations with NOBs in two laboratory-scale wastewater treatment systems to monitor the growth of NOBs and the NO$_2$ oxidation rate.

In addition, granular sludge could contain several bacterial types, such as nitrifying and denitrifying bacteria, which could enhance N removal. In particular, the denitrifying bacteria, such as *Meganema, Thauera, Paracoccus,* and *Zoogloea* were available in granular sludge. These bacteria helped to improve N removal (Szabó et al., 2017b).

Nitrifying bacteria have been found in granular sludge, which could enhance nitrification. Therefore, NO$_3$ produced was transferred denitrification. Finally, N removal increased. An examination of the important operational conditions to help stabilize granular sludge would be interesting. This could provide a good strategy to adjust the working conditions for different wastewater treatment plants.

In addition, influent characteristics could affect the stability of the granular sludge. In addition, granular sludge stability could depend on the feeding strategy, variation of substrate, and sludge growth rate within the granules, and waste sludge discharge. Granular sludge could help to reduce the environmental footprint and improve system performance. Therefore, it has potential applications in wastewater treatment. However, under certain conditions, granular sludge can be unstable. The mechanisms related to this should be examined in detail. This could help to explain how the granular sludge could affect the effluent quality. In addition, it could provide a deeper understanding of the bacterial interactions inside the granular sludge (Wilén et al., 2018).

In addition, it could provide a number of suitable conditions to optimize the treatment process. More importantly, this knowledge could determine the role of bacterial communities that exist in granular sludge, for their metabolism and activities to stabilize the granular sludge and to improve functionality. Based on this, similar conditions could be applied to full-scale wastewater treatment plants. These conditions could be adjusted to enhance variations in the organic loading rates, flow rates, working temperature, influent quality, and feeding sludge sources (Wilén et al., 2018). The examination of the major bacterial communities in granular sludge is important to determine the variation in them during reactor operation. These factors could be used when modeling or designing small or full-scale wastewater treatment plants. In addition, they could be used to study the kinetics of the nutrient removal processes.

A mass balance of C and nutrient removal, and bacterial growth, for example, nitrifiers and PAOs, should be conducted to confirm the efficiency of the treatment (Huang et al., 2014). When DO was >2 mg/L, *Nitrobacter* was dominant. However, when DO was controlled to <2 mg/L, the dominant bacteria were *Nitrospira* (Downing & Nerenberg, 2008). Based on the experimental results, a model could be developed to provide the correct conditions to control the dominance of *Nitrobacter* and *Nitrospira* in the system (Downing & Nerenberg, 2008). Shortcut nitrification could be carried out based on controlling the DO concentration. NO_2 accumulation increased at lower DO concentrations. However, nitrification rates increased at high DO concentrations.

6.8 CONCLUSIONS

Granular sludge is viewed as an alternative solution to upgrade the conventional ASP. An activated sludge system carries out nitrification well. However, granular sludge could help the conventional activated sludge system to reduce the volume of the bioreactor tank. It could be reduced by approximately 30% and, more importantly, the granular sludge could help to enhance the effluent quality. As reported in many studies, granular sludge is larger than the size of the conventional activated sludge flocs. Therefore, it could settle faster, and when discharged into the sludge storage tank, it could be thickened faster. Therefore, it could help to reduce the volume of the sludge tank.

Granular sludge settles and separates better than conventional activated sludge. This is because granular sludge is compacted and is larger than the size of the conventional activated sludge flocs. This would generate a higher sludge concentration in the same volume treatment system. Therefore, it could help to reduce the land required to build wastewater treatment plants. However, it is not easy to maintain the stability of granular sludge in the system over long operating times. This could make the system performance unstable. In addition, the system would be difficult to control for N removal. A detailed study into the bacteria in the granular sludge could provide suitable operational conditions to control wastewater treatment and improve nutrient removal.

Modeling could be used to predict the NH_4 oxidation rate and NO_2 formation based on the DO concentrations on the membrane surface. The kinetic parameters for *Nitrospira* and *Nitrobacter* in the reactor could be predicted by modeling based on the relationship between different NO_2 and O_2 concentrations.

REFERENCES

Adav, S. S., Lee, D. J., Show, K. Y., & Tay, J. H. (2008). Aerobic granular sludge: recent advances. *Biotechnol Adv, 26*, 411–423. doi:10.1016/j.biotechadv.2008.05.002

Ambuchi, J. J., Liu, J., Wang, H., Sha, L., Zhou, X., Mohammed, M. O. A., & Feng, Y. (2016). Microbial community structural analysis of an expanded granular sludge bed (EGSB) reactor for beet sugar industrial wastewater (BSIW) treatment. *Appl Microbiol Biot, 100*, 4651–4661. doi:10.1007/s00253-015-7245-2

Bassin, J. P., Pronk, M., Kraan, R., Kleerebezem, R., & van Loosdrecht, M. C. M. (2011). Ammonium adsorption in aerobic granular sludge, activated sludge and anammox granules. *Water Res, 45,* 5257–5265. doi.org/10.1016/j.watres.2011.07.034

Borowik, A., Wyszkowska, J., & Kucharski, J. (2020.) Impact of various grass species on soil bacteriobiome. *Diversity, 12*(6), 2–20. https://doi.org/10.3390/d12060212

Bratanis, E., Andersson, T., Lood, R., & Bukowska-Faniband, E. (2020). Biotechnological potential of *B. dellovibrio* and like organisms and their secreted enzymes. *Front Microbiol, 11.* doi:10.3389/fmicb.2020.00662

Coats, E. R., Brinkman, C. K., & Lee, S. (2017). Characterizing and contrasting the microbial ecology of laboratory and full-scale EBPR systems cultured on synthetic and real wastewaters. *Water Res, 108,* 124–136. doi:10.1016/j.watres.2016.10.069

de Kreuk, M. K., Pronk, M., & van Loosdrecht, M. C. M. (2005). Formation of aerobic granules and conversion processes in an aerobic granular sludge reactor at moderate and low temperatures. *Water Res, 39,* 4476–4484 doi.org/10.1016/j.watres.2005.08.031

de Sousa Rollemberg, S. L., Mendes Barros, A. R., Milen Firmino, P. I., & Bezerra dos Santos, A. (2018). Aerobic granular sludge: Cultivation parameters and removal mechanisms. *Bioresource Technol, 270,* 678–688. doi.org/10.1016/j.biortech.2018.08.130

Dolinšek, J., Goldschmidt, F., & Johnson, D. R. (2016). Synthetic microbial ecology and the dynamic interplay between microbial genotypes. *FEMS Microbiol Rev, 40,* 961–979. doi:10.1093/femsre/fuw024

Downing, L. S., & Nerenberg, R. (2008). Effect of oxygen gradients on the activity and microbial community structure of a nitrifying, membrane-aerated biofilm. *Biotechnol Bioeng, 101,* 1193–1204. doi.org/10.1002/bit.22018

Edwards, T. A., Calica, N. A., Huang, D. A., Manoharan, N., Hou, W., Huang, L., Panosyan, H., Dong, H., & Hedlund., B. P. (2013). Cultivation and characterization of thermophilic Nitrospira species from geothermal springs in the US Great Basin, China, and Armenia, *FEMS Microbiol Ecol, 85,* 283–292. doi:10.1111/1574-6941.12117

Feng, S., Tan, C. H., Constancias, F., Kohli, G. S., Cohen, Y., & Rice, S. A. (2017). Predation by *Bdellovibrio bacteriovorus* significantly reduces viability and alters the microbial community composition of activated sludge flocs and granules. *FEMS Microbiol Ecol, 93.* doi:10.1093/femsec/fix020

Franca, R. D. G., Pinheiro, H. M., van Loosdrecht, M. C. M., & Lourenço, N. D. (2018). Stability of aerobic granules during long-term bioreactor operation. *Biotechnol Adv, 36,* 228–246. doi.org/10.1016/j.biotechadv.2017.11.005

Gonzalez-Gil, G., & Holliger, C. (2011). Dynamics of microbial community structure of and enhanced biological phosphorus removal by aerobic granules cultivated on propionate or acetate. *Appl Environ Microbiol, 77,* 8041–8051. doi:10.1128/aem.05738-11

Gonzalez-Gil, G., & Holliger, C. (2014). Aerobic granules: Microbial landscape and architecture, stages, and practical implications applied and environmental. *Microbiology+, 80,* 3433–3441. doi:10.1128/aem.00250-14

Gruber-Dorninger, C., Pester, M., Kitzinger, K., Savio, D. F., Loy, A., Rattei, T., Wagner, M., & Daims, H., (2015). Functionally relevant diversity of closely related *Nitrospira* in activated sludge. *ISME J, 9,* 643–655. doi:10.1038/ismej.2014.156

He, Q., Zhou, J., Wang, H., Zhang, J., & Wei, L. (2016). Microbial population dynamics during sludge granulation in an A/O/A sequencing batch reactor. *Bioresource Technol, 214,* 1–8. doi.org/10.1016/j.biortech.2016.04.088

Hellinga, C., van Loosdrecht, M. C. M., & Heijnen, J. J. (1999). Model based design of a novel process for nitrogen removal from concentrated flows. *Math Comp Model Dyn, 5,* 351–371. doi:10.1076/mcmd.5.4.351.3678

Huang, P., Li, L., Kotay, S. M., & Goel, R. (2014). Carbon mass balance and microbial ecology in a laboratory scale reactor achieving simultaneous sludge reduction and nutrient removal. *Water Res, 53,* 153–167. doi.org/10.1016/j.watres.2013.12.035

Jena, J., Narwade, N., Das, T., Dhotre, D., Sarkar, U., & Souche, Y. (2020). Treatment of industrial effluents and assessment of their impact on the structure and function of microbial diversity in a unique Anoxic-Aerobic sequential batch reactor (AnASBR). *J Environ Manage, 261,* 1–13, doi.org/10.1016/j.jenvman.2020.110241

Kanimozhi, R., & Vasudevan, N. (2014). Effect of organic loading rate on the performance of aerobic SBR treating anaerobically digested distillery wastewater Clean. *Clean Technol Environ Policy, 16*, 467–476. doi:10.1007/s10098-013-0639-x

Kim, D-J., & Kim, S-H. (2006). Effect of nitrite concentration on the distribution and competition of nitrite-oxidizing bacteria in nitration reactor systems and their kinetic characteristics. *Water Res, 40*, 887–894. doi.org/10.1016/j.watres.2005.12.023

Kocaturk, I., & Erguder, T. H. (2016). Influent COD/TAN ratio affects the carbon and nitrogen removal efficiency and stability of aerobic granules. *Ecol Eng, 90*, 12–24. doi.org/10.1016/j.ecoleng.2016.01.077

Layer, M. Adler, A., Reynaert, E., Hernandez, A., Pagni, M., Morgenroth, E., Holliger, C., & Derlon, N. (2019). Organic substrate diffusibility governs microbial community composition, nutrient removal performance and kinetics of granulation of aerobic granular sludge. *Water Res, 4*, 1–16. doi.org/10.1016/j.wroa.2019.100033

Lemaire, R., Webb, R. I., & Yuan, Z. (2008). Micro-scale observations of the structure of aerobic microbial granules used for the treatment of nutrient-rich industrial wastewater. *ISME J, 2*, 528–541. doi:10.1038/ismej.2008.12

Liébana, R., Modin, O., Persson, F., Szabó, E., Hermansson, M., & Wilén, B-M. (2019) Combined deterministic and stochastic processes control microbial succession in replicate granular biofilm reactors. *Environ Sci Technol, 53*, 4912–4921. doi:10.1021/acs.est.8b06669

Lin, Y., de Kreuk, M., van Loosdrecht, M. C., & Adin, A. (2010). Characterization of alginate-like exopolysaccharides isolated from aerobic granular sludge in pilot-plant. *Water Res, 44*, 3355–3364. doi:10.1016/j.watres.2010.03.019

Lv, Y., Wan, C., Lee, D-J., Liu, X., Tay, J-H. (2014). Microbial communities of aerobic granules: Granulation mechanisms. *Bioresour Technol, 169*, 344–351. doi.org/10.1016/j.biortech.2014.07.005

Martin, V. N., Wei, Y., Zhang, J., Yu, D., & Tong, J. (2018). Characterization and dynamic shift of microbial communities during start-up, overloading and steady-state in an anaerobic membrane bioreactor. *Int J Environ Res Public Health, 15*, 1–20. doi:10.3390/ijerph15071399

Mosquera-Corral, A., de Kreuk, M. K., Heijnen, J. J, & van Loosdrecht, M. C. (2005). Effects of oxygen concentration on N-removal in an aerobic granular sludge reactor. *Water Res, 39*, 2676–2686. doi:10.1016/j.watres.2005.04.065

Paulo, A. M. S., Amorim, C. L., Costa, J., Mesquita, D. P., Ferreira, E. C., & Castro P. M. L. (2021). Long-term stability of a non-adapted aerobic granular sludge process treating fish canning wastewater associated to EPS producers in the core microbiome. *Sci Total Environ, 756*, 1–12. doi.org/10.1016/j.scitotenv.2020.144007

Pohlner, M., Dlugosch, L., Wemheuer, B., Mills, H., Engelen, B., & Reese, B. K. (2019). The majority of active Rhodobacteraceae in marine sediments belong to uncultured genera: A molecular approach to link their distribution to environmental conditions. *Front Microbiol, 10*. doi:10.3389/fmicb.2019.00659

Świątczak, P., & Cydzik-Kwiatkowska, A. (2018a). Performance and microbial characteristics of biomass in a full-scale aerobic granular sludge wastewater treatment plant. *Environ Sci and Pollut Res, 25*, 1655–1669. doi:10.1007/s11356-017-0615-9

Świątczak, P., & Cydzik-Kwiatkowska, A. (2018b). Performance and microbial characteristics of biomass in a full-scale aerobic granular sludge wastewater treatment plant, *Environ Sci Pollut Res, 25*, 1655–1669. doi:10.1007/s11356-017-0615-9

Świątczak, P., & Cydzik-Kwiatkowska, A. (2018c). Treatment of ammonium-rich digestate from methane fermentation using aerobic granular sludge. *Water Air Soil Pollut, 229*, 247. doi:10.1007/s11270-018-3887-x

Świątczak, P., Cydzik-Kwiatkowska, A., & Zielińska, M. (2019). Treatment of the liquid phase of digestate from a biogas plant for water reuse. *Bioresour Technol, 276*, 226–235. doi:10.1016/j.biortech.2018.12.077

Szabó, E., Liébana, R., Hermansson, M., Modin, O., Persson, F., & Wilén, B-M. (2017a). Comparison of the bacterial community composition in the granular and the suspended phase of sequencing batch reactors. *AMB Express, 7*, 168–168. doi:10.1186/s13568-017-0471-5

Szabó, E., Liébana, R., Hermansson, M., Modin, O., Persson, F., & Wilén, B-M. (2017b) Microbial population dynamics and ecosystem functions of anoxic/aerobic granular sludge in sequencing batch reactors operated at different organic loading rates. *Front Microbiol, 8*, 1–14. doi:10.3389/fmicb.2017.00770

Vázquez-Padín, J., Mosquera-Corral, A., Campos, J. L., Méndez, R., & Revsbech N. P. (2010). Microbial community distribution and activity dynamics of granular biomass in a CANON reactor. *Water Res, 44*, 4359–4370. doi.org/10.1016/j.watres.2010.05.041

Vázquez-Padín, J. R., Figueroa, M., Mosquera-Corral, A., Campos, J. L., & Méndez, R. (2009). Population dynamics of nitrite oxidizers in nitrifying granules. *Water Sci Technol, 60,* 2529–2536. doi:10.2166/wst.2009.602

Wei, D., Shi, L., Yan, T., Zhang, G., Wang, Y., & Du, B. (2014). Aerobic granules formation and simultaneous nitrogen and phosphorus removal treating high strength ammonia wastewater in sequencing batch reactor. *Bioresour Technol, 171,* 211–216. doi.org/10.1016/j.biortech.2014.08.001

Weissbrodt, D. G., Neu, T. R., Kuhlicke, U., Rappaz, Y., & Holliger, C. (2013). Assessment of bacterial and structural dynamics in aerobic granular biofilms. *Front Microbiol, 4,* 175–175. doi:10.3389/fmicb.2013.00175

Weissbrodt, D. G., Shani, N., & Holliger, C. (2014). Linking bacterial population dynamics and nutrient removal in the granular sludge biofilm ecosystem engineered for wastewater treatment. *FEMS Microbiol Ecol, 88,* 579–595. doi.org/10.1111/1574-6941.12326

Wilén, B-M., Liébana, R., Persson, F., Modin, O., & Hermansson, M. (2018). The mechanisms of granulation of activated sludge in wastewater treatment, its optimization, and impact on effluent quality. *Appl Microbiol Biotechnol, 102,* 5005–5020. doi:10.1007/s00253-018-8990-9

Xie, F., Zhao, B., Cui, Y., Ma, X., Li, D., & Yue, X. (2021). Enhancing nitrogen removal performance of anammox process after short-term pH and temperature shocks by coupling with iron-carbon micro-electrolysis. *J Clean Prod, 289,* 1–13. doi.org/10.1016/j.jclepro.2020.125753

Xu, M., Chen, X., Dai, R., Xinyi Xiang, X., Li, G., Cao, J., Xue, Z., & Shang, K., (2017). Performance and bacterial community change during the start-up period of a novel anaerobic bioreactor inoculated with long-time storage anaerobic granular sludge. *J Environ Eng, 143,* 413–423. doi:10.1061/(ASCE)EE.1943-7870.0001222

Ye, L., Shao, M-F., Zhang, T., Tong, A. H, Y., & Lok, S. (2011). Analysis of the bacterial community in a laboratory-scale nitrification reactor and a wastewater treatment plant by 454-pyrosequencing. *Water Res, 45,* 4390–4398. doi.org/10.1016/j.watres.2011.05.028

Zhou, D., Niu, S., Xiong, Y., Yang, Y., & Dong, S. (2014a). Microbial selection pressure is not a prerequisite for granulation: Dynamic granulation and microbial community study in a complete mixing bioreactor. *Bioresour Technol, 161,* 102–108. doi.org/10.1016/j.biortech.2014.03.001

Zhou, J., Wang, H., Yang, K., Ma, F., & Lv, B. (2014b). Optimization of operation conditions for preventing sludge bulking and enhancing the stability of aerobic granular sludge in sequencing batch reactors. *Water Sci Technol, 70,* 1519–1525. doi:10.2166/wst.2014.406

7 Microbial Degradation of Azo Dyes Present in Textile Industry Wastewater

Pushpa C. Tomar[1] and Praveen Dahiya[2]*

[1]Department of Biotechnology, Faculty of Engineering and Technology, Manav Rachna International Institute of Research and Studies, Faridabad, Haryana, India

[2]Amity Institute of Biotechnology, Amity University Uttar Pradesh, Gautam Buddha Nagar, Sector-125, Noida, Uttar Pradesh, India

*Corresponding author: pdahiya@amity.edu, praveen_sang@yahoo.com

CONTENTS

7.1 INTRODUCTION

Environmental contamination is one of the vital and greatest concerns of the modern world. The emission of effluents from different industries, for example, distilleries, textile, dyestuff, tanneries, and paper and pulp industries contribute an enormous amount of pollution. Among these industries, the textile industry contributes a large amount of contamination, which includes a wide range of dyes and chemicals, to the water dyes are used in the manufacturing of products in different industries, such as textile, paints, printing, cosmetics, food, plastics, photography, and pharmaceutical (Kammradt, 2004; Oliveira, 2005). Dyes are categorized into natural, synthetic, food, organic,

DOI: 10.1201/9781003130932-7

leather, and laser dyes. Among these, natural dyes are mainly obtained from plant sources, which includes the leaves, roots, twigs, stems, flowers, bark, wood, fruits, husks, and hulls of plants. Fungi, lichens, minerals, and invertebrates are considered natural sources of dyes. Most colors today contain synthetic dyes. They are in everything from paper to clothes, and from food to wood. This is because, they can easily be applied to the textile or material. In addition, they are indelible, brighter, and, most importantly, cost-effective. Synthetic dyes are further grouped into azo, acid, basic, and mordant dyes. Azo dyes account for ≤70% of the total organic dyes that are manufactured globally. Therefore, they are a significant part of the effluent from textile industries and are considered the most harmful of the synthetic dyes. Due to the complex chemical structure of azo dyes, which are composed of an aromatic ring, amino groups, and azo bonds; they remain in the aquatic environment for a longer time (Gurses et al., 2016; Shah, 2014; Lipskikh et al., 2018; Berradi et al., 2019; Ventura-Camargo & Marin-Morales, 2013).

The total volume generated and the effluent composition of the wastewater generated from textile mills are categorized as the most contaminated among all the manufacturing industries (Correia et al., 1994). The production of dyes has increased due to the increased demand for textile products. All the previous factors have contributed to the increase in dye wastewater, and therefore, are a substantial source of the acute pollution problems today (Stolz, 2001). As mentioned previously, the dyes are stable in the environment for a long time, which could be because they have high thermal and photo stability, which helps them to resist biodegradation (Kunz et al., 2002; Pearce, 2003). In aquatic environments, the dyes keep reflect and absorb sunlight and disrupt photosynthetic activity in algae, which harms the food chain (Petsas & Vagi, 2017). Most of the dyes and their by-products are detrimental to life due to their carcinogenicity and mutagenicity. They can cause skin, liver, kidney, and bladder cancer in laborers that work in dye industries. In addition, synthetic dyes can cause different types of skin and eyes allergies and respiratory disease (Chung, 2016). In effluent treatment plants, reverse osmosis (RO) techniques can be used to treat the discharge. Then, after primary and secondary treatments the water can be used for different purposes. Different textile industries could move to modern technologies, such as cleaner production technologies. In this technique, dyes that have a low salt concentration are used along with membrane filters and steady flow machines, which efficiently decrease the utilization of water ≤50%. Moreover, the addition of activated carbon (C) has demonstrated a significant reduction of pollutants in wastewaters (Gita et al., 2016).

This chapter will discuss the types and impact of azo dyes along with the physiological, chemical, and biological methods for wastewater treatment. The focus will be on the use of different bacterial strains to decolorize and degrade azo dyes along with the underlying mechanisms and roles of the bacterial enzymes in degradation.

7.2 TYPES OF AZO DYES

Based on Ventura-Camargo & Marin-Morales (2013), 1979 classified the dyes were classified into the following types by Kirk-Othmer (1979):

1. Acid dyes: Water-soluble and anionic, that are composed of molecules of carboxylic or sulfonic acid. Chemically they are composed of azo compounds, anthraquinones, aminoacetone, triarylmethanes, nitro, nitrous, and quinoline. They are used in the wool, nylon, silk, paper, cosmetic, and food industries.
2. Basic dyes: Water-soluble and cationic. They are powerful coloring agents. The chemical structure is composed of the azo compound, triarylmethane, anthraquinone, methane, acridine, thiazine, quinoline, and oxazine. Basic dyes are used in the modification of nylon, polyesters, papers, and are used as an antiseptic in pharmaceutical companies.

3. Direct dyes: Directly applied to the substrate, they are water-soluble and anionic. The chemical structure is composed of azo compounds, phthalocyanines, thiazoles, and oxazines. Direct dyes are used to color cotton, linen, rayon, silk, wool, and leather.

4. Fluorescent dyes: Organic compounds and nonproteins. Technically, they are not dyes; however, but due to their extensive use in coloring industries, they are placed in the table of dyes. They contain a xanthene compound and are manufactured by heating phthalic anhydride and resorcinol with a zinc catalyst. Then, they are crystallized to a deep red powder.

5. Reactive dyes: These dyes have a reactive group. The reactive group forms the covalent bond with the polymer of the fibers. They are composed of azo compounds, phthalocyanines, and anthraquinones. The reactive dyes give the most permanent color compared with the other dyes. They are used to dye cotton, wool, and cellulosic fibers.

6. Sulfurous dyes: Their chemical structure is composed of a di-sulfide linkage, and they are water-insoluble. An electrolyte can be incorporated to increase the exhaustion of the dye. They are used to color cotton, acrylic, nylons, polyester, paper, and silk. They produce limited and dull shades.

7. Vat dyes: Vat stands for the vessel. These dyes use sodium dithionite as a reducing agent, and they are converted into a soluble form. After attaching to the fabric, they change into an insoluble state. They are known as leuco dyes.

8. Dye precursors: These are known as color precursors. They are pigments with a small molecular size and can penetrate the hair deeply. The chemical structure includes naphthalene and benzene groups.

7.3 STRUCTURE OF AZO DYES

Azo dyes are composed of one or more azo groups and are part of a commercially predominant and large family of synthetic or organic dyes. To synthesize azo dyes, a diazonium salt and coupling component are required. During diazotization, the aromatic amines are converted into diazo forms. Then, these components react with phenol, naphthol, and aromatic amines to form the azo dyes. This is known as the coupling reaction.

According to Sudha et al. (2014), azo dyes are easily soluble in water due to the formation of the chromophore complex, which contributes to the color of the dye. It is composed of azo ($-N=N-$) groups and sulfonic (SO_3^-) groups. They are in demand in the trading sector (Barragán et al., 2007). The azo dyes are categorized into 30 groups, which depends on the chromophore's chemical structure. The most prominent are monoazo, diazo, triazo, polyazo, anthraquinone, triarylmethane, and phthalocyanine dyes (Mohammad, 2005; Sudha et al., 2014). The addition of side groups is important for color development and helps to generate a variety of shades (Zollinger, 1991). Research has determined that azo dyes are composed of autochrome, chromophoric, and solubilizing groups. These determine the color of the azo dyes (Benkhaya et al., 2016; Gürses et al., 2016).

Sarkar et al. (2017) discussed the classification of azo dyes according to hydrophobicity. The difference between hydrophobic and hydrophilic azo dyes is that the former dye is consumed and reduced inside the bacterial cells and later is reduced outside the bacterial cells. There are approximately 2,000 different azo dyes available commercially. During dyeing, the unbound dye merges with the effluent and pollutes the water. Umbuzeiro et al. (2005) described that following the reduction of azo bonds some of the azo dyes caused mutagenesis. In contrast, the carcinogenicity of aromatic amines depends on the chemical structure. It is important to evaluate the toxicity of textile dyes, because they cause different effects in the environment and to the organisms that are exposed to them (Ventura-Camargo & Marin-Morales, 2013). Despite the similarities between the structures, all dyes have disparate biological activities. Therefore, it is very difficult to understand the mutagenic properties by only considering one chemical group (Marechal et al., 1997).

7.4   IMPACT OF AZO DYES

As discussed previously, during dyeing in the textile and allied industries, a substantial amount of surplus chemical contamination and unprocessed material is discharged into the aquatic system and causes water pollution (Correia et al., 1994; Stolz, 2001; Kunz et al., 2002). Azo dyes have a detrimental effect on living organisms. Therefore, a strategy is required for the development of successful techniques or processes for the removal of or degradation of harmful dyes to a safe or useful form (Elshaarawy et al., 2017; Benkhaya et al., 2019). Benkhaya et al. (2020) reported that azo dyes of higher molecular weight disseminated rapidly in aquatic systems compared with lower molecular weights, because due to the increase in molecular weight the number of azo bonds increases, and therefore, decreases the degradation rate. During dyeing, a covalent bond is formed between the fabric and the azo dyes, and the amount of energy required to break these bonds approximately equals its degradation. Because the dyes persist in the environment for several years, the polluting effects prevail for a long time. Even the breakdown products from the azo dyes have carcinogenic, mutagenic, and toxic effects on life (Gita et al., 2016).

The presence of azo dyes in textile wastewater enhances the pH, chemical oxygen demand (COD), and biological oxygen demand (BOD). In addition, these dyes will cause an imbalance in the environmental organic and inorganic chemical levels. The dye will impart color to the water, which decreases the penetration of light in the water bodies and ultimately affects the flora, fauna, and the entire ecosystem of the water bodies. Toxic substances from the dye will enter and slowly concentrate in aquatic organisms, which will subsequently reach humans by entering the food chain. This can cause cramps, hypertension, sporadic disorders, and various other problems. They have been reported to cause cancer and tumors in animals. Because some azo dyes are soluble in water they are easily mixed in water and can cause allergies, irritation to the skin and eyes, edema on the face, respiratory distress, and cancers. Table 7.1 lists the various types of azo dyes, their uses, and their harmful effects.

7.5   PHYSICOCHEMICAL TREATMENT OF AZO DYES

Numerous physicochemical methods, such as coagulation, adsorption, filtration, coagulation and flocculation, precipitation, and oxidation have been used for the removal of textile dyes. However, due to several disadvantages and limitations, these methods have not been successfully implemented. Therefore, it is vital to develop effective, feasible, and profitable methods to degrade and decolorize dyes from industrial effluents (Alhassani et al., 2007). Extensive research has focused on various combinations of physicochemical treatments, which have been employed by industries in simple and isolated examples (Shah, 2014). Recently, Sarkar et al. (2017) suggested microbial degradation to treat different types of synthetic dyes. These methods are environmentally friendly, sustainable, cost-effective, and could be a cost-effective substitute for physicochemical processes.

Membrane filtration techniques use membranes for the segregation of dyestuffs, and therefore, decreases the color, BOD, and COD of the textile wastewater. This technique could potentially decrease the amount of dye waste and help in the recovery of salts. This process is beneficial due to the use of fast and reusable membranes that require smaller spaces. However, it has some disadvantages, such as membrane fouling and the high cost of the membranes. For the removal of suspended solids from the effluent, flocculation and coagulation techniques are mainly used. Flocculation includes binding together small and large particles to form a floc that can easily be separated. Coagulation involves the addition of chemical coagulating agents that will neutralize the charge on the suspended solids and the particles will join together. The coagulants include iron and aluminum salts (Shah, 2018). Ion exchange is not, very suitable for the removal of dyes from industrial effluents, because they cannot remove a wide range of dyes and many diverse additives are present in the effluents. The RO technique is highly efficient in the removal of reactive dyes, salts, and other chemicals that are present in industrial wastewater using a single-step process.

TABLE 7.1

Different Types of Azo Dyes, Their Uses, and Harmful Effects

Types of Azo Dyes	Uses of Dyes	Harmful Effects	References
Reactive Brilliant Red X 3b	Used for coloring cellulosic fibers, cotton and wool	Obstruct the function of human serum albumin and can cause a disturbance in the function of enzymes and proteins	Li et al. (2010)
Acid Violet 7	Used for silk, wool, wood, soap, medicine, cosmetics, leather, fiber, polyamide, and biological dyeing	Chromosomal aberration, inhibit the activity of acetylcholinesterase and membrane lipid peroxidation	Mansour et al. (2010)
Disperse Red 1 and 13, Disperse Orange 1	Used for dyeing polyester fibers due to high hydrophobicity	Mutagenic in nature; tendency to raise the micronuclei in lymphocytic cells of humans. Harmful effect on microbes	Chequer et al. (2015); Mahmood et al. (2016)
Disperse Blue 291	Used for dyeing and painting polyester and cotton fibers	Genotoxic, mutagenic, and cytotoxic they have severe effects on human hepatoma cells	Tsuboy et al. (2007)
Reactive Black 5	Approximately 90% dyestuffs bind to the fibers. Mainly used for cotton and other cellulosic fibers	Severe effects on the urinary tract, redness in the eyes, skin irritation, long term exposure can cause pneumoconiosis	Topac et al. (2009)
Direct Black 38	Used for dyeing silk, cotton, viscose, polyamide, wood, leather, biological, and plastic materials	Cancer in the bladder of humans	Cerniglia et al. 1986
Direct Blue 15	Used for dyeing cellulose, silk, wool, leather, paper, cotton, silk, wool, and biological materials; they are used to tint cinematographic film	Carcinogenic and mutagenic	Reid et al. 1984
Congo Red	Used to dye cotton fibers and staining tissues for microscopic examination in the laboratory	Mutagenic and carcinogenic	Gopinath et al. (2009)
Malachite Green	Used to dye silk, wool, jute, leather, and cotton	Carcinogenic, mutagenic, chromosomal fractures, respiratory toxicity, and teratogenicity	Srivastava et al. (2004)

Filtration techniques using nano, micro, and ultrafiltration are used for the removal of dyes from textile wastewaters. Ultrafiltration can remove the particles and macromolecules. In addition, they can remove, dyes but only between 31% and 76%. Microfiltration can be used for the removal of dye colors. It can be used in combination with RO and nanofiltration techniques. Similarly, nanofiltration is used for the removal of dye color in wastewaters with concentrated and complex solutions. Recently, adsorption techniques (physical and chemical) were used in the removal of textile dyes from wastewaters due to their high removal efficiency and application to a diverse range of dyes. Physical adsorption includes weak interactions, such as hydrogen (H) bonds and Van der Walls forces. Chemical adsorption involves stronger interactions between the adsorbent and adsorbate. This technique is cost-efficient and highly effective for the treatment of wastewaters that contain dyes (Shah, 2018).

7.6   AZO DYE DEGRADATION AND DECOLORIZATION BY BACTERIAL STRAINS

Textile and allied industries are releasing diverse carcinogenic, mutagenic, and allergic chemicals and dyes, which is putting pressure on these industries to minimize the harmful substances that are released in their effluents. Global regulatory bodies have prepared stringent guidelines for effluent treatment and to limit the dye color that is present in effluents. Textile dye effluent treatment includes the removal of color (decolorization) and the degradation and mineralization of dye molecules. Decolorization requires the removal of molecules from the solution or the chromophore to break down, where the molecule and the major fragment remain intact. Various methods have been developed to eliminate these synthetic dyes from industrial wastewater including physical, chemical, and biological treatments, which help to minimize their harmful effects on aquatic life and humans. Figure 7.1 shows different treatments processes and mechanisms for the degradation or reduction of azo dyes.

Textile and allied industries release various azo dyes that have a highly toxic effect on aquatic organisms, the ecosystem, and human health. Therefore, the treatment of wastewater from the textile industry is of utmost importance. There are various physical, chemical, and biological treatment techniques; however, the physicochemical treatments do not remove the synthetic dye efficiently. Therefore, there is more focus is more on biological treatment strategies that could remove the carcinogenic and mutagenic dyes that are present in the wastewater. Various bacteria, fungi, yeast,

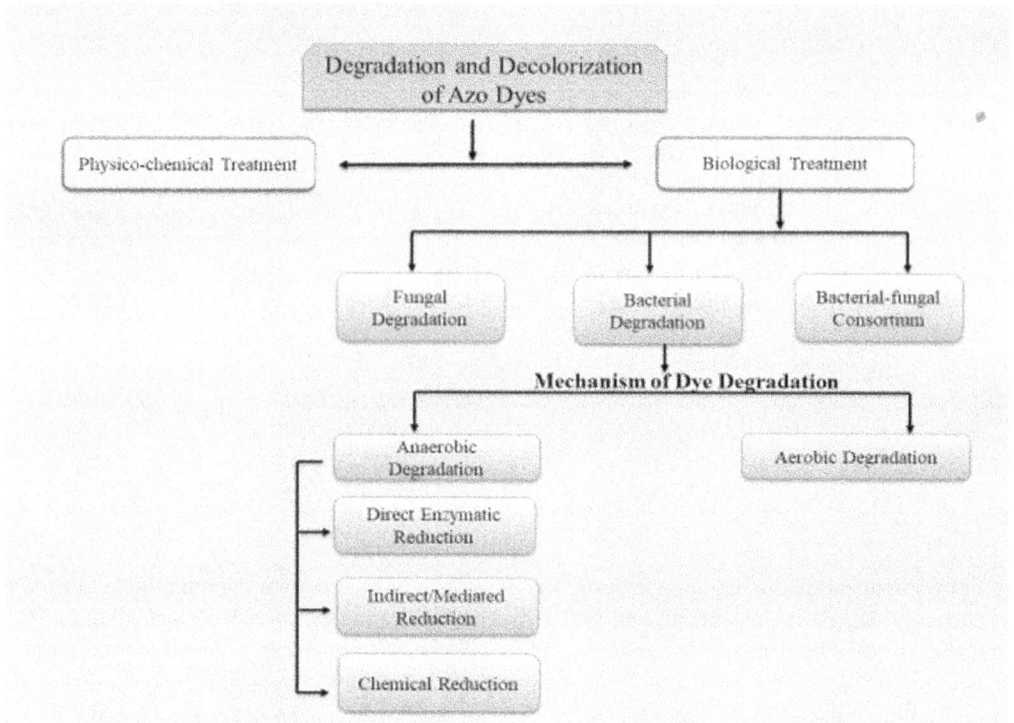

FIGURE 7.1   Treatment methods and mechanisms for azo dye degradation and decolorization.

and actinomyces have been used to biologically treat dyes in effluents (Wesenberg et al., 2003; Olukanni et al., 2006) where bacterial decolorization is nonspecific and faster. The identification of bacterial cultures that efficiently degrade azo dyes was initiated in 1970, with the successful selection of *Aeromonas hydrophila* and two species of a Bacillus (*Bacillus subtilis* and *Bacillus cereus*) (Wuhrmann et al., 1980; Dave et al., 2015). Then, scientists and researchers screened and selected various strains of bacteria for the efficient degradation or reduction of dyes present in wastewater. They included *Corneybaterium sp., Pseudomonas sp., Enterococcus sp., Staphylococcus sp., Escherichia coli, Dermacoccus s., Xenophilus sp., Lactobacillus sp., Alcaligenes sp., Bacillus sp., Rhizobium sp.,* and *Proteus sp.* (Imran et al., 2016; Kolekar et al., 2013; Haghshenas et al., 2016). The microorganisms can decrease the dyes by secretion or the production of azo reductase, hydrogenase, laccase, and peroxidase enzymes, which are further mineralized.

In contrast to the unspecified mechanisms for bacterial degradation of toxic azo dyes, aerobic bacteria are used for their reduction potential. This includes the growth of bacterial culture with simple azo compounds under aerobic conditions for a long time using a continuous culture system. These conditions produce enzyme azoreductases, which can degrade azo compounds and other similar compounds under aerobic conditions (Stolz, 2001). For a significantly higher reduction level, various parameters need to be optimized including the catalytic efficiency of an enzyme, cell density, and permeability. Kalyani et al. (2009) reported rapid decolorization by bacteria compared with fungal cultures and they demonstrated a better decolorization and biodegradation potential at a basic pH range. Various researchers investigated the metabolism of azo dyes and reported that the dyes were difficult to metabolize in aerobic environments, and therefore, lead to the formation of an intermediate compound formation, which did not undergo mineralization. However, these dyes might undergo complete degradation when the treatment process is a coupled aerobic–anaerobic process (McMullan et al., 2001). Under anaerobic conditions the bond in azo dyes cleaves, which forms aromatic amines that undergo further mineralization using nonspecific enzymes under aerobic conditions via ring cleavage. Therefore, toxic azo dyes are efficiently biodegraded by coupled treatments, which include anaerobic treatments followed by aerobic treatments (Feigel et al., 1993). Some conditions showed a significant bacterial growth rate under aerobic conditions; however, color removal was best observed under anaerobic conditions.

7.6.1 Decolorization by Single or Mixed Bacterial Cultures

Decolourization of azo dyes is possible using aerobic, anaerobic, and coupled aerobic–anaerobic methods. In addition, bacterial degradation is categorized into the use of single bacterial culture or a combination approach that uses consortia. Microbial or bacterial consortia have advantages for use in the reduction of synthetic dyes compared with a single or pure culture (Sudha et al., 2014). Mixed bacterial cultures support higher synergistic metabolic activities; therefore, they have a high level of mineralization. In mixed cultures, each culture has an important role. Each culture can attack dye molecules at specific positions, which results in the formation of metabolic end products that are utilized by another culture as a source of nutrients. The other cultures present can help to stabilize the ecosystem if they do not have a direct role in bioremediation. In addition, they will not allow any harmful intermediate substance to be released into the ecosystem. Moreover, mixed bacterial cultures are more stable under various environmental conditions compared with single cultures. The main bacterial culture used in azo dye decolorization is *Pseudomonas sp.* It can degrade a wide range of azo dyes, such as Reactive Red 2 and 22, Orange I, Orange II, Reactive Blue 172, and Red HE7B. Similarly, *Pseudomonas entomophila* BS1 significantly degraded (≤93%) of the azo dye, Reactive Black 5 was present in higher concentration after 5 days incubation (Khan & Abdul, 2015). A consortium of *Pseudomonas sp.* SUK1 and *P. rettgeri* HSL1 could significantly decolorize and degrade azo dyes (e.g., Direct Red 81, Reactive Black 5, Reactive Orange16, and Disperse Red 78) that were used in the study with slight differences in duration. Khan et al. (2014) selected a novel mix of bacterial culture (RkNb1) that showed a maximum dye decolorization rate. The bacterial consortium was analyzed using 16S rRNA analysis and it included *Arthrobacter crystallopoietes, Kocuria flavus, Bacillus beijingensis, Ochrobactrum intermedium, Ochrobactrum intermedium* M16-10-4, *Enterococcus faecalis,* and *Citrobacter freundii.* The bacterial consortium was efficient at decolorizing Reactive Violet 5 dye by 98.17% in 8 h. A higher rate of decolorization was observed under optimum conditions at 42°C and pH 7.0. Suad and Ahmed (2014) selectively used *Bacillus* sp., *Klebsiella* sp., *E. coli,* and *Pseudomonas aeruginosa* to study the decolorization of azo dyes (e.g., Direct Orange, Disperse Brown, and Reactive Green). The highest decolorization efficiencies of 94%, 88%, and 72% were reported when *P. aeruginosa* was used and the lowest decolorization was reported for *Klebsiella sp.* which were 80%, 45%, and 29% for Direct Orange, Disperse Brown, and Reactive Green, respectively. A similar observation for the decolorization of Acid Maroon V was reported by Patel et al. (2012). The bacterial consortium (EDPA) contained *Enterobacter dissolvens* AGYP1 and *P. aeruginosa* AGYP2. A maximum decolorization rate of 93% was reported at a pH from 6 to 9 with an optimum at pH 7.0 after 20 h of incubation. The EDPA consortium was efficient against higher concentrations of Acid Marron V and can degrade many other textile dyes.

7.7   FACTORS THAT INFLUENCE THE DEGRADATION OF DYES

Wastewater that is generated from different textile industries varies in chemical composition. This wastewater can be treated using diverse physicochemical and biological processes. In the biological processes, various factors might influence the reduction or removal of dye, such as pH, temperature, concentration, the structure of the dye, C and nitrogen (N) sources, and redox mediators as shown in Figure 7.2.

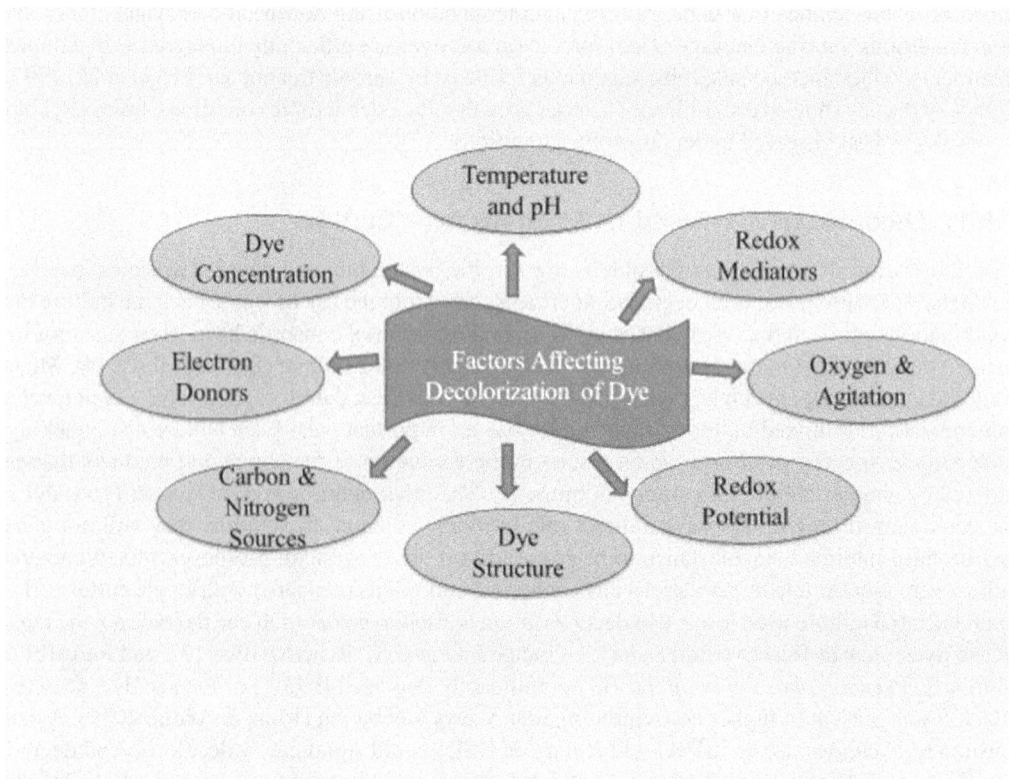

FIGURE 7.2   Factors that influence azo dye decolorization.

The pH influences the rate of dye decolorization with an optimum pH observed at neutral or weak alkaline pHs, which occurs mainly in reactive azo dyes. Dye color removal was maximum at the optimum pH, which decreased with an increase or decrease in the pH toward strongly acidic or alkali. During biological reduction, the azo bond breaks and increases the pH, because aromatic amines are formed, which are more basics compared with azo dyes (Kalyani et al. 2016). Temperature influences the microorganisms-based dye decolorization process, because each culture requires a specific optimum temperature for maximum efficiency. The optimum range was between 30°C and 45°C, which resulted in maximum azo dye color removal and further increases or decreases in temperature lead to a reduced rate of color removal.

Other factors that influence the efficiency of dye color removal include the structure of the dye and various functional groups that are present that influence its reduction capability. The reduction of color is maximum in azo dyes that are structurally simple and have low molecular weights compared with high molecular weight dyes that have a complex structure. Therefore, the reduction of color is fast in monoazo dyes compared with diazo and triazo dyes (Hu, 2001). The dye concentration in the wastewater is another important factor that influences the dye decolorization rate. Dye present in high concentrations will take longer to be reduced. Dye concentration in the range of 1–10 μM can be decolorized easily; however, with further increases (30 μM), a sharp decrease in the reduction was observed. Dye concentration might have a toxic impact on the single or consortium of bacteria used, it might block the active sites of azo reductase. The toxic impact of dyes at higher concentrations was minimized when a bacterial consortium was used (Saratale et al., 2009). This was because of the synergistic effects of the bacterial cultures used that could reduce the toxic concentrations and their effects.

Oxygen (O_2) has a vital function in the reduction of dye color. In anaerobic environments, there is high reductive enzymatic activity; however, a low O_2 concentration is required for the oxidative enzymes. Oxidative enzymes are included in dye reduction. Electron donors, such as sodium acetate, sodium citrate, glucose, and sodium pyruvate are involved to cleave the azo bonds. They act as external substrates that can increase the reduction of the dye color. Wastewater from textile and allied industries contain an insufficient quantity of these electron donors. These donors are required to support bacterial growth (aerobic and anaerobic), which results in the complete removal of dye color (Kalyani et al., 2016).

7.8   MECHANISM OF DYE DEGRADATION

Color removal from textile industry wastewater is initially carried out using bacterial cultures by the adsorption of dyes onto the bacterial biomass. Dye adsorption onto the bacterial biomass appears faster than various physical adsorption techniques. Adsorption is unsuitable when used for a long time for color removal from textile effluent due to the concentration of dyes on the biomass. Degradation of dyes by bacterial cultures involves two stages. First, the dye azo bond (–N=N–) undergoes reductive cleavage that leads to the formation of mainly colorless and hazardous aromatic amines. Second, the aromatic amines undergo degradation in the presence of O_2.

The methods for dye degradation are categorized as aerobic and anaerobic degradation strategies. The anaerobic degradation of azo dye involves three different mechanisms: (1) direct enzyme-based reduction processes; (2) indirect or mediated based reduction; and (3) chemical compound-based reduction (e.g., organic and inorganic). The direct enzyme-based reduction and indirect or mediated reduction use biologically produced enzymes and electron carriers. The chemical reduction includes chemical reactions that use biogenic reductants.

7.8.1   Bacterial Enzymes in Azo Dye Degradation

Industrial dyes are decolorized and degraded or reduced by microbial enzymes, such as azoreductase, laccase, lignin peroxidase, manganese peroxidase, and hydroxylases. To degrade azo, laccase and azoreductase have great potential and enzymes, such as peroxidase and oxidase are involved in the metabolic process and could degrade the azo dyes to a certain degree. The enzymes act intra and extracellularly. These enzymes have been found in bacteria, fungi, yeast, and plants. However, microbial enzymes have advantages due to their cost-effective production, easy processing and purification, and availability throughout the year. Some research has identified fungal cultures and their role in the removal of azo dyes, because they provide a good source of enzymes for dye degradation. In comparison, bacterial enzymes are faster to produce, possess wider substrate specificity, are easy to immobilize and are more stable; therefore, they can be used effectively for textile wastewater treatment. To reduce the cost of treatment and to speed up the reaction rate in dye industries, the enzymes can be immobilized onto suitable support, which provides them more stability and can be reused.

7.8.1.1 Azo Dye Degradation by Laccase

The laccase enzyme (EC 1.10.3.2) has been studied extensively for its potential in azo dye degradation. It is a small enzyme with lower specificity toward substrates but can degrade various xenobiotic compounds, aromatics, and nonaromatic substrates. Most laccases are obtained from fungus or plants and there are a few examples of the bacterial source of laccases. However, fungal laccases have a limited role in the industry, because they are quite unstable at high temperatures and pHs. In 1993, the first bacterial laccase (*Azospirillum lipoferum*) was isolated from rice rhizosphere. Bacteria can produce laccase extracellularly; however, they do not secrete it outside the cell (Givaudan et al., 1993). Most of the bacterial strains that produce laccase are Gram-positive, such as various species of *Azospirillum, Bacillus, Geobacillus, Streptomyces, Rhodococcus*, and *Staphylococcus* (Narayanan et al., 2015). Some examples of Gram-negative bacteria that secrete laccase have been reported, which include *Alteromonas sp., Delfia sp., Enterobacter sp., Pseudomonas sp.,* and *Proteobacterium sp.* (Solano et al., 1997).

Enzyme laccase-based degradation of azo dye via a free radical mechanism, which is nonspecific, leads to the production of phenolic compounds, and therefore, prevents aromatic amine formation (Chivukula et al. 1995). The enzyme laccase includes H^+ removal and the removal of amino groups from the hydroxyl group of the ortho and para-substituted phenolic substrates and aromatic amines. This enzyme can oxidize the dye phenolic groups and nucleophilic attack by water on the phenolic ring azo bonds that contain C (Camarero et al., 2005). In the reaction, the reactive species cross-coupled and leads to C–C and C–O bond formation between the phenolic compounds and C–N and N–N bond formation for aromatic amines (Zille et al., 2005). First, laccase oxidizes the phenolic ring to form a phenoxy radical, which is further oxidized to generate carbonium ions. Due to the nucleophilic attack of water, two compounds are generated, which are not stable under aerobic conditions. These are 4-sulfophenyldiazene and benzoquinone. Therefore, under aerobic conditions, 4-sulfophenyldiazene is oxidized and forms a phenyldiazene radical, which is followed by a sulfonyl radical and finally forms sulfophenyl hydroperoxide, scavenged by O_2 (Singh et al., 2015).

7.8.1.2 Azo Dye Degradation by Azoreductase

Azoreductases (EC 1.7.1.6) are cytoplasmic, or membrane-bound enzymes that can reduce azo dyes into colorless amines via a reductive cleavage mechanism. Both the enzyme systems (cytoplasmic and membrane-bound) for azoreductases include two different systems. The enzyme can catalyze the reaction using low molecular weight reducing equivalents, such as FADH and NADH (Robinson et al., 2001). Depending on the coenzyme utilized, there are three types of azoreductases that: (1) only use NADH: (b) only use NADPH; and (c) use NADH and NADPH. Azo reductases are sensitive to O_2; therefore, extracellular reduction is inhibited due to the availability of O_2.

Azoreductase can cleave the N=N azo bond, which transfers four electrons as reducing equivalents. At every stage, two electrons are transferred to the dye, which acts as an electron acceptor and results in a reduction of the dye. Some toxic intermediates (aromatic amines) are produced during reduction, which is subsequently reduced via aerobic or microaerophilic processes (Pandey et al., 2007). Redox mediators are used as electron shuttles in membrane-bound azoreductases. The redox mediator-based membrane-bound azoreductase mechanism is different compared with the cytoplasmic azoreductase mechanism. Nonsulfonated azo dyes undergo degradation via a cytoplasmic azoreductase (Chacko & Subramaniam, 2011), which passes through the cell membrane. Azoreductase is sensitive to O_2; therefore, reduction and decolorization are more efficient under anaerobic compared with aerobic conditions. In an aerobic environment, the azoreductase combines with O_2, which reduces the redox mediator rather than reducing the azo dye (Singh et al., 2015).

7.8.2 INDIRECT OR MEDIATED BIOLOGICAL DYE DEGRADATION

Sulfonated azo dyes cannot cross the cell membrane; therefore, they are reduced via redox mediators. These mediators have an important role in the bacterial reduction of azo dyes. Field and Brady (2003) reported a significant enhancement in the reduction of Mordant Yellow 10 when a small quantity of riboflavin was present. The removal of the color in many azo dyes was enhanced when anthraquinone-2-6-disulphonate, a synthetic electron carrier was added (Van der Zee et al., 2001). Acid Orange 7 enhanced the decolorization rate via the transfer of reducing equivalents (Mendez-Paz et al., 2005). *Sphingomonas sp.* BN6 cultivated with 2-naphthyl sulfonate in an aerobic environment showed an enhanced (10–20 fold) rate of azo dye (amaranth) decolorization. Dye color removal can be increased by the generation of redox intermediates during the aerobic degradation of aromatic compounds (Keck et al., 1997). Similar reports are available on the increased rate of azo dye decolorization by the addition of the supernatant from a culture of *E. coli* NO3 culture (Chang et al., 2004).

7.8.3 DYE DECOLORIZATION BY CHEMICAL REDUCTION

Dye decolorization can be enhanced by chemical reduction using various organic and inorganic compounds. Reactions with inorganic compounds, such as sulfide and ferrous ions can result in decolorization. Sulfate-reducing bacteria produce hydrogen sulfide (H_2S), which can lead to extracellular azo dye degradation (Yoo et al., 2000; Diniz et al., 2002). When sulfur compounds are not available, decolorization of the dye occurs in the granular sludge. There are reports on the kinetics of color removal at the laboratory and batch reactor stage that indicate the important role of chemical-based dye reduction mechanisms in anaerobic reactors (Van der Zee et al., 2003). Dawkar et al. (2008) studied many inducers and stabilizing agents of oxidoreductive enzymes that included calcium carbonate, indole, o-toluidine, veratrole, and vanillin that could enhance the decolorization rate of the dye.

7.9 CONCLUSIONS

The textile and allied industries are releasing a large amount of effluent into water bodies, which consist of colored waste, organic chemicals, and synthetic dyes. These pose a major threat to the environment and human health. The textile industry mainly uses azo dyes that are highly toxic and carcinogenic. Several industries including food, pharmaceutical, cosmetics, paper, and leather are using azo dyes due to their diverse application. These dyes can reach the groundwater and result in a change in the salinity, and the water becomes unfit for use and affects biological diversity. In addition, it affects aquatic organisms and microflora. Therefore, to avoid these issues, the effluents that contain toxic dyes from the textile and related industries must be treated properly. Various techniques have been utilized to overcome this problem, such as physical, chemical, and biological treatments. Biological treatments are efficient at removing the dye color and degrade the azo dyes. Aerobic or anaerobic bacterial cultures either as single strains or consortia can efficiently reduce the azo dyes. Diverse bacterial enzymes that attack the dyes can reduce the complex structure of the dye, which makes degradation by enzymes an important mechanism. Enzymes are eco-friendly, faster in action, highly stable, and can easily be immobilized to improve their catalysis and stability, and therefore, can be used efficiently in the textile wastewater treatment process. Further detailed studies are required in this area for the degradation and removal of azo dye color. In addition, based on the increased application of these dyes in various industries, new dyes need to be developed that have lower or no toxicity.

REFERENCES

Alhassani, H. A., Rauf, M. A., & Ashraf, S. S. (2007). Efficient microbial degradation of Toluidine Blue dye by *Brevibacillus* sp. *Dyes Pigm, 75*, 395–400.

Barragán, B. E., Costa, C., & Marquez, M. C. (2007). Biodegradation of azo dyes by bacteria inoculated on solid media. *Dyes Pigm, 75*(1), 73–81.

Benkhaya, S., Achiou, B., Ouammou, M., Bennazha, J., Younssi, S. A., M'rabet, S., & Harfi, A. E. (2019). Preparation of low-cost composite membrane made of polysulfone/polyetherimide ultrafiltration layer and ceramic pozzolan support for dyes removal. *Mater Today Commun, 19*, 212–219.

Benkhaya, S., Cherkaoui, O., Assouag, M., Mrabet, S., Rafik, M., & Harfi, A. E. (2016). Synthesis of a new asymmetric composite membrane with bi-component collodion: Application in the ultra filtration of baths of reagent dyes of fabric rinsing/padding. *J Mater Environ Sci, 7*(12), 4556–4569.

Benkhaya, S., M'rabet, S., & Harfi, A. E. (2020). Classifications, properties, recent synthesis and applications of azo dyes. *Heliyon, 6*, 1–26.

Berradi, M., Hsissou, R., Khudhair, M., Assouag, M., Cherkaoui, O., Bachiri, A. E., & Harfi, A. E. (2019). Textile finishing dyes and their impact on aquatic environs. *Heliyon, 5*, e02711.

Camarero, S., Ibarra, D., Martinez, M. J., & Angel, T. M. (2005). Lignin derived compounds as efficient laccase mediators for decolorization of different types of recalcitrant dyes. *Appl Environ Microbiol, 71*(4), 1775–1784.

Cerniglia, C. E., Zhuo, Z., Manning, B. W., Federle, T. W., & Heflich, R. H. (1986). Mutagenic activation of the benzidine based dye Direct Black 38 by human intestinal microflora. *Mutat Res, 175*(1), 11–16.

Chacko, J. T., & Subramaniam, K. (2011). Enzymatic degradation of azo dyes-a review. *Int J Environ Sci, 1*(6), 1250.

Chang, J. S., Chen, B. Y., & Lin, Y. S. (2004). Stimulation of bacterial decolorization of an azo dye by extracellular metabolites from *Escherichia coli* strain NO3. *Bioresour Technol, 91*(3), 243–248.

Chequer, F. M. D., Lizier, T. M., De Felício, R., Zanoni M. V. B., Debonsi, H. M., Lopes, N. P., & De Oliveira, D. P. (2015). The azo dye Disperse Red 13 and its oxidation and reduction products showed mutagenic potential. *Toxicol In Vitro, 29*(7), 1906–1915.

Chivukula, M., & Renganathan, V. (1995). Phenolic azo dye oxidation by laccase from *Pyricularia oryzae*. *Appl Environ Microbiol, 61*, 4347 4377.

Chung, K. T. (2016). Azo dyes and human health: A Review. *J Environ Sci Health, 34*(4), 233–261.

Correia, V. M., Stephenson, T., & Judd, S. J. (1994). Characterization of textile wastewaters: A review. *Environ Technol, 15*, 917–929.

Dave, S. R., Patel, T. L., & Tipre, D. R. (2015). Bacterial degradation of azo dye containing wastes. In S. N., Singh (Ed.) *Microbial degradation of synthetic dyes in wastewaters* (pp. 57–83). Switzerland: Springer International Publishing.

Dawkar, V. V., Jadhav, U. U., Jadhav, S. U., & Govindwar, S. P. (2008). Biodegradation of disperse textile dye Brown 3REL by newly isolated *Bacillus sp. VUS*. *J Appl Microbiol, 105*, 14–24.

Diniz, P. E., Lopes, A. T, Lino, A. R., & Serralheiro, M. L. (2002). Anaerobic reduction of a sulfonated azo dye, Congo Red, by sulfate-reducing bacteria. *Appl Biochem Biotechnol, 97*, 147.

Elshaarawy, R. F., Sayed, T. M., Khalifa, H. M., & El-Sawi, E. A. (2017). A mild and convenient protocol for the conversion of toxic acid red 37 into pharmacological (antibiotic and anticancer) nominees: organopalladium architectures. *Compt Rendus Chem, 20*, 934–941.

Feigel, B. J., & Knackmuss, H. J. (1993). Syntrophic interactions during degradation of 4-aminobenzenesulfonic acid by a two species bacterial culture. *Arch Microbiol, 159*, 124.

Field, J. A., & Brady, J. (2003). Riboflavin as a redox mediator accelerating the reduction of the azo dye Mordant Yellow 10 by anaerobic granular sludge. *Water Sci Technol, 48*(6), 187–193.

Gita, S., Hussan, A., & Choudhury, T. G. (2016). Impact of textile dyes waste on aquatic environments and its treatment. *Environ Ecol, 35*(3C), 2349–2353.

Givaudan, A., Effosse, A., Faure, D., Potier, P., Bouillant, M. L., & Bally, R. (1993). Polyphenol oxidase in *Azospirillum lipoferum* isolated from rice rhizosphere: evidence for laccase activity in non-motile strains of *Azospirillum lipoferum*. *FEMS Microbiol, 108*, 205–210.

Gopinath, K. P., Murugesan, S., Abraham, J. & Muthukumar, K. (2009). *Bacillus sp.* mutant for improved biodegradation of Congo red: Random mutagenesis approach. *Bioresour Technol, 100*(24), 6295–6300.

Gürses, A., Açıkyıldız, M., Günes, K., & Gürses, M. S. (2016). Classification of dye and pigments, in: Dyes and Pigments 31–45.

Haghshenas, H., Kay, M., Dehghanian, F. & Tavakol, H. (2016). Molecular dynamics study of biodegradation of azo dyes via their interactions with AzrC azoreductase. *J Biomol Struct Dyn, 34*(3), 453–462.

Hu, T. L. (2001). Kinetics of azoreductase and assessment of toxicity of metabolic products from azo dye by *Pseudomonas luteola*. *Water Sci Technol, 43*, 261–269.

Imran, M., Negm, F., Hussain, S. Ashraf, M,. Ahmad, Z., Arshad, M., & Crowley, D. E. (2016). Characterization and purification of membrane-bound azoreductase from azo dye degrading *Shewanella sp.* strain IFN4. *Clean, 44*(11), 1523–1530.

Kalyani, D. C., Telke, A. A., Dhanve, R. S., & Jadhav, J. P. (2009). Ecofriendly biodegradation and detoxification of reactive red 2 textile dye by newly isolated *Pseudomonas sp.* SUK1. *J Hazard Mater, 163*, 735–742.

Kalyani, P., Hemalatha, V., Vineela, K., & Hemalatha, K. P. J. (2016). Degradation of toxic dyes: A Review, *Int J Pure App Biosci, 4*(5), 81–89.

Kammradt, P. B. (2004). *Color removal of dye from industrial effluents by oxidation process Advanced.* (Unpublished master's dissertation). Paraná University, Curitiba, Brazil.

Keck, A., Klein, J., Kudlich, M., Stolz, A., Knackmuss, H.J., & Mattes, R. (1997). Reduction of azo dyes by redox mediators originating in the naphthalenesulfonic acid degradation pathway of *Sphingomonas sp.* strain BN6. *Appl Environ Microbiol, 63*(9), 3684–90.

Khan, R., Khan, Z., Nikhil, B., Devecha, J., & Datta, M. (2014). Azo dye decolorization under microaerophilic conditions by a bacterial mixture isolated from anthropogenic dye-contaminated soil. *Bioremediat J, 18*, 147–157.

Khan, S., & Abdul, M. (2015). Degradation of Reactive Black 5 dye by a newly isolated bacterium *Pseudomonas entomophila* BS1. *Can J Microbiol, 62*(3), 220–232.

Kirk-Othmer. (1979). *Encyclopedia of Chemical Technology* (3rd ed.) New York: Wiley-Inter Science.

Kolekar, Y. M., Konde, P. D., Markad, V. L., Kulkarni, S. V., Chaudhari, A. U. & Kodam, K. M. (2013). Effective bioremoval and detoxification of textile dye mixture by *Alishewanella sp.* KMK6. *Appl Microbiol Biotechnol, 97*(2), 881–889.

Kunz, A., Peralta-Zamora, P., Moraes, S. G., & Durán, N. (2002). Degradation of reactive dyes by the system metallic iron hydrogen peroxide. *Química Nova, 25*, 78.

Li, W. Y., Chen, F. F., & Wang, S. L. (2010). Binding of reactive brilliant red to human serum albumin: insights into the molecular toxicity of sulfonic azo dyes. *Protein Pept Lett, 17*(5): 621–629.

Lipskikh, O. I., Korotkova, E. I., Khristunova, Y. P., Barek, J., & Kratochvil, B. (2018). Sensors for voltammetric determination of food azo dyes: a critical review. *Electrochim Acta, 260*, 974–985.

Mahmood, S., Azeem, K., Muhammad, A., Mahmood, T., & Crowley, D. E. (2016). Detoxification of azo dyes by bacterial oxidoreductase enzymes. *Crit Rev Biotechnol, 36*(4), 639–651.

Mansour, H. B., Ayed-Ajmi, Y., Mosrati, R., Corroler, D., Ghedira, D., Barillier, D., & Ghedira, L. C. (2010). Acid violet 7 and its biodegradation products induce chromosome aberrations, lipid peroxidation, and cholinesterase inhibition in mouse bone marrow. *Environ Sci Pollut Res, 17*(7): 1371–1378.

Marechal, A. M., Slokar, Y. M., & Taufer, T. (1997). Decoloration of chlorotriazine reactive azo dyes with H_2O_2/UV. *Dyes Pigm, 33*, 281–298.

McMullan. G., Meehan, C., Conneely, A., Kirby, N., Robinson, T., Nigam, P., … Smyth, W. F. (2001). Microbial decolorization and degradation of textile dyes. *Appl Microbial Biotechnol, 561*(2), 81–87.

Méndez-Paz, D., Omil, F., Lema, J. M. (2005). Anaerobic treatment of azo dye acid orange 7 under fed-batch and continuous conditions. *Water Res, 39*, 771–778

Mohammad, S. (2005). *HPLC determination of four textile dyes and studying their degradation using spectro-photometric technique* (Unpublished Master's Dissertation). An-Najah National University. Palestine.

Narayanan, M. P., Murugan, S., Eva, A. S., Devina, S. U., & Kalidass, S. (2015). Application of immobilized laccase from *Bacillus subtilis* MTCC 2414 on decolourization of synthetic dyes. *Res J Microbiol, 10*, 421–432.

Oliveira, D. P. (2005). *Dyes as important class of environmental contaminants: A case study* (Unpublished doctoral thesis). São Paulo University, São Paulo, Brazil.

Olukanni, O. D., Osuntoki, A. A., & Gbenle, G. O. (2006). Textile effluent biodegradation potentials of textile effluent adapted and nonadapted bacteria. *Afr J Biotechnol, 5*, 1980–1984.

Pandey, A., Sing, P., & Iyengar, L. (2007). Bacterial decolorization and degradation of azo dyes. *Int Biodeterior, 59*(2), 73–84.

Patel, Y., Mehta, C., & Gupte, A. (2012). Assessment of biological decolorization and degradation of sulfonated di-azo dye Acid Maroon V by isolated bacterial consortium EDPA. *Int Biodeter Biodegrad, 75*, 187–193.

Pearce, C. I., Lloyd, J. R., & Guthrie, J. T. (2003). The removal of color from textile wastewater using whole bacterial cells: a review. *Dyes Pigm, 58*, 179–196.

Petsas, A. S., & Vagi, M. C. (2017). Effects of photosynthetic activity of algae after exposure to various organic and inorganic pollutants. *Chlorophyll*: Intech Open.

Reid, T. M., Morton, K. C., Wang, C. Y., & King, C.M. (1984). Mutagenicity of azo dyes following metabolism by different reductive/oxidative systems. *Environ Mutagen, 65*, 705–717.

Robinson, T., McMullan, G., Marchant, R., & Nigam, P. (2001). Remediation of dyes in textile effluent: a critical review on current treatment technologies with a proposed alternative. *Bioresour Technol, 77*, 247–255.

Saratale, R. G., Saratale, G. D., Kalyani, D. C., Chang, J. S., & Govindawar, S. P. (2009). Enhanced decolorization and biodegradation of textile azo dye Scarlet R by using developed microbial consortium-GR. *Bioresour Technol, 100*, 2493–2500.

Sarkar, S., Banerjee, A., Halder, U., Biswas, R., & Bandopadhyay, R. (2017). Degradation of synthetic azo dyes of textile industry: a sustainable approach using microbial enzymes. *Water Conserv Sci Eng, 2*, 121–131.

Shah, M. P. (2014). Effective treatment systems for azo dye degradation: a joint venture between physicochemical and microbiological process. *J Environ Biorem Biodegrad, 2*, 231–242.

Shah, M. P. (2018). Azo dye removal technologies. *Austin J Biotechnol Bioeng, 5*(1), 1090.

Singh, R. L., Singh, P. K., & Singh, R. P. (2015). Enzymatic decolorization and degradation of azo dyes-a review. *Int Biodeterior, 104*, 21–31.

Solano, F., Garcia, E., Perez, D., & Sanchez-Amat, A. (1997). Isolation and characterization of strain MMB-1 (CECT 4803), a novel melanogenic marine bacterium. *Appl Environ Microbiol, 63*, 3499–3506.

Srivastava, S., Sinha, R., & Roy, D. (2004). Toxicological effects of malachite green. *Aquat Toxicol, 66*(3), 319–329.

Stolz, A. (2001). Basic and applied aspects in the microbial degradation of azo dyes. *Appl Microbiol Biotechnol, 56*, 69–80.

Suad, G. K., & Ahmed, A. L. (2014). Some aerobic bacterial degradation and decolorization of different azo dyes. *J Biol Agric Healthc, 4*(2), 72–81.

Sudha, M. A., Saranya, G., Selvakumar, G., Sivakumar, N. (2014). Microbial degradation of azo dyes: A review. *Int J Curr Microbiol Applied Sci, 3*(2), 670–690.

Topac, F. O., Dindar, E., Ucaroglu, S., & Baskaya, H. S. (2009). Effect of a sulfonated azo dye and sulfanilic acid on nitrogen transformation processes in soil. *J Hazard Mater, 1702*(3), 1006–1013.

Tsuboy, M. S., Angeli, J. P. F., Mantovani, M. S., Knasmuller, S., Umbuzeiro, G.A., & Ribeiro, L. R. (2007). Genotoxic, mutagenic and cytotoxic effects of the commercial dye CI Disperse Blue 291 in the human hepatic cell line HepG2. *Toxicol In Vitro, 21*(8), 1650–1655.

Umbuzeiro, G. A., Freeman, H., Warren, S. H., Kummrow, F., & Claxton, L. D. (2005). Mutagenicity evaluation of the commercial product C.I. Disperse Blue 291 using different protocols of the Salmonella assay. *Food Chem Toxicol, 43*, 49–56.

Van der Zee, F. P., Bisschops, I. A. E., Blanchard, V. G., Bouwman, R. H. M., Lettinga, G., & Field, J. A. (2003). The contribution of biotic and abiotic processes during azo dye reduction in anaerobic sludge. *Water Res, 37*, 3098–3109.

Van der Zee, F. P., Lettinga, G., & Field, J. A. (2001). Azo dye decolourisation by anaerobic granular sludge. *Chemosphere, 44*, 1169–1176.

Ventura-Camargo, B. C., & Marin-Morales, M. A. (2013). Azo dyes: Characterization and toxicity: A review. *Textiles and Light Industrial Science and Technology, 2*(2), 103.

Wesenberg, D., Kyriakioles, I., & Agathos, S. N. (2003). White rot fungi and their enzymes for the treatment of industrial dye effluents. *Biotechnol Adv, 22,* 161–187.

Wuhrmann, K., Mechsner, K. L., & Kappeler, T. H. (1980). Investigation on rate-determining factors in the microbial reduction of azo dyes. *Appl Microbiol Biotechnol, 9*(4), 325–338.

Yoo, E. S., Libra, J., & Adrian, L. (2001). Mechanism of decolorization of azo dyes in anaerobic mixed culture. *J Environ Eng, 127,* 844–849.

Zille, A., Gornacka, B., Rehorek, A., & Cavaco-Paulo, A. (2005). Degradation of azo dyes by *Trametes villosa* Laccases over long periods of oxidative conditions. *Appl Environ Microbiol, 71*(11),6711–6718.

Zollinger, H. (1991). *Color chemistry: Synthesis, properties and application of organic dyes and pigments.* New York: VCH.

8 The Use of Microorganism for the Degradation of Azo Dyes

Pranjal Tripathi[1], Sonam[2], Devendra Mohan[2], and R.S. Singh[1]*

[1]Department of Chemical Engineering and Technology, IIT (BHU) Varanasi, U.P., India

[2]Department of Civil Engineering, IIT (BHU) Varanasi, U.P., India

*Corresponding author: rssingh.che@itbhu.ac.in

CONTENTS

8.1 INTRODUCTION

Due to the high rate of industrial revolution, some industrial pollutants, such as dyes and heavy metals are the major pollutants that are discharged into the environment [1]. Azo dyes are the oldest man-made dyes that contain ≥ 1 azo bonds (-N=N-) [2,3]. Azo dyes are used in various industries, such as food, textile, paper, cosmetics, and leather [4,5]. It has been estimated that globally, 280,000 t of dyes are released into the environment annually due to dyeing processes [6]. In the textile industries,

DOI: 10.1201/9781003130932-8

it has been estimated that approximately 80% of the total dyes are azo dyes for dyeing purposes, azo dyes are a very mutagenic and recalcitrant category of dye on the commercial scale [7–9].

Azo dyes are electron deficient xenobiotic compounds that have their electron removal groups, which adds electron deficiency to molecules [10]. The release of these synthetic azo dyes in large quantities is a serious concern because it causes various adverse effects on all forms of life [11–13].

Several physical and chemical methods, such as reverse osmosis (RO), ion exchange, advanced oxidation, ozonation, photo-Fenton, coagulation, and adsorption have been used to degrade azo dyes [14–17]. These processes are very expensive and produce a large amount of sludge after treatment [18].

Bioremediation is effective and environmentally friendly because it removes pollutants from the effluents. In addition, the sludge produced is less harmful to the environment [19,20].

8.2 CLASSIFICATION OF DYES

There are several methods to classify dye molecules. They can be classified by chemical structure, color, or application method. Chromophore and auxochrome components play vital roles in dyes molecule [21]. The color of a dye is due to the presence of a chromophores group. Important chromophores are –C=C-, -C=N-, -N=N-, -NO$_2$, -NO, and -C=O. Based on the structure of chromophores, there are 20–30 varieties of dyes. Azo, anthraquinone, phthalocyanine, and triarylmethane dyes are an important group [21–23].

The auxochrome is an electron-withdrawing component that intensifies the color of the chromophore by increasing the overall energy of the electron system. The most common auxochromes are -NH$_3$, -COOH, -OH, NR$_2$, NHR, and -SO$_3$H [22,24].

The majority of commercial important azo dyes belong to the following classes: acid, basic, direct, disperse, reactive, and solvent dyes.

8.3 METHODS FOR THE DECOLORIZATION AND DEGRADATION OF AZO DYES

Various physical, chemical, physicochemical, biological, enzymatic, and a combination of two or more methods have been used for the treatment of industrial effluents, which depend on the type of pollutant to be removed and the quality of the water to be discharged (see Figure 8.1).

8.3.1 Physicochemical Methods for the Removal of Dye from Wastewater

The physicochemical methods used for the treatment of textile effluents include adsorption, ion exchange, coagulation-flocculation, oxidation process, membrane filtration including RO, nanofiltration, electrodialysis. The major drawbacks of the physical methods are that they are only effective when there is a small volume of effluent, and they only transfer the dye molecules to another phase instead of degrading the pollutants [25]. One of the major disadvantages of membrane techniques includes the fouling of membranes, which occurs in a very short time, and therefore, requires the replacement of the membranes, which makes them expensive. Chemical methods have been widely used for the removal of dyes from wastewater; however, several disadvantages are associated with the application of chemical methods. The major drawbacks include cost, sludge accumulation, problems in disposal, and the generation of secondary sludge due to the excessive use of chemicals. Recently, emerging techniques, such as advanced oxidation processes (AOPs) involve the production of strong oxidizing agents, such as hydroxyl radicals (OH), which have been successfully used to degrade pollutants in the wastewater. Despite being effective, these methods are not preferred, since they require a large amount of electrical energy and are expensive, in addition to the consumption of chemicals. The conventional physicochemical processes will be discussed in brief.

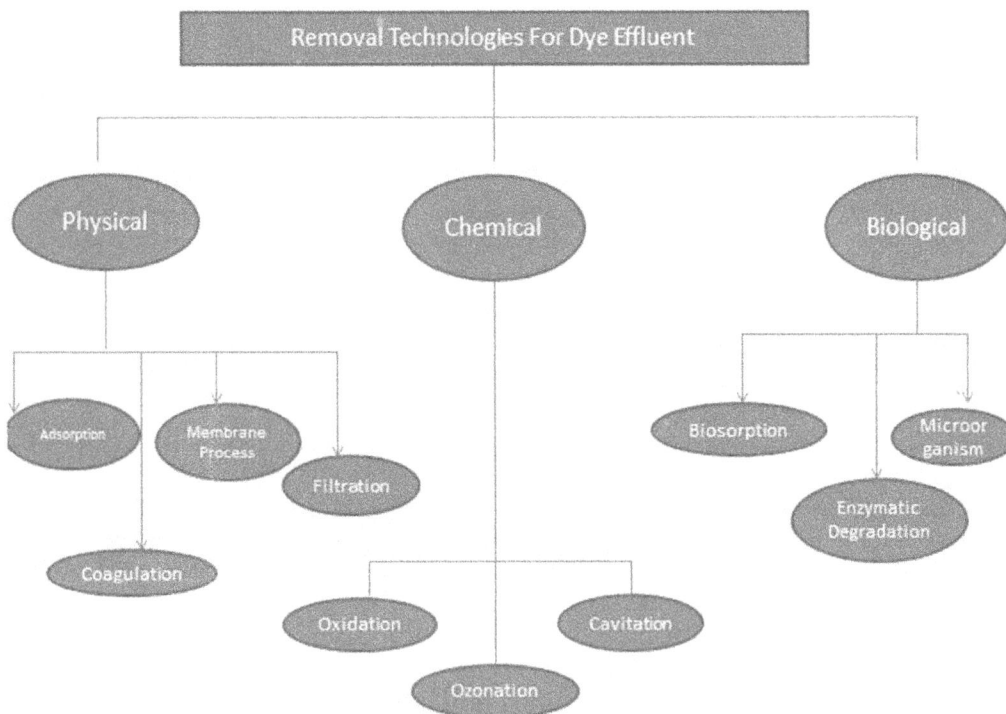

FIGURE 8.1 Removal technologies for dye effluents.

8.3.1.1 Adsorption

Adsorption is a surface phenomenon, which is used to efficiently separate chemical compounds from wastewater by adsorbing organic and inorganic contaminants on the surface of adsorbents. It is an attractive alternative for the remediation of wastewater, mainly if inexpensive sorbents are used and pretreatment is not required. The advantages of adsorption include low initial cost, design simplicity, insensitivity to toxic contaminants, flexibility, and ease of use. During adsorption, transfer of dye molecules occurs at the interface of two immiscible phases that are in contact. One of the most widely used methods for the removal of color includes liquid phase adsorption since it can produce high-quality effluent. The use of granular activated carbon (C) for dye removal is economical and used extensively. The removal of color from wastewater depends on various physicochemical parameters, such as surface area of the sorbent, temperature, particle size, interaction between adsorbate and adsorbent, contact time, and pH [26].

Some adsorbents, such as activated C have been used extensively as adsorbents in wastewater treatment since they are highly porous and have a very large surface area. It has been implemented for the removal of organic pollutants from textile effluents. Adsorption does not generate any harmful products [27].

Various low cost, easily available adsorbents, such as mango tree bark, locust beans bark, and neem tree bark have been studied for the adsorption of dyes from water. Novel materials have frequently been explored for the development of cost-effective and more efficient coagulants and adsorbents. Chemicals, such as ferric sulfate, aluminum chloride, and aluminum sulfate are commonly used in coagulation. Novel natural coagulants have been used as an alternative to inorganic coagulants, for the treatment of textile effluents using a coagulation–flocculation process, because natural coagulants possess numerous advantages over chemical coagulants, such as low toxicity, low cost, less sludge production, and biodegradability [28].

8.3.1.2 Ion Exchange

Ion exchange is a mechanism where mobile ions exchange from an external solution with ions that are electrostatically bound to the functional groups that are present in a solid matrix. Ion exchangers can be naturally occurring, synthetic, or composite. Ion exchange processes are operated in batches, continuous loops, columns, or in combination with membranes. An advanced version of an ion exchange process, which is known as electrochemical ion exchange, is considered to be a hybrid of conventional ion exchange and electrodialysis. It has been used for the treatment of textile wastewaters [29]. Chemical or electrochemical techniques, followed by ion exchange, to recycle textile dye wastewater have been studied. The ion exchange method could be used in the treatment of textile wastewater in combination with other methods, such as biological treatment or electrochemical techniques.

8.3.1.3 Coagulation–Flocculation

Coagulation–flocculation is a physicochemical treatment method, which is used to separate suspended solids from wastewater. Coagulation occurs due to the addition of coagulants, which are classified as primary coagulants (i.e., neutralize the charge on particles and clump together) and coagulant aids (i.e., add density and toughness to flocs). The first step in coagulation is the neutralization of the surface charge, followed by the formation of micro flocs. In the second step, the particle size increases with gentle mixing, as the micro flocs come into contact and are known as flocculation. After the flocs attain optimum size, sedimentation occurs. *Moringa oleifera* is a natural coagulant that has been used in wastewater treatment. *Moringa oleifera* seed has been used to extract to remove azo dyes from textile wastewater and it removed 99% of the dye.

Electrocoagulation is an alternative method for the treatment of wastewater. In this method, coagulating metal cations are released from the anode, which is made from iron or aluminum, when an electric current is applied. The metal plates produce metal hydroxides, which destabilize the colloids and allow them to coagulate.

Electrochemical methods are preferred alternative technologies for the treatment of wastewater, because no chemicals are used. It is similar to an electrocoagulation process, the only difference being the addition of hydrogen peroxide (H_2O_2) before electrolysis. In electrochemical processes, ferrous iron or H_2O_2 are produced, which allows the production of OH radicals. The application of iron in four different processes for the decolorization of wastewater was reported in a recent study. These methods include electro Fenton, peroxicoagulation, electrochemical Fenton, and electrocoagulation [30]. The Fenton process is a chemical oxidation process that occurs through OH radicals and is used for the degradation of refractory chemicals. The oxidation performance of OH radicals is better at lower pH values. In this study, 94.4% decolorization was reported at pH 3.

8.3.1.4 Ozonation and Chemical Coagulation

Dye molecules with a complex structure can be decomposed into smaller molecules using ozonation. Then, the smaller molecules can be degraded biologically using the activated sludge process. This method has been used in the degradation of azo dyes. Chemical methods can be used to oxidize organic compounds in wastewater via the use of oxidizing agents, such as H_2O_2, ultraviolet (UV) light, ozone (O_3), or a combination of oxidants [31].

8.3.1.5 Membrane Processes

Membrane techniques are effective methods, which are used for the removal of larger molecules and ions. In this process, the movement of particles occurs across a permeable or nonpermeable membrane, which is specific to the pollutants to be removed. The commonly used membrane technologies include RO and electrodialysis. RO is used for the removal of total dissolved solids from wastewater. For the removal of dyes from textile wastewater, membranes are used in combination with other techniques, and biological processes and are known as membrane bioreactors. The long term use of membranes for dye removal might result in clogging of the membranes. Fouling of the membranes can be avoided by pretreatment of the water that contains suspended solids, turbidity, trace organics, and colloids. Compared with the chemical treatments for textile wastewater, the key benefit of membrane processes is that they do not use chemicals. Therefore, the quality of the water discharged through the membranes decreases the total performance of the membrane process due to the fouling [32].

8.3.1.6 Cavitation

Cavitation is a physicochemical method that is used for the oxidization of volatile organic compounds in wastewaters. In this method, cavities form in water bubbles through the energy provided by an ultrasonic source. The cavities grow and collapse, which provides the conditions for strong oxidation to occur due to the production of OH radicals and H_2O_2. The bubbles formed by cavitation act as a microreactor for the treatment of organic compounds in textile wastewaters [33]. Cavitation is an AOP and can be used in combination with other methods, such as photo-oxidation, where energy sources are used to produce OH radicals. For textile wastewater treatment, the source of energy can be hydrodynamic, photolytic, or acoustic, which could be used in combination with ultrasonic sources for cavitation. This source can be acoustic, photolytic, or hydrodynamic energy for the treatment of textile wastewater. The increased interaction between dye molecules and OH radicals results in the efficient removal of textile effluents [33].

8.3.1.7 AOP

The conventional treatment processes, such as activated carbon adsorption, solvent extraction, and O_3 oxidation produce a large amount of solid wastes and hazardous byproducts, which adds to the cost of their safe disposal or secondary treatment. Therefore, complete oxidation of the pollutants into nontoxic products is required, which could be achieved through AOPs [34]. AOPs provide a promising alternative for the treatment of textile effluents. The AOPs that are driven by UV make use of UV light and oxidizers, such as O_3 or H_2O_2 to produce OH radicals, which attack the organic compounds at a greater rate. Studies have confirmed that H_2O_2 and UV can be used to degrade dyes even in dilute aqueous solutions, which contain azo dyes at approximately 20 mg/L [35].

Photochemical oxidation has been used as an alternative to conventional treatment methods. This method involves the production of OH radicals, which are highly reactive and can be successfully used in the removal of recalcitrant organic compounds that are present in the textile effluents. In addition, solar radiation can be used for the enhanced production of OH radicals at a lower cost [36].

8.3.1.8 Photocatalytic Degradation

Photocatalytic methods have been used for the degradation of dyes from textile effluents. In a study [37], a commonly used dye in the textile industry, Acid Red 14 (AR14), was photo catalytically degraded using a titanium dioxide (TiO_2) suspension irradiated by a UV-lamp (30 W). TiO_2 and UV light had an insignificant impact. Without TiO_2, photodegradation efficiency was less and without UV, it was negligible. The degree of degradation depends on various parameters, such as pH, initial dye concentration, and the amount of TiO_2 used.

8.3.1.9 Hybrid Processes

The effluents from textile industries contain a wide range of pollutants including nonbiodegradable compounds. The conventional methods used for the degradation of dyes have some advantages and limitations; therefore, hybrid systems have been used to enhance the degradation of pollutants. For example, the coagulation–flocculation process can precipitate some nonbiodegradable components in the textile effluent: however, the sludge produced contains a high amount of chemicals. To overcome this limitation, it could be used in combination with some other techniques.

In a study [38], an AOP was used for the degradation of biodegradable compounds. For the degradation of some of the pollutants, a membrane photoreactor was used, where UV lamps were used as a source of photons and TiO_2 as the catalyst. The bioreactor, when used in combination with photocatalytic reactions, could successfully treat water to consumable levels. The textile effluent characteristics include color, organic matter, low biodegradability, and high alkalinity. To increase biodegradability, an increase in the biological oxygen demand:chemical oxygen demand (BOD_5/COD) ratio is required, which can be achieved through coagulation using ferrous sulfate ($FeSO_4$), followed by biological treatment. For the degradation and removal of recalcitrant pollutants from textile effluents, the application of chemical oxidation is required before or after biological treatment, to enhance the degradability of some compounds. Ozonation has been used combined with anaerobic treatment [39], which resulted in 99% color removal and 85%–90% COD from synthetic wastewater that contained 100–1,000 mg/L azo dye.

8.4 BIOLOGICAL METHODS FOR THE TREATMENT OF DYE EFFLUENTS

Physical and chemical methods such as, adsorption, coagulation, oxidation and ozonation have some disadvantages, which are high cost and a large amount of sludge is generated, which needs to be disposed of. Biological methods appear to be economically feasible and produce environmentally friendly sludge compared with other techniques [40,41]. The bioremediation of textile dyes can be achieved using microorganisms and biocatalyst. Different strains of microorganisms can secrete an enzyme that can remove hazardous pollutants from contaminated sites [42].

8.4.1 BACTERIAL DEGRADATION

The mechanism involved in the bacterial degradation of azo dye includes the cleavage of azo bonds (-N=N-) under anaerobic conditions (see Table 8.1), which results in the formation of a colorless solution that contains potential hazardous aromatic amines, which are then degraded aerobically[43]. Bacterial degradation is faster compared with fungal degradation for the decolorization and mineralization of azo dyes. Microbial consortia are more useful than pure bacterial cultures for their capability to degrade azo dyes [41,44,45]. *Aspergillus oryzae* was used to biosorb reactive dye Procion Red HE7B and Procion Violet H3R at different pH values (2.5, 4.5, and 6.5) [41]. In another study, a consortium of *Pseudomonas sp. SUK1* and *Pseudomonas rettgeri HSL1* were used to degrade different types of azo dye (e.g., Reactive Black 5, Direct Red 81, Disperse Red 78, and Reactive Orange 16) [46]. Actinomycetes can decolorize textile dyes [47]. These are *Streptomyces sp.*, which produce extracellular peroxidase that has a key role in the biodegradation of lignin.

8.4.2 DECOLORIZATION BY FUNGI

Among microorganisms, bacteria and fungi play vital roles in the treatment of dye wastewater. The white-rot fungus *Phanerochaete Chrysosporium* has studied. They produce extracellular lignin enzymes. such as lignin peroxidase, laccase, and manganese peroxidase, which are responsible for the degradation of azo dyes [43,50]. In addition, *Aspergillus niger*, *Rigidoporus lignosus*, and *Tarametes versicolor* have been reported as dye degrading fungi [51,52]. *Penicillium oxalicum SAR* [53] can biodegrade some azo dyes, such as Direct Blue15, Acid Red 183, and Direct Red 75. In another study, a red azo dye was successfully decolorized using *A. niger* pH 9 [54].

TABLE 8.1
Bacterial Degradation of Azo Dyes

Serial Number	Name of Dye	Organism Used	Conditions (pH, temp, time)	References
1	Condo Red	*Bervibacillus parabervis*	7.0, 30, 72	[15]
2	Methylene blue	*Acaligenes faecalis*	7.0, 30, 60	[11]
3	Trypan blue	*Bacillus sp.*	7.0, 35, 48	[48]
4	Reactive red	*E.coli* and *Pseudomonas sp.*	7.0, 28, 25	[49]
5	Acid blue 125	Hybrid adsorbent	5.0, 27.5, 4.0	[16]
6	Acid Red 151	*Aspergillus Fumigatus fresenius*	5.5, 30, 80	[29]

8.4.3 DECOLORIZATION BY ALGAE

Algae are a diverse group of organisms that can photosynthesize. In total, >30 azo compounds were biodegraded by *Chlorella vulgaris*, *Chlorella pyrenoidosa,* and *Oscillateria trnuis* into simpler aromatic amines [55]. In addition, researchers have found that algae can use aniline, which is a degradation product of azo dye breakdown [56,57].

8.4.4 DEGRADATION OF AZO DYES IN A BIOREACTOR

Biodegradation of azo dye can be carried out using aerobic and anaerobic methods or a combination of both. Biodegradation should be operated in specifically designed bioreactors to increase the overall efficiency. Several researchers reported on the different types of bioreactors that are used in the degradation of azo dyes, such as, anaerobic–aerobic sequential bioreactors, fixed-film bioreactors, microbial fuel cells (MFC) bioreactors, air pulsed bioreactors, and hybrid bioreactor[58–60]. Fixed film processes are more suitable for the removal of xenobiotic compounds and offer several advantages over suspended growth reactors [61].

8.4.5 CONVENTIONAL BIOREACTOR

Several types of conventional bioreactors are in use, each has benefits. Combined or sequential bioreactors have been used by several researchers for the degradation of azo dye. These combinations include: (1) anaerobic expanded granular sludge blanket (EGSB) and an aerobic reactor [62]; (2) ozonation and an up-flow biological aerated filtration (UBAF) [63]; and (3) anaerobic biofilter and a fixed-film bioreactor [64]. If only ozonation is applied, it might lead to the production of a carcinogenic byproduct from the wastewater treatment: however, when combined with a biological process these problems diminished. The main benefit of a combined process is that the degradation and decolorization processes occur simultaneously.

Airlift bioreactors have been used by several researchers for microbial processes, because they have some advantages (e.g., mechanically simple, low energy consumption, and low operating cost) [65]. They are not suitable for a viscous liquid, because they do not provide sufficient mixing. A 5 L airlift bioreactor was used to decolorize Indigo carmine. 25 mg/L dye was dissolved in tap water with an immobilized laccase activity of 6×10^4 IU. The optimal air flow rate was 4 L/min, and 100% dye degradation was achieved [66].

Stirred tank bioreactors (STBR) are the most widespread reactors that are used to culture fungi due to several merits (e.g., low operating cost, easy scale-up, and O_2 transfer ability). A laboratory-scale STBR of 5,000 cm^3 capacity operated at room temperature, pH 5.0, and in continuous flow mode was used for the biological removal of Sulfur Black [67]. In total, 84.53% color removal was achieved. Using an AOP in a fluidized bed reactor was beneficial for dye degradation. Due to turbulence and high mass transfer rate, a fluidized bed reactor combined with an AOP is an effective solution, because it increases the reaction rate, has low energy consumption, and has an increased degradation capacity [68].

8.4.6 Hybrid Bioreactor

As mentioned previously, hybrid treatment is more beneficial than a single conventional process. Azo dyes are electron deficient xenobiotic compounds; therefore, they require an organic cosubstrate that can donate an electron to cleave the azo bond. Hybridization could provide an alternative solution since it amplifies microbial activity.

Recently, an anaerobic study was performed using an anaerobic hybrid reactor (e.g., upflow anaerobic sludge blanket (UASB) and aerobic filter) for the biodegradation of methylene blue. They successfully achieved 90% efficiency at 70 mg/L dye [69].

In another study, a hybrid acidogenic bioreactor (HAB) was used to treat an azo dye (acid red) contained in wastewater. A HAB is an acidogenic bioreactor used in conjunction with a biocatalyzed electrolysis module. They reported 93.5% Acid red G (ARG) removal at a hydraulic retention time (HRT) of 6 h [70].

Ultrasonic pretreatment of synthetic wastewater can improve the removal efficiency of a hybrid bioreactor. For a hybrid bioreactor (sequential anaerobic–aerobic) with ultrasonically pretreated wastewater, a maximum of 92% color removal efficiency was obtained for the diazo dyes Reactive Black 5 and Reactive Red [71].

Microbial fuel cells (MFCs) can be used as potential bioreactors for dye degradation, they simultaneously produce electricity with the help of microorganisms [72]. Researchers have used dual-chamber MFCs to generate electricity from the anaerobic treatment of wastewater. For the treatment of azo dyes, the anaerobic–aerobic sequential reactor and an MFC coupled system have been used [73]. Electricity can be produced from azo dye treatment processes. Researchers have constructed an MFC that used an azo dye as the cathode oxidant, which effectively removed Methyl Orange [74].

8.5   CONCLUSIONS

Azo dyes are one of the most widely used classes of dyes in the textile industry. The effluents that are discharged from these industries contain dye residues, which enter the environment and harm the environment and human health. Various physicochemical methods have been applied for the degradation and decolorization of dyes contained in wastewaters. However, the effectiveness of these processes is limited due to the high operating cost, sludge generation, and hazardous byproduct formation. Recently emerging techniques, such as AOP, could be used for the degradation of pollutants. Despite being effective, these methods are not preferred due to their high operating costs. Biological processes could offer a solution to these problems, because they have several advantages over conventional processes (eg. economical operation, environmentally friendly, and no generation of sludge). Different biological processes for the effective degradation of azo dyes have been highlighted in this chapter. In addition, the role of different microorganisms in the degradation of various dyes, under diverse operating conditions have been discussed. The biological methods are environmentally friendly and effective when used in combination with other techniques.

REFERENCES

1.   Lucas-Abell C. Pulsed light for a cleaner dyeing industry: Azo dye degradation by an advanced oxidation process driven by pulsed light. J Clean Prod. 2019; 217:757–66.
2.   Mohammed W, Alabdraba S, Burhan M, Albayati A. Biodegradation of azodyes a review biodegradation of azo dyes: A review. Int J Environ Eng Nat Resour. 2018; 1(4):179–89.

3. Srinivasan S, Sadasivam SK. Exploring docking and aerobic-microaerophilic biodegradation of textile azo dye by bacterial systems. J Water Process Eng. 2018; 22:180–91.
4. Singh RP, Singh PK, Singh RL. Bacterial decolorization of textile azo dye acid orange by *Staphylococcus hominis* RMLRT03. Toxicol Int. 2014; 2:160–7.
5. Chung YC, Chen CY. Mineralization. Degradation of azo dye reactive violet 5 by TiO_2 photocatalysis. Environ Chem Lett. 2009; 7:347–52.
6. Drumond Cheques FM, de Oliveira GAR, Ferraz ERA, Carvalho Cardoso J, Boldrin Zanoni MV et al. Textile dyes: Dyeing process and environmental impact, in Gunay M, editor. *Eco-Friendly Textile Dyeing and Finishing*. IntechOpen. doi: 10.5772/53659.
7. Bhatia D, Sharma NR, Singh J, Kanwar RS. Biological methods for textile dye removal from wastewater: A review. *Crit Rev Environ Sci Technol.* 2017. doi: 10.1080/10643389.2017.1393263
8. Sarkar S, Banerjee A, Halder U, Biswas R. Degradation of synthetic azo dyes of textile industry: A sustainable approach using microbial enzymes. Water Conserv Sci Eng. 2017; 2:121–31.
9. Arul S, Gopal S, Natesan S. Microbial degradation of azo dyes: A review. Int J Curr Microbiol App Sci. 2014; 3:670–90.
10. Singh RL, Singh PK, Singh RP. Enzymatic decolorization and degradation of azo dyes. Int Biodeterior Biodegrad. 2015; 104:21–31.
11. Bharti V. Vikrant K, Goswami M, Tiwari H, Sonwani RK, Lee J. et al. Biodegradation of methylene blue dye in a batch and continuous mode using biochar as packing media. Environ Res. 2019; 171: 356–64.
12. Taylor P, Singh K, Arora S. Removal of synthetic textile dyes from wastewaters: A critical review on present treatment technologies. Crit Rev Environ Sci. 2013; 14:807–78.
13. Asad S, Amoozegar MA, Pourbabaee AA, Sarbolouki MN, Dastgheib SMM. Decolorization of textile azo dyes by newly isolated halophilic and halotolerant bacteria. Bioresour Technol. 2007; 98(11):2082–8.
14. Gutie MC, Lo V. Decolourisation of simulated reactive dyebath effluents by electrochemical oxidation assisted by UV light. 2006; Chemosphere. 62:106–12.
15. Talha A, Goswami M, Giri BS, Sharma A, Rai BN, Singh RS. Bioremediation of Congo red dye in immobilized batch and continuous packed bed bioreactor by *Brevibacillus parabrevis* using coconut shell bio-char. Bioresour Technol. 2017; 252:37–43.
16. Ravikumar K, Pakshirajan K, Swaminathan T, Balu K. Optimization of batch process parameters using response surface methodology for dye removal by a novel adsorbent. Chem Eng J. 2005; 105:131–8.
17. Gupta VK. Application of low-cost adsorbents for dye removal: A review. J Environ Manag. 2009; 90(8):2313–42.
18. Gao Y, Bo Y, Wang Q. Biodegradation and decolorization of dye wastewater: A review. Earth and Environmental Science 2018; 178: 012013 doi :10.1088/1755-1315/178/1/012013.
19. Ledakowicz S. Zylla R, Pazdzior K, Wrebiak J, Sojka-Ledakowicz J. Integration of ozonation and biological treatment of industrial wastewater from dyehouse. Ozone Sci Eng. 2017; 39(5):357–65.
20. Sarker R, Chowdhury M, Deb AK. Reduction of color intensity from textile dye wastewater using microorganisms: A review. J Curr Microbiol Appl Sci. 2019; 8(2):3407–15.
21. Al Prol AE. Study of environmental concerns of dyes and recent textile effluents treatment technology: A review. Asian J Fish Aquat Res. 2019; 3(2):1–18.
22. Mohammad S, Azeez A. HPLC determination of four textile dyes and studying their degradation using spectrophotometric technique [Masters dissertation]. An-Najah National University, Palestine; 2005.
23. Benkhaya S, El Harfi S, El Harfi A. Classifications, properties and applications of textile dyes: A review. Appl J Environ Eng Sci. 2017; 3: 311–20.
24. Amran M, Salleh M, Khalid D, Azlina W, Abdul W, Idris A. Cationic and anionic dye adsorption by agricultural solid wastes: A comprehensive review. Desalination. 2011; 280(1–3):1–13.
25. Nigam TRP, Marchant IMBR. Microbial decolourisation and degradation of textile dyes. Appl Microbiol Biotechnol. 2001; 56(1–2):81–7.
26. Rupainwar DC. A comparative evaluation for adsorption of dye on Neem bark and Mango bark powder. Indian J Chem Technol. 2011; 8 :67–75.
27. Rashed MN. Adsorption technique for the removal of organic pollutants from water and wastewater. Organic pollutants – Monitoring, risk and treatment. London: IntechOpen; 2013.

28. Renault F, Sancey B, Badot P, Crini G. Chitosan for coagulation/flocculation processes: An eco-friendly approach. Eur Polym J. 2009; 45(5):1337–48.
29. Sharma P, Singh L, Dilbaghin N. Response surface methodological approach for the decolorization of simulated dye effluent using *Aspergillus fumigatus fresenius*. J Hazard Mater. 2009; 161:1081–6.
30. Ghanbari F, Moradi M. Textile wastewater decolorization by zero valent iron activated peroxymonosulfate: Compared with zero t copper. J Environ Chem Eng. 2014; 2:1845–51.
31. Choi J, Song HK, W. Lee, Koo K, Han C, Na B. Reduction of COD and color of acid and reactive dye-stuff wastewater using ozone. Koream J Chem Eng. 2004; 21(2):398–403.
32. Amini M, Arami M, Mohammad N, Akbari A. Dye removal from colored textile wastewater using acrylic grafted nanomembrane. DES. 2011; 267(1):107–13.
33. Reddy DR, Dinesh GK, Anandan S, Sivasankar T. Process intensification sonophotocatalytic treatment of Naphthol Blue Black dye and real textile wastewater using synthesized Fe doped TiO_2. Chem Eng Process. 2016; 99:10–18.
34. Mahadwad O, Parikh P, Patil C, Jasra RV. Photocatalytic degradation of Reactive Black-5 dye using TiO_2 impregnated ZSM-5. Bullet Mater Sci. 2014; 34: 551–556.
35. Galindo C, Kalt A. UV ± H_2O_2 oxidation of monoazo dyes in aqueous media: A kinetic study. Dyes Pigm.1999; 40(1):27–35.
36. Oller I, Malato S, Sanchez-Perez JA. Combination of advanced oxidation processes and biological treatments for wastewater decontamination: A review. Sci Total Environ. 2011; 409(20):4141–66.
37. Daneshvar N, Salari D, Khataee AR. Photocatalytic degradation of azo dye acid red 14 in water: investigation of the effect of operational parameters. J Phytochem Photobiol A: Chemistry. 2003; 157:111–16.
38. Lee K, Beak H, Choo K. Membrane processes. Water Environ Res. 2015;88:1050–1124.
39. Punzi M, Nilsson F, Anbalagan A, Svensson BM, Jonsson K, Mattiasson B, et al. Combined anaerobic–ozonation process for treatment of textile wastewater: Removal of acute toxicity and mutagenicity. Hazard Mater. 2015; 292:52–60.
40. Ali N, Ikramullah G, Lutfullah A, Hameed A, Ahmed S. Decolorization of Acid red 151 by *Aspergillus niger* SA1 under different physicochemical conditions. World J Microbiol Biotechnol. 2008; 24(7):1099–105.
41. Corso CR, Carolina A, De Almeida M. Bioremediation of dyes in textile effluents by *Aspergillus oryzae*. Microbial Ecol. 2009; 57(2):384–390.
42. Xiang X. Anaerobic digestion of recalcitrant textile dyeing sludge with alternative pretreatment strategies. Bioresour Technol. 2016; 222: 252–260.
43. Saratale RG, Saratale GD, Chang JS, Govindwar SP. Bacterial decolorization and degradation of azo dyes: A review. J Taiwan Inst Chem Eng. 2011; 42(1):138–57.
44. Shinkafi MS, Mohammed IU, Hayatu JM, Audu AA. Microbial biotechnology for the decolourization and mineralization of organic components of textile wastewater by single and mixed microbial consortium isolated from effluent treatment plant of African Textiles Industry Kano, Nigeria. J Environ Sci Toxicol Food Tech. 2016; 10(4):32–9.
45. Ahmad A, Mohd-Setapar SH, Chuong S, Khatoonm A. Recent advances in new generation dye removal technologies: Novel search for approaches to reprocess wastewater. RSC Adv. 2015; 30801–18.
46. Lade H, Kadam A, Paul D, Govindwar S. Biodegradation and detoxification of textile azo dyes by bacterial consortium under sequential microaerophilic/aerobic processes. EXCLI J. 2015; 14:158–74.
47. Robinson T, Mcmullan G, Marchant R, Nigam P. Remediation of dyes in textile effluent: A critical review on current treatment technologies with a proposed alternative. Biosour Technol. 2001; 77(3): 247–55.
48. Jeevitha P, Manjula D, Ramya I, Hemapriya J. Bioremediation and detoxification of Trypan Blue by *Bacillus sp.* isolated from textile effluents. Int J Curr Microbiol App Sci. 2018; 7(7):4381–91.
49. Chen J, Lin Y. Decolorization of azo dye by immobilized *Pseudomonas luteola* entrapped in alginate–silicate sol–gel beads. Process Biochem. 2007; 42:934–42.
50. Gou M, Qu Y, Zhou J, Ma F, Tan L. Azo dye decolorization by a new fungal isolate, *Penicillium sp. QQ* and fungal-bacterial cocultures. J Hazard Mater. 2009; 170:314–19.
51. Nilsson I, Mattiasson B, Rubindamayugi MST, Welander U. Decolorization of synthetic and real textile wastewater by the use of white-rot fungi. Enzyme Microb Technol. 2006; 38:94–100.
52. Karimi A, Vahabzadeh F, Bonakdarpour B. Use of *Phanerochaete chrysosporium* immobilized on Kissiris for synthetic dye decolourization: involvement of manganese peroxidase. World J Microbiol Biotechnol. 2006; 22: 1251–1257.

53. Saroj S, Kumar K, Pareek N, Prasad R, Singh RP. Biodegradation of azo dyes Acid Red 183, Direct Blue 15 and Direct Red 75 by the isolate *Penicillium oxalicum* SAR-3. Chemosphere. 2014; 107:240–8.
54. Mahmoud MS, Mostafa MK, Mohamed SA, Sobhy NA, Nasr M. Bioremediation of red azo dye from aqueous solutions by *Aspergillus niger* strain isolated from textile wastewater. J Environ Chem Eng. 2017; 5:547–54.
55. Daneshvar M. Ayazloo A, Khataee R, Pourhassan M. Biological decolorization of dye solution containing Malachite Green by microalgae *Cosmarium sp.* Bioresour Technol. 2007; 98:1176–82.
56. El-Sheekh MM, Gharieb MM, Abou-el-Souod GW. Biodegradation of dyes by some green algae and cyanobacteria. Int Biodeterior Biodegrad. 2009; 63(6):699–704.
57. Jinqi L, Houtian L. Degradation of azo dyes by algae. Environ Pollut. 1991; 75:273–8.
58. Vikrant K. Recent advancements in bioremediation of dye: Current status and challenges. Bioresour Technol. 2018 Nov; 253:355–67.
59. Balapure K, Bhatt N, Madamwar D. Mineralization of reactive azo dyes present in simulated textile waste water using down flow microaerophilic fixed film bioreactor. Bioresour Technol. 2015; 175:1–7.
60. Van Der Zee FP, Villaverde S. Combined anaerobic-aerobic treatment of azo dyes: A short review of bioreactor studies. Water Res. 2005; 39:1425–40.
61. Keharia H, Madamwar D. Bioremediation concepts for treatment of dye containing wastewater: A review. Indian J Exp Biol. 2003; 4:1068–75.
62. Tan NCG, Borger A, Slenders P, Svitelskaya A, Lettinga G, Field JA. Degradation of azo dye Mordant Yellow 10 in a sequential anaerobic and bioaugmented aerobic bioreactor. Water Sci Technol. 2018; (42):337–344.
63. Lu X, Yang B, Chen J, Sun R. Treatment of wastewater containing azo dye reactive brilliant red X-3B using sequential ozonation and upflow biological aerated filter process. J Hazard Mater. 2009; 161(1): 241–245.
64. Spagni A, Grilli S, Casu S, Mattioli D. Treatment of a simulated textile wastewater containing the azo-dye reactive orange 16 in an anaerobic-bio film anoxic aerobic membrane bioreactor. Int Biodeterior Biodegrad. 2010; 64(7):676–81.
65. Sodaneath H, Lee J, Yang S, Jung H, Ryu HW, Cho K. Toxic/hazardous substances and engineering decolorization of textile dyes in an air-lift bioreactor inoculated with *Bjerkandera adusta.* J Environ Sci Health A. 2017; 52(11):1099–11.
66. Teerapatsakul C, Parra R, Keshavarz T, Chitradon L. Repeated batch for dye degradation in an airlift bioreactor by laccase entrapped in copper alginate. Int Biodeterior Biodegrad. 2017; 120:52–7.
67. Andleeb S, Atiq N, Ali MI, Shafique M. Biological treatment of textile effluent in stirred tank bio-reactor. Int J Agric Biol. 2010; 12:256–60.
68. Farshchi ME, Aghdasinia H, Khataee A.Modeling of heterogeneous Fenton process for dye degradation in a fluidized-bed reactor: Kinetics and mass transfer. J Clean Prod. 2018; 182:644–53.
69. Farooqi IH, Basheer F. Biodegradation of methylene blue dye by sequential treatment using anaerobic hybrid reactor and submerged aerobic fixed film bioreactor. J Inst Eng Ser A. 2017; 98(4):397–403.
70. Wang H, Cheng H, Wang S, Cui D, Han J. Efficient treatment of azo dye containing wastewater in a hybrid acidogenic bioreactor stimulated by biocatalyzed electrolysis. J Environ Sci. 2015; 39:1–10.
71. Verma AK, Nath D, Bhunia P, Dash RR. Application of ultrasonication and hybrid bioreactor for treatment of synthetic textile wastewater. J Hazard Toxic Radio Waste. 2017; 21(2) 3–10.
72. Fang Z, Song H, Cang N, Li X. Electricity production from azo dye wastewater using a microbial fuel cell coupled constructed wetland operating under different operating conditions. Biosens Bioelectron. 2014; 68:135–41.
73. Li Z, Zhang X, Lin J, Han S, Lei L. Azo dye treatment with simultaneous electricity production in an anaerobic-aerobic sequential reactor and microbial fuel cell coupled system. Bioresour Technol. 2010; 101(12):4440–5.
74. Liu L, Li F, Feng C. Microbial fuel cell with an azo-dye-feeding cathode. Appl Microbiol Biotechnol. 2009; 85(1):75–83.

9 Lignolytic Enzymes and Their Role in the Bioremediation of Environmental Pollutants
Prospects and Challenges

Ajay Kumar Chauhan[1] and Bijan Choudhury[1]*
[1]Department of Biological science and Bioengineering Indian Institute of Technology, Roorkee, Uttarakhand 24667, India
*Corresponding author: bijan.choudhury@bt.iitr.ac.in

CONTENTS

9.1 INTRODUCTION

Recalcitrant xenobiotic chemicals in industrial effluents cause severe health hazards and environmental pollution, which are currently a major concern. These effluents contain various organic and inorganic wastes that are the major environmental pollutant (Bharagava, Saxena, Mulla, & Patel, 2018). In general, the organic pollutants in effluents contain: azo dyes, phenols, various pesticides, polyaromatic hydrocarbons (PAH), chlorinated and polychlorinated phenols (Zango et al., 2020). The organic content in wastes can be degraded by microorganisms; however, a high concentration in the waste stream leads to poor biodegradation. However, the inorganic content in textile, pulp and paper, tanneries, and pharmaceuticals waste effluent is composed of various toxic heavy metal ions such as; chromium (Cr), arsenic (Ar), cadmium (Cd), mercury (Hg), lead (Pb), manganese (Mn), Copper (Cu), and sodium (Na) (Aprianti, Miskah, Selpiana, Komala, & Hatina, 2018; Yaseen &

Scholz, 2019). The severe impact posed by the organic and inorganic content in effluents has a major effect on the environmental safety of all living organisms. Therefore, a suitable treatment method for industrial effluents needs to be developed that is environmentally friendly, cost-competitive, and sustainable.

Synthetic dyestuff that is released from various industries, such as textile, food, leather, paper, pharmaceutical and cosmetics causes various health hazards. According to the World Health Organization (WHO) data revealed that 17%–20% of industrial water pollution contained synthetic dyes (80% azo dyes) from textile industry effluents (Sudha, Saranya, Selvakumar, & Sivakumar, 2014). These synthetic dyes are mainly azo and have a chemical structure with ≥1 azo group (-N=N-) (Chang, Kuo, Chao, Ho, & Lin, 2000). Azo dyes contain benzene or naphthalene groups with different substituents such as, methyl, carboxyl, nitro, amino, hydroxyl, carboxyl, and chloro, which gives different types of azo dyes (Langhals, 2004). In addition, industrial effluents from pharmaceutical industries generally contain various types of carbamazepine, which was treated and converted into lower molecular compounds by lignolytic enzymes (Naghdi et al., 2018a). Wastewaters from various industries contain aromatic compounds, synthetic dyes, hydrocarbons, pesticides, and endocrine-disrupting substances. Various conventional techniques (e.g., filtration, precipitation, coagulation, adsorption, and ozonation) are available to treat these recalcitrant pollutants. However, secondary sludge formation and the generation of secondary waste are major limitations. In contrast, suitable bioremediation processes solve the problem of secondary waste generation and sustainably clean the environment. The bioremediation of recalcitrant material is dependent on microbial or enzymatic processes to degrade, detoxify, and biotransform pollutants into less toxic forms. Bioremediation that uses the microbial population is an environmentally friendly but long process. Recently, enzymes from various microbial species have been discovered, which are involved in the degradation of xenobiotic compounds. For effective bioremediation either the environmental conditions must favor microbial growth and activity, or a process could be designed that allowed microbial growth and degradation of waste compounds. Lignolytic enzymes are well known in the treatment of synthetic dyes and effluents: they gained attention due to the formation of low molecular weight and less toxic compounds. In addition, the similarity between the aromatic structural configuration of synthetic dyes, lignin, pharmaceutical waste, and coal means that lignolytic enzymes can degrade them into simpler forms. Enzymatic degradation is influenced by various physicochemical parameters, such as pH, solvent, metal ions, and temperature.

Therefore, due to cost-competitiveness, lignolytic enzymes need to be produced economically and a novel bioprocess technique needs to be developed that effectively bioremediates pollutants from the waste streams under operational conditions. Therefore, research in bioprocess technology could significantly reduce toxic pollutants into useful products. The chapter will provide descriptive information on the lignolytic enzyme-mediated bioremediation of environmental pollutants, research opportunities, prospects, challenges, and future outlook.

9.2 MICROBIAL ENZYMES INVOLVED IN BIOREMEDIATION

Enzymes are biological catalysts that convert substrates into desired products under favorable conditions. These enzymes are composed of protein and glycoproteins that contain polypeptides. The active site in the enzymatic subunit performs the catalytic activity. Various hydrolytic enzymes, such as lipase, protease, and cellulase can treat polluted soil, industrial wastes, and hydrocarbons from petrochemical industries. In addition, other than hydrolytic enzymes (lignolytic enzymes) are used in bioremediation. The enzymatic treatment provides bioremediation technologies in a safe and environmentally friendly manner. These lignolytic enzymes are produced by various species of plants (*Acorus calamus, Carica papaya,* and *Musa paradisiaca*), bacteria (*Bacillus sp., Pseudomonas sp.,* and *Citrobacter sp.*), fungi (*Streptomyces sp., Basidiomycete sp.,* and *Phanerochaete chrysosporium*) (Bansal & Kanwar, 2013). Wide applications of lignolytic enzymes achieve biodegradation of various environmental pollutants makes a sustainable form of waste valorization. These lignolytic enzymes are majorly produced of white-rot fungus (WRF). Lignolytic enzymes can degrade synthetic dyes, lignin from lignocellulosic biomass (LCB), and bioremediate various industrial effluents. In enzymatic catalysis, the reaction system is more complicated and depends on the various physicochemical process parameters, such as pH, temperature, and inhibitors. However, due to cost-competitiveness, the extremophiles perform better under extreme process conditions. These extremophile originated lignolytic enzymes can separately or synergistically remove lignin from LCB and degrade various similar aromatic structures. Enzymes were produced through solid-state/submerged fermentation (SSF/SmF), such as peroxidases [e.g., laccase and manganese peroxidase (MnP)], and lignin peroxidases (LiP) to remove lignin components) to produce biofuel (Paramjeet, Manasa, & Korrapati, 2018).

9.3 LIGNOLYTIC ENZYMES

Lignolytic enzymes are ubiquitous groups of enzymes widely produced by different microbes, such as bacteria, archaea, fungi, insects, and plants. These lignolytic enzymes are used in lignin depolymerization and synthetic dye decolorization. WRF is the primary producer of lignolytic enzymes (Torres-Farradá et al., 2017). In this study, a fungal strain *Basidiomycete sp.* PV 002, produced laccase and MnP to degrade Ranocid Fast Blue by 96% and Acid Black 210 by 70%, on days 5 and 9 respectively. In another study, recombinant lignolytic enzymes (LiP, MnP, and laccase) from *Pichia pastoris* were used for the degradation and detoxification of various azo dyes (e.g., Congo red, bromophenol blue, malachite green and methyl orange) (Liu, Xu, Kang, Xiao, & Liu, 2020). In another study, lignolytic enzymes of nonhalophilic origin did not survive under elevated pH, temperature, and salinity, which are major limitations in industrial applications. In a previously reported study, laccase of WRF (*Tramates villosa, Tramates versicolor,* and *Lentinus edodes*) significantly reduced effluent color by 40%–44%, total phenols by 30%–51%, and chemical oxygen demand (COD) by 37%–43% (Souza, Souza, Silva, & Paiva, 2014). However, industrial effluent waste streams consist of a high concentration of inorganic salts that provides favorable conditions for the growth of halophiles and the activation of their enzymes. Therefore, in previous reports, a strain of *Klebsiella pneumonia* produced more stable forms of lignolytic enzymes (e.g., MnP and laccase) (Gaur, Narasimhulu, & Y, 2018). In another report, laccase from *Chromohalobacter sp.* had optimal activity at pH 8, 45°C, and at 3 M NaCl (Rezaei, Shahverdi, & Faramarzi, 2017). A halophilic strain *Haloferax volcanii* that produces laccase showed stable activity at 0.1 M–1.4 M NaCl, 55°C and depolymerized organic substances, such as 2,2'- azino-bis(3-ethylbenzothiazoline-6-sulfonic acid (ABTS), 2,6dimethoxyphenol (DMP), and syringaldazine (Uthandi, Saad, Humbard, & Maupin-Furlow, 2010).

9.3.1 LACCASES

Laccases are polyphenol oxidases. Their structure contains at least four Cu atoms in their catalytic sites. Therefore, laccase is a metalloprotein, which is known as a multicopper oxidase (Baldrian & Valášková, 2008). Laccase (EC 1.10.3.2) are mainly used for the degradation of a wood structure known as lignin (Morozova, Shumakovich, Gorbacheva, Shleev, & Yaropolov, 2007). The structures of lignin are composed of high molecular weight phenylpropanoid alcohol units of monolignols [e.g., p-coumaryl (H type), coniferyl (G type), and sinapyl alcohol (S type)]. The metallic components' presence acts as a substrate for laccase; these metal ions are iron (Fe^{2+} to Fe^{3+}) and Mn (III). Laccase is a widely distributed enzyme found in plants, some bacteria, and fungi. In plants, laccases are present intracellularly and help in the formation of lignin, which provides rigidity to the plant (Baldrian & Valášková, 2008). In previous research, intracellular laccase that was produced from various strains of fungi and bacteria transformed the aromatic compounds (low molecular weight) present in plant cell lignin (Janusz et al., 2017). In addition, laccase production for biotechnological processes has received attention due to its extracellular production, which can be induced, does not need any cofactors for its conversion, and has low substrate specificity (Plácido & Capareda, 2015). Laccase production can be induced using Cu, aromatic compounds such as dyes, and another recalcitrant compound (lignin) (Minussi, Pastore, & Durán, 2007). In a previous reported study, the bacterial laccase from *Streptomyces ipomoeae* CECT degraded Reactive Black 5 by 94%, Acid black by 22%, Acid orange 63 by 14%, and Orange II by 92% at neutral pH and 60°C (Blánquez et al., 2019). In another study, soil contaminants, such as 2,4-dichlorophenol were bioremediated by biochar-immobilized laccase, and 65% degradation was achieved (Wang et al., 2019). Therefore, laccase could degrade various aromatic structures that are present in various industrial effluents that originate from textiles, pulp and paper, soil contaminants, and pharmaceuticals wastes. In addition, it could be used in coal biosolubilization to improve the quality of coal and the pretreatment of LCB that originates from agro and municipal wastes, and various woods for biofuel generation. The major advantage of laccase is that it does not add any additional cost for a cosubstrate for enzymatic biocatalysis and its attraction toward substrate depolymerization helps to increase substrate accessibility.

9.3.2   LiP

LiP (EC 1.11.1.14) was first discovered in a carbon and nitrogen source limited culture of *P. chrysosporium* (Tien & Kirk, 1984). LiP is known as a diaryl propane oxygenase, which contains Fe as a prosthetic group. LiP has a high redox potential (1.4 V) and oxidizes a wide range of nonphenolic structures in lignin and similar types of structures that are present in other types of waste (Martínez et al., 2005). In addition, it oxidizes various aromatic phenolic compounds by using hydrogen peroxide (H_2O_2)as a cosubstrate (Baciocchi, Fabbrini, Lanzalunga, Manduchi, & Pochetti, 2001). In some studies, the high redox potential of approximately 0.7–1.4 V made LiP capable of oxidizing substrates, which were not oxidized by other types of peroxidases. Recently, LiP from *P. chrysosporium* was used for the successful degradation of various azo dyes (e.g., Amido black 10 B, Evans blue, and Guinea green) (Ilić Đurđić et al., 2020). In general, LiP mediated reactions require veratryl alcohol (VA) as a mediator to increase the rate of the reactions and form products faster. However, the major drawback of LiP is that it does not work well with high molecular weight compounds during degradation, and the use of LiP causes the repolymerization of the substrates. Therefore, LiP catalyzed reactions use a capping agent to overcome repolymerization, which prevents the formation of undesirable degraded products from side reactions. In addition, a higher cosubstrate concentration leads to the inactivation of LiP mediated degradation reactions and significantly contributes to the release of VA radicals that are responsible for veratraldehyde formation and release from the enzyme binding site. The use of H_2O_2 as the cosubstrate added extra costs to degradation. At the molecular level, LiP mediated catalytic oxidation forms aryl cation radical (VA*) for a very short time, which unable to degrade high molecular weight of compounds. Therefore, its essentially required to maintain both LiP* and VA* radicals during enzymatic catalysis.

9.3.3   MnP

MnP was first discovered in 1984 in batch cultures of *P. chrysosporium* (Kuwahara, Glenn, Morgan, & Gold, 1984). MnP (EC 1.11.1.13) is known as the lignin modifying enzyme, which is composed of glycoproteins that have a molecular weight of 38–62.5 kDa and are made up of approximately 350 amino acid residues (43% similarity sequence with LiP). However, one of the disulfide bridges participates in the Mn-binding site, which helps during the enzyme-catalyzed degradation of phenolic substrates. Therefore, this important characteristic of MnP differentiates it from other peroxidases (Sundaramoorthy, Kishi, Gold, & Poulos, 1994). In the MnP catalyzed reactions Mn^{3+} ions are a substrate to degrade phenolic substrate. Similar to LiP catalyzed reactions, MnP requires a cosubstrate to start the MnP catalyzed reactions.

9.3.4 VERSATILE PEROXIDASE

Versatile peroxidase (VP) is a newly discovered lignolytic enzyme. VP (EC 1.11.1.16) is composed of the molecular architecture of LiP and MnP, which typically oxidize phenolic components (VA), and nonphenolic model lignin compounds (Hofrichter, Ullrich, Pecyna, Liers, & Lundell, 2010). VP is composed of glycoproteins, which have a molecular mass of 40–45 kDa with an isoelectric point of 3.4–3.9 (Mester & Field, 1998). However, the Mn^{2+} binding site is structurally similar to MnP (Pérez-Boada et al., 2005).

9.3.5 DYE OXIDIZING ENZYMES

These enzymes catalyze the oxidation reaction of various organic and inorganic substrates by utilizing H_2O_2 as a cosubstrate. These dye oxidizing enzymes are Dyp-A type and peroxidase Dyp-B type peroxidases (EC 1.11.1.19), and haloperoxidases (Torres-Farradá et al., 2017).

9.4 INDUSTRIAL WASTE CHARACTERISTICS AND THE ROLE OF HALOPHILIC LIGNOLYTIC ENZYMES IN THEIR VALORIZATION

Lignolytic enzymes have wide applications and can be used to treat phenolic waste present in the effluent that is generated from various industries. For instance, carbamazepine (a pharmaceutically active compound) that is present in pharmaceutical wastewater was suitably degraded ≤95% into less toxic low molecular weight compounds 10,1-dihydro-10,11-dihydroxy-carbamazepine and 10,11-dihydro-10,11-epoxy-carbamazepine using 60 IU/L laccases in the presence of 18 µM ABTS at pH 6 and 35°C (Naghdi et al., 2018a). In another study, LiP (Sigma Aldrich, MO) was used for the degradation of carbamazepine (60%±8%), diclofenac (59%±8%), and paracetamol (9%±5%) at pH 5, 55°C, within 72 h (Pylypchuk, Daniel, Kessler, & Seisenbaeva, 2020). Naghdi et al. (2018b) reported that various oxidoreductase enzymes (LiP, MnP, laccase, VP, and Dyp) could degrade various pharmaceutical compounds present in wastewater (e.g., acetaminophen, diazepam, and sulfapyridine) (Naghdi et al., 2018b). Industrial effluents have various characteristics and chemical compositions that depend on the raw material that was processed. The textile industry, pulp, paper mill, tanneries, and molasses-based distillery waste streams contain a high concentration of various metals and salts, which means that haloalkaliphiles are the most suitable for their treatment (Table 9.1) (Amoozegar, Safarpour, Noghabi, Bakhtiary, & Ventosa, 2019). The presence of a high NaCl concentration in pulp and paper mill waste could play an essential role in supporting the haloalkaliphile growth, which could be applied to biobleaching and biopulping. In addition, haloalkaliphiles can be used to remove phenol from the waste stream, and they reduce biological oxygen demand (BOD) and COD in the industrial effluent.

9.4.1 TEXTILE INDUSTRIES

Textile industries discharge problematic groups of synthetic dyes in their effluent, which are not easily biodegradable. Synthetic dyes cause severe health hazards and affect aquatic life. Water is consumed by humans for drinking, food preparation, and household purposes. Therefore, the presence of synthetic dye in water causes mutagenic activity, which results in cancer or even death. Textile industries usually operate at an elevated pH, high concentrations of salt (2.5–3.5 M), and higher temperatures (e.g., washing at 60°C–70°C, textile desizing at 80°C–90°C, and starch gelatinization at 100°C) (Silva, Martins, Jing, Fu, & Cavaco-Paulo, 2018). In addition, at a sodium salt concentration >3 g/L, the presence of metal ions, synthetic dyes, and organic solvents adversely affects fungal-originated enzymes (Silva et al., 2018). The enzymes derived from extremophiles, such as haloalkaliphiles, can be useful as substitutes for lignolytic enzymes that are derived from other

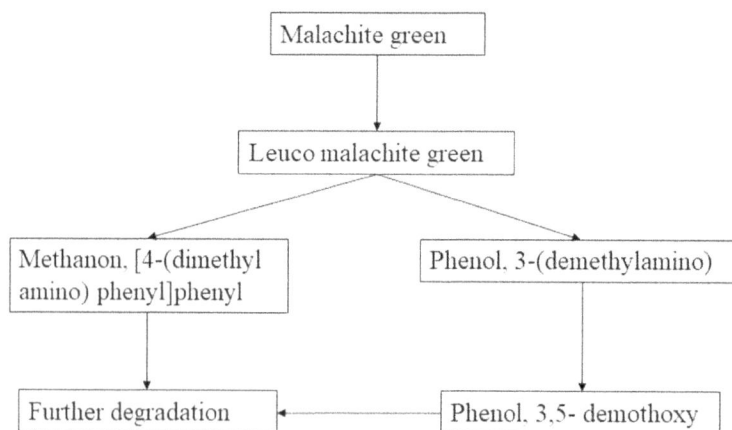

FIGURE 9.1 Microbial biodegradation pathway of MG.

microbial species. In a study, a *Pichia kudriavzevii* strain, which is a halotolerant strain, reported the microbial degradation of Red B ≤94.2% at 100 mg/L in 35 h (Feng, Fang-yan, & Yu-bin, 2014). In another study on halophilic archaea, *Halogeometricum borinquese sp.* degraded Remazol black B (400 mg/L) up to more than 70% at pH 7.0, 45°C and 2.5–3.4 M NaCl (Kiadehi, Amoozegar, Asad, & Siroosi, 2018). These studies demonstrated the potential of extremophiles for the degradation of synthetic dyes.

Azo dyes are a major textile industrial effluent and have severe health and environmental impacts. Therefore, the biological degradation method is preferred over physiochemical methods due to their sustainability and environmentally cleaner approach. Chaturvedi et al. (2015) first reported the degradation of triphenylmethane (MG), a cationic azo dye using *Ochrobactrum pseudogrignonense* (Chaturvedi & Verma, 2015). In addition, 400 ppm MG was decolorized within 96 h in the presence of 50 mM $CuSO_4$ in a supported minimal medium, and the degradation pathway followed is shown in Figure 9.1. *Basidiomycete sp.* PV 002 decolorize Ranocid Fast Blue by 96% (day 5) and Acid Black by 70% (day 9) using lignolytic enzymes and microbial utilization. In addition, it was very difficult to predict whether decolorization (of azo dyes) was achieved solely through microbial utilization or enzymatic degradation (Verma & Madamwar, 2005). Various dye decolorization studies by lignolytic enzymes are listed in Table 9.2.

9.4.2 PULP AND PAPER MILL

Pulp mill wastewater contains black liquor components and chlorinated aromatic compounds in their waste streams. These chlorinated organic compounds are mainly organic halides, such as tetrachlorocatechols, tetrachloro guaiacols, and pentachlorophenol, and have been treated using fungal lignolytic enzymes (Ellouze & Sayadi, 2016). Similar to textile operational conditions, pulp bleaching and washing occur at 55°C–70°C, combined with the use of a high concentration of salt (2.5–3.5 M) (Silva et al., 2018). In addition, waste streams from pulp and paper industries are enriched with waste wood fiber, aromatic lignin, and various chlorinated toxic components (Karn, Chakrabarty, & Reddy, 2010). Therefore, it provides favorable conditions for halophiles to grow and supplies all the necessary conditions that are required to activate their enzymes (Table 9.1). In a previously reported study, *H. volcanii,* a halophilic archaeon that produces laccase, was stable at 1.4 M NaCl at 55°C and degraded various organic substances (Uthandi et al., 2010). In another study, the laccase from *Chromohalobacter sp.* showed their optimal activity at 3 M NaCl, 45°C, and pH 8 (Rezaei et al., 2014).

TABLE 9.1
Characteristics of Various Industrial Wastes

Parameters	Textile Wastewater (Furusjö & Pettersson, 2016)	Black Liquor and Wastewater from Pulp Mill (Mehmood et al., 2019; Yaseen & Scholz, 2019)	Tanneries Wastewater (Chowdhury, Mostafa, Biswas, Mandal, & Saha, 2015)
Color	Dye 10–70 (mg/L)	Brownish	Yellowish brown
pH	6–10	12.5	7.5
Temperature (°C)	35–45	35–50	34
Total suspended solids (ppm)	24.9–3,950	784	6,880
Total dissolved solids (ppm)	90.7–5,980	2710	8,100
Total solids (ppm)		3494	9,340
Alkalinity (ppm)		NA	1,900
BOD_5 (ppm)	80–6,000	975	920
COD (ppm)	150–12,000	2820	3,980
SO_4^- (ppm)	NA	NA	4,000
Cl⁻ (ppm)	1,000–6,000	1900	5,000
Na⁺ (ppm)	7,000	206000	3,840
Zn, Ni, Mn, Fe, Cu, B, Ar, Hg, Fl, Cr (ppm)	10	Pb, Cd, Cu, Zn 7.7 mg/g of lignin	NA

9.4.3 TANNERIES

Tanneries provide suitable conditions for the growth of halophiles and the activation of their enzymes (Table 9.1). Various salt-tolerant bacteria have been isolated from tannery wastewater, such as *Bacillus flexus, Staphylococcus aureus, Exiguobacterium homiense,* and *Pseudomonas aeruginosa,* which grew at NaCl 8% (w/v), 37°C, pH 8, and degraded tannery wastewater and reduced COD by 80% (Sivaprakasam, Mahadevan, Sekar, & Rajakumar, 2008).

9.5 OTHER BIOREMEDIATION APPLICATIONS FOR LIGNOLYTIC ENZYMES

9.5.1 COAL DEPOLYMERIZATION

Brown coal, which is known as lignite coal, could be solubilized using lignolytic enzymes. After biosolubilization of brown coal, the efficiency significantly improved. According to previous studies, the calorific value of brown coal is low, and after combustion, it produces nitrogen and sulfur oxide gases (NOx and SOx) (Igbinigie et al., 2008). By liquefaction, coal quality can be improved, and it can be converted into cleaner fuel, which is an energy-intensive process. Therefore, another alternate and sustainable biological approach is required. The biosolubilization of coal yields humic and fulvic acids and other organic compounds as primary products. The potential applications for biosolubilized products were sorbents, which were used to remove heavy metals. *Fusarium oxysporium,* which produces laccase, could be used with a suitable mediator system for biofertilizer production (humic acid) (Kwiatos et al., 2020; Strzelecki, & Bielecki, 2018). Recently, Kwaitos et al. (2020) demonstrated that an engineered strain of *Saccharomyces cerevisiae* produced higher levels of laccase, which was further utilized for coal modification and biosolubilization of coal, which released more fulvic acid (Kwiatos et al., 2020). In a previously reported study, 8 N nitric acid pretreated brown coal was further biosolubilized and the microbial biosolubilization increased by 89% using *Gordonia alkanivorans,* and *Bacillus mycoides.* Therefore, the implementation of the biochemical approach significantly

TABLE 9.2
Bioremediation Applications for Lignolytic Enzymes

Microbial Strain	Source	Growth Conditions	Lignolytic Enzymes	Substrate Support Growth	Applications	References
Pseudomonas sp.	Soil	Aerobic	Lignolytic enzymes	Waste effluents	Increased degradation of BTX compounds	Kumar and Chandra (2020)
Aneurinibacillus-neurinlyticus	Pulp mill effluent	Microaerophilic/facultative anaerobes	Lignolytic enzymes	Kraft lignin	Degradation into low molecular weight kraft lignin and its decolorization	Kumar and (2020)
Azotobacter	Soil	Aerobic	Lignolytic enzymes	-	Solubilize and decolorize lignin	Kumar & Chandra (2020)
Bacillus sp.	Pulp and paper sludge	Facultative	Lignolytic enzymes	Kraft lignin	Degradation into low molecular weight kraft lignin and its decolorization	Kumar & Chandra (2020)
H. aswanensis	Sambhar lake	Aerobic	Lignolytic enzymes	Wheat bran	Degradation of azo dye	Chauhan & Choudhury (2021)
Basidiomycete sp.	Cinnamomum Camphora tree	Aerobic	MnP	-	Dye and lignin biodegradation	Rekik et al. (2019)
Pseudomonas fluorescens	-	Aerobic	Dyp	-	Textile dye degradation	Lončar et al. (2019)
P. chrysosporium	-	Aerobic	LiP	-	Textile dye degradation	(Ilić Đurđić et al. (2020)
P. tenuiculus	-	Aerobic	Laccase	-	Atrazine (herbicide degradation)	da Silva Coelho-Moreira et al. (2013)
Commercial	-	-	Laccase	-	Pharmaceutical compounds	Naghdi et al. (2018b)

Note: BTX, benzene, toluene, and xylene.

contributed to improving coal quality and its efficiency (Romanowska, Strzelecki, & Bielecki, 2015). The presence of coal in the water is responsible for increased salinity and metal ions; thus, it creates a favorable environment to grow various halophilic microbes.

9.5.2 Pesticide and Herbicide Degradation

For improved productivity of crops, pesticides and herbicides are used to kill various pests and unwanted plants. In pesticides, their chemical structure contains polychlorinated biphenyls (PAHs). When humans consume these indirectly by drinking water and food, they cause chronic health hazards and problems, such as skin diseases, cancers, neurological disorders, and memory loss (Bansal & Kanwar, 2013). In addition, pesticides contain polychlorinated bisphenols as an aromatic structure, which is similar to dyes and lignin. PAHs were degraded by the LiP and MnP. Herbicides contain similar polychlorinated aromatic structures in their chemical makeup, which is toxic. For instance, in a recent study, atrazine (a chlorinated herbicide) was biodegraded by *Basidiomycete sp.* (*Pluteus cubensis* and *Gloelophyllum striatum*) by 30%–38% within 20 d. In addition, *Polyporous tenuiculus* produces laccase and contributed 13.9 % to the degradation of atrazine (Henn, Monteiro, Boscolo, da Silva, & Gomes, 2020). Lignolytic enzymes are used to degrade other environmental pollutants, such as polycyclic aromatic hydrocarbons [pentachloronitrobenzene (PCNB)] and chloroaniline (Renner, 1980). Therefore, when lignolytic enzymes of haloalkaliphilic origin are used, under suitable reaction conditions they degrade herbicides and pesticides. In Table 9.2, various bioremediation applications for lignolytic enzymes are listed.

9.6 CONCLUSION

Lignolytic enzymes are well known for the bioremediation and detoxification of various pollutants; however, they have not been commercially implemented. Versatility and availability help when accessing a wide range of substrates by lignolytic enzymes. Lignolytic enzymes from a fungal origin did not withstand industrial applications. Therefore, the commercial implementation of fungal-originated lignolytic enzymes has challenges to increase stability (under industrial conditions), the economical production of enzymes, and an increase in redox potential. In addition, it is very important to achieve higher productivity and determine the degradation pathways for xenobiotic compounds. This could help in the implementation of lignolytic enzymes in various industries for the degradation of effluents, such as textile, pulp mill, and pharmaceutical. In conventional techniques for lignolytic enzymes, mediated catalytic reactions are slow and cost-intensive. Simultaneously, lignolytic enzymes could be altered with the organic solvents and ionic liquids to address the limitations in the generation of biofuel and lignin valorization. Due to the potential roles of lignolytic enzymes and new applications, there could be an environmentally friendly and sustainable approach in selective biological catalysis.

REFERENCES

Amoozegar, M. A., Safarpour, A., Noghabi, K. A., Bakhtiary, T., & Ventosa, A. (2019). Halophiles and their vast potential in biofuel production. *Frontiers in Microbiology, 10*(1895). doi:10.3389/fmicb.2019.01895

Aprianti, T., Miskah, S., Selpiana, Komala, R., & Hatina, S. (2018). Heavy metal ions adsorption from pulp and paper industry wastewater using zeolite/activated carbon-ceramic composite adsorbent. *AIP Conference Proceedings, 2014*(1), 020127. doi:10.1063/1.5054531

Baciocchi, E., Fabbrini, M., Lanzalunga, O., Manduchi, L., & Pochetti, G. (2001). Prochiral selectivity in H_2O_2 promoted oxidation of aryl alkanols catalysed by chloroperoxidase. *European Journal of Biochemistry, 268*(3), 665–672. doi:10.1046/j.1432-1327.2001.01924.x

Baldrian, P., & Valášková, V. (2008). Degradation of cellulose by basidiomycetous fungi. *FEMS Microbiology Reviews, 32*(3), 501–521. doi:10.1111/j.1574-6976.2008.00106.x

Bansal, N., & Kanwar, S. S. (2013). Peroxidase(s) in environment protection. *The Scientific World Journal, 2013*, 714639. doi:10.1155/2013/714639

Bharagava, R. N., Saxena, G., Mulla, S. I., & Patel, D. K. (2018). Characterization and identification of recalcitrant organic pollutants (ROPs) in tannery wastewater and its phytotoxicity evaluation for environmental safety. *Archives of Environmental Contamination and Toxicol*ogy, *75*(2), 259–272. doi:10.1007/s00244-017-0490-x

Blánquez, A., Rodríguez, J., Brissos, V., Mendes, S., Martins, L. O., Ball, A. S., … Hernández, M. (2019). Decolorization and detoxification of textile dyes using a versatile *Streptomyces* laccase-natural mediator system. *Saudi Journal of Biological Sciences, 26*(5), 913–920. doi.org/10.1016/j.sjbs.2018.05.020

Chang, J.-S., Kuo, T.-S., Chao, Y.-P., Ho, J.-Y., & Lin, P.-J. (2000). Azo dye decolorization with a mutant Escherichia coli strain. *Biotechnology Letters, 22*(9), 807–812. doi:10.1023/A:1005624707777

Chaturvedi, V., & Verma, P. (2015). Biodegradation of malachite green by a novel copper-tolerant *Ochrobactrum pseudogrignonense* strain GGUPV1 isolated from copper mine waste water. *Bioresources and Bioprocessing, 2*(1), 42. doi:10.1186/s40643-015-0070-8

Chauhan, A. K., & Choudhury, B. (2021). Synthetic dyes degradation using lignolytic enzymes produced from *Halopiger aswanensis* strain ABC_IITR by solid state fermentation. *Chemosphere, 273*, 129671. doi.org/10.1016/j.chemosphere.2021.129671

Chowdhury, M., Mostafa, M. G., Biswas, T. K., Mandal, A., & Saha, A. K. (2015). Characterization of the effluents from leather processing industries. *Environmental Processes, 2*(1), 173–187. doi:10.1007/s40710-015-0065-7

da Silva Coelho-Moreira, J., Maciel, G. M., Castoldi, R., da Silva Mariano, S., Inácio, F. D., Bracht, A., & Peralta, R. M. (2013). Involvement of lignin-modifying enzymes in the degradation of herbicides. In Herbicides-Advances in Research: InTechOpen.

Ellouze, M., & Sayadi, S. (2016). White-rot fungi and their enzymes as a biotechnological tool for xenobiotic bioremediation. *Management of Hazardous Wastes* (pp. 103–120). London: InTechOpen.

Feng, C., Fang-yan, C., & Yu-bin, T. (2014). Isolation, identification of a halotolerant acid red B degrading strain and its decolorization performance. *Apcbee Procedia, 9*, 131–139.

Furusjö, E., & Pettersson, E. (2016). Mixing of Fast Pyrolysis Oil and Black Liquor: Preparing an Improved Gasification Feedstock. *Energy & Fuels, 30*(12), 10575–10582. doi:10.1021/acs.energyfuels.6b02383

Gaur, N., Narasimhulu, K., & Y, P. (2018). Biochemical and kinetic characterization of laccase and manganese peroxidase from novel *Klebsiella pneumoniae* strains and their application in bioethanol production. *RSC Advances, 8*(27), 15044–15055. doi:10.1039/C8RA01204K

Henn, C., Monteiro, D. A., Boscolo, M., da Silva, R., & Gomes, E. (2020). Biodegradation of atrazine and ligninolytic enzyme production by *Basidiomycete* strains. *BMC Microbiology, 20*(1), 266. doi:10.1186/s12866-020-01950-0

Hofrichter, M., Ullrich, R., Pecyna, M. J., Liers, C., & Lundell, T. (2010). New and classic families of secreted fungal heme peroxidases. *Applied Microbiology and Biotechnology, 87*(3), 871–897. doi:10.1007/s00253-010-2633-0

Igbinigie, E. E., Aktins, S., van Breugel, Y., van Dyke, S., Davies-Coleman, M. T., & Rose, P. D. (2008). Fungal biodegradation of hard coal by a newly reported isolate, *Neosartorya fischeri*. *Biotechnology Journal, 3*(11), 1407–1416. doi:10.1002/biot.200800227

Ilić Đurđić, K., Ostafe, R., Prodanović, O., Đurđević Đelmaš, A., Popović, N., Fischer, R., … Prodanović, R. (2020). Improved degradation of azo dyes by lignin peroxidase following mutagenesis at two sites near the catalytic pocket and the application of peroxidase-coated yeast cell walls. *Frontiers of Environmental Science & Engineering, 15*(2), 19. doi:10.1007/s11783-020-1311-4

Janusz, G., Pawlik, A., Sulej, J., Świderska-Burek, U., Jarosz-Wilkołazka, A., & Paszczyński, A. (2017). Lignin degradation: Microorganisms, enzymes involved, genomes analysis and evolution. *FEMS Microbiology Reviews, 41*(6), 941–962. doi:10.1093/femsre/fux049

Karn, S. K., Chakrabarty, S. K., & Reddy, M. S. (2010). Pentachlorophenol degradation by *Pseudomonas stutzeri* CL7 in the secondary sludge of pulp and paper mill. *Journal of Environmental Sciences, 22*(10), 1608–1612. doi.org/10.1016/S1001-0742(09)60296-5

Kiadehi, M. S. H., Amoozegar, M. A., Asad, S., & Siroosi, M. (2018). Exploring the potential of halophilic archaea for the decolorization of azo dyes. *Water Science and Technology, 77*(6), 1602–1611. doi:10.2166/wst.2018.040

Kumar, A., & Chandra, R. (2020). Ligninolytic enzymes and its mechanisms for degradation of lignocellulosic waste in the environment. *Heliyon, 6*(2), e03170. doi.org/10.1016/j.heliyon.2020.e03170

Kuwahara, M., Glenn, J. K., Morgan, M. A., & Gold, M. H. (1984). Separation and characterization of two extracelluar H_2O_2-dependent oxidases from ligninolytic cultures of *Phanerochaete chrysosporium*. *FEBS Letters, 169*(2), 247–250. doi.org/10.1016/0014-5793(84)80327-0

Kwiatos, N., Jędrzejczak-Krzepkowska, M., Krzemińska, A., Delavari, A., Paneth, P., & Bielecki, S. (2020). Evolved *Fusarium oxysporum* laccase expressed in *Saccharomyces cerevisiae*. *Scientific Reports, 10*(1), 1–11.

Kwiatos, N., Jędrzejczak-Krzepkowska, M., Strzelecki, B., & Bielecki, S. (2018). Improvement of efficiency of brown coal biosolubilization by novel recombinant *Fusarium oxysporum* laccase. *AMB Express, 8*(1), 133–133. doi:10.1186/s13568-018-0669-1

Langhals, H. (2004). Color Chemistry. Synthesis, Properties and Applications of Organic Dyes and Pigments [Review of the book (3rd revised ed.) by H. Zollinger]. *Angewandte Chemie International Edition, 43*(40), 5291–5292. doi:10.1002/anie.200385122

Liu, S., Xu, X., Kang, Y., Xiao, Y., & Liu, H. (2020). Degradation and detoxification of azo dyes with recombinant ligninolytic enzymes from *Aspergillus sp.* with secretory overexpression in *Pichia pastoris. Royal Society Open Science, 7*(9), 200688. doi:doi:10.1098/rsos.200688

Lončar, N., Drašković, N., Božić, N., Romero, E., Simić, S., Opsenica, I., … Fraaije, M. W. (2019). Expression and characterization of a dye-decolorizing peroxidase from *Pseudomonas fluorescens* Pf0-1. *Catalysts, 9*(5), 463.

Martínez, Á. T., Speranza, M., Ruiz-Dueñas, F. J., Ferreira, P., Camarero, S., Guillén, F., … Río Andrade, J. C. D. (2005). Biodegradation of lignocellulosics: microbial, chemical, and enzymatic aspects of the fungal attack of lignin.

Mehmood, K., Rehman, S. K. U., Wang, J., Farooq, F., Mahmood, Q., Jadoon, A. M., … Ahmad, I. (2019). Treatment of Pulp and Paper Industrial Effluent Using Physicochemical Process for Recycling. *Water, 11*(11), 2393.

Mester, T., & Field, J. A. (1998). Characterization of a novel manganese peroxidase-lignin peroxidase hybrid isozyme produced by *Bjerkandera* species strain BOS55 in the absence of manganese. *Journal of Biological Chemistry, 273*(25), 15412–15417.

Minussi, R. C., Pastore, G. M., & Durán, N. (2007). Laccase induction in fungi and laccase/N-OH mediator systems applied in paper mill effluent. *Bioresource Technology, 98*(1), 158–164. doi:10.1016/j.biortech.2005.11.008

Morozova, O., Shumakovich, G., Gorbacheva, M., Shleev, S., & Yaropolov, A. (2007). Blue laccases. *Biochemistry (Moscow), 72*(10), 1136–1150.

Naghdi, M., Taheran, M., Brar, S. K., Kermanshahi-pour, A., Verma, M., & Surampalli, R. Y. (2018a). Biotransformation of carbamazepine by laccase-mediator system: Kinetics, by-products and toxicity assessment. *Process Biochemistry, 67*, 147–154.

Naghdi, M., Taheran, M., Brar, S. K., Kermanshahi-Pour, A., Verma, M., & Surampalli, R. Y. (2018b). Removal of pharmaceutical compounds in water and wastewater using fungal oxidoreductase enzymes. *Environmental pollution (Barking, Essex: 1987), 234*, 190–213. doi:10.1016/j.envpol.2017.11.060

Paramjeet, S., Manasa, P., & Korrapati, N. (2018). Biofuels: Production of fungal-mediated ligninolytic enzymes and the modes of bioprocesses utilizing agro-based residues. *Biocatalysis and Agricultural Biotechnology, 14*, 57–71. doi.org/10.1016/j.bcab.2018.02.007

Pérez-Boada, M., Ruiz-Dueñas, F. J., Pogni, R., Basosi, R., Choinowski, T., Martínez, M. J., … Martínez, A. T. (2005). Versatile peroxidase oxidation of high Redox potential aromatic compounds: Site-directed mutagenesis, spectroscopic and crystallographic investigation of three long-range electron transfer pathways. *Journal of Molecular Biology, 354*(2), 385–402. doi.org/10.1016/j.jmb.2005.09.047

Plácido, J., & Capareda, S. (2015). Ligninolytic enzymes: a biotechnological alternative for bioethanol production. *Bioresources and Bioprocessing, 2*(1), 23.

Pylypchuk, I. V., Daniel, G., Kessler, V. G., & Seisenbaeva, G. A. (2020). Removal of diclofenac, paracetamol, and carbamazepine from model aqueous solutions by magnetic sol–gel encapsulated horseradish peroxidase and lignin peroxidase composites. *Nanomaterials, 10*(2), 282.

Rekik, H., Zaraî Jaouadi, N., Bouacem, K., Zenati, B., Kourdali, S., Badis, A., … Jaouadi, B. (2019). Physical and enzymatic properties of a new manganese peroxidase from the white-rot fungus *Trametes pubescens* strain i8 for lignin biodegradation and textile-dyes biodecolorization. *International Journal of Biological Macromolecules, 125*, 514–525. doi.org/10.1016/j.ijbiomac.2018.12.053

Renner, G. (1980). Metabolic studies on pentachloronitrobenzene (PCNB) in rats. *Xenobiotica, 10*(7–8), 537–550. doi:10.3109/00498258009033788

Rezaei, S., Shahverdi, A. R., & Faramarzi, M. A. (2014). An extremely halophilic laccase from the saline water isolate *Chromohalobacter sp.*

Rezaei, S., Shahverdi, A. R., & Faramarzi, M. A. (2017). Isolation, one-step affinity purification, and characterization of a polyextremotolerant laccase from the halophilic bacterium *Aquisalibacillus elongatus* and its application in the delignification of sugar beet pulp. *Bioresour Technol, 230*, 67–75. doi:10.1016/j.biortech.2017.01.036

Romanowska, I., Strzelecki, B., & Bielecki, S. (2015). Biosolubilization of Polish brown coal by *Gordonia alkanivorans* S7 and *Bacillus mycoides* NS1020. *Fuel Processing Technology, 131*, 430–436. doi.org/10.1016/j.fuproc.2014.12.019

Silva, C., Martins, M., Jing, S., Fu, J., & Cavaco-Paulo, A. (2018). Practical insights on enzyme stabilization. *Critical Reviews in Biotechnology, 38*(3), 335–350.

Sivaprakasam, S., Mahadevan, S., Sekar, S., & Rajakumar, S. (2008). Biological treatment of tannery wastewater by using salt-tolerant bacterial strains. *Microbial Cell Factories, 7*, 15–15. doi:10.1186/1475-2859-7-15

Souza, É. S., Souza, J. V., Silva, F. T., & Paiva, T. C. (2014). Treatment of an ECF bleaching effluent with white-rot fungi in an air-lift bioreactor. *Environmental Earth Sciences, 72*(4), 1289–1294.

Sudha, M., Saranya, A., Selvakumar, G., & Sivakumar, N. (2014). Microbial degradation of azo dyes: A review. *International Journal of Current Microbiology and Applied Sciences, 3*(2), 670–690.

Sundaramoorthy, M., Kishi, K., Gold, M. H., & Poulos, T. L. (1994). The crystal structure of manganese peroxidase from *Phanerochaete chrysosporium* at 2.06-A resolution. *Journal of Biological Chemistry, 269*(52), 32759–32767.

Tien, M., & Kirk, T. K. (1984). Lignin-degrading enzyme from *Phanerochaete chrysosporium*: Purification, characterization, and catalytic properties of a unique H_2O_2-requiring oxygenase. *Proceedings of the National Academy of Sciences of the United States of America, 81*(8), 2280–2284.

Torres-Farradá, G., Manzano León, A. M., Rineau, F., Ledo Alonso, L. L., Sánchez-López, M. I., Thijs, S., … Vangronsveld, J. (2017). Diversity of ligninolytic enzymes and their genes in strains of the genus Ganoderma: Applicable for Biodegradation of Xenobiotic Compounds? *Frontiers in Microbiology, 8*(898). doi:10.3389/fmicb.2017.00898

Uthandi, S., Saad, B., Humbard, M. A., & Maupin-Furlow, J. A. (2010). LccA, an archaeal laccase secreted as a highly stable glycoprotein into the extracellular medium by *Haloferax volcanii*. *Applied and Environmental Microbiology, 76*(3), 733–743.

Verma, P., & Madamwar, D. (2005). Decolorization of azo dyes using *Basidiomycete* strain PV 002. *World Journal of Microbiology and Biotechnology, 21*(4), 481–485. doi:10.1007/s11274-004-2047-1

Wang, Z., Ren, D., Zhao, Y., Huang, C., Zhang, S., Zhang, X., … Guo, H. (2019). Remediation and improvement of 2,4-dichlorophenol contaminated soil by biochar-immobilized laccase. *Environmental Technology*, 1–14. doi:10.1080/09593330.2019.1677782

Yaseen, D. A., & Scholz, M. (2019). Textile dye wastewater characteristics and constituents of synthetic effluents: a critical review. *International Journal of Environmental Science and Technology, 16*(2), 1193–1226. doi:10.1007/s13762-018-2130-z

Zango, Z. U., Jumbri, K., Sambudi, N. S., Ramli, A., Abu Bakar, N. H. H., Saad, B., … Sulieman, A. (2020). A critical review on metal-organic frameworks and their composites as advanced materials for adsorption and photocatalytic degradation of emerging organic pollutants from wastewater. *Polymers, 12*(11), 2648.

10 Biodegradation of Synthetic Dyes from the Textile Industry by Microbes

Yashsvi Raval[1] and Anupama Shrivastav[1]*

[1]Department of Microbiology, Parul Institute of Applied Sciences, Parul University, Vadodara, Gujarat, India

*Corresponding author: yashsviraval1903@gmail.com, anupamashrivastav@gmail.com

CONTENTS

10.1 INTRODUCTION

Dyes can be prepared from natural sources, such as flowers, vegetables, roots, woods, and insects. However, due to increased demands, industries are more dependent on dyes that are manufactured from petrochemicals, for instance, artificial dyes. These dyes are used in different industries, such as textiles, paper, printing, cosmetics, pharmaceutical, color photography, and petroleum. Compared with natural dyes, synthetic dyes are more soluble in water, easily absorbed, give fast coloration, and provide a large range of colors. A large number of dyes are used in the textile industry. Based on the dyeing process, textile processing is water-intensive and releases a large volume of wastewater. The released wastewater contaminates the water and soil, which results in environmental pollution, and has an impact on the water quality, color, pH, and recalcitrant synthetic compounds are present. Several impacts on human health are associated with residual dyestuff including irritation, respiratory problems, cancers, skin allergies, infections, and other effects on the immune system.

One of the most obvious indicators of water pollution is the colored wastewaters from the textile industry. Colored dye wastewater has a severe effect on aquatic environments. Apart from the color, the discharged dye wastewater contains other pollutants, such as degradable organics, pH altering agents, salts, nutrients, toxicants, and refractory organics. Therefore, the decomposition of the dye molecules depends on the complexity of their structure. Dyes are classified based on their chemical composition, they can be acid, basic, pigment, sulfur, and azo dyes. Azo dyes have complex structures, and they are the most popular synthetic dyes because of their cost-effectiveness and ease of use.

Considering the overall load of the effluent from textile industries, dye wastewater is treated using physicochemical and biological methods. The physical and chemical methods are very expensive, and they produce a lot of sludge and release toxic substances. The biological treatment of dye wastewater is an alternative method. It is cost-effective, nonhazardous, environmentally friendly, produces less sludge, and the ecological nature of biological systems means that they are preferable compared with the physicochemical methods to treat textile wastes.

10.2   DECOLORIZATION AND DEGRADATION METHODS

The wastewater from textile industries contains color, dyes, and oily chemicals. Various physical, chemical, and biological methods are available to treat wastewater. These techniques can be employed to remove color from the dye contained in wastewater. In general, every technique has advantages and limitations, and dye removal strategies consist of a mixture of techniques (Figure 10.1).

10.2.1   Microbial Methods for Dye Removal from Textile Wastewater

Several physicochemical methods are used for the treatment of textile effluents. However, using microorganisms or microbial enzymes or combining them with a physicochemical method provides better results for economic viability. The use of microbes could decolorize very complex synthetic dyes.

Textile dyes are decolorized by microorganisms using three methods, either by microbial biomass adsorption or by cell biodegradation or by enzymes. If the effluent is highly toxic and cannot support the growth and maintenance of microbial cells, biomass might be used. Bacteria, fungi, microalgae, and plants are used as adsorbents, and the dye is not degraded into fragments during the process. Therefore, biodegradation is a more practical option (Table 10.1).

10.2.1.1   Biodegradation of Synthetic Dyes by Fungi

Fungi have been classified as white-brown or soft-rot fungi based on the technical decay and descriptions, regardless of their taxonomic classification. The enzyme systems and metabolic pathways involved in the breakdown of carbohydrates and lignin are very distinct in these fungi. The important physiological characteristic of decay in a fungal culture is the production of extracellular enzymes, phenoloxidases and peroxidases. Fungi produce various oxidoreductases that degrade lignin and related aromatic compounds. Fungi are used in mycoremediation and are used in biotechnological applications for biodegradation. The nonspecific enzyme systems include lignin peroxidase (LiP), manganese peroxidase (MnP), and laccase. The production of laccase by *Phanerochate chrysosporium* and *Neurospora crassa* has been studied for the removal of pigments and phenol from wastewater. *P. chrysosporium* could decolorize synthetic dyes, such as the azo dyes

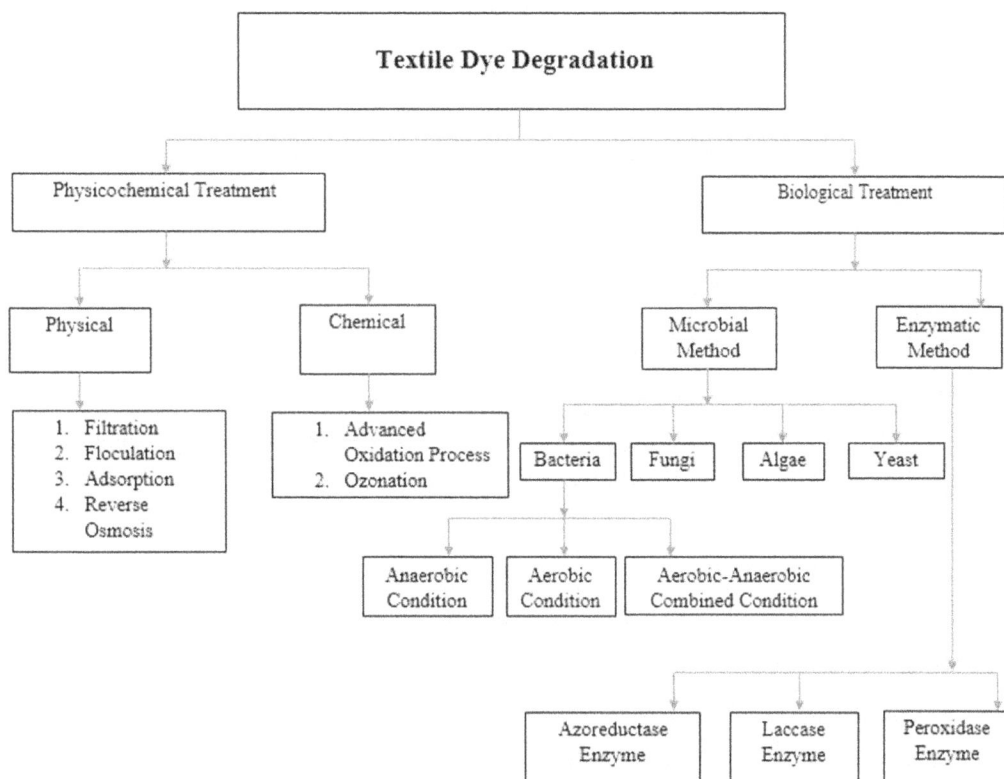

FIGURE 10.1   Degradation methods of dyes.

Sources: Jamee and Siddique (2019); Satyendra *et al.* (2016).

Acid orange 6, Orange 2, and Congo red. Recently, more potent fungal strains, such as *Trametes versicolor, Dichomitus squalens, Phlebia fascicularia, Irpex flavus,* and *Polyporus sanguineus,* have been discovered (Roderiguez *et al.*, 1999; Pointing *et al.*, 2000; Gill *et al.*, 2002; Chander *et al.*, 2004).

10.2.1.2   Biodegradation of Dyes by Yeasts

Biodegradation of dyes by yeasts occurs through three mechanisms: biosorption, bioaccumulation, and biodegradation. Biosorption is defined as the binding of solutes to biomass when there is no metabolic energy or transport is not involved, although these processes might occur at the same time when live biomass is used. Biosorption takes place in either living or dead biomass. Bioaccumulation is the uptake of toxins by living cells and their transport into cells. It is a growth-dependent process mediated by living biomass. The mechanism is intracellular and more complex than biosorption and is not fully understood. Biodegradation is an energy-dependent process that involves the breakdown of dyes into different by-products through the action of different enzymes. When biodegradation is complete, it is known as mineralization. During biodegradation, yeast biomass removes the color of the dye, then dye removal occurs through biosorption or bioaccumulation.

Compared with bacteria and white-brown or soft-rot fungi, yeasts have many advantages. Yeasts are an inexpensive and readily available source of biomass. Yeasts can survive in unfavorable environments, such as at low pH and temperatures. Some yeast can efficiently treat high strength organic wastewaters, such as food industry effluents that are highly colored.

10.2.1.3 Biodegradation of Dyes by Algae and Plants

Algae are photosynthetic organisms, which have been used to remove color via biosorption and reductive mechanisms. Therefore, the species of algae used and the molecular structure of the dye that is taken up as their sole supply of C and N. There are approximately 30 azo compounds that can be biodegraded and decolorized by *Chlorella pyrenoidosa, Chlorella vulgaris,* and *Oscillateria tenius,* the azo dyes are decomposed into simpler aromatic amines (Yan *et al,* 2004). Algae play an important role in the removal of azo dyes and aromatic amines in stabilization ponds (Banat *et al.,* 1996).

The exploitation of plants to remove contaminants from polluted water is phytoremediation. It is an *in situ* method, eco-friendly, decrease soil erosion, improves soil fertility by increasing organic matters in the soil, and uses plants for degradation, transformation, extraction, and detoxification of chemical contamination (Organum and Bacon, 2006). Phytoremediation technology has been used for the use of growing aquatic plants such as *Phragmites australis* (Hussein and Scholz, 2017) and using free-floating plants like *Lemna minor* (Uysal *et al.,* 2014). *Brassica juncea, Sorghum vulgare,* and *Phaseolus mungo* have successfully decolorized textile effluents. *Blumea malcommi* and *Tagetes patula* have had positive results in the literature.

10.2.1.4 Biodegradation of Synthetic Dyes by Bacteria

Several species of Gram-negative bacteria from different genera such as *Escherichia, Citrobacter, Aeromonas, Pseudomonas,* and *Sphingomonas* can decolorize dye solutions and effluents (Forgacs *et al.,* 2004). Gram-positive bacteria, such as *Bacillus, Paenibacillus, Nocardia, Streptomyces, Micrococcus,* and *Clostridium* can degrade synthetic dyes. Bacterial oxidoreductive enzymes are important in the degradation of synthetic dyes. The dynamic metabolism of bacteria enables them to use the complex xenobiotic compounds in the dyestuff as substrates. The compounds are broken down into less complex metabolites. Obtaining bacteria from the wastewater disposal sites increases the probability of enzyme activation, which helps to decompose the dyes.

10.2.1.4.1 Bacterial Methods for the Degradation of Dyestuffs

The biodegradation of dyes using bioremediation removes synthetic dyes from textile effluents with minimum costs and optimum operating times. Compared with physicochemical methods, biological methods are economically feasible, environmentally friendly, and generates less sludge. Synthetic dyes are degraded into less toxic inorganic compounds due to the breakdown of chemical bonds that help to remove them. The degradation of synthetic dyes, for instance, azo dyes involves two steps (1) the azo bonds in the dyes are cleaved and amines are formed; and (2) the aromatic amines degraded into smaller nontoxic, molecules under aerobic conditions. Some bacteria can survive under aerobic and anaerobic conditions that help with the degradation of the azo bonds within the dyes.

10.2.1.4.1.1 Decolorization of Synthetic dyes Under Aerobic Conditions   The degradation of dyes under aerobic conditions is catalyzed by specific enzymes, mainly azo reductase. In the presence of azo reductase, some aerobic microorganisms can reduce azo dyes and produce aromatic amines, which are then degraded by microorganisms under aerobic conditions for the complete mineralization of the azo dyes. Aerobic bacteria, such as *Pigmentiphaga kullae K*24 and *Xenophilus azovorans* KF 46 can cleave -N=N- azo bonds and utilize amines as the C source and energy for their growth. These types of organisms are substrate sensitive. Bacterial strains, such as *Vibrio logei* and *Pseudomonas nitroreducens* are aerobic bacteria species that utilize Methyl red as a sole C source. No amines were detected in the growth medium, which highlighted the degradation of the dye.

10.2.1.4.2 Decolorization of Synthetic Dyes Under Anaerobic Conditions Dye decolorization and degradation under anaerobic condition involves the simple and nonspecific reduction of azo dyes. This type of azo reduction relies on redox intermediates. The main redox intermediates are flavin adenine dinucleotide (FAD) and nicotinamide adenine dinucleotide (NADH). In direct enzymatic catalysis, azo reductases in microorganisms directly catalyze the reduction of the azo dye. The existence of this enzyme has not been directly verified. The reduction reaction depends on the redox intermediates, such as FAD and NAD(P)H for the degradation of the azo dye under anaerobic conditions. The redox intermediates receive electrons from an azo reductase and then transfer the electrons to the azo dye. Then azo bonds (–N=N–) are cleaved to produce aromatic amines, which decolorize the azo dye. The reduction reaction between the azo dyes and redox intermediates depends on the redox potential. Under the anaerobic conditions, a redox potential of 50 mV can cause the effective decolorization of azo dyes. The cleavage of the chromosphere groups in the dyes results in colorless, odorless, and toxic intermediates of aromatic amines, which are commonly mineralized under aerobic conditions. Anaerobic azo dye reduction is an efficient and economical method of color removal from textile wastewater.

10.2.1.4.3 Decolorization of Synthetic Dyes Under Anaerobic–Aerobic Conditions In general, few azo dyes can be biodegraded completely; however, under anaerobic conditions, the aromatic amines in azo dyes often cannot be degraded further under anaerobic conditions, which results in the accumulation of toxic substances. These aromatic amines are biodegradable under aerobic conditions; therefore, a combined aerobic–anaerobic process to treat azo dye wastewater could be a good method to completely degrade the dye. This method is costs less and is suitable for various dyes. However, the main limitation of biological treatment is low biodegradability, lower flexibility in design and operation, the larger surface area required, and longer times required for decolorization. Therefore, a method is required to remove dyes from effluents continuously in liquid-state fermentation.

10.2.2 ENZYMATIC METHODS TO DEGRADE DYES

Enzymes, such as azo reductase, laccases, LiP, MnP, and hydroxylases are used to degrade dyes using enzymatic methods. Laccase and azoreductase can degrade azo dyes (Rodriguez *et al.*, 1999; Reyes *et al.*, 1999). Dye molecules have a variety of structures and are degraded by a few enzymes (Table 10.2).

10.2.2.1 Laccases

Laccases have been studied for their degradation of azo dyes. (Chivukula *et al.*, 1995; Kirby *et al.*, 2000; Peralta *et al.*, 2003; Blánquez *et al.*, 2004; Novotny *et al.*, 2004). Laccases work by removing H^+ ions from the hydroxyl and amino groups of the ortho and para-substituted phenolic substrates and aromatic amines.

Laccases oxidize the phenolic group of the azo dye, and there is a nucleophilic attack by water on the phenolic ring C that contains the azo bonds (Camarero *et al.*, 2005). Cross-coupling between the reactive species results in the formation of C–C and C–O bonds between phenolic molecules and C–N and N–N bonds between aromatic amines (Zille *et al.*, 2005). By phenolic cross-coupling, an electron is removed from the hydroxy group, which generates an alkoxy radical. The alkoxy radical forms dimers within the ortho and para positions with the hydroxy groups. Phenolic radicals then oxidized to produce oligomeric products. The C–C dimers formed might participate in the coupling reaction to form extended quinines under certain conditions.

10.2.2.2 Peroxidases

Peroxidases are hemoproteins that catalyze reactions in the presence of hydrogen peroxide (H_2O_2) (Duran and Esposito, 2002). Lignin and MnPs have identical reaction mechanism that starts with enzyme oxidation via H_2O_2 to an oxidized state during their catalytic cycle. The first ligninolytic peroxidase was isolated from *P. chrysosporium* and was LiP (Glenn *et al.*, 1983; Tien *et al.*, 1983) MnP

TABLE 10.1
Microbial Decomposition of Azo Related Industrial Dyes

Strain	Organisms	Dye
Bacteria	Galactomyces geotrichum	Yellow 84A
	Enterobacter sp	CI Reactive Red 195
	Bacillus subtilis	Acid blue 113
	Brevibacillus laterospous MTCC2298	Navy blue 3G
	Bacillus Fusiformis kmk 5	Acid Orange 10 and Disperse Blue 79
	Enterobacter agglomerans	Methyl Red
Fungi	Geotrichum sp	Reactive black 5, Reactive red 158 and Reactive yellow 27
	Shewanella sp. NTOVI	Crystal violet
	Phanaerochaete chrysosporium	Orange II
	Aspergillus ochraceus NCIM -1146	Reactive blue 25
Algae	Spirogyra Rhizopus	Acid red 247
	Cosmarium sp	Triphenylmethane dye and Malachite green
Yeast	Saccharomyces cerevisiae MTCC463	Methyl red
	Kluyveromyces marxianus IMB3	Ramazol black B
Actinomycetes	Streptomyces ipomoea	Orange II

(Kamlesh Shah, et al., 2014)

(Kuwahara *et al.*, 1984; Wariishi *et al.*, 1988). LiP is used to catalyzes the oxidation of nonphenolic aromatic compounds. MnP oxidizes Mn^{2+} to Mn^{3+}, and Mn^{3+} is responsible for the oxidation of many phenolic compounds (Glenn *et al.*, 1986). Mn^{2+} is required for the completion of the catalytic cycle of MnP. LiP or MnP are directly involved in the degradation of various xenobiotic compounds and dyes (Paszczynski *et al.*, 1991).

10.2.2.3 Azoreductases

Azo reductase is a membrane-bound enzyme that catalyzes the reaction in presence of reducing equivalents, such as FADH and NADH (Robinson *et al.*, 2001). Therefore, the reduction process occurs in bacterial cells with intact cell membranes (Russ, Rau, and Stolz., 2000). In Gram-negative bacteria, the enzyme might be in direct contact with either the azo dye substrate or a redox mediator at the cell surface (Myers and Myers, 1992). In addition, a few low molecular weight redox mediator compounds can act as electron shuttles between the azo dye and an NADH-based azo reductase, which is located in the outer membrane (Gingell and walker, 1971). These enzymes are oxygen (O_2) sensitive; therefore, in the extracellular environment, this reduction mechanism is inhibited by O_2. The mechanisms for redox reactions were under anaerobic conditions, and the redox mediators depend on cytoplasmic reducing enzymes to supply electrons (Yoo *et al.*, 2001).

10.3 ISOLATION AND SCREENING OF BACTERIA THAT DEGRADE SYNTHETIC DYES

Currently, industrial effluent that contains textile dyes is regarded as the major environmental globally. Isolation of bacteria started by screening azo dye degrading bacteria that were isolated from the textile effluents. Effluent samples must be collected from different areas in the textile effluent. Dye degrading bacteria have been isolated from textile wastewater using serial dilution techniques,

TABLE 10.2
Enzymes Mediated Decolorization of Some Dyes

Substrates	Enzyme	Source of enzyme
3-(4dimethyl amino- 1 phenylazo) Benzene sulfonic acid	Laccase	Trametes villosa
Acid Orange 6, Acid Orange 7, Methyl Orange and Methyl Red	Mixture of Bacterial Oxidoreductases	Sludge Methonogens
Acid Blue	Laccase	Cladosporium cladosporioides
Tertazine and Ponceau	Azoreductase	Green algae
Direct Yellow	Horse radish peroxidases	Amorracia rusticana
Reactive Yellow, Reactive Black, Reactive Red and Direct Blue	Azoreductase	Staphylococcus arlettae

(Kamlesh Shah, et al., 2014)

agar plating, and enrichment culture techniques. Experimental dyes should be stored at room temperature. A 1% solution of the experimental stock solution was prepared by dissolving 1 g of each powder dye into 100 ml of autoclaved distilled water, which was followed by filtration. The isolation process was carried out in Bushnell-Haas (BH) medium.

Then, bacteria were grown and selected that could decolorize and degrade the synthetic dyes. Several processes can be used for this.

10.3.1 GROWTH AND SELECTION OF BACTERIA

10.3.1.1 Enrichment Culture for the Sample

The tested sample should be grown in a nutrient medium that is supplemented with the experimental dye. The choice of the media depends on the growth and requirements of the bacteria in the sample. Nutrient broth supplemented with 1% dye is suitable for several bacteria. It is composed of peptone and a beef extract. The peptone provides organic N as amino acids and long-chain fatty acids. The beef extract provides additional vitamins, carbohydrates, salts and different natural N compounds. BH medium, mineral salts medium, synthetic wastewater medium could be used for the enrichment batch culture to select the decolorizing bacteria. BH medium is composed of 0.1% K_2HPO_4, 0.02% $MgSO_4 \cdot 7H_2O$, 0.002% $CaCl_2 \cdot 2H_2O$, 0.1% NH_4Cl, 0.1% NH_4NO_3, 0.01% $NaCl$, and 0.005% $FeCl_3 \cdot 6H_2O$ at pH 7 might be used. Minerals salts medium containing $NaHPO_4$ (3.6 g), $(NH_4)_2SO_4$ (1.0 g) K_2HPO_4 (1.0 g), $MgSO_4$ (1.0 g), Fe (NH_4) citrate (0.01 g), $CaCl_2 \cdot 2H_2O$ (0.10 g) and 10.0 mL of trace element solution per liter might be. The basic composition of the synthetic wastewater medium was (g/L): $(NH_4)_2SO_4$, 0.28; NH_4Cl, 0.23; KH_2PO_4, 0.067; $MgSO_4 \cdot 7H_2O$, 0.04; $CaCl_2 \cdot 2H_2O$, 0.022; $FeCl_3 \cdot 6H_2O$, 0.005; $NaCl$, 0.15; $NaHCO_3$, 1.0; and 1 mL/L of a trace element solution containing (g/L): $ZnSO_4 \cdot 7H_2O$, 0.01; $MnCl_2 \cdot 4H_2O$, 0.01; $CuSO_4 \cdot 5H_2O$, 0.392; $CoCl_2 \cdot 6H_2O$, 0.248; and $NiCl_2 \cdot 6H_2O$, 0.2). To enrich the dye decolorizing bacteria, the effluent and dye aggregate were added to a conical flask that contained the appropriate media, and incubated under the appropriate conditions (e.g., 37°C, 100–150 rpm). Media must be supplemented with glucose and yeast extract if the sample of microorganisms cannot use the dye as a C source. Following incubation, the solution is monitored for decolorization at predetermined intervals. For positive results with a sample for a certain dye, the results for individual colonies are confirmed using a plating method on a screening medium.

10.3.1.2   Serial Dilution of Sample

A textile effluent sample or soil affected by the effluent should be obtained from a contaminated site. The sample should be diluted with sterile distilled water or a sterile salt solution. Each dilution should be plated onto nutrient agar plates at an optimum temperature for 24 h. Pure colonies are isolated, and the isolates must be tested for their ability to decolorize dyes in media supplemented with dyes (Khalid *et al.*, 2016).

The bacteria should be identified by biochemical tests, genomic DNA, or 16S rRNA.

10.4   QUANTIFICATION OF DYE DECOLORIZATION

Decolorization of the respective dyes should be determined by measuring the absorption of the supernatants from the cultures at absorbance maxima for the dyes after centrifugation at 10,000 rpm for 15 min and absorbance was measured at λ max of the dye. The decolorization in percent for dyes was determined, according to the equation

$$\text{Decolorization}\ (\%) = (I–F) \times 100/I$$

Where I is the initial absorbance of untreated dye solution (control), and F is the absorbance of dye solution.

10.5   SCREENING OF BACTERIA THAT DECOLORIZE DYES

10.5.1   INITIAL SCREENING OF DYE DECOLORIZING BACTERIAL ISOLATES USING MICROTITER PLATE TECHNIQUE

Bacterial isolates were initially screened using a microtiter plate technique (Lucas *et al.*, 2008) using selected dyes. Isolated cultures were cultivated in nutrient broth for 24 h before screening was carried out in mineral salt media. For initial screening, 10% (v/v) aliquots of each isolated strain were inoculated into a 96 well microtiter plate, that contained 200 µL of individual dye solution. Decolorization of the dye solution was confirmed visually after 48 h incubation at 32°C.

10.5.2   FINAL SCREENING AND DYE DECOLORIZING EFFICIENCY OF BACTERIAL ISOLATES

The final screening was carried out using selected dyes in mineral salts (MS) media. Each selected strain was cultivated for 24 hs in nutrient broth. A 5% (v/v) of the inoculum was then transferred into 250 mL Erlenmeyer flasks that contained 50 mL of MS media. A final concentration of 100 mg/L of dye was added to each flask and the absorbance was measured at their absorbance maxima after 24 h or ≤72 h at 32°C±2 °C under static (anoxic) and shaking (aerobic) conditions at 150 rpm. Based on the reduction method in absorbance, the percent decolorization was calculated.

10.6   FACTORS THAT INFLUENCE BACTERIA DURING DEGRADATION

Biodegradation of synthetic dyes and other chemicals in textile effluents is influenced by a variety of operational parameters, which includes intrinsic factors, such as bacterial strain characteristics, and external factors, such as pH, temperature, C and N sources, the availability of O_2, dye concentration and structure, and metal ions.

1. **Dye Structure**

   The structure of dyes strongly influences their degradability by pure cultures and the enzymes produce. Variations that concern dye decolorization could be feasible because of the complicated chemical structure, for example, diazo dyes have more complicated structures than monoazo dyes. In addition, the chemical structure of dyes influences their decolorization. Dyes with simpler structures and low molecular weights have higher rates of color removal. The components of the aromatic ring influence enzymatic oxidation. Electron donating methyl and methoxy substituents increase the enzymatic degradation of azo phenols and electron removing chloro, fluoro, and nitro substitutes inhibit oxidization.

2. **Dye Concentration**
   An increase in dye concentration gradually decreases the rate of decolorization, probably because of the toxic effect of dyes on the individual bacteria and insufficient biomass concentration, dye molecules of various structures might block the active sites of azo reductase.

3. **C and N sources**
   Dyes are deficient in C and N sources, and the biodegradation of dyes without these supplements is very difficult. Microbial cultures need complex organic sources, such as yeast extract, peptone, or a combination of both, for dye decolorization and degradation.

4. **O_2 and Agitation**
   Environmental conditions can affect azo dyes degradation and decolorization directly, which depends on the reductive or oxidative environmental conditions, which influences microbial metabolism. Different groups of bacteria decolorize dyes under anaerobic, facultatively anaerobic, and aerobic conditions. Under anaerobic conditions, reductive enzyme activities are higher than in anoxic conditions, which breaks down the structures of the synthetic dyes. However, a small amount of O_2 is required for the oxidative enzymes that are involved in the degradation of azo dyes. Since both molecules serve as electron acceptors and O_2 is a much stronger oxidant, dissolved O_2 inhibits dye reduction.

5. **Temperature**
   Temperature is a very important factor for all processes that are associated with microbial applications, which includes the remediation of wastewater and soil. In addition, the decolorization rate of azo dyes increases up to the optimum temperature, then there is a marginal reduction in the decolorization activity.

6. **pH**
   pH has a major effect on the efficiency of decolorization, the optimum pH for color removal in bacteria is often between pH 6.0 and 10.0. The tolerance to a high pH is important, especially for industrial processes that use reactive azo dyes, which are usually performed under alkaline conditions. This factor is important for aquatic because most fish only survive at a pH range between 6 and 9.

7. **Electron Donor**
   The addition of electron donors, which include glucose or acetate ions, helps the reductive cleavage of azo bonds. The type and availability of electron donors are important to achieve a good color under anaerobic conditions.

8. **Redox Mediator**
   Redox mediators enhance many reductive processes under anaerobic conditions, which includes the reduction of azo dyes.

10.7 CONCLUSIONS

Textile wastewaters require appropriate treatment before their release into the environment. Residual dyes have complex structural compositions that are toxic to soil and water organisms in the affected areas. These industries need a cost-efficient and feasible method to treat their effluents; therefore, when it is released into the water bodies, it will have a small impact on the environment.

Among the treatment methods for dye wastewaters, dye degradation is important. The physical and chemical treatment processes for dye wastewaters are inefficient and expensive and produce refractory pollutants, which results in secondary pollution. The biological methods are eco-friendly and efficient, inexpensive and widely applicable. In microbial methods, the bacterial strains from many habitats have been explored and utilized for dye removal from dyes contained in wastewater. Aerobic, anaerobic, and a combination of aerobic–anaerobic systems have their advantages but possess certain disadvantages. Enzymatic degradation methods could be considered an excellent molecular weapon to prevent the problems associated with environmental pollution that is caused by toxic textile sludge. It is significant in industrial wastewater treatment because they are eco-friendly, inexpensive, and produce less sludge.

REFERENCES

Banat, I.M. *et al.* (1996) 'Microbial decolorization of textile dye containing effluents: A review', *Bioresource Biotechnology*, 58, pp. 217–227.

Blanquez, P. *et al.* (2004) 'Mechanisms of textile metal dye biotransformation by Trametes versicolour', *Journal of Water Research*, 38, pp. 2166–2172.

Camarero S., Ibarra, D., Marti'nez, M.J., and Angel, T.M. (2005) 'Lignin-derived compounds as efficient laccase mediators for decolourization of different types of recalcitrant dyes', *Applied Environmental Microbiology*, pp. 1775–1784.

Chander, M., Arora, D.S., and Bath, H.K. (2004) 'Biodecolorization of some industrial dyes by white-fungi', *Journal of Industrial Microbiology & Biotechnology*, 31, pp. 94–97.

Chivukula, M., and Renganathan, V. (1995) 'Phenolic azo dye oxidation by laccase from *Pyricularia oryzae*', *Applied Environmental Microbiology*, 61, pp. 4347–4377.

Duran, N., and Esposito, E. (2002) 'Potential applications of oxidative enzymes and phenoloxidase-like compounds in wastewater and soil treatment: A review, *Applied Catalysis B: Environmental*, 28, pp. 83–99.

Forgacs, E., Cserhatia, T., and Oros, G. (2004) 'Removal of synthetic dyes from wastewater: A review', *Environment International*, 30, pp. 953–971.

Garg, S.K., and Tripathi M. (2016) 'Microbial strategies for decolorization and detoxification of azo dyes from textile effluents,' *Research Journal of Microbiology*, 12(1), pp. 1–19.

Gill, P.K., Arora, D.S., and Chander, M. (2002) 'Biodecolourization of azo and tripphenylmethane dyes by Dichomitus squalens and Phlebia spp', *Journal of Industrial Microbial Biotechnology*, 12, pp. 869-872.

Gingell, R., and Walker, R. (1971) 'Mechanisms of azo reduction by *Streptococcus faecalis*: The role of soluble flavins', *Xenobiotica*, 1(3), pp. 232–239.

Glenn, J.K. *et al.* (1983) 'An extracellular requiring H_2O_2 enzyme preparation involved in lignin biodegradation by the white-rot basidiomycete *Phanerochaete chrysosporium*', *Biochemical and Biophysical Research Communications*. 114, pp. 1077–1083

Glenn, J.K., Akileswaran, L., and Gold, M.H. (1986) 'Mn (II) oxidation is the principal function of the extra-cellular Mn-peroxidase from *Phanerochaete chrysosporium*', *Archives of Biochemistry and Biophysics*, 251, pp. 688–696.

Hussein, A., and Scholz, M. (2017) 'Dye wastewater treatment by vertical-flow constructed wetlands', *Ecological Engineering*, 101, pp. 28–38.

Jamee, R., and Siddique, R. (2019) 'Biodegradation of synthetic dyes of textile effluent by microorganisms: An environmentally and economically sustainable approach', *European Journal of Microbiology and Immunology*, 9(4), pp. 114–118.

Khalid, T. *et al.* 'Microbial decolorization of textile effluent', *RADS Journal of Biological Research & Applied Sciences*, 2016, pp. 28–34.

Kirby, N., Marchan, R., and McMullan, G. (2000) 'Decolourization of synthetic textile dyes by *Phlebia tremellosa*', *FEMS Microbiology Letters* 188, pp. 93–96.

Kuwahara, M., Glenn, Morgan, K.J.M. A., and Gold, M.H. (1984) 'Separation and characterization of two extracellular H_2O_2 dependent oxidizes from ligninolytic cultures of *Phanerochaete chrysosporium*', *FEBS Letters*, 169, pp. 247–250.

Lucas, M. *et al.* (2008) 'Synthetic dye decolorization by white rot fungi: Development of original microtitre plate method and screening', *Enzyme and Microbial Technology*, 42, pp. 97–106.

Myers, C.R., and Myers, J.M. (1992) 'Localization of cytochromes to the outer membrane of anaerobically grown *Shewanella putrefaciens* MR-1', *Journal of Bacteriology*, 174(11), pp. 3429–3438.

Novotny, C. *et al.* (2004) 'Biodegradation of synthetic dyes by *Irpex lacteus* under various growth conditions', *International Biodegradation and Biodegradation*, 54, pp. 215–223.

Organum, N., and Bacon, F. (2006) 'Bioremediation technologies', in, Alvarez, P.J.J., and Illman, W.A. (eds.) *Bioremediation and natural attenuation*. New Jersey: John Wiley and Sons, pp. 351–455.

Paszczynski A, Crawford RL. (1991) Degradation of azo compounds by ligninase from Phanerochaete chrysosporium: involvement of veratryl alcohol. *Biochem Biophys Res Commu.* 178(3), pp. 1056–1063.

Pointing, S. B., Bucher, V.V.C., and . Vrijmoed, L.L.P., (2000) *World Journal of Microbiology and Technology*, 16, pp. 199–205.

Reyes, P., Pickard, M.A., and Vazquez-Duhal, R. (1999) 'Hydroxybenzotriazole increases the range of textile dyes decolourized by immobilized laccase', *Biotechnology Letters*, 21, pp. 875–880.

Robinson, T. *et al.* (2001) 'Remediation of dyes in textile effluent: a critical review on current treatment technologies with a proposed alternative', *Bioresource Technology*, 77, pp. 247–255.

Rodriguez, E., Pickard, M.A., and Vazquez-Duhal, R. (1999) 'Industrial dye decolorization by laccases from ligninolytic fungi', *Current Microbiology*, 38, pp. 27–32.

Russ, R., Rau, J., and Stolz, A., (2000) 'The function of cytoplasmic flavin reductases in the reduction of azo dyes by bacteria', *Applied Environmental Microbiology*, 66(4), pp. 1429–1434.

Shah, K. (2014) 'Biodegradation of azo dye compounds', *International Research Journal of Biochemistry and Biotechnology*, 1(2), pp. 5–3.

Tien, M., and Kirk, T.K. (1983) 'Lingin- degrading enzyme from the hymenomycete *Phanerochaete chrysosporium* Burds', *Science*, 221, pp. 661–663.

Uysal, Y., Aktas, D., and Caglar, Y. (2014) 'Determunation of colour removal efficiency of *Lemna minor* L. from industrial effluents', *Journal of Environmental Protection and Ecology*, 15, pp. 1718–1726.

Wariishi, H., Akileswaran, L., and Gold, M.H. (1988) 'Manganese peroxidase from the basidiomycete *Phanerochaete chrysosporium*: Spectral characterization of the oxidized states and the catalytic cycle', *Biochemistry*, 27, pp. 1736–1744.

Yan, H. *et al.* (2004) 'Study on the extraction and purification of microcystins', *Acta Scientiae Circumstantiae,* 24(2), pp. 355–359.

Yoo, E.S., Libra, J., and Adrian, L. (2001) 'Mechanism of decolourization of azo dyes in anaerobic mixed culture', *Journal of Environmental Engineering and Science*, 127(9), pp. 844–849.

11 Removal of Emerging Contaminants in Water and Wastewater by Microbes

Shubhangi Parmar[1], Sagar Daki[1], Sourish Bhattacharya[2], and Anupama Shrivastav[1]

[1]Department of Microbiology, Parul Institute of Applied Sciences, Parul University, Vadodara, Gujarat, India

[2]Process Design and Engineering Cell, CSIR, Central Salt and Marine Chemicals Research Institute, Bhavnagar, 364002, India

*Corresponding author: anupama.shrivastav82045@paruluniversity.ac.in

CONTENTS

DOI: 10.1201/9781003130932-11

11.1 INTRODUCTION

Many different compounds, such as pharmaceuticals, personal care products (PCPs), and synthetic compounds are being used by millions of people and have become an indispensable component of society. Modern society is continuously increasing its demand for organic and inorganic chemical compounds in daily use. Recently, scientific research has focused on studying the occurrence, fate, and consequences of these anthropogenic substances, which could have harmful effects on terrestrial as well as aquatic ecosystems with chronic exposure. The concentration of persistent organic pollutants and heavy metals that are released into the environment reduced drastically in the 1970s when legislation forced the reduction of these compounds, because there were concerns about the environment. The priority pollutants were defined and intensive monitoring combined with control programs were implemented. Since then, a broad range of manmade chemicals, synthetic drugs, agricultural chemicals, industrial compounds, and consumer products along with some by-products of production processes are continuously emitted. The major concerns that are associated with these anthropogenic compounds are bioaccumulation in human and animal tissue, biomagnification at different levels of the food chain, and the consequences on human health and the environment.

Numerous definitions have been presented for ECs globally by different researchers. Some of the most relevant terms, which cover an introduction to all aspects of ECs, are given below.

1. A contaminant whose new origin, an alternate route to humans, or techniques for treatment has been under constant innovation is termed as "emerging". They pose probable or actual risk to human health (US EPA, 2012).
2. The term "emerging pollutants" is apt for the substances that are released in the environment to which no regulations are present currently, for their environmental monitoring (Thomaidis *et al.*, 2012).
3. ECs are synthetic or naturally occurring chemicals or any microbes, which are not commonly monitored in the environment, but they enter the environment and are suspected to have eco-toxicological or impact on human health (Rosenfeld *et al.*, 2011).

Current water and wastewater treatment facilities do not have mechanisms for the effective elimination of these compounds, which might occur in minute concentrations. The question remains, what would be the effects of chronic exposure to these contaminants on health and the environment. The constant addition of these compounds via wastewater or sewage into the aquatic environment remains a threat, because the exact impacts remain unknown. (Oulton *et al.*, 2010).

ECs are not regulated in drinking water and the environment. The traditional systems used in the treatment of wastewaters in all urban wastewater treatment (UWTPs) plants consists of three parts: conventional primary treatment, secondary treatment, and tertiary treatment. In addition, disinfection is achieved in some UWTPs by ultraviolet (UV) radiation or chlorination treatment before the discharge of the treated water into the environment. However, these treatments do not ensure the complete elimination of all the ECs that are present in wastewater, because they are usually present in wastewater in ng/L to µg/L concentrations; therefore, additional techniques combined with conventional methods need to be applied (Batt *et al.*, 2007). Terms, such as contaminants and pollutants have been applied without differentiation, but there is a slight difference in the meaning of both words. Contaminant refers to any substance that enters the environment and may or may not result in any adverse biological effects on the system. However, all pollutants are contaminants that mostly adversely affect biological systems. In summary, pollutants have adverse biological effects and contaminants pose a potential biological and environmental risk, which needs to be confirmed from chemical analysis and toxicological data. Following the precautionary public and environmental health issues, biodegradation of ECs is the most sustainable and eco-friendly process. Microorganisms are important to mankind and can be applied to various advanced oxidation processes, which can degrade the ECs.

Biodegradation is one of the most complex processes that occur during biological treatment and remarkable characteristic of biodegradability in microbes makes them ideal candidates for applications in bioremediation. This chapter discusses various ECs, their sources, effects, and different bioremediation approaches that are used to eliminate them from wastewater treatment plants (WWTPs).

11.2 SOURCES OF WASTEWATER AND TYPES OF ECS PRESENT

Knowledge of the different types of ECs could help to predict different sources from which they could potentially be added to the environment. According to different researchers, the substances that could be considered as ECs vary. The most frequently discussed ECs are given in Table 11.1 (Petrović *et al.*, 2003).

Every human is directly or indirectly responsible for the addition of ECs to various ecosystems. A summary of the contaminants that are extensively studied as characteristic major ECs are discussed in the following sections.

11.2.1 PHARMACEUTICALS

Pharmaceuticals are referred to as a group of chemical compounds that have medicinal properties and includes all prescription, nonprescription, and all other over the counter therapeutic and veterinary drugs. Pharmaceuticals are produced worldwide, on a scale of 100,000 t per year due to increasing usage globally. These compounds possess anthropogenic properties, and they continuously reach water supplies. They are formulated with a molecular weight <500 kDa along with high bioavailability. Of note, they are produced to have specific pharmacological and physiological effects at low doses, and they could have unintended outcomes for wildlife and ecosystems. With expanding potential markets, their production rate will not decrease in the future. Following administration, they are never 100% absorbed in the body and the parent compounds combine with different metabolites in the body and become more hydrophilic in nature. After they have achieved their aim, a broad range of compounds are eliminated, either unchanged as by-products, which are not eliminated by traditional WWTPs (Halling *et al.*, 1998). Antibiotics are major pharmaceuticals that are included as ECs. The increased and unneeded use of antibiotics combined with the spread of resistance in the environment poses a serious risk to humans.

TABLE 11.1
List of Most Commonly Studied ECs

Algal toxins	(EDCs)
Biocides	Synthetic musk
Disinfection by-products	Per-fluorinated substances
Drugs of abuse	Veterinary products
Flame retardants	Food additives
Preservatives	Microplastics
Nanomaterials	Steroids
Pharmaceuticals	Surfactants
PCPs	Plasticizers
Plant protection products	Microplastics
Trace metals	Anticorrosive agents

Sources: Mandaric *et al.* (2015) and Lapworth *et al.* (2012).

11.2.2 PCPs

PCPs are another class of ECs that incorporate all consumer chemicals that are found, for example, in fragrances, lotions, shampoos, cosmetic products, steroids, and sunscreens.

11.2.2.1 Synthetic Musk Compounds

Synthetic musk compounds are bioaccumulative and persistent xenobiotic compounds that are used in detergents, cosmetics, perfumes, and PCPs to add fragrances. Musks are classified into four major categories that depend on their physical and chemical properties: (1) nitro musks, which includes musk ketone; (2) musk xylene; (3) musk tibetan; and (4) musk moskene. Polycyclic musks are another category that includes galaxolide, tonalide, celestolide, cashmeran, and traseolide. Macrocyclic musks include compounds, such as ambrettolide, muscone, brassilite, and globalide, and acyclic musks include romandolide and helvetolide. Synthetic musk compounds are predominantly used in PCPs with different chemical structures. The most common examples include tonalide (AHTN), galaxolide (HHCB), musk xylene, and musk ketones (Clara *et al.*, 2011).

11.2.2.2 Preservatives with Antimicrobial Activity

Antimicrobial preservatives are chemicals that inhibit or kill microbes in consumer products to remove contamination. Parabens are the most common ingredients listed in products, such as shampoos, commercial moisturizers, shaving gels, makeup, lubricants, and toothpaste. They have excellent bactericidal and fungicidal properties and benzyl, butyl, isobutyl, isopropyl, methyl, propyl compounds are currently used. These compounds are used as food additives. Preservatives are becoming controversial because of their increased microbial and algal toxicity, estrogenicity, and their contribution toward antimicrobial resistance when added to the environment (Golden *et al.*, 2005).

11.2.2.3 UV Filters

UV filters are organic compounds in sunscreen and cosmetics that protect against harmful UV radiation from the sun. Organic compounds that are incorporated into UV filters include benzophenones, 2-phenylbenzimidazole-5-sulfonic acid (PBSA), isoamyl methoxycinnamate (IAMC), and octocrylene (OC). These compounds are relatively stable against biotic degradation (Fent *et al.*, 2008).

11.2.3 Per-fluorinated Compounds

Per-fluorinated compounds (PFCs) are fully fluorinated hydrophobic linear carbon (C) compounds with hydrophilic heads. Common PFCs include perfluorooctanoic acid (PFOA), perfluoro-octane sulfonic acid (PFOS), and perfluoro-nonanoic acid (PFNA). Because these compounds repel oil and water they are used in surface treatments. The use of these compounds includes surfactants in emulsion polymerization, the semiconductor industry, and to manufacture Teflon using fluoropolymers. Industrial applications of these compounds include the textile, paint, electronics, and food packaging industries. PFCs are resistant to break down and might accumulate in different levels of the food chain (Clara *et al.*, 2008).

11.2.4 Nanomaterials

Nanomaterials are small substances and materials that are scaled at nano levels and are <100 μm in ≥1 dimension. Nanomaterials have different categories, such as organic and inorganic nanomaterials. Organic nanomaterials include all the carbon-based substances, such as fullerenes, graphene, and C nanotubes and the latter includes zinc oxide, silver chloride, and titanium dioxide. Due to their unique size, nanomaterials have unique physical, chemical, and biological properties. They can travel long distances compared with particles of the same size suspended in air and water (Klaine *et al.*, 2008).

11.2.5 OTHER ECs

Natural and synthetic steroids are the most potent EDCs and are widely used to treat hormonal disorders and cancer. They affect aquatic organisms and increase estrogenic responses even at low concentrations. Another ubiquitous emerging contaminant is plasticizers, which are a large group of compounds represented by dimethyl, diethyl, dibutyl, and butyl benzyl. Applications of plasticizers include resins, adhesives, and cellulose film coating and polyvinyl chloride resins and compounds. Anticorrosive agents, such as benzotriazole and tolytriazole are extensively used to inhibit corrosion and as silver polishing agents in dishwasher powders and hydraulics fluids. However, they have a very low sorption capacity and are not readily degradable. Surfactants are another major emerging contaminant that is used to reduce the surface tension of water and form micelles in solvents. They are among the most produced organic chemicals used by households that are generated by the use of detergents, soaps, shampoos, and laundry aids. Industries that used surfactants include textile, leather, petroleum, pesticide, and emulsions (Mandaric *et al.*, 2015).

11.2.6 SOURCES OF ECs

Of note, there is a list of compounds that could be included in ECs. After becoming familiar with the different types of Ecs, their sources can be easily predicted. The major sources of these compounds are water and there are different water systems, such as groundwater, industrial wastewater, agricultural waters, and sewage. The WWTPs are a crucial destination where all such water travels to undergo traditional treatment. However, traditional WWTPs are designed for the elimination of pathogens and suspended matters and were not developed for the removal of these microcontaminants (Mandaric *et al.*, 2015). Municipal WWTPs are one of the major sources for the release of ECs, because most of the urban WWTPs still use secondary biological treatments, which only remove a fraction of ECs. Drinking water has been affected due to the high solubility of some ECs, which means that they are not susceptible to manmade treatments. Remarkable work has been carried out on the performance of wastewater technologies for the removal of nutrients: however, the ability to remove ECs and the toxicological impacts of these compounds on surface and drinking waters needs to be determined. ECs can enter the environment from landfill sites, aquaculture, power stations, farms, leaching of fertilizers and spills from oil pipelines. (Daughton *et al.*, 2004).

11.3 THREATS OF EMERGING POLLUTANTS ON THE ENVIRONMENT AND HEALTH

Rapid industrialization, large-scale manufacturing, improper management of waste, and its disposal has produced newly identified contaminants that pose threats to terrestrial and aquatic ecosystems. ECs mainly have adverse effects on human health and the environments to which they are added. The toxicity of these contaminants is associated with many factors including the chemical structure of compounds, and the different sections, such as soil, sediment, air, ground or surface waters that receive the compound, different physicochemical properties, such as solubility, adsorption properties, volatility and its biological targets including fish, birds, and humans. (de Oliveira *et al.*, 2013). The toxicity profiles and the threats from major ECs to the environment will be discussed further. One of the major threats from the increased discharge of ECs is the pollution of aquatic ecosystems and environments. Evidence of the exposure to ECs and their association with reproductive and health effects in humans and other living organisms have been noted by various researchers. Rivers,

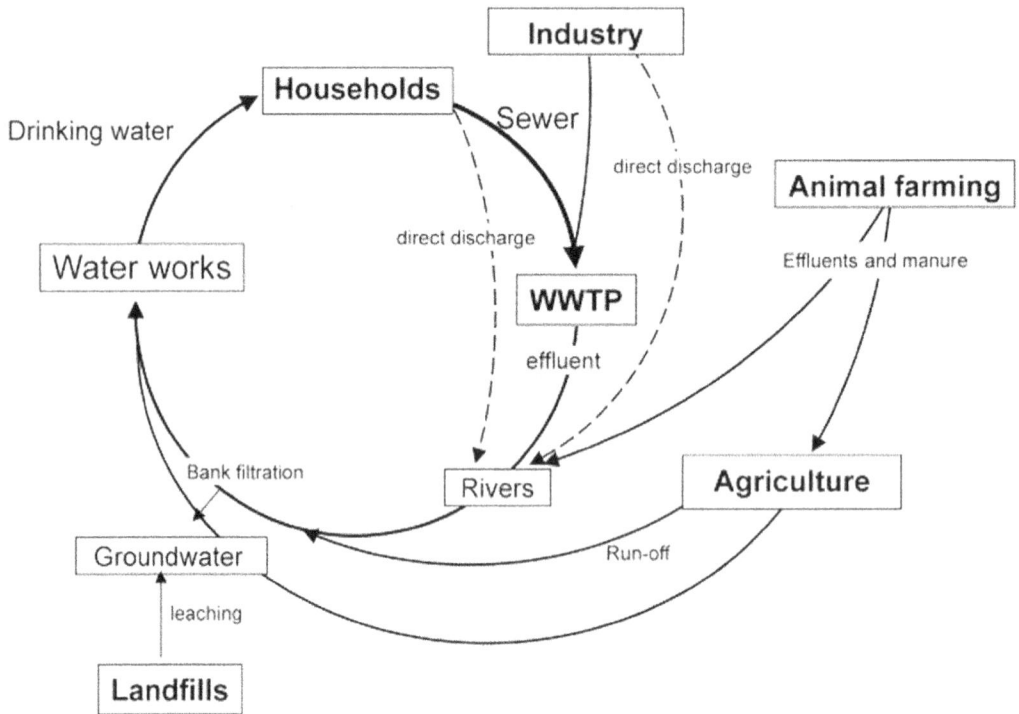

FIGURE 11.1   Different sources and migration of ECs in the environment.

Source: (Petrović *et al.*, 2003).

streams, groundwater, and marine environments are all susceptible to water pollution. Water pollution occurs from several specific sources, which includes municipal sewage, industrial wastewaters, farm effluents, domestic drainage, and nonspecific sources, such as wash off from agriculture and roads.

Endocrine disruption is another threat that is associated with ECs. ECs can disrupt the endocrine system by mimicking and blocking hormone functions or by interfering with normal hormone functions. Aquatic ecosystems are the most affected by these EDCs, because of bioaccumulation and biomagnification. Various studies highlighted the abnormalities observed in aquatic species, such as the abnormal ratio of androgens including estrogen and testosterone, abnormalities in reproductive tissues, and sexual abnormalities in fish that lived near WWTPs. Exposure to EDCs in humans can have effects including lower sperm count, and increases in prostate, ovarian and breast cancers combined with reproductive malfunctions (Bolong *et al.*, 2009).

11.4   CONVENTIONAL METHODS FOR THE TREATMENT OF WASTEWATERS

Wastewater treatment involves the breakdown of organic compounds into simpler compounds; therefore, the adverse environmental impacts are minimized when this water is discharged into the groundwater. The contaminants in the wastewater are eliminated by physical, chemical, and biological methods. Physical methods include treatment methods in which physical forces are involved. Screening, flocculation, sedimentation, filtration, and adsorption and are examples of physical methods. Physical methods include an adsorption process in which a substance on the material is concentrated at a solid surface and separated from its liquid or gaseous surroundings. Filtration includes varied approaches, such as microfiltration, ultrafiltration, and nanofiltration. Methods that include the addition of chemicals or chemical reactions for the removal of contaminants are chemical techniques. Chemical precipitation is accomplished by producing a precipitate that settles that contains the chemicals added and the contaminants that are removed along with the precipitate. Ozonation is another chemical method that is based on oxidation where hydrogen peroxide (H_2O_2) combined with an iron catalyst (known as Fenton's reagent) is used for the treatment of hazardous wastewater. A photocatalytic method is applied to complex contaminants by exposing them to UV treatment in the presence of H_2O_2, which produces hydroxyl radicals. Biological treatments are used to remove biodegradable compounds from wastewater. Removal of these contaminants is achieved via biological processes and biochemical pathways. Biological treatments are easily reproducible and eco-friendly strategies. Activated sludge was designed to remove C and was then applied to remove other ECs. Biological methods use the natural characteristics of various microbes, such as algae, fungi, and bacteria to detoxify ECs (Adin *et al.*, 1998).

Typical wastewater treatment has steps including preliminary, primary, secondary, and tertiary treatments and advanced nutrient removal procedures. The preliminary treatment removes any wastewater constituents that could cause operational problems by becoming stuck in the system. Screening, grit removal, and skimming are used as preliminary treatments for the removal of dead animals, rags, oils and tree branches. In addition, primary treatment is a precursor for secondary treatment, because a fraction of the suspended solids and organic matter is removed during this treatment via sedimentation techniques. When primary treatment is complete, further secondary treatment of the effluent occurs. Filters, aerated tanks, and oxidation ponds are part of the secondary treatment where biological decomposition of the organic matter occurs under aerobic or anaerobic conditions via bacteria. In anaerobic secondary treatment, anaerobic lagoons and treatment tanks are used. Tertiary treatments are advanced processes that are carried out after the secondary process to remove toxic compounds, nutrients, and to kill pathogenic bacteria. Processes, such as coagulation sedimentation, filtration, and chlorination are employed in the advanced tertiary treatment. Coagulation, flocculation, sedimentation, and filtration techniques are used to reduce toxic substances and heavy metals (Topare *et al.*, 2011).

11.5 MICROBIAL DETOXIFICATION AND ITS ADVANTAGES

Recently, the amount of research into the use of efficient processes to clean up and minimize pollution in water bodies has increased. Bioremediation has gained considerable attention due to its characteristic environmental compatibility and eco-friendliness. Bioremediation is composed of the natural processes that use biological agents, such as microorganisms and green plants, which are applied to remove toxic compounds from the environment and turn toxic compounds into a natural nonhazardous state. Bioremediation that uses microbes has been effective even for very low concentrations of ECs where most of the physicochemical methods cannot operate efficiently. Microorganisms can adapt to harsh environmental conditions and function even if present in very dilute samples. Commonly used microbes in bioremediation include individuals or consortia of bacteria, algae, and fungi (Coelho *et al.*, 2015). The action of microbes, under specific conditions, can achieve contaminant mineralization, transformation, or immobilization. Various physicochemical methods used require large capital and high maintenance costs and after investment that they are not as efficient as bioremediation. Microbial detoxification has low operating costs, and no additional chemicals or catalysts are required and no hazardous by-products are formed when the process is complete. Microbial mediated detoxification has gained significant public acceptance because they predominantly depend on natural biochemical and metabolic pathways in microbes without any additional modifications. In addition, the contaminants are not transferred from one environment to another but are eliminated by detoxification. It is a cheap process compared with the other methods and requires minimal space and equipment to operate. The ability of microorganisms to use different substrates means that they are the most appropriate organisms that can be exploited for bioremediation. The characterization of microbes can be achieved from the wastewater that contains ECs, because microbes have developed specific metabolic processes to use ECs as energy sources with time. The process could be stimulated by controlling parameters, such as temperature, pH, moisture content, oxygen (O_2) availability, and interaction time with ECs. The bioremediation of wastewaters using microbes is a very promising method due to its efficiency, cost-effectiveness, and ease of operation. Microbial detoxification could be applied in processes, such as biofiltration, phytoremediation, bioadsorption, and biotransformation and is gaining attention because it converts ECs into less toxic forms without the generation of secondary pollutants. Based on these advantages, microbes are excellent tools to restore the natural environments and prevent pollution (Coelho *et al.*, 2015; Kensa *et al.*, 2011; Shah *et al.*, 2020).

11.5.1 OPERATIONAL PARAMETERS THAT AFFECT MICROBIAL ACTIVITY DURING BIOREMEDIATION

Bioremediation can be influenced by various factors, such as the chemical components, the concentration of pollutants, and the availability of microorganisms. These are grouped into two different categories: (1) factors that affect microbial growth conditions; and (2) the characteristics of ECs in wastewater. Biological factors include competition between microbes for limited sources and phenomena, such as antagonism and predation by phages and protozoans. The expression of specific enzymes by microbial cells has a direct relationship with the rate of degradation. Different interactions, mutations, horizontal gene transfer, and population size influence bioremediation. The growth and activity of microbes are affected by environmental factors including pH, temperature, moisture, the solubility of contaminants in the water, nutrients, redox potential, contaminant concentration, type, chemical structure, and toxicity. The toxic contaminants can inhibit the metabolic activities of the microorganisms and slow down degradation (Abatenh *et al.*, 2017).

11.6 DIVERSITY OF MICROORGANISMS USED IN WASTEWATER TREATMENT

A variety of different microorganisms are found in different environments. Microbial technologies help in the reduction of contamination by adapting the pathways that microbes developed due to exposure to toxic waste. Numerous naturally resistant strains could be used to biotransform toxic chemical compounds into less hazardous forms (Saratale *et al.*, 2011).

The biodegradation of ECs includes pathways for the removal of pharmaceutical and other associated compounds from wastewater and surface water. The microbial species diversity that inhabits the earth is estimated at approximately 1 trillion, in which bacteria are abundant. In addition, bacteria are known to be the oldest life forms on earth. Because they play a dominant role on the planet, their diverse metabolic characteristics can be used for all types of biotechnological applications. (Locey *et al.*, 2016).

Pharmaceuticals including alprenolol, bisoprolol, metoprolol, propranolol, venlafaxine, and norfluoxetine present in an aerobic granular sludge reactor had bacterial diversity with the most abundant phyla including Proteobacteria, Bacteroidetes, Actinobacteria, Chlamydiae, Chloroflexi, Cyanobacteria, Epsilonproteobacteria, Firmicutes, Gemmatimonadetes, Temericutes, and Verrucomicrobia (Amorim *et al.*, 2018).

Studies indicated that the removal efficiencies of pharmaceuticals, such as atenolol, atorvastatin, azithromycin, caffeine, carbamazepine, ciprofloxacin, clarithromycin, diclofenac, erythromycin, fluoxetine, ibuprofen, ketoprofen, metoprolol, naproxen, and paracetamol significantly increased with biostimulation that provided nutrients to sludge the phyla Verrucomicrobia, Nitrospirae, Chlorobi, Acidobacteria, Chloroflexi, Actinobacteria, and Firmicutes. Compounds, such as methylamine and trimethylamine dimethylamine were removed by ≤96% by microbes including *Paracoccus sp.* and *Arthrobacter sp*. Fungal species employed for the removal of pharmaceuticals and other recalcitrant analgesics, anti-inflammatory, antibiotics, and psychotropic drugs included *Trametes versicolor* (Ramírez-Durán *et al.*, 2017).

PCPs typically contain compounds, such as triclosan, carbamazepine, ibuprofen, and sulfamethoxazole. These compounds can be effectively eliminated by microbes including *Pseudomonas aeruginosa, Acinetobacter sp., Firmicutes sp., Betaproteobacteria, Bacillus subtilis, Pseudomonas putida, Sphingomonas sp., Rhizobium sp., and Nitrosomonas europaea* under the different operational conditions (Prasad *et al.*, 2019). The interaction between microbes and the environment and contaminants significantly affects bioremediation and its efficiency. Significant species of microbes employed for the bioremediation of different compounds are discussed further.

P. putida is a common gram-negative organism employed for the bioremediation of dyes and toluene containing compounds. In addition, it is used for the biodegradation of contaminants from the petroleum industries including naphthalene (Petersen *et al.*, 1996). *Dechloromonas aromatica* can oxidize aromatic compounds via a coupling reaction with reducing chlorate and nitrate. It is a rod-shaped bacterium that can oxidize benzene under anaerobic conditions. Surface waters and groundwater contain aromatic compounds, because substances that contain these compounds are used in households. However, *D. aromatica* could be used for bioremediation (Chakraborty *et al.*, 2005).

Industrial effluents have a variety of nitro compounds in their wastewater. Microbial activities are mainly used for the elimination of mineral nitro compounds including ammonia, nitrites, and nitrates. Microbial detoxification of nitro compounds occurs via nitrification and denitrification. Organisms, such as *N. europaea* are used for the oxidation of ammonium to nitrite. Nitrite is oxidized further to nitrate by *Nitrobacter sp.*, such as *Nitrobacter hamburgensis*. Following the process, if there are anaerobic conditions, the nitrate produced is used as a terminal electron acceptor by microbes, such as *Paracoccus denitrificans* and nitrogen gas is produced.

Extremophiles, such as *Deinococcus radiodurans* are applied for the bioremediation of heavy metals and solvents in wastewaters. This microbe is resistant to radiation and can eliminate ionic mercury substances from wastewater (Brim *et al.*, 2000). Bioremediation of compounds such as methyl tert butyl ether (MTBE) is achieved by *Methylibium petroleiphilum*, which uses MTBE as the sole C and energy source (Hanson *et al.*, 1999). Environments that are contaminated by hydrocarbons, grease, or oil are successfully cleaned using *Alcanivorax borkymensis*, which is a marine bacteria. Its successful application was noted in the Gulf of Mexico to clean gallons of oil from an oil spill (Biello *et al.*, 2010).

Archaea have potential in the bioremediation of ECs. Archaea have greater catabolic diversity than expected. Archaea bacteria, such as *Halobacterium, Haloferax, and Halococcus* coccus are widely applied for the biodegradation of hydrocarbons and oil spills. Mycoremediation uses fungi for the elimination of toxic compounds. White-rot fungi are extensively used, because of their ability to degrade lignin extracellularly by their hyphal extensions. Strains, such as *Phanerochaete chrysosporium* have the potential for the elimination of hydrocarbon pollutants, remediating metals, and pesticide removal (Margesin *et al.*, 2001).

Before analyzing different bioremediation approaches that occur in WWTPs for the removal of ECs, it is important to understand the different types of transformations and reactions that occur within these systems. Transformations occur via different types of biochemical reactions that include hydrolysis, oxidation, reduction, and the total mineralization of compounds into compounds including C, nitrogen (N), and O_2. In microbe mediated transformations of ECs in wastewater plants, enzymes play a vital role in enzyme-dependent degradation. The exact process of degradation remains unclear; however, enzyme-dependent pathways can be divided into the following steps.

1. Bioavailability and transport of contaminants to the microbe.
2. Uptake of contaminants and diffusion across the cell walls by intracellular enzymes.
3. Formation of a contaminant–enzyme complex after binding to the reaction site.

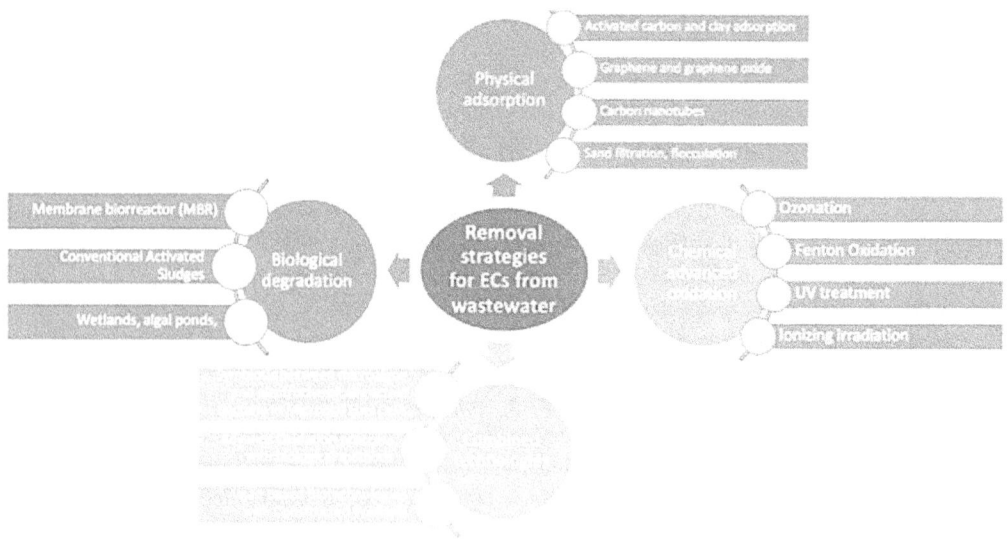

FIGURE 11.2   Different removal strategies for ECs from wastewater.

Source: Tolboom *et al.* (2019).

4. Activation of the complex with the help of cofactors and coenzymes.
5. Release of the transformed product into the environment.

The previously mentioned steps are affected by different factors, because enzymes are specific for the type of contaminants they bind combined with the polarity of the contaminants and their transport across microbial membranes (Christensen *et al.*, 2014).

Photochemical reactions result in the transformation of ECs into other molecules that are more biodegradable or hydrolysable by the cleavage of the covalent bonds between molecules in presence of solar radiation or photosensitized species. Photochemical transformation may be direct through the absorption of direct solar radiation or occur indirectly with the help of photosensitized species. Photochemical transformations are widely applied in sewage treatment plants for the degradation of ECs (Richard *et al.*, 2005). Apart from this many physical, chemical, and biological approaches are used in the removal of ECs, which are shown in Figure 11.2.

11.7   DIFFERENT BIOREMEDIATION APPROACHES FOR ECS IN WASTEWATER TREATMENTS

Bioremediation techniques depend on biomass for their successful implementation. Biomass refers to any material that is directly or indirectly produced by living microorganisms, animals, or plants. Different efficient and sustainable methods that use biomass are discussed further.

11.7.1   BIOSORPTION-BASED REMEDIATION

Adsorption means the mass transfer of substances that occurs between any two states of matter. The main principle that applies is the absorption of any contaminants on the surface of an adsorbent due to the presence of intermolecular forces. Two major events that occur during adsorption are physisorption and chemisorption. Physisorption is used when weak physical forces are responsible for adsorption. Chemisorption includes the formation of chemical bonds between a solid surface and the adsorbents. Various types of biomass have been used instead of commercially available synthetic adsorbents, which could help in the reduction of operational costs and environmental pollution. Materials used as bioadsorbents include plants, bacteria, algae, and fungi. In addition, several agricultural wastes, such as bamboo chips, eucalyptus bark, rice husk and straws, coconut husks, coir dust, wood chips, and fruit peels have been employed for bioadsorption. Bioadsorbents are easily available and are economical; therefore, they are viable alternatives for the removal of ECs. Of note, the selection of a suitable bioadsorbent is a crucial step for the successful implementation of the process. Factors that commonly affect the process include particle size, temperature, pH, contact time, ionic strength selectivity, and the concentration of the adsorbent available (Ismail *et al.*, 2020).

11.7.2 PHYTOREMEDIATION AND MICROBIAL REMEDIATION

Phytoremediation is sustainable for the removal of contaminants from various environments. The term means using plants, plant microbes, and the total microbiota of plants to remove contaminants from soil and water. It occurs as a natural process in the ecosystem, and therefore, is considered a low-cost remediation method. The mechanism of phytoremediation occurs via the uptake of contaminants by plants through their roots and the absorption of water nutrients combined with that. The contaminants are stored in various parts of plants and are converted into less harmful, nontoxic substances during transpiration. Phytoremediation occurs in many different ways and the different categories of phytoremediation are listed in Table 11.2. Phytoremediation is used for the treatment of wastewaters, surface waters, and groundwaters (Lakshmi *et al.*, 2017).

11.7.3 MEMBRANE-BASED BIOREACTORS

Membranes have been used for several decades for aerobic and anaerobic treatments. Membrane-based bioreactor technologies are emerging as suitable processes for different municipal wastewater treatments. The reactors operate similarly to activated sludge processes, but the tertiary stages are not required. In membrane-based bioreactors, feed water that contains the contaminants is mixed with biomass. Then, this mixture undergoes filtration across the membrane, and then the biomass is separated from the treated effluent.

There are two different types of membrane alternatives: submerged membrane bioreactors and side-stream membrane bioreactors, which can be used during the operation. The submerged membrane bioreactor operates under vacuum conditions and the side-stream operates under pressure. In the submerged process, the membrane is placed directly in liquid, and a pump is used in the side-stream

---

TABLE 11.2
Subsets of Different Phytoremediation Methods

Technique	Description
Phyto extraction	Accumulation of pollutants in harvestable biomass (i.e., shoots)
Phyto filtration	Sequestration of pollutants from contaminated waters by plants
Phyto stabilization	Limit the mobility and bioavailability of pollutants in soil by plant roots
Phyto volatilization	Conversion of pollutants into volatile form and their subsequent release into the atmosphere
Phyto degradation	Degradation of organic xenobiotics by plant enzymes within plant tissues
Phyto desalination	Removal of excess salts from saline soils by halophytes
Rhizo degradation	Degradation of organic xenobiotics in the rhizosphere by rhizosphere microorganisms

Source: Ali *et al.* (2013).

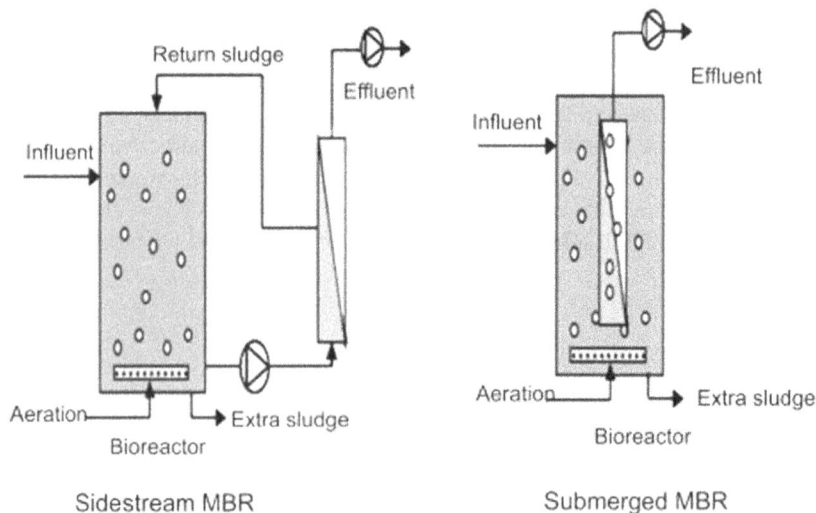

FIGURE 11.3 Schematic diagram of side-stream membrane bioreactor and submerged membrane bioreactor.

Source: Deowan *et al.* (2015).

process for access of the effluent to the membrane. The difference between these processes is shown in Figure 11.3. The removal of contaminants is based on many factors and the experimental setup. These systems have been applied efficiently for the removal of microcontaminants in WWTPs and other industrial effluent treatment plants (Ismail *et al.*, 2020; Mert *et al.*, 2018).

11.7.4 CONSTRUCTED WETLANDS

Constructed wetlands are manmade systems that are generated for wastewater treatment. The design setup includes the arrangement of plants, microorganisms, and soil combined with natural wetland ecosystems and hydrology. They utilize wetland vegetation combined with microbes for the treatment of waters from domestic, industrial, and agricultural runoff sources. Constructed wetlands are promising options for the removal of contaminants combined with nutrients from WWTPs. The characteristics of constructed wetlands are similar to natural wetlands and the elimination of contaminants occurs via a combination of biochemical processes that mimic wetlands in the natural environments Numerous types of wetlands have been constructed, which depend on the contaminants to be degraded. The wetlands are surface flow, hybrid constructed wetlands, and subsurface wetlands, which are based on vegetation flow and direction (Narayan *et al.*, 2018).

Emerging methods, such as microbial fuel cells (MFCs) have demonstrated great potential in the removal of ECs, which is discussed in the following sections of this chapter. Activated sludge is one of the numerous methods that can be employed for the bioremediation of ECs; however, it is one of the most conventional methods used for use for treatment.

11.8 SUCCESSFUL APPLICATIONS FOR THE REMOVAL OF ECS FROM WASTEWATERS USING BIOREMEDIATION

Numerous studies have been carried out on the removal of ECs from wastewaters by employing bioremediation approaches, some of which are highlighted in the following section.

11.8.1 Removal of ECs Using MFCs

MFCs are used to enhance bioremediation for the removal of contaminants. It is a device that uses bacteria for oxidation at the anode and eliminates contaminants and generates electricity. The microbes that are added to the anodic chamber are provided with a substrate. The substrate provided is anaerobically degraded by the microbes and electrons are released. The chemical energy is transformed into electrical energy via the oxidation of a variety of wastes that are consumed by electrochemically active microorganisms. Studies indicated that MFCs have an elimination efficiency ≤84% for 50 mg/L chloramphenicol within 12 h of MFC operation.

Sulfamethoxazole is the most common broad-spectrum antibiotic that was successfully degraded ≤85% at 20 ppm within 12 hours. MFCs that contain *B. subtilis* have shown promising results for the degradation of recalcitrant toxic compounds, such as phenols. MFCs are coupled with constructed wetlands and applied to azo dye decolorization with *Ipomoea aquatica* plants. This system achieved a 91.24% decolorization rate with a voltage output of 610 mV. In addition, they have been successfully applied for the elimination of substances including antibiotics, synthetic dyes, nitro compounds, ethyl acetate, pesticides, sulfur compounds, organic substances, and other ECs (Mandal *et al.*, 2021).

11.8.2 Removal of ECs by Spent Mushroom Compost

Novel techniques, such as mycoremediation that uses extracellular enzymes from white-rot fungi are employed to degrade ECs including acetaminophen and sulfonamides. Spent mushroom composts from nine different mushrooms were tested and *Pleurotus eryngii* had the highest removal rate for both compounds (Chang *et al.*, 2018).

11.8.3 Microalgal Removal of Pharmaceuticals from Wastewater

Compounds including paracetamol and salicylic acid were eliminated by the application of *Chlorella sorokinia* and the results concluded that nutrients, such as nitrates were removed ≤70% and 89% of phosphates were removed by the application of semicontinuous culture. In addition, semicontinuous culture achieved ≤93% elimination efficiency for salicylic acid (Escapa *et al.*, 2015).

11.8.4 Bioremediation of Agro-Industrial Effluent by Fungi

Studies were carried out on the elimination of pesticides including thiabendazole, imazalil, thiophanate, and diphenylamine from agricultural effluent by using three white-rot fungi *P. chrysosporium*, *T. versicolor*, and *Pleurotus ostreatus*. *T. versicolor* and *P. ostreatus* degraded all pesticides except for thiabendazole and were the most efficient strains. Pesticides, such as diphenyl-amine, ortho-phenyl phenol, and imazalil ≤50 mg/L were fully or partially degraded by *T. versicolor* (Karas *et al.*, 2011).

11.9 CONCLUSIONS AND FUTURE PERSPECTIVES

Different anthropogenic substances are constantly added to the environment due to their increased use globally. A large number of ECs are not reported due to a lack of knowledge. Limited research has been carried out on the elimination of these recalcitrant compounds and sufficient data is not available on this topic. Bioremediation technologies have demonstrated remarkable results for the elimination of these compounds. Bioremediation techniques are green strategies and can help in sustainable development. The proper characterization of the physicochemical properties of ECs and the specific biochemical pathways that occur in microbes requires analysis under different environmental conditions. Most of the elimination approaches employed are in the early stages of development and research into large-scale developments are required. Research is required in this field and integrative approaches combined with omics technologies for the maximum elimination of ECs via large-scale applications need to be developed. Future studies should focus on the characterization of novel microorganisms and their ability to remove these compounds. ECs should be extensively studied and new methods for their removal should be developed when their interaction pathways have been characterized.

REFERENCES

Abatenh, E., Gizaw, B., Tsegaye, Z. and Wassie, M. (2017) 'The role of microorganisms in bioremediation: A review', *Open Journal of Environmental Biology*, 2(1), pp. 30–46.

Adin, A. and Asano, T. (1998) 'The role of physical-chemical treatment in wastewater reclamation and reuse. *Water Science and Technology*', 37(10), pp. 79–90.

Ali, H., Khan, E. and Sajad, M.A. (2013) 'Phytoremediation of heavy metals: concepts and applications. *Chemosphere;*, 91(7), pp. 869–881.

Amorim, C.L. *et al.* (2018) Bacterial community dynamics within anaerobic granular sludge reactor treating wastewater loaded with pharmaceuticals', *Ecotoxicology and Environmental Safety*, 147, pp. 905–912.

Batt, A.L., Kim, S. and Aga, D.S. (2007) 'Comparison of the occurrence of antibiotics in four full-scale wastewater treatment plants with varying designs and operations', *Chemosphere*, 68(3), pp. 428–435.

Biello, D. (2010) 'Slick solution: how microbes will clean up the Deepwater Horizon Oil Spill', *Scientific American*, 25, pp. 1–4.

Bolong, N., Ismail, A.F., Salim, M.R. and Matsuura, T. (2009) 'A review of the effects of emerging contaminants in wastewater and options for their removal. *Desalination*', 239(1–3), pp. 229–246.

Brim, H. *et al.* (2000) 'Engineering *Deinococcus radiodurans* for metal remediation in radioactive mixed waste environments', *Nature Biotechnology*, 18(1), pp. 85–90.

Chakraborty, R., O'Connor, S.M., Chan, E. and Coates, J.D. (2005) 'Anaerobic degradation of benzene, toluene, ethylbenzene, and xylene compounds by *Dechloromonas* strain RCB', *Applied and Environmental Microbiology*, 71(12), pp. 8649–8655.

Chang, B.V. *et al.* (2018) 'Removal of emerging contaminants using spent mushroom compost. *Science of the Total Environment*', 634, pp. 922–933.

Christensen, E.R. and Li, A. (2014) Physical and Chemical Processes in the Aquatic Environment. Hoboken: Wiley & Sons, Inc, pp. 255–259.

Clara, M. *et al.* (2008) 'Emissions of perfluorinated alkylated substances (PFAS) from point sources: identification of relevant branches', *Water Science and Technology*, 58(1), pp. 59–66.

Clara, M. *et al.* (2011) 'Occurrence of polycyclic musks in wastewater and receiving water bodies and fate during wastewater treatment,' *Chemosphere*, 82(8), pp. 1116–1123.

Coelho, L.M. *et al.* (2015) 'Bioremediation of polluted waters using microorganisms', *Advances in Bioremediation of Wastewater and Polluted Soi*, 10, p.60770.

Daughton, C.G. (2004) 'Non-regulated water contaminants: Emerging research', *Environmental Impact Assessment Review*, 24(7–8), pp. 711–732.

Deowan, S.A., Bouhadjar, S.I. and Hoinkis, J. (2015) 'Membrane bioreactors for water treatment', in Basile, A., Cassano, A., and Rastogi, N.K., (eds.) *Advances in Membrane Technologies for Water Treatment: Materials, Processes and Applications* 1st UK edn. Cambridge: Elsevier, pp. 155–184.

Escapa, C. *et al.* (2015) 'Nutrients and pharmaceuticals removal from wastewater by culture and harvesting of *Chlorella sorokiniana*', *Bioresource Technology,* 185, pp. 276–284.

Fent, K., Kunz, P.Y. and Gomez, E. (2008) 'UV filters in the aquatic environment induce hormonal effects and affect fertility and reproduction in fish', *CHIMIA International Journal for Chemistry,* 62(5), pp. 368–375.

Golden, R., Gandy, J. and Vollmer, G. (2005) 'A review of the endocrine activity of parabens and implications for potential risks to human health', *Critical Reviews in Toxicology*, 35(5), pp. 435–458.

Halling-Sørensen, B.N.N.S. *et al.* (1998) 'Occurrence, fate and effects of pharmaceutical substances in the environment: A review', *Chemosphere*, 36(2), pp. 357–393.

Hanson, J.R., Ackerman, C.E. and Scow, K.M. (1999) 'Biodegradation of methyl tert-butyl ether by a bacterial pure culture', *Applied and Environmental Microbiology*, 65(11), pp. 4788–4792.

Ismail, W.N.W. and Mokhtar, S.U. (2020) 'Various Methods for Removal, Treatment, and Detection of Emerging Water Contaminants', in Nuro, A. (ed.) *Emerging Contaminants.* IntechOpen.

Karas, P.A. *et al.* (2011) Potential for bioremediation of agro-industrial effluents with high loads of pesticides by selected fungi. *Biodegradation,* 22(1), pp. 215–228.

Kensa, V.M., (2011) 'Bioremediation-an overview', *Journal of Industrial Pollution Control*, 27(2), pp.161–168.

Klaine, S.J. *et al.* (2008) 'Nanomaterials in the environment: behavior, fate, bioavailability, and effects', *Environmental Toxicology and Chemistry: An International Journal*, 27(9), pp. 1825–1851.

Lakshmi, K.S., Sailaja, V.H. and Reddy, M.A. (2017) 'Phytoremediation-a promising technique in waste water treatment', *International Journal of Scientific Research and Management*, 5(6), pp. 5480–5489.

Lapworth, D.J., Baran, N., Stuart, M.E. and Ward, R.S. (2012) 'Emerging organic contaminants in groundwater: a review of sources, fate and occurrence', *Environmental Pollution*, 163, pp. 287–303.

Locey, K.J. and Lennon, J.T. (2016) 'Scaling laws predict global microbial diversity', *Proceedings of the National Academy of Sciences*, 113(21), pp. 5970–5975.

Mandal, S.K. and Das, N. (2021) 'Application of microbial fuel cells for bioremediation of environmental pollutants: An overview', *Journal of Microbiology, Biotechnology and Food Sciences*, 7(4), pp. 437–444.

Mandaric, L., Celic, M., Marcé, R., Petrovic, M. (2015) 'Introduction on emerging contaminants in rivers and their environmental risk', in Petrovic, M., Sabater, S., Elosegi, A., and Barceló, D. (eds.) *Emerging contaminants in river ecosystems: occurrence and effects under multiple stress conditions.* Cham: Springer International, pp. 3–25.

Margesin, R. and Schinner, F. (2001) 'Biodegradation and bioremediation of hydrocarbons in extreme environments', *Applied Microbiology and Biotechnology*, 56(5–6), pp. 650–663.

Mert, B.K., Ozengin, N., Dogan, E.C. and Aydıner, C. (2018) *Efficient Removal Approach of Micropollutants in Wastewater Using Membrane Bioreactor.* London: IntechOpen, pp. 42–69.

Narayan, M., Solanki, P., Srivastava, R.K. and Zhu, I.X. (2018) 'Treatment of sewage (domestic wastewater or municipal wastewater) and electricity production by integrating constructed wetland with microbial fuel cell', in Zhu, I.X. (ed) *Sewage.* IntechOpen.

Oliveira, I., Blöhse, D. and Ramke, H.G. (2013) 'Hydrothermal carbonization of agricultural residues', *Bioresource Technology*, 142, pp. 138–146.

Oulton, R.L., Kohn, T. and Cwiertny, D.M. (2010) 'Pharmaceuticals and personal care products in effluent matrices: A survey of transformation and removal during wastewater treatment and implications for wastewater management', *Journal of Environmental Monitoring*, 12(11), pp. 1956–1978.

Petrović, M., Gonzalez, S., and Barceló, D., (2003) 'Analysis and removal of emerging contaminants in wastewater and drinking water', *Trends in Analytical Chemistry*, 22(10), pp. 685-696.

Petersen, G. (1996). U.S. Patent No. 5,536,407. Washington, DC: U.S. Patent and Trademark Office.

Prasad, M.N.V., Vithanage, M., & Kapley, A. (eds.) (2019) *Pharmaceuticals and personal care products: waste management and treatment technology: emerging contaminants and micro pollutants.* Oxford: Butterworth-Heinemann.

Ramírez-Durán, N., Moreno-Pérez, P.A. and Sandoval-Trujillo, A.H. (2017) 'Bacterial Treatment of Pharmaceutical Industry Effluents', in Gómez-Oliván, L.M. (ed.) Ecopharmacovigilance. Cham: Springer, pp. 175–187.

Richard, C. and Canonica, S. (2005) 'Aquatic phototransformation of organic contaminants induced by coloured dissolved natural organic matter', In Robertson, P.K.J. (ed.) *Environmental Photochemistry Part II*. Berlin: Springer, pp. 299–323.

Rosenfeld, P.E. and Feng L.G.H. (2011) Emerging contaminants. In Rosenfeld, P.E. Feng, L.G.H. (eds.) *Risks of hazardous wastes*, 1st edn. William Andrew Publishing, Boston, pp. 215–222.

Saratale, R.G., Saratale, G.D., Chang, J.S. and Govindwar, S.P. (2011) 'Bacterial decolorization and degradation of azo dyes: A review'. *Journal of the Taiwan Institute of Chemical Engineers,* 42(1), pp.138–157.

Shah, M.P (ed.). (2020*). Microbial Bioremediation & Biodegradation.* New York: Springer.

Thomaidis, N.S., Asimakopoulos, A.G. and Bletsou, A.A. (2012) 'Emerging contaminants: A tutorial mini-review', *Global NEST Journal*, 14(1), pp. 72–79.

Tolboom, S.N. *et al.* (2019) 'Algal-based removal strategies for hazardous contaminants from the environment: A review', *Science of The Total Environment*, 665, pp. 358–366.

Topare, N. S., Attar, S. J., & Manfe, M. M. (2011) 'Sewage/wastewater treatment technologies: A review', *Scientific Reviews and Chemical Communications*, 1(1), pp. 18–24.

US EPA (2012) Water: Contaminant Candidate List 3. Washington, DC: US Environmental Protection Agency. http://water.epa.gov/scitech/drinkingwater/dws/ccl/ccl3.cfm

12 Systems Biology Aided Functional Analysis of Microbes that Have Rich Bioremediation Potential for Environmental Pollutants

Jyotsna Choubey[1,2], Jyoti Kant Choudhari[1], Mukesh Kumar Verma[1,3], Tanushree Chaterjee[2], and Biju Prava Sahariah[1]*

[1]Chhattisgarh Swami Vivekanand Technical University, Bhilai (C.G), India 491107

[2]Raipur Institute of Technology Raipur (C.G), India 491102

[3]National Institute of Technology Raipur (C.G), India 492010

*Corresponding author: biju.sahariah@gmail.com

CONTENTS

12.1 INTRODUCTION

The quality of life on earth is related to the overall quality of the environment. The reciprocal ratio of increasing human population and unnatural activities with decreasing natural resources, such as forests globally, the contamination levels of the earth with a large number of toxic pollutants from multiple sources are increasing. Advances in science and technology, the destruction of natural

DOI: 10.1201/9781003130932-12

resources, rapid industrialization, and lifestyle changes have generated increased amounts of wastes in various forms, such as raw sewage, toxic organic and inorganic chemicals, heavy metals, and nuclear waste that have potential health hazards to humans and the environment. Over time, the scenario of pollutant diversity in agricultural, industrial, and municipal waste is becoming complex with the addition of more complex and new pollutants, such as carbon (C) nanomaterials (e.g., C nanotubes, graphene, metals, and metal oxide nanoparticles) and responsible for significant adverse effects on the environment and human health. The pollution of the fundamental components of the environment, for example, the air, water, and soil are potential environmental hazards to human health. Continuous recycling and the condensation of air pollutants has led to the presence of contaminants in fog, rain, and snow, which has detrimental effects on biochemical cycling and noxious effects on all parts of the environment. This makes it a global issue; however, soil and water pollutants produce local and geographical issues (Boehler *et al.*, 2017). Most countries' populations have challenges with airborne and waterborne diseases, as well as diseases caused by a polluted environment. Approximately 80% of the world's population is estimated to be at high risk for water security, which required the removal of pollutants removal and treatment at its source for the survival of the environment and all the species on earth.

To sustain and achieve safeguard humans and the environment from environmental pollution, new feasible and advanced treatment processes need to be designed that are environmentally feasible and sustainable. The bioremediation of organic and inorganic pollutants has been used since ancient times. With the development of new technologies for the treatment of pollution [e.g., physicochemical processes, ozonation, base-catalyzed chlorination, and ultraviolet (UV) oxidation] bioremediation is unique due to its advantages, such as cost-effectiveness, eco-friendly, complete mineralization of pollutants, and the sustainability of bioremediation agents. Plants and a wide variety of microorganisms, such as bacteria, fungi, and algae that have efficient bioremediating properties have been employed for the efficient removal of toxins from polluted environments. The bioremediation of pollutants describes the utilization of agents of biological origin that are alive or dead, such as plants, algae, fungi, and various bacteria that have a high neutralizing ability through the process, such as biodegradation bioaccumulation, biosorption, or bioleaching to remove or neutralize the environmental pollutants (Eevers *et al.*, 2017). The remediation techniques in organisms are influenced by their structure, components, cell properties, for example, the extracellular and intracellular membranes and the enzymes produced by them that participate in the metabolic pathways. These organisms act on the pollutants and can partially or completely mineralize or convert them into less harmful products (Megharaj *et al.*, 2011).

Many strains of microbes are only effective at bioremediation under laboratory conditions and growth is limited under extreme environments. Most microbes prefer optimal conditions where pH, temperature, oxygen (O_2), soil structure, moisture, nutrients levels, low bioavailability of contaminants, and the presence of other toxic compounds are controlled. Bioremediation of pollutants by microbes under aerobic, aerobic, and anoxic conditions has been investigated for simple and recalcitrant pollutants. Laboratory studies have helped to determine the bioremediating agents; however, the ratio of studied microbes toward the noncultured and noninvestigated microbes is enormous, and therefore, bioinformatic tools, such as a systems biology (SB) approach could help to gain an insight into the microbial world.

An SB approach uses omics techniques (e.g., genomics, proteomics, transcriptomics, and metabolomics) to explore and characterize microbial communities their composition, cellular and molecular activity, and complex networks and interconnections during various biological processes at the molecular, cellular, population, community, and ecosystem levels (Pavlopoulos *et al.*, 2015). The normal behavior of a microbial community is influenced by the presence of toxic chemicals. During bioremediation or the detoxification of toxic compounds, the influence of environmental variables and cell–cell interactions could be investigated by the proper selection of SB tools. The selection process for the methods to use in an SB study involves the consideration of cost, time frame, personnel, and the objectives of the process.

If the focus is to determine the microbial community composition, DNA-based omics tools, such as a 16S rRNA clone library, PhyloChip, or sequencing could be used. If the aim is to understand cellular pathways and to identify the functional genes involved in microbially mediated reactions, tools that identify ribonucleic acids (RNAs), such as GeoChip, RNAseq, and various mass spectrometry methods are used. Nuclear magnetic resonance, desorption electrospray ionization, and matrix-assisted laser desorption or ionization are generally used for the characterization of microbially secreted small molecules. Concomitant monitoring of limiting nutrients, electron donors, electron acceptors, and hydrology is crucial for an SB conceptual model to be useful.

To identify and follow the microbial community during bioremediation, metagenomic analysis that includes16S rRNA-based clone libraries have been studied for metals (Cardenas *et al.*, 2008), and hydrocarbons and chlorinated solvents (Militon *et al.*, 2010). In combination, these metagenomics techniques have reiterated that the microbial diversity that exists in most environments is larger than expected.

Recently, high throughput microarrays, such as PhyloChip and GeoChip have received attention in metal and organics bioremediation studies to quickly characterize the microbial community and functions. PhyloChip, a 16S rRNA-phylogenetic microarray, characterizes and monitors microbial community dynamics and GeoChip, a functional gene microarray tracks the functional gene activity changes in microbes in the environment (DeSantis *et al.*, 2007). Microbial community proteomics and metabolomics are breakthroughs that provide a deeper insight into microbial cellular function and gene products that interact in the environment. A novel application of immunomagnetic separation to target and monitor specific microorganisms during *in situ* bioremediation has the potential to enable transcriptomics, proteomics, or metabolomics-based studies directly on cells collected from the field. The integration of all of these techniques using the latest advances in bioinformatics and modeling will enable breakthroughs in environmental biotechnology.

This chapter aims is to highlight SB approaches with an emphasis on functional genomics during the bioremediation of organic and inorganic wastes in the environment that are generated from various sources. Various types of organic pollutants are released into the environment; therefore, a variety of microbes are required for the effective bioremediation of these pollutants.

12.2   ROLE OF SB IN BIOREMEDIATION

The SB approach for bioremediation requires the identification and characterization of microbial communities and the molecular processes involved; however, it becomes difficult because the toxic contaminants present to influence the normal activity of the microbial communities. In general, SB provides information on gene expression, enzymes, biosynthetic pathways, secondary metabolites in microbes, and alterations in the existing pathways under stress caused by various pollutants.

The data from the omic tools provide significant insights into complex microbial metabolic pathways. Information from the characterization of cellular processes, microbial community composition, and metabolic activity under stress induced by toxic compounds, which alters the typical behavior of the microbial communities, provides the opportunity to upgrade the bioremediation efficiency of bioremediating agents. The target objective, cost, time and workforce must be considered when selecting the appropriate SB tools. For example, to explore the composition of a microbial community, DNA-based omics approaches, such as genomics, metagenomics, and phenomics could be used. If the focus is to identify the functional genes involved in cellular interactions and to understand cellular gene expression, metabolic pathways, and proteins, techniques, such as transcriptomics, proteomics, and mass spectroscopy are employed (Chakraborty *et al.*, 2012). However, functional genomics microarrays, proteomics, and metabolomics provide insights into the key microbial reactions that are employed to identify the functional activity of enzymes in a complex microbial community.

12.3   SB TECHNIQUES FOR BIOREMEDIATION STUDIES

The continuous monitoring of onsite bioremediation to record physiological and metabolic changes and their interactions with the microbial population helps to generate a conceptual model. The monitoring technology involves functional genomic microarrays, phylogenetic analysis, metabolomics, proteomics, and quantitative polymerase chain reactions (q-PCR) sometimes at ionic level changes. To identify the microbial community structure, the DNA-based genomics tool, such as a 16S rRNA clone library, PhyloChip, or sequencing can be useful. If the aim is to identify the functional genes involved and to understand the associated cellular pathways via microbial bioremediation, various tools, such as RNAseq, GeoChip (for RNA), and several mass spectroscopy methods (for proteins) can be used.

12.3.1   16S rRNA

The 16S rRNA is highly conserved and widely used as identification criteria for different taxonomic units that have a significant effect on the relative community structural profile, indicative changes in diversity, consistency, and relative densities of certain groups. This tool can indicate the presence or absence of certain microbes and identifies the community structural changes that are advantageous or detrimental toward bioremediation. However, 16S rRNA analysis cannot provide information about the microbes' activities (e.g., metabolically active or not).

12.3.2   PhyloChip

The PhyloChip arrays technique helps to identify the abundance and presence or absence of a known prokaryotic species within a sample. The chip was developed by Affymetrix and uses a microarray that consists of 25mer single-stranded oligonucleotides that are called probe groups, which are complementary to a certain region on the 16S rRNA gene of a particular species. The 16S rRNA codes an RNA component of the small ribosomal subunit and uses it as a convenient taxonomic marker present in all bacteria and archaea. The conserved regions allow the easy design of PCR primers from variable regions to differentiate species. PhyloChip contains >1 million different probes and allows the discrimination of > 50,000 operational taxonomic units. PhyloChip helped to quantify the relative abundance of microbial species in deepwater samples from the Gulf of Mexico, and differences in the relative abundance of the microbial populations (i.e., Gammaproteobacteria that are efficient in the biodegradation of hydrocarbon) in samples that were collected from within an oceanic oil plume versus those that were collected outside of the oil plume was observed (Hazen *et al.*, 2010).

12.3.3   GeoChip

The GeoChip functional gene microarray approach is based on the use of a DNA microarray evaluation that contains oligonucleotide probes. These gene probes are focused on the biogeochemical cycles of certain metals and important nutrients including C, nitrogen (N), phosphorus, and sulfur (S). This helps to distinguish resistance to antibiotics, viruses, energy production, the ability to degrade organic contaminants and *gyrB* gene based phylogenetic markers of slow-growing bacteria, such as Mycobacterium. The advantages of GeoChip are that no previous knowledge of the microbial community being sampled is required, and it works on microorganisms from environments, such as soil, water, air, human and animal sources. It has successfully detected low levels of microorganisms, in addition, it prevents annotation bias, it is relatively quick, and can receive data nearly every day from either DNA or RNA. A disadvantage of GeoChip is that it cannot detect novel gene families, because it can only detect the genes present on the probes. GeoChip 5.0 contains 167,044 probes that can cover 395,894 coding sequences and 1,500 gene families. It has been used to show functions that were inhibited during the Deepwater Horizon oil spill and predicted the geochemistry in a mixed waste site.

12.3.4 Phospholipid Fatty Acids

Phospholipid fatty acids (PLFAs) are the main component of cell membranes, and their analysis is a useful tool in microbial community analysis. These lipids generally degrade quickly on cell death, which allows their analysis to target viable cells, unlike many other microbiology techniques. However, PLFA has disadvantages associated with its measurements. PLFA analysis is highly sensitive and easily reproducible, fast and comparatively inexpensive. However, PLFA analysis requires a large number of samples to make the available data statistically significant. For certain techniques, such as fingerprinting, PLFA requires ≤10 times more samples than the analyses based on the examination of fatty acid methyl esters, which is another type of fatty acid analysis. Four major types of analyses conducted using PLFAs are total biomass, physiological indicators, fingerprinting, and taxonomic biomarkers. When examining the Deepwater Horizon spill plume, the PLFA data supported the 16S rRNA pattern analysis results and provided additional biomass measurements (Hazen *et al.*, 2010).

12.3.5 Functional Gene Clone Libraries

The use of functional gene clone libraries is a method that can help determine gene functionality based on nucleotide sequence data. This technique is often used at the population level of SB, in particular, when the sequence of a gene of interest is known and the corresponding function is not. The advantages of this technique include its culture-independent nature, which bypasses the need for microbial isolation. For example, a gene can be PCR-amplified and cloned into a plasmid, such as pUC19, and then transformed into a model heterologous host, such as *Escherichia coli*.

Another advantage is the ability to selectively engineer a regulated promoter to drive the transcription of a gene; therefore, controlling the expression of the phenotype. A commonly used regulated promoter in *E. coli* is the T5-lac promoter. Limitations of this technique include the functional heterologous expression of a gene in a foreign host might be considered an unexpected event, at best. For example, a gene of interest might require a specific chaperone to produce key folds in the protein structure that are critical for enzymatic activity. If the heterologous host does not have the genes that encode encoding for these chaperones, then the resulting protein might not fold correctly, and therefore, produce no enzymatic activity. Even if protein folding occurs independently of the chaperones, the enzyme in question might require additional cofactors that would need to be supplied for the process to function, such as vitamin B12. For example, the Dehalococcoides rely on reductive dehalogenases to perform organohalide respiration; however, they cannot produce vitamin B12 that is essential to reductive dehalogenase function. (Yan *et al.*, 2013) If the heterologously expressed enzyme is functional, the level of its activity might not reflect the activity that occurs in the gene native organism. This might occur because the activity of a gene might be regulated by single or even multiple regulatory networks in the native host and that regulation might be absent in the foreign host. In addition, the expression of the target gene might be toxic to foreign hosts, which might compromise host cell viability. These limitations must be acknowledged and addressed when employing functional gene cloning in SB.

12.3.6 GENOMICS

Genomics is based on the study of entire genomes of organisms and incorporates genetic elements. Genomics uses a combination of recombinant DNA sequencing methods and bioinformatics to sequence, assemble, and analyse genome structure and function. Genomics provides an overall view of genetic material, such as DNA and RNA expression in microorganisms when exposed to pollutants where initial genome sequencing is conducted followed by bioinformatic analysis that uses a group of tools and algorithms. A recent study found that the genomes of 270,567 organisms have been sequenced and >46,000 genome sequencing projects are being undertaken worldwide. The analysis of the genome sequence of *Pseudomonas sp.* KT2440 showed the presence of group enzymes or proteins that encoded several enzymes or proteins, such as dehydrogenase, oxidoreductases, oxygenases, ferredoxin, cytochromes, glutathione-S transferase, sulfur-metabolizing proteins, and efflux pumps. This information helps to understand the degradation of chemicals in industrial effluents (Belda *et al.*, 2016). Sequencing the entire genome of *Arthrobacter sp.* LS16 and YC-RL1 revealed that they contain metabolic networks involved in the bioremediation of aromatic compounds, such as naphthalene, 1,2,3,4-tetrachlorobenzene, fluorine, 4-nitrophenol, bisphenol A, biphenyls, and p-xylenes (Ren *et al.*, 2016). The Alkaliphila JAB1 P genomic sequence indicated the presence of various groups of *bph* genes that coded the degradation pathways of biphenyls and their derivatives (Ridl *et al.*, 2018). In addition, several scientific reports suggested the role of other microorganisms in the biorestoration of heavy metals and dyes based on the results obtained from their entire genome sequencing.

12.3.7 METAGENOMICS IN BIOREMEDIATION

Metagenomics is a genomics-based approach, in which genomic DNA that is directly isolated from an environmental microbial sample, is sequenced and analyzed in the laboratory without obtaining pure cultures of each member of the bacterial community (Baweja *et al.*, 2016). Metagenome analysis via next-generation sequencing provides allows the extensive analysis of the environmental genomes where a metagenomic approach could investigate poorly characterized or unknown diverse degradation pathways that present in diverse microbes.

Studies that investigated the microbial communities from diverse environments, such as sediments and marine water (DeLong *et al.*, 2006; Yooseph *et al.*, 2007), the human gut (Turnbaugh *et al.*, 2007), soils (Smets and Barkay, 2005), and acid mine drainage (Tyson *et al.*, 2005), provided novel insights into microbial systems and functions Table 12.1.

Metagenomic bioremediation offers more positive results and complete information to enhance degradation ratios compared with other approaches to bioremediation (Kosaric, 2001). With an increased understanding of the structural and functional attributes of microbes toward the degradation of xenobiotic compounds, information is generated about microbes from contaminated and undisturbed sites. This helps to identify key microbial processes and establishes suitable species for specific sites.

Metagenomics could provide an appropriate metagenomic database that offers a library of genes for the construction of novel microbial strains for targeted use in bioremediation efforts. Microbiologists consider the metagenomics-based bioremediation approach to be one of the most important and potent tools for the eradication of pollutants from the environment (Chandran and Das, 2011).

12.3.8 TRANSCRIPTOMICS

Transcriptomics studies all RNA molecules within a cell, which is known as the transcriptome. Several studies of transcripts focused on messenger RNA molecules. It is primarily used to study differential gene expression that is regulated and degraded in response to environmental pollutants. In addition, it might help in determining the function of previously unknown genes. Techniques, such as microarray and RNA sequencing could be applied to quantify a set of predetermined

TABLE 12.1

Genomic Studies on Microbial Bioremediation of Different Contaminants

Number	Microorganisms	Approaches	Toxic Compounds	References
1	*Brevibacterium epidermidis EZ-K02*	Genomics	Benzoate, p-hydroxybenzoate, acetophenone, catechol, gentisate, arsenic, cobalt, and cadmium	Ziganshina *et al.* (2018)
2	*Microbacteria moleivorans*	Genomics	Nitroacetate and nitriloacetate	Miller *et al.* (2016)
3	*Irpexlacteus*	Genomics	Aromatic dye	Sun *et al.* (2016)
4	*Bacillus subtilis HUK15*	Genomics	Hexachlorocyclohexane	Hwang *et al.* (2015)
5	*Anaeromyxobacter sp. Fw109-5*	Genomics	Nitrate and uranium	Birolli *et al.* (2018)
6	*Mycobacterium, Rhodococcus wratislaviensis strain 9*	Genomics	PAHs, organophosphate, para-nitrophenol, and phenanthrene compounds of pesticides	
7	*Bacillus, Coprothermobacter; Rhodobacter, Pseudomonas, Achromobacter, Desulfitobacter; Desulfosporosinus, T78, Methanobacterium, Methanosaeta*	Metagenomics	Sulfate-reducing, CO_2-assimilating, hydrocarbon-rich petroleum refinery waste	Roy *et al.* (2018)
8	*Proteobacteria and Firmicutes*	Metagenomics	Degradation of fatty acids, chloroalkanes, and chloroalkanes	Roy *et al.* (2018)
9	*Naegleria, Vorticella, Arabidopsis, Asarum and Populus*	Metagenomics South African petroleum-contaminated water aquifer sites	Hydrocarbons	Kachienga *et al.* (2018)
12	*Thiobacillus sp.*	Metagenomics	Thiocyanate	Rahman *et al.* (2017)
16	*Acinetobacter venetianus RAG-1*	Transcriptomics	Alkanol (dodecanol)	Kothari *et al.* (2016)
17	*Pseudomonas putida KT2440, Sphingobium sp. 1017-1*	Transcriptomics	Organophosphates, pyrethroids, and carbamates	Gong *et al.* (2018)
18	*Pedobactersteynii DX4*	Transcriptomics	crude oil	Chang *et al.* (2017)
19	*Penicillium oxalicum*	Proteomics	Polycyclic aromatic hydrocarbons (anthracene)	Lucero Camacho-Morales *et al.* (2018)
21	*Miscanthus sinensis*	Proteomics	Antimony (SB)	Xue *et al.* (2015)
23				
18	*Acinetobacter guillouiae SFC 500–1 A*	Proteomics	phenol and chromium (VI)	Ontañon *et al.* (2018)
19	Microbial sediment sample	Metaproteomics and metabolomics	Crude oil hydrocarbons	Bargiela *et al.* (2015)

sequences. Analyzing transcriptome microbial communities requires :(1) RNA isolation; (2) cDNA synthesis; (3) library preparation; (4) library purification; (5) DNA sequencing; and (6) transcriptome data analysis (Zhu *et al.*, 2017). In a study, RNA sequencing was used to assess the physiology of clean environmental crops. and to monitor the catabolic gene expression profile in mixed microbial communities. In total, >100 genes were identified that were affected by the growth of cis-dichloroethene (cDCE) in *Polaromonas sp.* JS666. The comprehensive genomic microarray techniques for *Polaromonas sp.* JS666 were synthesized and used for hybridization with cDNA obtained from *Polaromonas sp.* during growth on cDCE (Jennings *et al.*, 2009). In a separate study, a transcriptomic approach was used to compare the expression profile of *Pseudomonas extremaustralis* under aerobic and microaerophilic conditions. Genes involved in alkaline degradation, including *alkB*, were overexpressed under microaerobic conditions without hydrocarbon compounds (Tribelli *et al.*, 2018). The transcriptomic profile of calmodulin encoding genes in *Talaromyces sp.* in response to 2, 2-dichlorovinyl dimethyl phosphate (DDVP) and c-hexachlorocyclohexane (lindane) were considered. The amplification of *ITS* (e.g., *ITS1* and *ITS4* combinations) was performed; two of the most frequent fungi were characterized and subjected to in vitro DDVP and lindane tolerance testing at different concentrations. Fungi were screened using reverse transcription-PCR (RT-PCR) techniques for the presence and expression of the calmodulin gene (*cam*). Two *Talaromyces* strains were identified as *Talaromyces astroroseusasemo G* and *Talaromyces purogenumasemo N* (GenBank scan numbers KY488464 and KY488468, respectively). In addition, transcriptomic analysis of *Pedobacter steynii* DX4 showed the mechanism of low-temperature crude oil degradation. During the study, researchers predicted several hydrocarbon oxygenases, chemokinetic proteins, and biosurfactants syntheses in the bacterial strain.

12.3.9 PROTEOMICS

In addition to genomics and transcriptomics, proteomics is a valuable and more complex technique than genomics, because an organism's genome is quite constant, and proteomics differs from cell to cell and from time to time (Nzila *et al.*, 2018). Proteomics is the branch of science that deals with the study of proteomes. A proteome is a group of proteins produced within a cell, tissue, organ, or body. This technique is important, because the observed phenotype is based on protein expression and interactions and not on the genomic sequence. Proteomics allows researchers to monitor and analyze the overall expression of induced proteins in microorganisms that live in polluted sites, because of their anthropogenic activities. The analysis of the suite of proteins produced by bacterial cultures (proteomics) and in environmental samples (metaproteomics) could be used to determine variations in the composition and production of proteins and the detection of many proteins that are important in the physiological response of microbes in the presence of pollutants. In addition, metaproteomics has recently been used to detect the protein expression profiles of microbial communities in environmental samples without members of the bacterial community being cultured in the laboratory (Aydin *et al.*, 2017). Therefore, this approach reflects the actual functional activities of microbial populations in a particular ecosystem. This approach accounts for the real functional activities of microbial populations within a specific ecosystem. The development of two-dimensional gel electrophoresis (2D-E) combined with mass spectrometry and protein sequencing and improved protein structure databases facilitated its use (Wang *et al.*, 2016). *Mycobacterium vanbaalenii* PYR-1 cultured in the presence of high molecular weight polycyclic aromatic hydrocarbons (HMW-PAHs) produced several PAH induced proteins detected by 2D-E.

12.3.10 METABOLOMICS

Metabolomics is the application of metabolome-based approaches to environmental samples and develops models that consider microbial activities under different bioremediation strategies.

Therefore, it enhances the understanding of the dynamic operations in microbial communities and their functional contributions to their habitat. Metabolites under normal and stress conditions can help to identify the optimization of bioavailability and bioremediation of pollutants and microbial interactions.

The metabolomics of microbes is generally investigated fusing metabolite profiling, target analysis for the identification and quantification of several cellular metabolites, metabolic flux analysis, and metabolic fingerprinting. Gas chromatography, high-performance liquid chromatography, Fourier-transform infrared spectroscopy (FTIR), capillary electrophoretic mass spectrometry (EC), and direct injection mass spectrometry are used for metabolite analysis. The metabolomic approach is mainly limited to the investigation of the functional roles of low molecular weight metabolites. Several recent studies used microbial metabolomic analysis to study the biodegradation of human pollutants (Table 12.1).

12.4 LIMITATIONS AND CHALLENGES OF SB

SB is a valuable tool for the characterization of functional molecules from various signaling pathways in biomedical research; however, only a few laboratories currently apply it to environmental issues due to its cost and the requirement for highly specialized facilities and qualified personnel to perform the analyses. Proteomics technology needs to be developed to make the environment cleaner at a more reasonable cost. However, proteomics studies are currently used in environmental biotechnology laboratories. It has applications that are beyond SB and has been successfully applied in various areas; However, some limitations are associated with this approach. SB provides large amounts of data on genes, proteins, and metabolites, but adequate and efficient calculation tools are required for the better annotation of genes, pathways, and metabolites (Yooseph et al., 2007). In addition, the calibration of the calculation tools, models, and algorithms is challenging. The use of SBs for bioremediation has been limited to date, due to the high cost of sample processing and the requirement for specialized instruments. Therefore, only a few laboratories currently use SB in bioremediation studies. In addition, the integration of omics techniques, in particular with metabolomics, is a major challenge. Low molecular weight primary and secondary metabolites are equally important in bioremediation (Kumar et al., 2017).

12.5 COMPUTATIONAL TOOLS USED FOR ANALYSIS IN SB

Omics techniques that are used in SB, generate a large amount of data; therefore, it is necessary to integrate and analyse this complex data (Zhang et al., 2010). In the last decade, various computational tools, such as software, web resources, pipelines, and algorithms have been designed and developed for the complete analysis or the interpretation of the data (Zhang et al., 2010); however, but tools that specifically apply SB to bioremediation are insufficient. A Biocatalysis/Biodegradation Database (EAWAG-BBD), which was initially developed by the University of Minnesota, MIN, US and hosted by EAWAG (Swiss Federal Institute of Aquatic Science and Technology, Switzerland), is the first of its kind to contain information on large numbers of microbial biodegradation reactions and pathways for recalcitrant chemicals (Gao et al., 2010). The new updated version of this database contains lists of 219 pathways, 1,503 reactions, 1,396 compounds, 993 enzymes, and 543 types of microbial entries. Another freely available web resource, MetaRuter, contains information on various biochemical compounds, enzymes, reactions, and organisms that are related to biodegradation and bioremediation in an integrated framework (Pazos et al., 2005). In addition, it is helpful to find pathways between two sets of compounds. Several other tools and software used for the analysis of SB data are described in Table 12.2. These resources give valuable preliminary information that could be used as a starting point in metabolic pathway engineering for bioremediation.

TABLE 12.2
Databases Used for SB in Bioremediation

Number	Name	Database	URL
1	1000 Genomes Project	DNA	www.internationalgenome.org/
2	Arrayexpress	RNA	www.ebi.ac.uk/arrayexpress/
3	NCBI databases, Ensambl, IntAct	DNA, RNA, proteins	www.ncbi.nlm.nih.gov/gquery/ www.ensembl.org/index. html www.ebi.ac.uk/intact/
4	Negatome 2.0 PAUDA	Proteins	http://mips.helmholtz-muenc hen.de/proj/ ppi/negatome/ https://ab.inf.uni-tuebingen. de/software/pauda
5	BioRadBase Bioremediation of radioactive	Waste generated from nuclear plants	http://biorad.igib.res.in/
6	AromaDeg	Phylogenomics for bioremediation of aromatic compounds	http://aromadeg.siona. helmholtz-hzi.de/
7	OxDBaseRHObase	Oxygenases involved in biodegradation	www.imtech.res.in/raghava/ oxdbase/ http://bicresour ces.jcbose.ac.in/ ssaha4/ Rhobase/
8	Bionemo	Provides molecular information about biodegradation metabolism	https://omictools.com/ bionemo-tool
9	BioSurfDB	Contains algorithms for biosurfactants and biodegradation studies	www.biosurfdb.org/
10	PBT profiler	Identification of toxic chemicals	www.pbtprofiler.net/
11	BSD	Biodegradative Strain Database	http://bsd.cme.msu.edu/
12	KBase	Search and organize microbial communities, plants, genomics and SB data	https://kbase.us/
13	BioFNet	database for analysis and synthesis of biosystems	http://kurata22.bio.kyutech. ac.jp/db/pub/ pub_main. php?Ver ¼3.8
	MEMOSys 2.0	Genome-scale models (GEMs) database	https://memosys.i-med.ac.at/ MEMOSys/ home.seam
14	SABIO-RK	The database for biochemical reactions kinetic properties	http://sabio.h-its.org/
15	MetaCyc	Database describe metabolic pathways and enzymes from all domains of life	http://metacyc.org/
16	BioCyc	Contains full information about organism-specific pathway/ genome databases	http://biocyc.org/
17	SYSTOMONAS	Genome database of *Pseudomonas*	http://systomonas.tu-bs.de/
18	Genevestigator V3	Reference expression database for transcriptomes	https://genevestigator.com/gv/

TABLE 12.2 (Continued)
Databases Used for SB in Bioremediation

Number	Name	Database	URL
19	University of Minnesota Biocatalysis/ Biodegradation Database (UMBBD)	university-minnesota-biocatalysis-andbiodegradation-database Give information about molecular mechanisms involved in biodegradation pathways and biotransformation rules, enzymes, genes, and reactions involved in microbial degradation of xeno pesticidal compounds	www.msi.umn.edu/content/
20	Biodegradation Network-Molecular Biology Database (Bionemo)	Dynamic regulation of metabolic pathways and transcription factors in degradation pathways	http://bionemo.bioinfo.cnio.es
21	Pesticide Target Interaction Database (PTID)	Interaction of pesticides with their target	http://lilab.ecust. edu.cn/ptid
22	Microbial Genome Database (MBGD)	Comparative analysis of microbial genome	http://mbgd.genome. ad.jp
23	Metarouter	MetaRouter Maintain diverse information related to biodegradation	http://pdg.cnb.uam.es/
24	Pesticide Action Network (PAN)	Give informative data on the toxicity of pesticides	http://pesticideinfo.org/Index. html
25	The Environmental Contaminant Biotransformation Pathway (EAWAGBBD/ PPS)	Give informative from bulk data of multi-omics approaches	https://envipath.org/

12.6 CONCLUSIONS AND FUTURE PERSPECTIVES

Microorganisms play important role in the decomposition of organic materials, recycling nutrients, and the remediation of environmental pollutants. However, it requires more research to achieve the effective and reliable clean-up of environmental contaminants. It is very challenging, because as most of the microorganisms in the environment are not easy to culture; therefore, their functional biology has not been studied fully. The SB and functional genomic approach has revolutionized this field. It integrates various bioremediation approaches, which is very useful in understanding the complex process of bioremediation. Today, different omics-based approaches are being extensively used for bioremediation. New advances in molecular techniques, such as genomics, transcriptomics, metatranscriptomics, and proteomics offer individual qualities as tools to gather information, starting from genes to proteins and metabolites. To achieve breakthroughs in the application of bioremediation to environmental clean-ups, these approaches need to be mapped and integrated from the theoretical to practical stages. Although significant progress has been made in the development of different in silico databases and software and computational models for the study of microbial processes, the major problems are encountered when analyzing the results using suitable, user-friendly, and simplified bioinformatics tools to make draw conclusions. Therefore, more advanced computational tools are required to fully exploit the omics derived data for a better anno-tation of genes, understanding metabolic networks, and metabolites. A single user-friendly platform is required, which could provide all the bioinformatics tools and databases for data analysis and metabolic pathway reconstruction models. This platform could provide all the information related to bioremediation research, such as data, analytical methods, and pipelines. This requires coord-ination between researchers working in different laboratories to share data, update, and maintain databases. By using this information, better *in situ* and *ex situ* models for bioremediation could be proposed. Genomics and metagenomics have potential applications in environmental bioremedi-ation. This technology could be applied to the effective eradication of contaminants, diversification of microorganisms at the contaminated site, and the identification of new enzymes and pathways. The applications of these approaches are still in development; however, the data that is being generated by the current omic technology needs to be organized within the informative databases. Omics approaches demonstrate a good ability to predict organism's metabolism in polluted envir-onments and to determine the microbially mediated attenuation of the pollutants to help bioremedi-ation. The study of the molecular mechanisms behind microbial transformations of toxic pollutants using omic approaches to bioremediation could help to identify the responsible organisms and in the efficient elimination of the contaminants from the environment.

REFERENCES

Aydin, S. *et al.* (2017) 'Aerobic and anaerobic fungal metabolism and omics insights for increasing polycyclic aromatic hydrocarbons biodegradation', *Fungal Biology Reviews*, 31, pp. 61–72.

Bargiela, R. *et al.* (2015) 'Metaproteomics and metabolomics analyses of chronically petroleum-polluted sites reveal the importance of general anaerobic processes uncoupled with degradation,' *Proteomics*, 15, pp. 3508–3520. doi.org/10.1002/pmic.201400614

Baweja, M., Nain, L., Kawarabayasi, Y., and Shukla, P. (2016) 'Current technological improvements in enzymes toward their biotechnological applications,' *Frontiers in Microbiology*, 7, 965. doi.org/10.3389/fmicb.2016.00965

Belda, E. *et al.* (2016) 'The revisited genome of *Pseudomonas putida* KT2440 enlightens its value as a robust metabolic chassis', *Environmental Microbiology*, 18, pp. 3403–3424.

Birolli, W.G. *et al.* (2018) 'Biodegradation of anthracene and several PAHs by the marine-derived fungus *Cladosporium sp.* CBMAI 1237. *Marine Pollution Bulletin*, 129, pp. 525–33. doi.org/10.1016/j.marpolbul.2017.10.023

Boehler, S. *et al.* (2017) 'Assessment of urban stream sediment pollutants entering estuaries using chemical analysis and multiple bioassays to characterise biological activities,' *Science of the Total Environment*, 593, pp. 498–507.

Cardenas, E. *et al.* (2008) 'Microbial communities in contaminated sediments, associated with bioremediation of uranium to submicromolar levels', *Applied and Environmental Microbiology*, 74, pp. 3718–3729.

Chakraborty, A., Tripathi, S.N., and Gupta T. (2017) 'Effects of organic aerosol loading and fog processing on organic aerosol volatility,' *Journal of Aerosol Science*, 105, pp. 73–83.

Chakraborty, R., Wu, C.H., and Hazen, T.C. (2012) 'Systems biology approach to bioremediation,' *Current Opinion in Biotechnology*, 23, pp. 483–490.

Chandran, P., and Das, N. (2011) 'Characterization of sophorolipid biosurfactant produced by yeast species grown on diesel oil,' *International Journal of Science and Nature*, 2, pp. 63–71.

Chang, S. *et al.* (2017) 'The complete genome sequence of the cold adapted crude-oil degrader: *Pedobacter steynii* DX4,' *Standards in Genomic Sciences*, 12(45).

DeLong, E.F. *et al.* (2006) 'Community genomics among stratified microbial assemblages in the ocean's interior,' *Science*, 311, pp. 496–503. doi.org/10.1126/science.1120250

DeSantis, T.Z. et al. (2007) 'High-density universal 16S rRNA microarray analysis reveals broader diversity than typical clone library when sampling the environment,' *Microbial Ecology*, 53, pp. 371–383.

Eevers, N., White, J.C., Vangronsveld, J., and Weyens, N. (2017) 'Bio- and phytoremediation of pesticide-contaminated environments: a review,' *Advances in Botanical Research*, 83, pp. 277–318.

Gao, J., Ellis, L.B., and Wackett, L.P. (2010) 'The University of Minnesota biocatalysis/biodegradation database: improving public access,' *Nucleic Acids Research*, 38, D488–D491.

Gong, T., *et al.* (2018) 'An engineered *Pseudomonas putida* can simultaneously degrade organophosphates, pyrethroids and carbamates,' *Science of The Total Environment*, 628–629: doi.org/10.1016/j.scitotenv.2018.02.143

Hazen TC. *et al.* (2010) 'Deep-sea oil plume enriches indigenous oil-degrading bacteria,' *Science*, 330, pp. 204–208.

Hwang, C. *et al.* (2015) 'Complete genome sequence of *Anaeromyxobacter sp.* Fw109-5, an anaerobic, metal-reducing bacterium isolated from a contaminated subsurface environment,' *Genome Announcements,* 3. doi.org/10.1128/genomeA.01449-14

Jennings, L.K. *et al.* (2009) 'Proteomic and transcriptomic analyses reveal genes upregulated by cis-dichloroethene in *Polaromonas sp.* strain JS666,' *Applied and Environmental Microbiology*, 75, pp. 3733–3744.

Kachienga, L., Jitendra, K., and Momba, M. (2018) 'Metagenomic profiling for assessing microbial diversity and microbial adaptation to degradation of hydrocarbons in two South African petroleum-contaminated water aquifers. *Science Reports*, 8. doi.org/10.1038/s41598-018-25961-0

Kosaric, N. (2001) 'Biosurfactants and their application for soil bioremediation,' *Food Technology and Biotechnology*,' 39, pp. 295–304.

Kothari, A. *et al.* (2016) 'Transcriptomic analysis of the highly efficient oil-degrading bacterium *Acinetobacter venetianus* RAG-1 reveals genes important in dodecane uptake and utilization,' *FEMS Microbiology Letters*, 363. doi.org/10.1093/femsle/fnw224

Kumar, R. *et al.* (2017) 'Metabolomics for plant improvement: status and prospects,' *Frontiers in Plant Science*, 8, 1302.

Lucero Camacho-Morales, R. *et al.* (2018) 'Anthracene drives sub-cellular proteome-wide alterations in the degradative system of *Penicillium oxalicum*,' *Ecotoxicology Environmental Safety*, 159, pp. 127–35. doi.org/10.1016/j.ecoenv.2018.04.051

Megharaj, M. *et al.* (2011) 'Bioremediation approaches for organic pollutants: a critical perspective,' *Environment International*, 37, pp. 1362–1375.

Militon, C. *et al.* (2010) 'Bacterial community changes during bioremediation of aliphatic hydrocarbon-contaminated soil,' *FEMS Microbiology Ecology*, 74, pp. 669–681.

Miller, N.T. *et al.* (2016) 'Draft genome sequence of *Pseudomonas moraviensis* strain Devor implicates metabolic versatility and bioremediation potential,' *Genomics Data*, 9, pp. 154–159.

Nzila, A., and Ramirez, C.O. (2018) 'Pyrene biodegradation and proteomic analysis in *Achromobacter xylosoxidans* PY4 strain,' *International Biodeterioration & Biodegradation*, 130, pp. 40–47.

Ontañon, O.M. *et al.* (2018) 'What makes *A. guillouiae* SFC 500-1A able to co-metabolize phenol and Cr(VI)? A proteomic approach,' *Journal of Hazardous Material*, 354, pp. 215–24. doi.org/10.1016/j.jhazmat.2018.04.068

Pavlopoulos, G.A. et al. (2015) 'Visualizing genome and systems biology: technologies, tools, implementation techniques and trends, past, present and future,' *Gigascience*. 4. doi.org/10.1186/s13742-015-0077-2

Pazos, F., Guijas, D., Valencia, A., and De Lorenzo, V. (2005) 'MetaRouter: bioinformatics for bioremediation,' *Nucleic Acids Research*, 33, D588–D592.

Rahman, S.F. et al. (2017) 'Genome-resolved metagenomics of a bioremediation system for degradation of thiocyanate in mine water containing suspended solid tailings,' *MicrobiologyOpen*, 6, e00446.

Ren, L. *et al.* (2016) 'Complete genome sequence of an aromatic compound degrader *Arthrobacter sp.* YC-RL1,' *Journal of Biotechnology*, 219, pp. 34–35.

Ridl, J. *et al.* (2018) 'Complete genome sequence of *Pseudomonas alcaliphila* JAB1 (= DSM 26533), a versatile degrader of organic pollutants,' *Standards in Genomic Sciences*, 13:3.

Roy, A. *et al.* (2018) 'Biostimulation and bioaugmentation of native microbial community accelerated bioremediation of oil refinery sludge,' *Bioresource Technology*, 253, pp. 22–32. doi.org/10.1016/j.biortech.2018.01.004

Smets, B.F., and Barkay, T. (2005) 'Horizontal gene transfer: perspectives at a crossroads of scientific disciplines,' *Nature Reviews Microbiology*, 3, pp. 675–678. doi.org/10.1038/nrmicro1253

Sun, S. *et al.* (2016) 'Genomic and molecular mechanisms for efficient biodegradation of aromatic dye,' *Journal of Hazardous Materials*, 302, pp. 286–95. doi.org/10.1016/j.jhazmat.2015.09.071

Tribelli, P.M. *et al.* (2018) 'Microaerophilic alkane degradation in *Pseudomonas extremaustralis*: a transcriptomic and physiological approach,' *Journal of Industrial Microbiology & Biotechnology*, 45, pp. 15–23.

Turnbaugh, P.J. *et al.* (2007) 'The human microbiome project,' *Nature*, 449, pp. 804–810. doi.org/10.1038/nature06244

Tyson, G.W. *et al.* (2005) 'Genome-directed isolation of the key nitrogen fixer *Leptospirillum ferrodiazotrophum sp.* nov. from an acidophilic microbial community,' *Applied Environmental Microbiology*, 71, pp. 6319–6324. doi.org/10.1128/AEM.71.10.6319-6324.2005

Wang, D-Z., Kong, L-F., Li, Y-Y., and Xie, Z-X. (2016) 'Environmental microbial community proteomics: status, challenges and perspectives,' *International Journal of Molecular Sciences*, 17, 1275.

Xue, L. *et al.* (2015) 'Comparative proteomic analysis in *Miscanthus sinensis* exposed to antimony stress,' *Environmental Pollution*, 201, pp. 150–160. doi.org/10.1016/j.envpol.2015.03.004

Yan, J., Im, J., Yang, Y., and Löffler, F.E. (2013) 'Guided cobalamin biosynthesis supports *Dehalococcoides mccartyi* reductive dechlorination activity,' *Philosophical Transactions of the Royal Society: Biological Sciences*, 368, 20120320.

Yooseph, S. et al. (2007) 'The Sorcerer II global ocean sampling expedition: Expanding the universe of protein families,' *PLoS Biol*, 5. doi.org/10.1371/journal.pbio.0050016

Zhang, W., Li, F., and Nie, L. (2010) 'Integrating multiple 'omics' analysis for microbial biology: application and methodologies,' *Microbiology*, 156, pp. 287–301.

Zhu, Y,. *et al.* (2017) 'Genomic and transcriptomic insights into calcium carbonate biomineralization by marine actinobacterium *Brevibacterium linens* BS258,' *Frontiers in Microbiology*, 8, 602.

Ziganshina, E.E. *et al.* (2018) 'Draft genome sequence of *Brevibacterium epidermidis* EZ-K02 isolated from nitrocellulose-contaminated wastewater environments,' *Data in Brief*, 17, pp. 119–123.

13 Bioremediation of Azo Dyes

*Jyoti Kant Choudhari[1], Mukesh Kumar Verma[1,2],
Jyotsna Choubey[1], Anandkumar J. Sweta Singh[1,2],
and Biju Prava Sahariah[1]**

[1]Chhattisgarh Swami Vivekanand Technical University,
Bhilai, Chhattisgarh, India 491107

[2]National Institute of Technology Raipur, Raipur,
Chhattisgarh, India 492010

*Corresponding author: biju.sahariah@gmail.com

CONTENTS

13.1 INTRODUCTION

Color is a vital element that is used to enhance the attractiveness of an item to the human eye. The large number of natural colors that are available is not sufficient for human use and to address the large demand, various dyes have been synthesized for today's culture and for use in advanced accessories. Azo dyes are one of the largest synthesized and utilized organic dyes that contain >1 azo group (–N=N–) in their configuration. Due to the diversity of the dye components, a large quantity of azo dyes are consumed by the textile, leather, plastic, print, paint, and food industries. Simultaneously, a large quantity of nonbound dye is released into the wastewater, which deteriorates the aesthetic condition and is a threat to the environment and human health. Due to their high solubility, the dyes prevent light penetration, interfere with photosynthesis, and the dissolved oxygen (O_2) levels in aquatic ecosystems. The linking rings of phenyl and naphthyl in the azo group is frequently exchanged with another functional group, such as hydroxyl (OH), methyl, nitro, and sulfonate. Azo dyes have acute and chronic toxicity. The –N=N– linkage that is responsible for the depth of color is susceptible to the formation of an amino group ($-NH_2$), which is a potential mutagen and carcinogen (DeVito, 1993; Wang et al., 2020). If the unused azo dye that is released into the wastewater is not treated it could enter the trophic levels and exhibit biomagnification, with a negative impact on soil microorganisms and plant germination and growth (Imran et al., 2014; Rehman et al., 2018).

The complex aromatic molecular configurations of the dye mean that it is xenobiotic and persists in the environment. The desirable qualities of a commercial dye are its resistance to light, washing, and microbial activity for a long time.

It is difficult to follow natural treatment processes in an ecosystem. Therefore, the treatment techniques for azo dyes that include conventional and advanced techniques are being investigated to achieve satisfactory outcomes to preserve water resources and the environment.

DOI: 10.1201/9781003130932-13

13.2 TREATMENT PROCESSES FOR AZO DYES

The high solubility and persistence of azo dyes in water mean that their removal from wastewater is challenging. Decolorization is achieved when the connecting bonds in the azo dye are separated via physicochemical, biological agents or both. Advanced physicochemical and biological processes are employed carefully for the treatment of dye wastewater to achieve the treatment aims. Physicochemical processes, such as adsorption, coagulation, flocculation, ion exchange; and oxidation process, such as advanced oxidation, ozonation, the Fenton process, photo-Fenton, photocatalysis, and electrochemical oxidation, individually or in combination have been used for dye wastewater treatment. Various Fenton and advanced oxidation processes (AOPs) combined with different treatments are often used for decolorization due to their high efficiency. Table 13.1

TABLE 13.1
Treatment of Azo Dyes Using Physicochemical Processes

Number	Physicochemical Treatment	Remark	References
1.	Precipitation of monocarboxylic azo dye (MCD) using aluminum–magnesium hydroxycarbonate (LH)	LH was modified with azo dye at 15 % wt and 20 % wt concentrations efficient in pH 8–8.5, for 3 h. Acquired hybrid pigment dried at 80°C for 24 h under static air atmosphere New Peak arises in XRD (x-ray diffraction) study that does not belong to MCD or LH suggesting hybrid composite of the MCD, and LH attributed to strong interactions between the carboxylic group in the dye and the Mg^{2+} in the inorganic matrix	Szadkowski et al. (2018)
2.	Chitosan conjugation beads (CS–AL)	Aliquat-336 impregnated CS–AL were synthesized through the reaction of amino groups of chitosan with tricaprylylmethyl ammonium chloride Optimum adsorbent dosage of 2 g/L with high adsorption behavior and a wide pH range of 7–11	Ranjbari et al. (2019)
3.	Fenton oxidation	Optimal conditions for the decolorization and chemical oxygen demand (COD) removal of direct blue 71 (DB71): pH=3.0, Fe^{2+}=3 mg/L and H_2O_2=125 mg/L At optimal conditions, color and COD removal at initial dye 100 mg/L aqueous solution were 94% and 50.7%, respectively after 20 min of reaction	Ertugay et al. (2017)
4.	Fenton oxidation	Considered parameter: pH 2.5–9.0, catalyst loading (0.25–3.0 g/L), and BR18 dye concentration (0.1–0.3 mM) using powder magnetite nanoparticles (NPs) Color removal efficiencies were achieved as 44% at pH 9.0 and 76% at pH 3.5 for adsorption and Fenton oxidation of BR18 dye (0.1 mM)	Ozbey Unal et al. (2019)

TABLE 13.1 (Continued)
Treatment of Azo Dyes Using Physicochemical Processes

Number	Physicochemical Treatment	Remark	References
5.	Photo-Fenton Process	Adsorption of Acid Red 1 on an iron modified composite adsorbent coating Operating pH from 3 to 11 Performed adsorption tests using magnetic bar (350 rpm) and stirrer (RT-10 Power) Effective adsorption ≤10 adsorption–regeneration cycles via photo-Fenton process	Azha et al. (2019)
6.	Anodic oxidation, EF, and photoelectro-Fenton (PEF)	Degradation of Reactive Yellow 160 (RY160) azo dye from 100 cm^3 solution with 0.167 mmol/dm^3 dye in sulfate medium at pH 3.0	Bedolla-Guzman et al. (2016)
7.	Solar PEF (SPEF)	Allura Red AC azo dye is degraded by EF and SPEF with fast decolorization and almost total but slow mineralization by SPEF with a Pt/air-diffusion cell with 16 identified intermediates Optimum condition for cost-effective SPEF treatment for 460 mg L1 azo dye in 0.05 M Na_2SO_4 at 50 mA cm^2, which yielded 95% mineralization with 81% current efficiency and 8.50 kW h/m^3	Thiam et al. (2015)
8.	Electrochemical oxidation, OH radical, Fenton-like process	Degradation of Acid Yellow 36 solution by the electrooxidation process containing an Ir-Sn-Sb oxide anode and a stainless-steel cathode at pH 3.0 Quick decolorization in Cl^-/SO_4^{2-} as electrolyte but accumulation of chloro derivatives as intermediate Acid Yellow 36 yields maleic and acetic acids along with SO_4^{2-} and NO_3^- ions	Aguilar et al. (2017)
9.	Heterogeneous photo-Fenton process	Methyl Orange (MO) decolorization in heterogeneous photo-Fenton process using a natural clay as a photocatalyst under sun irradiation Direct photolysis achieved for a substrate of 10^{-4} mol/L, at optimal conditions [clay]=1 g/L and pH=3 in the presence of oxalic acid = 10^{-2} M	Khennaoui et al. (2017)
10.	Flax shives; anionic azo dyes; cationic surfactant modification; synchrotron infrared analysis; interface transport	Applied cetyltrimethylammonium bromide surfactant-modified flax shives for removal of anionic azo dyes Acid Orange 7 (AO7), Acid Red 18 (AR18) and Acid Black 1 (AB1) Adsorption of three anionic azo dyes influenced by adsorbent dose, pH, and ionic strength but spontaneous	Wang et al. (2017)

(continued)

TABLE 13.1 (Continued)
Treatment of Azo Dyes Using Physicochemical Processes

Number	Physicochemical Treatment	Remark	References
11.	AOP driven by pulsed light (PL)	Azo dye decolorization tested at PL/H_2O_2 and PL/ H_2O_2/ferrioxalate process. High rate of decolorization achieved at low dye dose and high H_2O_2 doses in the PL/H_2O_2 process with >50% decolorization after applying 54 J/cm^2 (25 light pulses) The efficacy of this process increased to >95% with addition of ferrioxalate in PL/H_2O_2/ferrioxalate process for MO at 10 pulses	Martínez-López et al. (2019)
12.	Electrochemical AOPs: anodic oxidation with electrogenerated H_2O_2 (EO-H_2O_2), EF, and PEF	For decolorization of the copper-phthalocyanine dye Reactive blue 15 dye in sulfate medium, PEF is the most powerful treatment followed by EF and EO-H_2O_2	Solano et al. (2016)
13.	Innovative hydrodynamic cavitation (HC) and hybrid process of HC as HC + UV and HC + photocatalytic and photocatalysis	Decolorization of RR180 dye solutions is high combined with hydrodynamic cavitation and photocatalytic processes at pH 6.9–7.1	Çalışkan et al. (2017)
13.	Homogeneous, heterogeneous, copper ions, alumina, H_2O_2 catalyst	The reaction rate for dyes Chromotrope 2B, Chromotrope 2R, and Chrysophenen increase with increasing pH and temperature, and it is entropy controlled during homogeneous and heterogeneous catalytic oxidation processes in presence of copper (II) ions, copper (II) ions supported on alumina and zinc oxide and a copper–ammonia complex supported on alumina	Salem et al. (2020)
14.	SiO_2-Co core-shell nanoparticles	MO is reductively degradable in aqueous solution within 1 min using 50 mg of SiO_2-Co core-shell nanoparticles at 0.076 mM at pH 2.5 and Congo Red requires more core-shell nanoparticles to achieve the same amount of degradation	Gao et al. (2016)

lists the significant physicochemical treatment processes that have been adopted for the treatment of azo dyes to date. Parameters, such as pH, reaction time, temperature, reactant dose, and initial dye concentration have a significant role in decolorization during physicochemical treatment. The bioremediation of azo dyes is used where biological agents, such as microbes, bacteria, fungi, and algae can decolorize azo dyes. In biological processes, azo dyes are rarely degradable under aerobic conditions; however, biodegradation is possible under anaerobic conditions, which results in the generation of aromatic amines as by-products that are more toxic than the original dyes. Therefore, it requires further treatment to convert these intermediates into nonharmful end products. Microbial agents, such as bacteria and fungi can be used in microbial fuel cells, along with various extracted enzymes and algae. The treatment unit might operate in an individual and or combined mode, which depends on the strength of wastewater and the required water quality. The following sections describe various biological treatment methods used in the bioremediation of azo dyes.

13.3 BIOREMEDIATION: AGENTS AND MECHANISMS

Bioremediation is often encouraged in the treatment of pollutants, such as organics, inorganics, and metals due to its cost-effectiveness, eco-friendly end products, and easy operation. Currently, thousands of azo dyes are used commercially. The variety of dyestuffs can be matched with biological agents, for example, bacteria, fungi, algae, and enzymes extracted from these agents for treatment. The principal mechanism in the bioremediation of a dye can be categorized as the biodegradation of a complex molecule to a simpler molecule and biosorption in the cell wall or cell. Biodegradation of azo dyes is an efficient method for decolorization and the microbes break down the azo bonds in the dyestuff that generates the energy required for their survival and growth. This is governed by various mechanisms based on the species and can be enhanced by the reductive or oxidative enzymes present in the organisms that promote biodegradation. Azoreductase is a prominent reductive enzyme, and the oxidative enzymes are polyphenol oxidases, manganese peroxidase, lignin peroxidase, laccase, tyrosinase, N-demethylase, dye decolorizing peroxidases, and cellobiose dehydrogenase. Biosorption of the dyestuff by microbial species incorporates the dye removal from the influent wastewater via adhesion of the dyestuff onto the surface or pores of a substrate of biological origin. Various functional groups, such as amino, OH, carboxyl, phosphate, and other charged groups are the basic units of heteropolysaccharides and lipids that are present in the cell wall. These functional groups generate strong attractive forces between the cell wall and azo dyes, and this is influenced by operational parameters, such as pH, temperature, ionic strength, time of contact, adsorbent, dye concentration, dye structure, and type of microorganism used.

The reduction of the azo bond ($-N=N-$) is the first step of azo dye biodegradation either intracellularly or extracellularly and involves three basic mechanisms: enzymes, low molecular weight redox mediators (RMs), and chemical reduction by biogenic reductants. These mechanisms might work individually or in combination based on the species and environment provided. The operating conditions have a significant influence on the performance of the microbes, and the efficiency could be enhanced with acclimatization and by providing optimum conditions, such as pH, temperature, O_2 availability and oxidation–reduction potential. Under anaerobic conditions, the azo dyes act as electron acceptors and azo bonds are reduced forming colorless aromatic amines, which are rarely mineralized anaerobically but can be degraded under aerobic conditions. Therefore, azo dye biodegradation is often conducted under an anaerobic–aerobic sequential system, anoxic and intermittent aeration systems to avoid the negative impact of toxic intermediates and for the complete mineralization of the dye species. As mentioned previously, persistent azo dye molecules with a complex configuration require efficient treatment for complete mineralization. Table 13.2 summarizes various bioremediation studies into azo dye treatment. The following sections discuss various biological agents that can be used for the bioremediation of azo dyes.

13.3.1 BIOREMEDIATION AGENTS FOR AZO DYES

Microbes with azo dye decolorization and degradation efficiency are naturally available and cover bacteria, fungi, and algae that have a degrading mechanism. Due to the variety and complexity of azo dyes, microbes possess specific responses and have different reactions to different dyes and the degree of dye decolorization by these agents is related to the molecular structures of the dyes.

TABLE 13.2
Bioremediation Techniques for Azo Dyes

Number	Dye and Treatment Agent	Review Comment	References
1.	**Dye:** Direct Black (DB) G **Species:** *Enterobacter, Pseudomonas* and *Morganella sp.* plus kaolin, bentonite and powdered activated carbon	Sequential processes that combine the anaerobic and aerobic steps and simultaneous processes utilizing anaerobic zones within aerobic bulk phases **Remark** Particle size and surface texture of carrier material with irregular solid surface and a proper particle size contains convenient pores to enhance dye biodegradation as substrate becomes concentrated near the cells (adsorption capacity of the solid) and promotes formation of microaerophilic niches and permits microbial growth	Barragán et al. (2006)
2.	**Dye:** MO **Species:** *Aeromonas sp.*	pH 3.0–7.0 and a broad temperature range (5°C–45°C) after incubation for 12 h **Remark** The main intermediates of MO degradation are N,N-dimethyl-p-phenylenediamine and 4-aminobenzenesulfonic acid. Enzymes such as laccase, NADH–DCIP reductase, and azo reductase are responsible for MO degradation	Du et al. (2015)
3.	**Dye:** Reactive Black 5 (RB 5), Reactive Orange 16 (RO16), Disperse Red 78 (DR 78) and Direct Red 81 (DR 81) **Species:** *Providencia rettgeri* HSL1 and *Pseudomonas sp.*	Sequential microaerophilic/aerobic processes **Remark** Efficient biodegradation of the azo dyes occurred with complete detoxification	Lade et al. (2015)
4.	**Dye:** Direct Fast Scarlet 4BS **Species:** *Penicillium funiculosum Thom* and *Sphingomonas sanguinis*	Microbes immobilized using polyvinyl alcohol (PVA) as the carrier. Optimum pH range (5–8), temperature range (25°C–40°C) under shaking culture of high DO level for high performance **Remark** The high decolorization activity (approximately 81.25 mg/L/h), which occurred at a dye concentration of 1,000 mg/L and high repeat reuse of carrier PVA beads are feasible	Fang et al. (2004)

#	Species/Dye	Remark	Reference
5.	DB22 **Species:** Mixed culture	Intermittent aeration strategy **Remark** Intermittent aeration enhances azo dye biodegradation and COD removal, but decolorization velocity decreases at high-frequency aeration cycle suggesting use of O_2 as an electron acceptor over azo dyes	Oliveira et al. (2020)
6.	Procion Red MX-5B **Species:** *Aspergillus niger* and *Aspergillus terreus*	Biosorption and biodegradation in Erlenmeyer flasks containing dye solution at 200 μg/mL, pH 4.0, and 3 mg/mL of fungal biomass **Remark** Biosorption provides decolorization and reducing the toxicity of the dye whereas biodegradation remain efficient only for decolorization but toxicity due to toxic metabolites exists	Almeida and Corso (2014)
7.	**Dye:** Alizarin yellow R (AYR) **Species:** *Enterobacter* and *Enterococcus*	Hybrid anaerobic digestion (AD) bioreactor with built-in BESs [electrodes installed in liquid phase (R1) and sludge phase (R2)] **Remark** Very high decolorization at high loading (at influent loading rate of 800 g AYR/ m^3/d) and BES enhanced performance of the reactor in AD compared with sludge phase	Cui et al.(2016)
8.	**Dye:** AO7, Reactive red 2, RB5, Direct yellow 12 (DY12) and DB71 **Species:** Methanogenic granular sludge	Lab-scale high rate anaerobic methanogenic bioreactor with influent azo dyes at 600 mg/L, and 50 μM anthraquinone-2-sulfonate as RM in the last 25 days of operation **Remark** Acetolactic methanogens are more susceptible to high concentration azo dyes than hydrogenotrophic methanogens and the high biological toxicity of azo dyes can be mainly attributed to enrichment effect in tightly bound-extracellular polymeric substances	Dai et al. (2016)
9.	**Dye:** From textile wastewater. **Species:** Sulfate-reducing bacteria	Potential effects of NPs zinc oxide nanoparticles (ZnO-NPs)) with batch biological sulfate-reducing reactor **Remark** ZnO-NPs suppressed various oxidoreductive enzyme activities involved in dye decolorization and biotransformation	Rasool and Lee (2016)

(continued)

TABLE 13.2 (Continued)
Bioremediation Techniques for Azo Dyes

Number	Dye and Treatment Agent	Review Comment	References
10.	**Dye:** Direct violet, DR. DB, Reactive blue, RR, RO. Alura red, Fast green, Crystal violet, Tartrazine, Trypan blue, Janus green, Alirazin yellow, Evans blue, Brilliant green, Safranin, Pararosaniline, Ponceau S, Naphthol blue black, and Methyl red **Species:** Five species of *Aspergillus* and *Lichtheimia* sp.	MSM media in tubes with dye and culture incubated at 150 rpm at 28°C for 5 days **Remark** Direct violet easily transformable azo dye and glucose supplementation enhances decolorization No able to produce lignin peroxidases by the fungal strains in absence of organic nitrogen source	Abd El-Rahim et al. (2017)
11.	**Dye:** Methylene blue (MB) **Species:** Brown algae (Phaeophyta, *S. polycystum*) Red algae (Rhodophyta, *E. spinosum K. striatum K. alverezii*) Green algae (Chlorophyta, *C. lentillifera*)	Biosorption in lab with 0.025 g of biosorbent made of dry algae biomass into 25 mL of MB solution with the initial dye at 100 mg/L at pH 7, and room temperature **Remark** All species of red, brown and green algae possess high efficiency for MB dye removal and *E. spinosum* show high biosorption efficiency among all considered	Mokhtar et al. (2017)
12.	**Dye:** Acid Violet 7, Acid Red 1, Allura Red AC, Orange G and Sunset Yellow FCF **Enzyme:** Laccase from *T. versicolor*	1-hydroxybenzotriazole used as RM **Remark** RM increased decolorization efficiency with laccase from *T. versicolor* and final products possess lower inhibition effect compared to initial selected azonaphthalene dyes	Legerská et al. (2018)
13.	**Dye:** 2-Naphthol Red dye (Acid red 88) - containing industrial wastewater **Species:** Mixed culture	A laboratory-scale of upflow anaerobic sludge bed followed by an aerobic process **Remark** Combined system provides high removal and mineralization efficiency at 6–12 h HRT (hydraulic retention time) at 100 mg/L dye and performance deteriorates at lower HRT	Gadow and Li, (2020)

No.	Dye/Enzyme/Species	Method and Remark	Reference
14.	Dye: RB5 **Enzyme**: Horseradish peroxidase	Laboratory batch reactor with 100 mL as reaction volume at 50 mg/L of dye at varied enzyme concentration, H_2O_2, and pH. **Remark** Horseradish peroxidases have high RB5 dye removal efficiency at pH 4 and pH influence dye removal reactions to a great extent, followed by the enzyme concentration	Ulson de Souza et al. (2018)
15.	Dye: AO7 Mixed species, combined system	Single-chamber BES with stacked modules **Remark** *Pseudomonas, Geobacter, Comamonas, Meniscus, Bellilinea, Achromobacter,* and *Paludibacter* were present as specific microbial communities of the biocathode, and *Geobacter* and *Acinetobacter* were enriched in the microbial community of the bioanode contributing the electron transfer and/or azo dye decolorization	Kong et al. (2018)
16.	Reactive Violet 5 (RV5) Species: *Trichosporon, Aspergillus* and *Clostridium sp.* in activated sludge treatment (ASP)	Sequential chemical (AOP–Fenton process) and ASP **Remark** Pretreatment with the Fenton process significantly enhanced the dye removal efficiency of ASP	Meerbergen et al. (2017)

Bacterial species including *Pseudomonas, Bacillus, Enterobacter, Enterococcus Klebsiella, Streptomyces, Lactobacillus, Lysinibacillus, Acinetobacter, Stenotrophomonas, Pelosinus, Morganella, Chryseobacterium,* and *Clostridium* can produce azo dye degrading enzymes that are used for azo dye treatment. However, the overall list of bacterial species is not discussed in this chapter.

Fungi species that are associated with azo dye treatment are species of *Aspergillus* and white-rot fungi (e.g., *Phanerochaete chrysosporium, Bjerkandera adusta, Trametes versicolor, Phlebia radiata, Lichtheimia, Abortiporus biennis, Bjerkandera fumosa, Cerrena unicolor, Fomitopsis pinicola, Geotrichum sp., Gloephyllum odoratum, Kuehneromyces mutabilis, Phlebia radiata, Stropharia rugoso-annulata, Agrocybe cylindracea, Rhizopus arrhizus, Flammulina velutipes, Ganoderma applanatum, Heterobasidion annosum, Lentinus edodes, Panus tigrinus, Pestalotia sp., Pholiota glutinosa, Pleurotus pulmonarius, Trametes san guinea, Bjerkandera fumosa, Kuehneromyces mutabilis, Stropharia rugoso-annulata, Coprinus micaceus, Pestalotia sp.* and *Diplomitoporus crustulinus*).

A few examples of algal species used for decolorization are *Euchema spinosum, Enteromorpha sp., Kappaphycus alvarezii, Oscillateria tenuis, Chlorella pvrenoidosa, Chlorella vulgaris,* and *Sargassum polycystum.* Algae have been used many times as biosorbents combined with the biodegradation of azo dyes and dead algae biomass is preferred for the biosorption of dye material. Similar to the bacterial species, the list of fungi and algae is long and based on azo dye molecule specific species that have varied performances.

13.4   FUTURE SCOPE

Currently, research with different treatment configurations, such as microbial fuel cells, bioelectrochemical systems (BESs), biological treatment combined with the Fenton process, photocatalysis, electro-Fenton (EF) processes, and enzymes is being carried out and many methods have been applied in the field. Further studies are required with individual dye components mixed with other inorganics, organics, and metal pollutants to efficiently address the dye wastewater pollution load and save water resources.

13.5   CONCLUSIONS

Azo dyes are constituents of various industrial wastewaters, such as textiles, tanneries, and personnel care and they are widely distributed. However, due to the rich biodiversity of bacteria, fungi, algae, and their combinations, the inbuilt morphological and functional properties of these microbes could help to address dye bioremediation. This efficiency could be enhanced by providing suitable parameters, such as pH, acclimatization of the microbes, and pre or post-treatment with some other treatment options. Currently, wastewater treatment requires rapid and cost-effective technologies to deal with the azo dyes in wastewater, and therefore, research to identifying the correct combination of techniques for the wastewater is essential.

REFERENCES

Abd El-Rahim, W. M., Moawad, H., Abdel Azeiz, A. Z., & Sadowsky, M. J. (2017). Optimization of conditions for decolorization of azo-based textile dyes by multiple fungal species. *Journal of Biotechnology, 260,* 11–17. doi.org/10.1016/j.jbiotec.2017.08.022

Aguilar, Z. G., Brillas, E., Salazar, M., Nava, J. L., & Sirés, I. (2017). Evidence of Fenton-like reaction with active chlorine during the electrocatalytic oxidation of Acid Yellow 36 azo dye with Ir-Sn-Sb oxide anode in the presence of iron ion. *Applied Catalysis B: Environmental, 206,* 44–52. doi.org/10.1016/j.apcatb.2017.01.006

Almeida, E. J. R., & Corso, C. R. (2014). Comparative study of toxicity of azo dye Procion Red MX-5B following biosorption and biodegradation treatments with the fungi *Aspergillus niger* and *Aspergillus terreus*. *Chemosphere*, *112*, 317–322. doi.org/10.1016/j.chemosphere.2014.04.060

Azha, S. F., Sellaoui, L., Engku Yunus, E. H., Yee, C. J., Bonilla-Petriciolet, A., Ben Lamine, A., & Ismail, S. (2019). Iron-modified composite adsorbent coating for azo dye removal and its regeneration by photo-Fenton process: Synthesis, characterization and adsorption mechanism interpretation. *Chemical Engineering Journal*, *361*, 31–40.doi.org/10.1016/j.cej.2018.12.050

Barragán, B. E., Costa, C., & Carmen Márquez, M. (2006). Biodegradation of azo dyes by bacteria inoculated on solid media. *Dyes and pigments*, *75*(1), 73–81. doi.org/10.1016/j.dyepig.2006.05.014

Bedolla-Guzman, A., Sirés, I., Thiam, A., Peralta-Hernández, J. M., Gutiérrez-Granados, S., & Brillas, E. (2016). Application of anodic oxidation, electro-Fenton and UVA photoelectro-Fenton to decolorize and mineralize acidic solutions of Reactive Yellow 160 azo dye. *Electrochimica Acta*, *206*, 307–316. doi.org/10.1016/j.electacta.2016.04.166

Çalışkan, Y., Yatmaz, H. C., & Bektaş, N. (2017). Photocatalytic oxidation of high concentrated dye solutions enhanced by hydrodynamic cavitation in a pilot reactor. *Process Safety and Environmental Protection*, *111*, 428–438. doi.org/10.1016/j.psep.2017.08.003

Cui, M. H., Cui, D., Lee, H. S., Liang, B., Wang, A. J., & Cheng, H. Y. (2016). Effect of electrode position on azo dye removal in an up-flow hybrid anaerobic digestion reactor with built-in bioelectrochemical system. *Scientific Reports*, *6*(1), 1–9. doi.org/10.1038/srep25223

Dai, R., Chen, X., Luo, Y., Ma, P., Ni, S., Xiang, X., & Li, G. (2016). Inhibitory effect and mechanism of azo dyes on anaerobic methanogenic wastewater treatment: Can redox mediator remediate the inhibition? *Water Research*, *104*, 408–417. doi.org/10.1016/j.watres.2016.08.046

DeVito, S. C. (1993). Predicting azo dye toxicity. *Critical Reviews in Environmental Science and Technology*, *23*(3), 249–324. doi.org/10.1080/10643389309388453

Du, L. N., Li, G., Zhao, Y. H., Xu, H. K., Wang, Y., Zhou, Y., & Wang, L. (2015). Efficient metabolism of the azo dye methyl orange by *Aeromonas sp.* strain DH-6: Characteristics and partial mechanism. *International Biodeterioration and Biodegradation*, *105*, 66–72. doi.org/10.1016/j.ibiod.2015.08.019

Ertugay, N., Kocakaplan, N., & Malkoç, E. (2017). Investigation of pH effect by Fenton-like oxidation with ZVI in treatment of the landfill leachate. *International Journal of Mining, Reclamation and Environment*, *31*(6), 404–411. doi.org/10.1080/17480930.2017.1336608

Fang, H., Wenrong, H., & Yuezhong, L. (2004). Investigation of isolation and immobilization of a microbial consortium for decoloring of azo dye 4BS. *Water Research*, *38*(16), 3596–3604. doi.org/10.1016/j.watres.2004.05.014

Gadow, S. I., & Li, Y. Y. (2020). Development of an integrated anaerobic/aerobic bioreactor for biodegradation of recalcitrant azo dye and bioenergy recovery: HRT effects and functional resilience. *Bioresource Technology Reports*, *9*, 100388. doi.org/10.1016/j.biteb.2020.100388

Gao, F., Wanjala, B., Li, Z., Zhang, Y., Cernigliaro, G., & Gu, Z. (2016). High Efficiency reductive degradation of a wide range of azo dyes by SiO$_2$-Co core-shell nanoparticles. *Applied Catalysis B: Environmental*, *199*, 504–513. doi.org/10.1016/j.apcatb.2016.06.030

Imran, M., Crowley, D. E., Khalid, A., Hussain, S., Mumtaz, M. W., & Arshad, M. (2014). Microbial biotechnology for decolorization of textile wastewaters. *Reviews in Environmental Science and Biotechnology*, *14*(1)73–92. doi.org/10.1007/s11157-014-9344-4

Khennaoui, B., Malouki, M. A., López, M. C., Zehani, F., Boutaoui, N., Salah, Z. R., & Zertal, A. (2017). Heterogeneous photo-Fenton process for degradation of azo dye: Methyl orange using a local cheap material as a photocatalyst under solar light irradiation. *Optik*, *137*, 6–16. doi.org/10.1016/j.ijleo.2017.02.081

Kong, F., Ren, H.-Y., Pavlostathis, S. G., Wang, A., Nan, J., & Ren, N.-Q. (2018). Enhanced azo dye decolorization and microbial community analysis in a stacked bioelectrochemical system. *Chemical Engineering Journal*, *354*, 351–362.doi.org/10.1016/j.cej.2018.08.027

Lade, H., Kadam, A., Paul, D., & Govindwar, S. (2015). Biodegradation and detoxification of textile azo dyes by bacterial consortium under sequential microaerophilic/aerobic processes. *EXCLI Journal*, *14*, 158–174. doi.org/10.17179/excli2014-642

Legerská, B., Chmelová, D., & Ondrejovič, M. (2018). Azonaphthalene dyes decolorization and detoxification by laccase from *Trametes versicolor*. *Nova Biotechnologica et Chimica*, *17*(2), 172–180. doi.org/10.2478/nbec-2018-0018

Martínez-López, S., Lucas-Abellán, C., Serrano-Martínez, A., Mercader-Ros, M. T., Cuartero, N., Navarro, P., Pérez, S., …Gómez-López, V. M. (2019). Pulsed light for a cleaner dyeing industry: Azo dye degradation by an advanced oxidation process driven by pulsed light. *Journal of Cleaner Production*, *217*, 757–766. doi.org/10.1016/j.jclepro.2019.01.230

Meerbergen, K., Crauwels, S., Willems, K. A., Dewil, R., Van Impe, J., Appels, L., & Lievens, B. (2017). Decolorization of reactive azo dyes using a sequential chemical and activated sludge treatment. *Journal of Bioscience and Bioengineering*, *124*(6), 668–673. doi.org/10.1016/j.jbiosc.2017.07.005

Mokhtar, N., Aziz, E. A., Aris, A., Ishak, W. F. W., & Mohd Ali, N. S. (2017). Biosorption of azo-dye using marine macro-alga of *Euchema spinosum*. *Journal of Environmental Chemical Engineering*, *5*(6), 5721–5731. doi.org/10.1016/j.jece.2017.10.043

Oliveira, J. M. S., de Lima e Silva, M. R., Issa, C. G., Corbi, J. J., Damianovic, M. H. R. Z., & Foresti, E. (2020). Intermittent aeration strategy for azo dye biodegradation: A suitable alternative to conventional biological treatments? *Journal of Hazardous Materials*, *385*, 121558. doi.org/10.1016/j.jhazmat.2019.121558

Ozbey Unal, B., Bilici, Z., Ugur, N., Isik, Z., Harputlu, E., Dizge, N., & Ocakoglu, K. (2019). Adsorption and Fenton oxidation of azo dyes by magnetite nanoparticles deposited on a glass substrate. *Journal of Water Process Engineering*, *32*, 100897. doi.org/10.1016/j.jwpe.2019.100897

Ranjbari, S., Tanhaei, B., Ayati, A., & Sillanpää, M. (2019). Novel Aliquat-336 impregnated chitosan beads for the adsorptive removal of anionic azo dyes. *International Journal of Biological Macromolecules*, *125*, 989–998. doi.org/10.1016/j.ijbiomac.2018.12.139

Rasool, K., & Lee, D. S. (2016). Effect of ZnO nanoparticles on biodegradation and biotransformation of co-substrate and sulphonated azo dye in anaerobic biological sulfate reduction processes. *International Biodeterioration and Biodegradation*, *109*, 150–156. doi.org/10.1016/j.ibiod.2016.01.015

Rehman, K., Shahzad, T., Sahar, A., Hussain, S., Mahmood, F., Siddique, M. H., … Rashid, M. I. (2018). Effect of Reactive Black 5 azo dye on soil processes related to C and N cycling. *PeerJ*, *2018*(5), e4802. doi.org/10.7717/peerj.4802

Salem, I. A., Shaltout, M. H., & Zaki, A. B. (2020). Homogeneous and heterogeneous catalytic oxidation of some azo dyes using copper(II) ions. *Spectrochimica Acta Part A: Molecular and Biomolecular Spectroscopy*, *227*, 117618. doi.org/10.1016/j.saa.2019.117618

Solano, A. M. S., Martínez-Huitle, C. A., Garcia-Segura, S., El-Ghenymy, A., & Brillas, E. (2016). Application of electrochemical advanced oxidation processes with a boron-doped diamond anode to degrade acidic solutions of Reactive Blue 15 (Turquoise Blue) dye. *Electrochimica Acta*, *197*, 210–220. doi.org/10.1016/j.electacta.2015.08.052

Szadkowski, B., Marzec, A., Rybiński, P., Maniukiewicz, W., & Zaborski, M. (2018). Aluminum-magnesium hydroxycarbonate/azo dye hybrids as novel multifunctional colorants for elastomer composites. *Polymers*, *11*(1), 43. doi.org/10.3390/polym11010043

Thiam, A., Sirés, I., Centellas, F., Cabot, P. L., & Brillas, E. (2015). Decolorization and mineralization of Allura Red AC azo dye by solar photoelectro-Fenton: Identification of intermediates. *Chemosphere*, *136*, 1–8. doi.org/10.1016/j.chemosphere.2015.03.047

Ulson de Souza, A. A., Baumer, J. D., Valério, A., Ulson de Souza, S. M. G., Erzinger, G. S., & Furigo, A. (2018). Remazol black dye (reactive black 5) decolorization by horseradish peroxidase enzyme. *Journal of Textile Engineering & Fashion Technology*, *4*(2). doi.org/10.15406/jteft.2018.04.00132

Wang, W., Huang, G., An, C., Xin, X., Zhang, Y., & Liu, X. (2017). Transport behaviors of anionic azo dyes at interface between surfactant-modified flax shives and aqueous solution: Synchrotron infrared and adsorption studies. *Applied Surface Science*, *405*, 119–128. doi.org/10.1016/j.apsusc.2017.01.311

Wang, Y., Jiang, L., Shang, H., Li, Q., & Zhou, W. (2020). Treatment of azo dye wastewater by the self-flocculating marine bacterium *Aliiglaciecola lipolytica*. *Environmental Technology and Innovation*, *19*, 100810. doi.org/10.1016/j.eti.2020.100810

14 Microbial Degradation of Azo Dyes Using Bacteria

Vandana Gupta[1]* and Mayur R. Raviya[2]

[1]Department of Chemical Engineering, National Institute of Technology Raipur, Raipur, 492010, Chhattisgarh, India

[2]Reverse Osmosis Department, CSIR-Central Salt and Marine Chemicals Research Institute, Council of Scientific and Industrial Research (CSIR). Gijubhai Badheka Marg, Bhavnagar, 364002, Gujarat, India

*Corresponding author: vgupta.che@nitrr.ac.in

CONTENTS

14.1 INTRODUCTION

Dyes are chemical compounds that are used as coloring agents. A large number of dyes are available that are natural or synthetic. The common natural dyes are hina, turmeric, saffron, and tea. Synthetic dyes are generally synthesized from organic or inorganic chemical compounds. Synthetic dyes can be classified as acid, basic, azoic, vat, nitro, reactive, and sulfur dyes (Kiernan, 2001). The global production of dye was estimated to be >1000,000 t by the Chemical Economics Handbook (2018). India is the second-largest producer of dyes and consumes a lot of dyes in domestic and industrial applications (Global Dyes Market Analysis Report, 2018). Azoic dyes account for 50% of the dyes produced and have high consumption levels due to their thermal and chemical stability, durability, and wide color range (Stolz, 2001).

Dyes have applications in numerous industries, such as textile, printing, leather, food and beverages, cosmetics, plastics, and inks. The textile industry is the major consumer of dyes. In the textile industry, dyes are used to provide permanent color to fibrous materials that can resist fading

DOI: 10.1201/9781003130932-14

of the material when exposed to sunlight, a harsh chemical environment, sweat, water, and micro-bial attack (Rai et al., 2005). The high consumption of dye bearing water in a dry processing mill and woven fabric finishing mill in the textile industry results in the discharge of wastewater that contains dyes into the environment. The loss of dyes into the wastewater usually depends upon its binding with the fabric. Basic dyes exhibit minimum loss (2%) and maximum loss has been reported for reactive dyes (O'Niel et al., 1999). The wastewater that contains dyes and compounds from the reactive dyes are toxic and contaminates the groundwater, surface water, and soil. In addition, the discharge of dyes into the environment can alter the pH, color, dissolved oxygen (O_2), biological oxygen demand (BOD), and chemical oxygen demand (COD) of the water bodies. Alterations in water quality can have a severe effect on soil fertility, ecosystems, natural resources, aquatic life or organisms, and human health. Therefore, the degradation and decolorization of dyes in effluents is the main concern before discharge into the environment.

14.2 AZO DYES

14.2.1 STRUCTURE AND SOURCE

A large quantity of azo dyes is consumed in industries due to their stability, cost-effectiveness, ease of synthesis, and variety of colors. Azo dyes are water-soluble aromatic compounds with ≥ 1 –N=N– (azo group) and $–SO^{3-}$ (sulfonic group) in their chemical structure (Barragan et al., 2007). Azo dyes can be classified based on the number of –N=N– groups in their chemical structure. Dyes with one –N=N– group is a monoazo dye and diazo and triazo dyes have two and three –N=N– groups, respectively (Zollinger, 1991). In addition, a variety of azo dyes can be synthesized by linking the azo group with benzene, phenyl, or naphthyl rings, which are usually substituted with different substituents such as nitro ($–NO_2$), chloro (–Cl), amino ($–NH_2$), methyl ($–CH_3$), carboxyl (–COOH), hydroxyl (–OH), sulfonate ($– SO_3^-$) groups (Zollinger, 1991; Bell et al., 2000). These side groups are responsible for the different shades, colors, intensity, and absorption spectrum. Structures of various azo dyes are represented in Table 14.1. In total, >3,000 azo dyes have been reported in the litera-ture, among which 80% are only utilized in the textiles industries. Other industries that contribute to the high consumptions of azo dye are the leather, food, paper, printing, pharmaceutical, and cos-metic industries (Telke et al., 2008; Elbanna et al., 2010). For the textile industry, approximately 10% of azo dyes do not bind to the fabrics during dyeing and are discharged into the environment as effluent. The release of dyes into the environment has resulted in water and soil contamination. Some dye compounds are highly toxic. The toxicity and its effect on the environment and living organisms are discussed in the following section.

14.2.2 TOXICITY

When an effluent that contains dyes is discharged into the environment it harms the flora and fauna. Textile industry effluent contributes the maximum discharge of dye at approximately around 10–200 mg/L. The improper discharge of a highly concentrated dye effluent into the water bodies affects aquatic diversity. The aquatic diversity is affected due to the reduction in sunlight penetration, which affects dissolved O_2 concentrations, photosynthesis activity, BOD, COD, total organic carbon, and water quality (Saratale et al., 2009a). Approximately 1 mg/L of azo dye can reduce the visibility in the water. The discharge of dye effluent onto land affects the soil quality in the germination rate and the biomass of plants, erosion, and soil infertility (Ghodake et al., 2009a). Some researchers have reported mutagenic activity due to azo dyes and their metabolites (Umbuziero et al., 2005). Many azo dyes are carcinogenic and cause allergic reactions in humans. The toxicity and carcinogenicity of azo dye increase with the increase in the number of benzene rings in their structure. In addition, carcinogenicity depends on the mechanism of degradation (Gičević et al., 2020).

TABLE 14.1
Structure of Azo Dyes

Name	Structure
Mono azo dye (Methyl red)	
Acid Red 88	
1-(4'-carboxyphenylazo)-2-naphthol	
1-(4'-carboxyphenylazo)-2-naphthol (carboxy orange I)	
Orange II	
Disperse orange 3	

(*continued*)

TABLE 14.1 (Continued)
Structure of Azo Dyes

Name	Structure
Acid orange 7	
Methyl orange	
4'carboxy-4'-sulfoazobenzene	
4,4'-Dicarboxyazobenzene	
Disperse yellow 3	
Diazo dye (Direct blue 15)	
Reactive black	

TABLE 14.1 (Continued)
Structure of Azo Dyes

Name	Structure
Congo red	
Methylene blue	

The degradation of azo dyes produces aromatic amines that have a different structure, such as arylamine and benzidine, which are carcinogenic (Sudha et al., 2014). Some azo dyes, for example, 1,4-diamino benzene, methylene blue, and benzidine can cause cancer, skin irritation, chemosis, permanent blindness, vomiting gastritis, vertigo, hypertension, drowsiness, discoloration of the oral mucosa, colds, redness or dryness, and bladder cancer (Gičević et al., 2020; Anirudhan et al., 2015). Toxic compounds in effluents that contain dyes are mixed with the surface and groundwater, which can reach humans and aquatic organisms and cause severe damage to the environment. The European Union banned the use of some azo dyes in early 2002 that broke down into carcinogenic compounds. In addition, several countries including China, Japan, India, and Vietnam have limited and restricted the use of some azo dyes that produce toxic, mutagenic, and carcinogenic compounds. Therefore, it is important to treat the effluent that contains effluents before being discharged into the environment.

14.3 REMEDIATION TECHNIQUES

Several remediation techniques have been reported for the treatment of dye bearing effluents. These remediation techniques are physical, chemical, and biological. The chemical treatment of effluent involves oxidation, ozonation, and electrolysis processes, and the physical methods include membrane separation, filtration, reverse osmosis, coagulation, flocculation, and adsorption (Pandey et al., 2007; Saratale et al., 2011). The biological methods involve the use of microorganisms, such as yeasts, fungi, algae, and bacteria, and some enzymes (Pandey et al., 2007; Saratale et al., 2011; Sudha et al., 2014).

Physical or chemical methods for dye remediation have major disadvantages; coagulation, flocculation, and electrolysis generate large amounts of sludge that might cause disposal problems. Adsorption has the problems of adsorbent loading, regeneration, and disposal (Pandey et al., 2007). Membrane filtration, oxidation, and ozonation are not economically feasible for energy requirements and chemical consumption and cannot completely remove azo dyes and their metabolites (Pandey et al., 2007; Zhang et al., 2004). Among all methods, biological treatment or the bioremediation of azo dyes have gained more attention because they are eco-friendly, completely mineralize organic pollutants, cost-effective, energy-efficient, have a fast degradation rate, and reduced water consumption (Saratale et al., 2011; Rai et al., 2005).

14.4   BIOREMEDIATION

Bioremediation involves microorganism and their degradation ability. When the microorganism comes into contact with a toxic pollutant they acclimatize themselves to it and new resistant strains develop. The breakdown of toxic pollutants into less harmful compounds is associated with the development of new resistant strains. Degradation and decolorization of azo dyes using bio-remediation techniques are performed under specific environments, such as aerobic or anaerobic conditions. Numerous microorganisms can degrade and decolorize azo dyes, such as bacteria, fungi, algae, and yeast, and some enzymes, such as laccase (Srinivasan et al., 2020; Ajaz et al., 2019). Fungal bioremediation of azo dyes is performed in fungal bioreactors. Several fungi that can degrade azo dyes include ligninolytic fungi, *Aspergillus niger*, *Pleurotus ostreatus*, *Penicillium geastrivous*, *Phanerochaete chrysosporium*, and white-rot fungi. The ligninolytic fungi have been reported widely for azo dye degradation and decolorization (Ajaz et al., 2019). Fungi produce enzymes as by-products during the mineralization of dyes. Enzymes play an important role in dye degradation and decolorization. Enzymes, such as laccase, manganese peroxidase, lignin perox-idase, and hydroxylase acts as azo dye reductases and medicate the microbial degradation of azo dyes. Laccases have been extensively studied. They utilize free radical mechanisms for the degradation and decolorization of an azo dye, which reduces the formation of toxic aromatic amines. Azoreductases are used in the form of a membrane-bound enzyme. Therefore, dye degradation occurs in bacterial cells with intact cell membranes (Ajaz et al., 2019). The ability of photosynthetic organisms, such as algae and cyanobacteria to degrade and decolorize azo dyes has been reported by some researchers. The mechanisms for algal remediation of azo dyes occurs in three stages: (1) creation of algal biomass; (2) alteration via hydrogen peroxide (H_2O_2) and carbon dioxide (CO_2) from colored to colorless molecules; and (3) biosorption or adsorption onto the algal biomass (Mahalakshmi et al., 2015). In this method, the algal biomass eliminates the toxic aromatic amines and colorants. Bioremediation of azo dye using yeasts has been examined by a few researchers. Yeast species such as *Saccharomyces cerevisiae*, *Galactomyces geotrichum*, *Trichosporon beigelli* and *Candida oleophila* have been utilized for the degradation and decolorization of methyl red, malachite green, azo-triphenyl methane, Navy blue HER and Reactive black, respectively (Saratale et al., 2009b; Jadhav et al., 2008a; Lucas et al., 2007). Phytoremediation or bioremediation using plants is an emerging technology for dye degradation and decolorization. Phytoremediation is an economical and efficient technique for the remediation of soil and groundwater. Phytoremediation involves planting autotrophic plants on contaminated sites or wetlands. This process is carried out with some specific plant species, such as *Sorghum vulgare* (great millet), *Brassica juncea* (brown mustard), *Phaseolus mungo* (black gram), *Blumea malcommi*, *Tagetes patula* (marigold), *Sesbania cannabina pers*, and *Medicago sativa*, which efficiently degraded the azo dyes and industrial effluent (Zhou & Xiang, 2013; Ghodake et al., 2009a, b; Patil et al.,2009). The disadvantages associated with phytoremediation are the contaminants tolerance limit and the availability of a grafting zone. Bacterial degradation of azo dyes uses bacterial strains or bacterial enzymes that work under a wide pH range, high concentrations of dye, have short production times, and produce less toxic by-products (Srinivasan et.al., 2020). Therefore, bacterial degradation leads to the eco-friendly degradation of azo dyes and is preferred above the other degradation techniques. Bacterial degradation is more efficient than other biodegradation techniques due to its thermal stability, high pH and –Cl tolerance (Fang et al., 2011).

14.5   BACTERIAL DEGRADATION AND DECOLORIZATION OF AZO DYES

The main advantages factors associated with the use of bacterial degradation for azo dyes are its cost-effectiveness, ability to degrade a wide range of azo dyes, and speed of growth with a shorter doubling time. Bacterial degradation and decolorization of azo dyes are facilitated with the –N= N– bond reductive cleavage as a first step. Azo dye degradation occurs in presence of diverse bacteria under aerobic, anaerobic, or anoxic conditions. The degradation of azo dyes under different conditions, types of bacteria, and mechanism of degradation are discussed in the following section.

14.5.1 TYPE OF BACTERIA

The identification of bacteria that could potentially degrade azo dyes was performed in the early 1970s with the isolation of three bacterial strains *Bacillus subtilis, Aeromonas hydrophila,* and *Bacillus cereus* (Dave et al., 2015). Bacteria can be broadly classified as aerobic and anaerobic. The aerobic and anaerobic bacteria include *Acinetobacter sp., Aeromonas sp., Alcaligenes sp., Alishewanella sp., Bacillus subtilis, Clostridium sp., Corneybaterium sp.,* and *Dermacoccus sp.,*

In addition, *Enterococcus sp., Escherichia coli, Geobacillus sp., Klebsiella sp., Lactobacillus sp., Micrococcus sp., Morganella sp., Proteus sp., Pseudomonas sp., Rhizobium sp., Rhodobacter sp., Staphylococcus sp., Shewanella sp.,* and *Xenophilus sp.* efficiently degrade azo dyes (Sarkar et al., 2017; Haghshenas et al., 2016; Imran et al., 2016; Sudha et al., 2014; Kolekar et al., 2013). Bacterial degradation of dyes can be specific or nonspecific and use a pure or mixed culture (consortium). Numerous bacterial strains that efficiently degrade and decolorize azo dyes are listed in Table 14.2.

TABLE 14.2
Bacteria Responsible for Efficient Degradation and Decolorization of Various Azo Dyes

Name of Bacteria	Dye	References
A. hydrophila	Reactive red 141, Reactive red 198, Reactive black 5, Reactive red 141, Reactive blue 171, Reactive yellow 84, Red RBN, Malachite green, Brilliant green, Crystal violet, Reactive yellow F3R, Joyfix Red RB	Chen et al., 2009; Hsueh et al., 2009; Chen et al., 2003; Mahmood et al., 2016; Srinivasan et al., 2017; Kumar et al., 2016
Aeromonas sp.	Reactive black	Shah, 2014
Acinetobacter calcoaceticus NCIM-2890	Direct brown MR	Ghodake et al., 2009a
Alishewanella sp. KMK6	Carmosine, Golden yellow HER, Reactive blue 59, Red BLI, Chocolate brown HT	Kolekar et al., 2013
Bacillus sp.	Red HE7B, Congo red, Navy blue 2GL	Thakur et al., 2014; Dawkar et al., 2009; Kannappan et al., 2009
B. cereus	Orange II,	Wuhrmann et al., 1980; Garg and Tripathi, 2013
Comamonas sp. UVS	Direct red 5B	Jadhav et al., 2008a
Enterococcus faecalis YZ66	Direct red 81	Sahasrabudhe et al., 2014
L. sphaericus MTCC 9523	Remazol yellow RR	Srinivasan, 2018
Micrococcus lutes	Acid black	Kanagaraj et al., 2015
Moraxella osloensis	Mordant black 17	Karunya et al., 2014
Morganella sp. HK1	Reactive black B (RB-B)	Pathak et al., 2014
Proteus sp.	Congo red	Barragan et al., 2007
Pseudomonas sp.	Reactive blue 13, Reactive black, Orange I, Orange II	Dave et al., 2015; Khan et al., 2015; Lade et al., 2015; Lin et al., 2010
Pseudomonas sp. SUK1	Reactive red 2; Methyl orange	Kalyani et al., 2009, 2008

(continued)

TABLE 14.2 (Continued)
Bacteria Responsible for Efficient Degradation and Decolorization of Various Azo Dyes

Name of Bacteria	Dye	References
P. aeruginosa	Acid red, acid blue, Remazol orange	Lucious et al., 2014; Sarayu and Sandhya, 2010
Pseudomonas sp. SU-EBT	Congo red, Acid red 119, Navy blue 2GL, Acid orange 10	Telke et al., 2009; Dave et al., 2015
Shewanellaxia menensis BC01	Congo red	Ng et al., 2014
Shewanella sp.	Reactive black 5, Direct red 81, Acid red 88, Orange 3. Reactive blue, Acid yellow,	Imran et al., 2016
Shewanella sp. IFN4	Acid red-88, direct red-88, reactive black-5	De Souza et al., 2019
Mutant Bacillus sp. (ACT1 and ACT2)	Congo red	Gopinath et al., 2009
Geobacillus stearothermophilus	Indigo carmine	Mehta et al., 2016
Bacillus pumilus HKG212 and *Zobellella taiwanensis* AT1–3	Reactive green-19	Das and Mishra, 2017
Brevibacillus laterosporus NCIM2298 and *GG-BL* *G. geotrichum* (MTCC *1360*)	Golden yellow HER	Waghmode et al., 2011
A. hydrophila MTCC 1739 and *L. sphaericus* MTCC 9523	Drimaren red CL5B	Srinivasan and Sadasivam, 2018

14.5.2 Mechanisms

Bacterial degradation of azo dyes starts with the cleavage of azo dye bonds (–N=N–), under aerobic or anaerobic conditions. Cleavage of –N=N– bonds occurs via a chemical reduction process that uses reductants, enzymes, or a redox mediator.

Direct enzymatic degradation includes the transfer of reduced equivalents to the azo dye via enzymes. Enzymes could be specific (i.e., only reduce the azo dye) or nonspecific (i.e., reduce the substrate). Transfer of electrons between the extracellular dye and the intracellular reductase occurs via redox mediators (Chacko & Subramaniam, 2011). This reaction is mediated by azoreductases in the presence of some reducing equivalent, for instance, nicotinamide adenine dinucleotide phosphate, or flavine adenine dinucleotide. (Van der Zee & Cervantes, 2009). Chemical reduction using azoreductases was first observed by Rafii et al. (1990) for the degradation of sulfonated azo dyes under anaerobic conditions. Azoreductase activity depends upon the microorganism and culture conditions and can involve more than one reductase. However, laccase producing bacteria, such as *Bacillus sp.* SF and *A. hydrophila* W11 can degrade azo dyes by direct oxidation without the cleaving of azo bonds. The intermediate product produced could be toxic or carcinogenic, which can be treated further under aerobic conditions (Jadhav et al., 2011).

14.5.3 Degradation and Decolorization Under Different Conditions

14.5.3.1 Aerobic Conditions

Azo dye degradation and decolorization under aerobic conditions requires substrates for bacteria; growth. A commonly used substrate is glucose, which has been reported by many researchers. For instance, Nachiyar and Rajkumar (2005) utilized glucose to degrade Navitan Fast blue S5R using *P. aeruginosa*. Some aerobic bacterial strains can grow on an azo compound as the sole carbon (C) source and utilize amines as a source of energy and C. These bacteria are substrate-specific. Kulla et al. (1983) and Zimmermann et al. (1982), reported the aerobic growth of *Pigmentiphaga kullae* K24 and *Xenophilus azovorans KF46* and on carboxy orange II and carboxy orange I, respectively. However, the growth of these strains on structurally analogous sulfonated dyes and acid orange was not observed. One of the commonly used aerobic processes is the activated sludge process for effluent treatment; however, it is quite ineffective in color reduction efficiency (Roy et al., 2018). Some pure aerobic bacterial stains are difficult to isolate. Therefore, bacterial consortia or mixed culture is preferred over the pure culture for the degradation of azo dyes, which could degrade and mineralize the aromatic amines formed during anaerobic degradation (Mishra & Maiti, 2018; Khalid et al., 2008).

14.5.3.2 Anaerobic Conditions

Anaerobic or methanogenic conditions for azo dye degradation and decolorization require the presence of methanogenic, acidogenic, or acetogenic bacteria, which require organic C or an energy source. Under these conditions, the azo bond in azo dyes undergoes cleavage with the formation of aromatic amines, which are hazardous and colorless. These aromatic amines can be further degraded in an aerobic step (Popli & Patel, 2015). Some studies have shown that degradation depends on the structure of the dye and source of organic C under anaerobic conditions (Bromley-Challenor et al., 2000; Stolz, 2001). Several researchers used simple and complex substrates, such as whey, tropica, glucose, acetate, and ethanol under methanogenic conditions (Van der Zee & Villaverde, 2005; Talarposhti et al., 2001). Efficient degradation and decolorization of azo dyes using methanogenic bacteria or a combination of acidogenic and methanogenic bacteria and a mixed population of bacteria were reported by Razo-Flores et al. (1997) and Bras et al. (2001). Tan et al. (2005) revealed that sulfonated aromatic amines could not be degraded under methanogenic conditions. Anaerobic azo dye degradation is a secondary process, sometimes the dye participates in the electron transport chain by acting as an electron acceptor or might be involved in nonspecific extracellular reactions (de Souza et al., 2019). It can be involved in a nonspecific extracellular reaction with the reduced anaerobic biomass (Van der Zee & Villaverde, 2005). Therefore, for complete degradation and decolorization of azo dyes, a combination of anaerobic and aerobic conditions is preferable.

14.5.3.3 Anoxic Conditions

Degradation and decolorization of azo dyes under anoxic conditions has been reported by many researchers using mixed aerobic and facultative anaerobic consortia. Under this condition, bacterial growth occurs under aerobic conditions followed by degradation under anaerobic conditions. Degradation and decolorization of azo dyes by bacterial cultures (e.g., consortia or pure) requires complex organic sources or a combination of carbohydrates. Yeast extract and peptone are the most common sources of complex organics (Khehra et al., 2005; Chen et al., 2003). The utilization of glucose as a substrate for bacterial growth is favorable under anaerobic conditions; however, it can form acids or cause catabolic repression, which contributes to the reduction of pH when operated under anoxic conditions (Chen et al., 2003). The aromatic amines formed during cleavage of the azo group under anaerobic conditions can be degraded under aerobic conditions. Therefore, efficient degradation of azo dyes requires anaerobic and aerobic conditions.

Degradation of azo dyes in anaerobic–aerobic conditions can be carried out simultaneously or sequentially. The sequential treatment system can be operated either in separate rectors or in a single reactor (at different times). In simultaneous operations, the anaerobic zone is maintained in a biofilm or gel matrix (Kudlich et al., 1996).

14.5.4 Factors that Affect Bacterial Degradation and Decolorization of Azo Dyes

Dye bearing effluent might consist of various components, for example, dyes, toxic compounds, salts, and sulfur compounds. The bacterial degradation of dye bearing effluent can be influenced by some physicochemical parameters, such as pH, temperature, electron donors, redox mediators, dye structure, and dye concentration. Therefore, knowledge of the effect of these parameters is essential to achieve fast, efficient, and feasible bacterial degradation of azo dyes.

The pH of the medium has a major influence on dye degradation efficiency. The optimum pH for efficient degradation and decolorization of azo dye is between pH 6 and 10 (Guo et al., 2007; Kilic et al., 2007). An increase in pH during degradation resulted in the reduction of azo bonds and the formation of alkaline aromatic amines. Azo dye degradation decreases with increasing or decreasing pH. In addition, the pH of the medium has an important role as a rate-limiting step for azo dye degradation by influencing the transport of dye molecules across cell membranes (Guo et al., 2007; Kilic et al., 2007).

An increase in dye concentration can reduce the rate of degradation due to the incorrect cell to dye ratio and blocking the active sites of azoreductase (Tony et al., 2009; Jadhav et al., 2008b). Inhibition of bacterial growth on the aromatic ring at high dye concentrations has been reported (Kalyani et al., 2008). Dye structure varies from simple to complex with a low or high molecular weight. Therefore, the degradation rate is high for a simple structure with a low molecular weight compared with a high molecular weight and complex structure (Hsueh et al., 2009). The production of azoreductase and the rate of electron transfer is related to dye structure and the associated functional groups. High electron density in the functional group can reduce the degradation rate due to the formation of dianion (Pearce et al., 2003).

The degradation rate of azo dyes can be enhanced by the provision of an external substrate or electron donor under anaerobic conditions (Telke et al., 2009). Different electron donors or acceptors have different effects on azo dye degradation. Anaerobic degradation participates in the electron transport chain by oxidizing the electron donor and provides electrons to electron acceptors via a multicomponent system. Some electron donors, such as sodium acetate, sodium citrate, sodium succinate, sodium pyruvate, and sodium formate enhance the degradation rate (Telke et al., 2009). The region of electron density in the azo bond has a key role in the degradation rate. Lower electron density could result in an enhanced degradation rate.

Transfer of reduced equivalents to electron acceptors can be accelerated by a redox mediator. Redox mediators are highly dependent on redox potential (from –200 to 350 mV). The degradation rate increases with a decrease in redox potential and vice versa (Liu et al., 2009). Some of the commonly used redox mediators are flavin and quinone based compounds.

The azoreductase responsible for bacterial cell growth can be active at temperature <60°C (Pearce et al., 2003). Therefore, temperature is an important parameter for azo dye degradation and microbial vitality. A small temperature change can cause an alteration in activation energy. In addition, a small temperature range has been reported for the degradation of azo dye using complex bacterial cultures. The degradation rate increases up to an optimum temperature (60°C) and then it decreases. The decrease in the rate of degradation at high temperatures could be due to the denaturation of azoreductase or the loss of cell viability (Saratale et al., 2009b)

14.6 RECENT ADVANCES

Recent studies into the bacterial degradation of azo dyes were carried out using novel bacterial strains and cultures. Currently, bacterial strains are either isolated from effluents or some domestic or food sources. In addition, bacterial strains can be linked with nanoparticles to enhance their degradation efficiency. Srinivasan and Sadasivam (2018) studied the degradation of Drimaren red CL5B by *A. hydrophila* MTCC 1739 and *Lysinibacillus sphaericus* MTCC 9523 and identified naphthalene, phenol, and 2-aminophenol as degraded metabolites. *A. hydrophila* SK16 degraded reactive yellow F3R into aniline and naphthalene (Srinivasan et al., 2017). Saravanan et al. (2017) isolated *Leuconostoc lactis* from idli batter and linked it with silver nanoparticles for the degradation of organic azo dyes (e.g., methyl orange and congo red). Bekhit et al. (2020) synthesized *Enterobacter cloacae* A3 coated with magnetic iron oxide nanoparticles for the degradation of basic red 46 (BS46). Degradation of disperse red 1 using *Microbacterium sp.* (FJ598006), *Leucobacter albus* (FJ598008), *Klebsiella sp.* and *Staphylococcus arlettae* (AB009933) was reported by Francisoon et al. (2015), which were isolated from textile effluent. In addition, recent advances in bioinformatics or computational approaches for methods and prediction are gaining more attention to save time and money. They can predict nature, structure, interactions, toxicity, mechanisms, pathways, functions, and contaminants, which could be utilized for real-time applications (Srinivasan et al., 2020).

14.7 CONCLUSIONS

Azo dyes and their metabolites are toxic and hazardous to aquatic and human life. In addition, they harm soil properties. Azo dye remediation can be assessed by various physicochemical and biological methods. Physicochemical processes generate secondary pollutants; however, complete degradation and decolorization of azo dyes can be achieved via bioremediation. Therefore, bioremediation could overcome the disadvantages of the physiochemical processes. Among all the microorganisms, bacterial degradation proved to be more efficient due to their rapid growth, superior hydraulic retention time, ability to degrade a wide spectrum of azo dyes, thermal stability, high pH tolerance, –Cl tolerance, and cost-effectiveness. Bacterial strains can be easily isolated from the environment. Therefore, well-characterized bacterial strains that have excellent dye degradation capability (e.g., either natural or genetically manipulated) could be a potential for azo dye degradation.

REFERENCES

Ajaz, A., Shakeel, S., & Rehman, A. (2019). Microbial use for azo dye degradation: a strategy for dye bioremediation. *International Microbiology, 23*(3), 1–11.

Anirudhan, T. S., & Ramachandran, M. (2015). Adsorptive removal of basic dyes from aqueous solutions by surfactant modified bentonite clay (organoclay): Kinetic and competitive adsorption isotherm. *Process Safety and Environmental Protection, 95*, 15–225.

Barragán, B. E., Costa, C., & Marquez, M. C. (2007). Biodegradation of azo dyes by bacteria inoculated on solid media. *Dyes and Pigments, 75*(1), 73–81.

Bekhit, F., Farag, S., & Attia, A. M. (2020). Decolorization and degradation of the Azo dye by bacterial cells coated with magnetic iron oxide nanoparticles. *Environmental Nanotechnology Monitoring Management, 14*(100376), 1–11.

Bell, J., Plumb, J. J., Buckley, C. A., & Stuckey, D.C. (2000). Treatment and decolorization of dyes in an anaerobic baffled reactors. *Journal of Environmental Engineering, 126*, 1026–1032.

Bras, R., Isabel, M., Ferra, A., Pinheiro, H. M., & Goncalves, I. C. (2001). Batch tests for assessing decolorization of azo dye by methanogenic and mixed cultures. *Journal of Biotechnology, 89*, 155–162.

Bromley-Challenor, C. A., Knapp, J. S., Zhang, Z., Gray, N. C., Hetheridge, M, J., & Evans, M. R. (2000). Decolorization of an azo dye by unacclimated activated sludge under anaerobic conditions. *Water Research, 34*(18), 4410–4418

Chacko, J., T, & Subramaniam, K. (2011). Enzymatic degradation of azo dyes: A review. *International Journal of Environmental Science, 1*(6), 1250–1260.

Chemical Economics Handbook. (2018). Retrieved from https://ihsmarkit.com/products/dyes-chemical-economics-handbook.html

Chen, B. Y., Lin, K. W., Wang, Y. M., & Yen, C. Y. (2009). Revealing interactive toxicity of aromatic amines to azo dye decolorizer *Aeromonas hydrophila. Journal of Hazardous Materials, 166*, 187–194.

Chen, K. C., Wu, J. Y., Liou, D. J., & Hwang, S. C. J. (2003). Decolorization of the textile dyes by newly isolated bacterial strains. *Journal of Biotechnology, 101*, 57–68.

Das, A., & Mishra, S. (2017). Removal of textile dye reactive green-19 using bacterial consortium: process optimization using response surface methodology and kinetics study. *Journal of Environmental and Chemical Engineering, 5*, 612–627.

Dave, S. R., Patel, T. L., & Tipre, D. R. (2015). Bacterial degradation of azo dye containing wastes. In: S. N. Singh (Ed.) *Microbial Degradation of Synthetic Dyes in Wastewaters*. Switzerland: Springer International Publishing. 57–83.

Dawkar, V. V., Jadhav, U. U., Ghodake, G. S., & Govindwar, S. P. (2009). Effect of inducers on the decolorization and biodegradation of textile azo dye Navy blue 2GL by *Bacillus sp*. VUS. *Biodegradation, 20*(6), 777–787.

de Souza, N. A., Ramaiah, N., Damare, S., Furtado, B., Mohandass, C., Patil, A., & De Lima, M. (2019). Differential protein expression in *Shewanella seohaensis* decolorizing azo dyes. *Current Proteomics, 16*, 156–164.

Elbanna, K., Hassan, G., Khider, M, & Mondour, R. (2010). Safe biodegradation of textile azo dyes by newly isolated lactic acid bacteria and detection of plasmids associated with degradation. *Journal of Bioremediation and Biodegradation, 1*(3), 110.

Fang, Z., Li, T., Wang, Q., Zhang, X., Peng, H., Fang, W., & Xiao, Y. (2011). A bacterial laccase from marine microbial metagenome exhibiting chloride tolerance and dye decolorization ability. *Applied Microbiology and Biotechnology, 89*(4), 1103–1110.

Franciscon, E., Mendonca, D., Seber, S., Morales, D. A., Zoco, G. J., Zanoni M. B. Z.,…Umbuzeiro, G. A. (2015). Potential of a bacterial consortium to degrade azo dye Disperse Red 1in a pilot scale anaerobic–aerobic reactor. *Process Biochemistry, 50*, 816–825

Garg, S. K., & Tripathi, M. (2013). Process parameters for decolorization and biodegradation of orange II (Acid Orange 7) in dye-simulated minimal salt medium and subsequent textile effluent treatment by *Bacillus cereus* (MTCC 9777) RMLAU1. *Environmental Monitoring and Assessment, 185*(11), 8909–8923.

Ghodake, G. S., Telke, A. A., Jadhav, J. P., & Govindwar, S. P. (2009a). Potential of *Brassica juncea* in order to treat textile effluent contaminated sites. *International Journal of Phytoremediation, 11*(4), 297–312.

Ghodake, G., Jadhav, S., Dawkar, V. S., & Govindwar, S. (2009b). Biodegradation of Diazo dye Direct Brown MR by *Acinetobacter calcoaceticus* NCIM 2890. *International Biodeterioration and Biodegradation, 63*, 433–439.

Gičević, A., Hindija, L., & Karačić, A. (2020). Toxicity of azo dyes in pharmaceutical industry. In: A. Badnjevic, R. Škrbić & L. Gurbeta Pokvić (Eds.) *Proceedings of the International Conference on Medical and Biological Engineering, 73*. Cham: Springer. doi.org/10.1007/978-3-030-17971-7_88.

Global Dyes Market Outlook to 2022. (2018). Retrieved from www.researchandmarkets.com/research/ll4f8h/global_dyes?w=5

Gopinath, K. P., Murugesan, S., Abraham, J. (2009). *Bacillus sp*. mutant for improved biodegradation of Congo red: random mutagenesis approach. *Bioresource Technology, 100*(24), 6295–6300.

Guo, J. B., Zhou, J. T., Wang, D., Tian, C. P., Wang, P., Uddin, M. S., & Yu, H. (2007). Biocatalyst effects of immobilized anthraquinone on the anaerobic reduction of azo dyes by the salt-tolerant bacteria. *Water Research, 41*(2), 426–432.

Haghshenas, H., Kay, M., Dehghanian, F. Tavakol, H., (2016). Molecular dynamics study of biodegradation of azo dyes via their interactions with AzrC azoreductase. *Journal of Biomolecular Structure & Dynamics, 34*(3), 453–462.

Hsueh, C. C., Chen, B. Y., & Yen, C. Y. (2009). Understanding effects of chemical structure on azo dye decolorization characteristics by *Aeromonas hydrophila. Journal of Hazardous Materials, 167*(1–3), 995–1001.

Imran, M., Negm, F., Hussain, S., Ashraf, M., Ahmad, Z., Arshd, M., & Crowley, D. E. (2016). Characterization and purification of membrane-bound azoreductase from azo dye degrading *Shewanella sp*. strain IFN4. *Clean – Soil, Air, Water, 44*(11), 1523–1530.

Jadhav, S. B., Phugare, S. S., Patil, P. S., & Jadhav, J. P. (2011). Biochemical degradation pathway of textile dye Remazol red and subsequent toxicological evaluation by cytotoxicity, genotoxicity and oxidative stress studies. *International Biodeterioration & Biodegradation, 65*(6), 733–743.

Jadhav, S. U., Dawkar, V. V., Ghodake, G. S., & Govindwar, S. P. (2008a). Biodegradation of direct Red 5B, a textile dye by newly isolated *Comamonas* sp. UVS. *Journal of Hazardous Materials, 158*(1–2), 507–516.

Jadhav, S. U., Jadhav, M. U., Kagalkar, A. N., & Govindwar, S. P. (2008b). Decolorization of Brilliant blue G Dye mediated by degradation of the microbial consortium of *Galactomyces geotrichum* and *Bacillus sp*. *Journal of the Chinese Institute of Chemical Engineers, 39*(6), 563–570.

Kalyani, D. C, Telke, A. A, Dhanve, R. S., & Jadhav, J. P. (2008). Ecofriendly biodegradation and detoxification of Reactive Red 2 textile dye by newly isolated *Pseudomonas sp.* SUK1. *Journal of Hazardous Materials, 163*, 735–742.

Kalyani, D. C, Telke, A. A., Govindwar, S. P., & Jadhav, J. P. (2009). Biodegradation and detoxification of reactive textile dye by isolated *Pseudomonas sp.* SUK1. *Water & Environmental Research, 81*(3), 298–307.

Kanagaraj, J., Senthilvelan, T., & Panda, R. C. (2015). Degradation of azo dyes by laccase: biological method to reduce pollution load in dye wastewater. *Clean Technologies & Environmental Policy, 17*(6), 1443–1456.

Kannappan, P. G., Hajamohideen, A. M. S., Karuppan, M., & Manickam, V. (2009). Improved biodegradation of Congo red by using *Bacillus sp. Bioresource Technology, 100*(2), 670–675.

Karunya, A., Rose, C., & Nachiyar, C. V. (2014). Biodegradation of the textile dye Mordant Black 17 (Calcon) by *Moraxella osloensis* isolated from textile effluent-contaminated site. *World Journal of Microbiol & Biotechnology, 30*(3), 915–924.

Khalid, A., Arshad, M., & Crowley, D. E. (2008). Decolorization of azo dyes by *Shewanella sp.* under saline conditions. *Applied Microbiology and Biotechnology, 79*, 1053–1059.

Khan, S., & Abdul, M. (2015). Degradation of Reactive Black 5 dye by a newly isolated bacterium *Pseudomonas entomophila* BS1. *Canadian Journal of Microbiology, 62*(3), 220–232.

Khehra, M. S., Saini, H. S., Sharma, D. K., Chadha, B. S., & Chimni, S. S. (2005), Comparative studies on potential of consortium and constituent pure bacterial isolates to decolorize azo dyes. *Water Research, 39*, 5135–5141.

Kiernan, J. A. (2001). Classification and naming of dyes, stains and fluorochromes. *Biotechnic & Histochemistry, 76*(5–6), 261–278.

Kilic, N. K., Nielsen, J. L., Yuce, M., & Donmez, G. (2007). Characterization of a simple bacterial consortium for effective treatment of wastewaters with reactive dyes and Cr(VI). *Chemosphere, 67*(4), 826–831.

Kolekar, Y. M., Konde, P. D., Markad, V. L., Kulkarni, S.V., Chaudhari, A. U., &Kodam, K. M. (2013). Effective bioremoval and detoxification of textile dye mixture by *Alishewanella sp.* KMK6. *Applied Microbiology & Biotechnology, 97*(2), 881–889.

Kudlich, M., Bishop, P., Knackmuss, H. J., & Stolz, A. (1996). Synchronous anaerobic and aerobic degradation of the sulfonated azo dye Mordant Yellow 3 by immobilized cells from a naphthalenesulfonate-degrading mixed culture. *Applied Microbiology & Biotechnology, 46*, 597–603.

Kulla, H. G., Klausener, F., Meyer, O., Ludeke, B., & Leisinger, T. (1983). Interference of aromatic sulfo groups in the microbial degradation of the azo dyes orange I and orange II. *Archive of Microbiology, 135*, 1–7.

Kumar, S. S., Shantkriti, S., Muruganandham, T., Murugesh, E., Rane, N,. & Govindwar, S. P. (2016). Bioinformatics aided microbial approach for bioremediation of wastewater containing textile dyes. *Ecological Informatics, 31*, 112–121.

Lade, H., Kadam, A., Paul, D. Govindwar, S. (2015). Biodegradation and detoxification of textile azo dyes by bacterial consortium under sequential microaerophilic/aerobic processes. *EXCLI Journal, 14*, 158.

Lin J, Zhang X, Li Z, Lei L (2010) Biodegradation of Reactive Blue 13 in a two-stage anaerobic/aerobic fluidized beds system with a *Pseudomonas sp.* isolate. *Bioresource & Technology, 101*(1), 34–40.

Liu, G., Zhou, J., Wang, J., Zhou, M., Lu, H., & Jin, R. (2009). Acceleration of azo dye decolorization by using quinone reductase activity of azoreductase and quinone redox mediator. *Bioresource & Technology, 100*(11), 2791–2795.

Lucas, M. S., Dias, A. A., Sampaio, A., Amaral, C., & Peres, J. A. (2007). Degradation of a textile reactive azo dye by a combined chemical–biological process: Fenton's reagent-yeast. *Water Research, 41*, 1103–1109.

Lucious, S., Reddy, E. S., Anuradha, V., Yogananth, N., Ali, M. Y. S., Vijaya P, … Parveen, P. M. K. (2014). Decolorization of acid dyes by *B. cereus* and *P. aeruginosa* isolated from effluent of dyeing industry. *International Journal of Pure and Applied Biosciences, 2*(3), 23–29.

Mahalakshmi, S., Lakshmi, D., & Menaga, U. (2015). Biodegradation of different concentration of dye (Congo red dye) by using green and blue-green algae. *International Journal of Environmental Research, 9*, 735–744.

Mahmood, S., Azeem, K., Muhammad, A. Mahmood, T., Crowley, D. E. (2016) Detoxification of azo dyes by bacterial oxidoreductase enzymes. *Critical Reviews in Biotechnology, 36*(4), 639–651.

Mehta, R., Singhal, P., Singh, H., Dmale, D., & Sharma, A. K. (2016). Insight into thermophiles and their wide-spectrum applications. *3 Biotech, 6*(1), 81–89.

Mishra, S., & Maiti, A. (2018). The efficacy of bacterial species to decolourise reactive azo, anthraquinone and triphenylmethane dyes from wastewater: a review. *Environmental Science and Pollution Research, 5,* 8286–8314.

Nachiyar, C. V., & Rajakumar, G. S. (2005). Purification and characterization of an oxygen insensitive azoreductase from *Pseudomonas aeruginosa*. Enzyme and *Microbial Technology, 36*(4), 503–509.

Ng, I., S., Chen, T., Lin, R., Zhang, X., Ni, C., & Sun, D. (2014) Decolorization of textile azo dye and Congo red by an isolated strain of the dissimilatory manganese-reducing bacterium *Shewanella xiamenensis* BC01. *Applied Microbiology & Biotechnology, 98*(5), 2297–2308.

O'Neill, C., Hawkes, F. R., Hawkes, D. L., Lourenco, N. D., Pinheiro, H. M., & Delee, W. (1999). Color in textile effluents sources, measurement, discharge consents and simulation: a review. *Journal of Chemical Technology & Biotechnology, 74,* 1009–1018.

Pandey, A., Singh, P., & Iyengar, L. (2007). Bacterial decolorization and degradation of azo dyes. *International Biodeterioration & Biodegradation, 59*(2), 73–84.

Pathak, H., Soni, D., & Chauhan, K. (2014). Evaluation of in vitro efficacy for decolorization and degradation of commercial azo dye RB-B by *Morganella sp.* HK-1 isolated from dye contaminated industrial landfill. *Chemosphere, 105,* 126–132.

Patil, P., Desai, N., Govindwar, S., Jadhav. J. P., & Bapat, V. (2009). Degradation analysis of reactive red 198 by hairy roots of *Tagetes patula L.*(Marigold). *Planta, 230,* 725–735.

Pearce, C. I., Lloyd, J. R., & Guthriea JT (2003). The removal of colour from textile wastewater using whole bacterial cells: A review. *Dyes and Pigments, 58*(3), 179–196.

Popli S, Patel UD (2015) Destruction of azo dyes by anaerobic–aerobic sequential biological treatment: a review. *International Journal of Environmental Science and Technology*, 12:405–420

Rafii, F., Franklin, W., & Cerniglia, C. E. (1990). Azoreductase activity of anaerobic bacteria isolated from human intestinal microflora. *Applied and Environmental Microbiology, 56,* 2146–2151.

Rai, H., Bhattacharya, M., Singh, J., Bansal, T. K., Vats, P., & Banerjee, U. C. (2005). Removal of dyes from the effluent of textile and dyestuff manufacturing industry: A review of emerging techniques with reference to biological treatment. *Critical Reviews in Environmental Science & Technology, 35*(3), 219–238.

Razo-Flores, E., Luijten, M., Donlon, B., Lettinga, G., & Field, J. (1997). Biodegradation of selected azo dyes under methanogenic conditions. *Water Science & Technology, 36,* 65–72.

Roy, U., Manna, S., Sengupta, S., Das, P., Datta, S., Mukhopadhyay, A., & Bhowal, A. (2018). Dye removal using microbial biosorbents. In: Gregorio Crini and Eric Lichtfouse (Eds), *Green adsorbents for pollutant removal: Innovative materials.* Switzerland: Springer Nature: 253–280.

Sahasrabudhe, M. M., Saratale, R. G., Saratale, G. D., & Pathade, G.R. (2014). Decolorization and detoxification of sulfonated toxic diazo dye CI direct red 81 by *Enterococcus faecalis* YZ 66. *Journal of Environmental Health Science & Engineering, 12*(151), 1–13.

Saratale, R. G., Saratale, G. D., Chang, J. S., & Govindwar, S. P. (2011). Bacterial decolorization and degradation of azo dyes: A review. *Journal of the Taiwan Institute of Chemical Engineers, 42,* 138–157.

Saratale, R. G., Saratale, G. D., Kalyani, D. C., Chang, J. S., & Govindwar, S. P. (2009a). Enhanced decolorization and biodegradation of textile azo dye Scarlet R by using developed microbial consortium-GR. *Bioresource Technology, 100,* 2493–2500.

Saratale, R. G., Saratale, G. D., Chang, J. S., & Govindwar, S. P. (2009b). Decolorization and biodegradation of textile dye navy blue HER by *Trichosporon beigelii* NCIM-3326. *Journal of Hazardous Materials, 166,* 1421–1428.

Saravanan, C., Rajendiran, R., Kaviarasan, T., Krishnan, M., Digambar, K., & Shetty, P. H. (2017). Synthesis of silver nanoparticles using bacterial exopolysaccharide and its application for degradation of azo-dyes. *Biotechnology Reports, 15,* 33–40.

Sarayu, K., & Sandhya, S. (2010). Aerobic biodegradation pathway for Remazol Orange by *Pseudomonas aeruginosa. Applied Biochemistry and Biotechnology, 160*(4), 1241–1253.

Sarkar, S., Banerjee, A., Halder, U., Biswas, R., & Bandopadhyay, R. (2017). Degradation of synthetic azo dyes of textile industry: a sustainable approach using microbial enzymes. *Water Conservation Science and Engineering, 2,* 121–131.

Shah, M. P. (2014). Environmental bioremediation: a low cost nature's natural biotechnology for environmental clean-up. *Journal of Petroleum & Environmental Biotechnology, 5,* 1–12.

Srinivasan, S. (2018). Biodegradation of textile azo dye, Remazol Yellow RR using non-autochthonous bacteria *Lysinibacillus sphaericus* MTCC 9523, supported by docking. In International Conference on Biodiversity & Sustainable Resource Management (242–255). University of Madras, Chennai, India.

Srinivasan, S., Parameswari, K. M., & Nagaraj, S. (2020). Latest innovations in bacterial degradation of textile azo dyes. In Kostas Marinakis (Ed.) *Emerging Technologies in Environmental Bioremediation.* Amsterdam: Elsevier. 285–309

Srinivasan, S., & Sadasivam, S. K. (2018). Exploring bacterial systems for docking and aerobic-microaerophilic biodegradation of textile azo dye. *Journal of Water Process Engineering, 22,* 180–191.

Srinivasan, S., Shanmugam, G., Surwase, S. V., Jadhav, J. P., & Sadasivam, S. K. (2017). In silico analysis of bacterial systems for textile azo dye decolorization and affirmation with wetlab studies. *Clean Soil, Air, Water, 45*(9), 1600734, 1–31.

Stolz, A. (2001). Basic and applied aspects in the microbial degradation of azo dyes. *Applied Microbiology & Biotechnology, 56,* 69–80.

Sudha, M., Saranya, A., Selvakumar, G., & Sivakumar, N. (2014). Microbial degradation of azo dyes: a review. *International Journal of Current Microbiology and Applied Sciences, 3*(2), 670–690.

Talarposhti, A. M., Donnelly, T., Anderson, G. K. (2001). Colour removal from a simulated dye wastewater using a two-phase anaerobic packed bed reactor. *Water Research, 35,* 425–432.

Tan, N. C. G., van Leeuwen, A., van Voorthuizen, E. M., Slenders, P., Prenafeta-Boldu, F. X., Temmink, H., … Field. J. A. (2005). Fate and biodegradability of sulfonated aromatic amines. *Biodegradation, 16,* 527–537.

Telke, A., Kalyani, D., Jadhav, J., & Govindwar, S. (2008). Kinetics and mechanism of Reactive Red 141 degradation by a bacterial isolate *Rhizobium radiobacter* MTCC 8161. *Acta Chimica Slovenica, 55,* 320–329.

Telke, A. A., Kalyani, D. C., Dawkar, V. V., & Govindwar, S. P. (2009). Influence of organic and inorganic compounds on oxidoreductive decolorization of sulfonated azo dye C.I. Reactive Orange 16. *Journal of Hazardous Materials, 172*(1), 298–309.

Thakur, J. K., Paul, S., Dureja, P., Annapurna, K., Padaria, J. C., & Gopal, M. (2014). Degradation of sulphonated azo dye Red HE7B by *Bacillus sp.* and elucidation of degradative pathways. *Current Microbiology, 69*(2), 183–191.

Tony, B. D., Goyal, D., & Khanna, S. (2009). Decolorization of textile azo dyes by aerobic bacterial consortium. *International Biodeterioration & Biodegradation, 63*(4), 462–469.

Umbuzeiro, G. D. A., Freeman, H. S., Warren, S. H., de Oliveira, D. P., Terao, Y., Watanabe, T.., & Claxton, L. D. (2005). The contribution of azo dyes to the mutagenic activity of the Cristais river. *Chemosphere, 60*(1), 55–64.

Van der Zee, F. P., & Cervantes, F. J. (2009). Impact and application of electron shuttles on the redox (bio) transformation of contaminants: A review. *Biotechnology Advances, 27*(3), 256–277.

Van der Zee, F. P., & Villaverde, S. (2005.) Combined anaerobic aerobic treatment of azo dyes: A short review of bioreactor studies. *Water Research, 39*(8), 1425–1440.

Waghmode, T. R., Kurade, M. B., Khandare, R. V., & Govindwar, S. P. (2011). A sequential aerobic/microaerophilic decolorization of sulfonated mono azo dye Golden Yellow HER by microbial consortium GG–BL. *International Biodeterioration & Biodegradation, 65,* 1024–1034.

Wuhrmann, K., Mechsner, K., & Kappeler, T. (1980). Investigations on rate determining factors in the microbial reduction of azo dyes. *European Journal of Applied Microbiology and Biotechnology, 9,* 325–338.

Zhang, F., Yediler, A., Liang, X., & Kettrup, A. (2004) Effects of dye additives on the ozonation process and oxidation by-products: A comparative study using hydrolyzed CI Reactive Red 120. *Dyes and Pigments, 60*(1), 1–7.

Zhou, X., & Xiang. X. (2013). Effect of different plants on azo-dye wastewater bio-decolorization. *Procedia Environmental Sciences, 18,* 540–546.

Zimmermann, T., Kulla, H., Leisinger, T. (1982). Properties of purified orange II-azoreductase, the enzyme initiating azo dye degradation by *Pseudomonas* KF46. *European Journal of Biochemistry, 129,* 197–203.

Zollinger, H. (1991). *Colour chemistry: synthesis, properties and applications of organic dyes and pigments* (5th ed.). Weinheim. VCH Publishers

15 Bioremediation
A Low-Cost and Clean Green Technology for Environmental Management

Ishani Lahiri[1], Hare Ram Singh[1], Sushil Kumar Shukla[2], and Santosh Kumar Jha[1]*

[1]Department of Bioengineering, Birla Institute of Technology, Mesra, Ranchi-835215, Jharkhand, India

[2]Department of Transport Science and Technology, Central University of Jharkhand, Brambe, Ranchi-835205, Jharkhand, India

*Corresponding author: skjha@bitmesra.ac.in

CONTENTS

DOI: 10.1201/9781003130932-15

15.1   INTRODUCTION

Rapid industrialization has increased various unwanted elements that have now accumulated in the biosphere in toxic levels, which cannot degrade in the natural environment. However, no proper waste treatment facilities are available. Therefore, words, humans are harming and endangering the environment. The aquatic and terrestrial ecosystems or the overall biodiversity are being exposed to ubiquitous contamination and pollution due to the industrial production of chemicals, excessive use of petroleum and its derivatives, polyethylene, pesticides and organic herbicides. The textile industry is one of the major polluters of surface and groundwater resources because it can consume millions of liters of water per day to dye tons of fabric. The wastewater released from the textile dyeing industry contains approximately 72 toxic chemicals, of which 30 chemicals cannot be removed by waste treatment processes. The chemical dyes are readily soluble in water and are heavy risk factors to the environment and humans. The chemical-based textile dyes are used due to their cost-effectiveness and high stability to various parameters, such as temperature and light and, in general, are extremely stubborn compounds. Due to its dark color, textile effluents usually block the penetration of sunlight in aquatic bodies and hinder the photosynthetic activity of plants. In addition, it disturbs the oxygen (O_2) levels in the waters (Ghaly *et al.*, 2014).

15.2   CLASSIFICATION OF TEXTILE DYES

The textile industry has been using >10,000 different dyes and pigments annually worldwide for fabric production. Dye classification and categorization mainly depend on the type of fabric they are used on and the chemical and physical nature of the dye. Most dyes are either made of chromophore molecules, which gives them their distinctive color or auxochromes that are color enhancers. Some of the functional groups present in chromophores are $-N=N-$, $-C=O$, $-NO_2$, $O=(C_6H_4)=O$ $-N=N-$, $-C=O$, $-NO_2$, and $O=(C_6H_4)=O$ and auxochromes consist of $-NH_3$, $-COOH$, $-OH$, and $-SO3H$ functional groups (Srinivasan *et al.*, 2010).

Azo and anthraquinone are two of the major dye categories out of the twelve different chromophores that have been reported. N=N is present in most azo dyes and can absorb light in the visible spectrum. Azo dyes account for 70% of all pigments that are used in the textile industry. Sometimes the azo group can be substituted with benzene or naphthalene groups, and it might contain many different substituents, such as chloro (–Cl), methyl (–CH3), nitro (–NO_2), amino (–NH_2), hydroxyl (–OH), and carboxyl (–COOH), which give rise to different types of azo dyes (Zollinger, 1991). In contrast, anthraquinone dyes are derived from anthraquinone, which consists of a quinoid ring as the chromophore. However, –OH or –NH_2 groups might be present, which makes them a good chromophore molecule. Most anthraquinone dyes are toxic, carcinogenic, and mutagenic because they can be converted into benzidine, which is harmful, and are not affected and are resistant to microbial degradation. Most of the dyes are extensively used in the textile, paper, food, leather, cosmetics, and pharmaceutical industries.

15.3   IMPACT ON THE ENVIRONMENT

Synthetic dyes, due to their complex structure, are difficult to remove from the environment (Figure 15.1). They cannot be removed by traditional wastewater or sludge treatment techniques, such as coagulation–flocculation or by adsorption and chemical degradation techniques. Therefore, the accumulation of different dyes in the land and waterbodies has occurred for a long time. The major reason is the lack of a cost-efficient and environmentally friendly method for dye treatment (Figure 15.2). The discharged wastewater is rich in toxic aromatic amines due to O_2 deficiency in the untreated effluent.

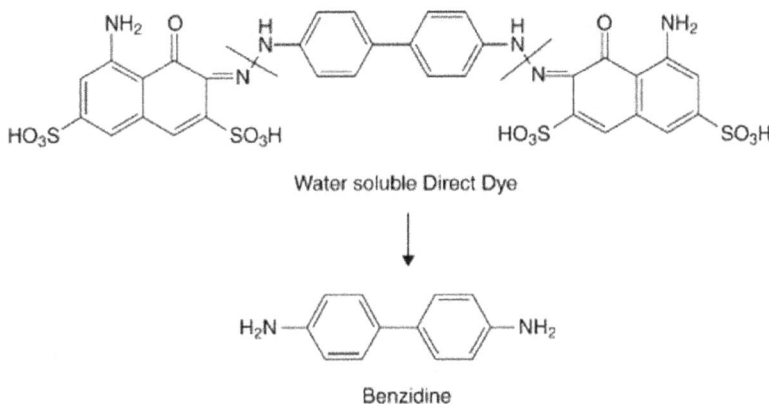

Water soluble Direct Dye

Benzidine

FIGURE 15.1 A typical water-soluble dye that can generate carcinogenic benzidine metabolite upon reduction in the animal body.

Source: Clark (2011).

Anthraquinone

Drimarene Blue K₂RL

Remazol Brilliant Blue R (RBBR)

Reactive Blue 4 (RB4)

Reactive Blue 19 (RB19)

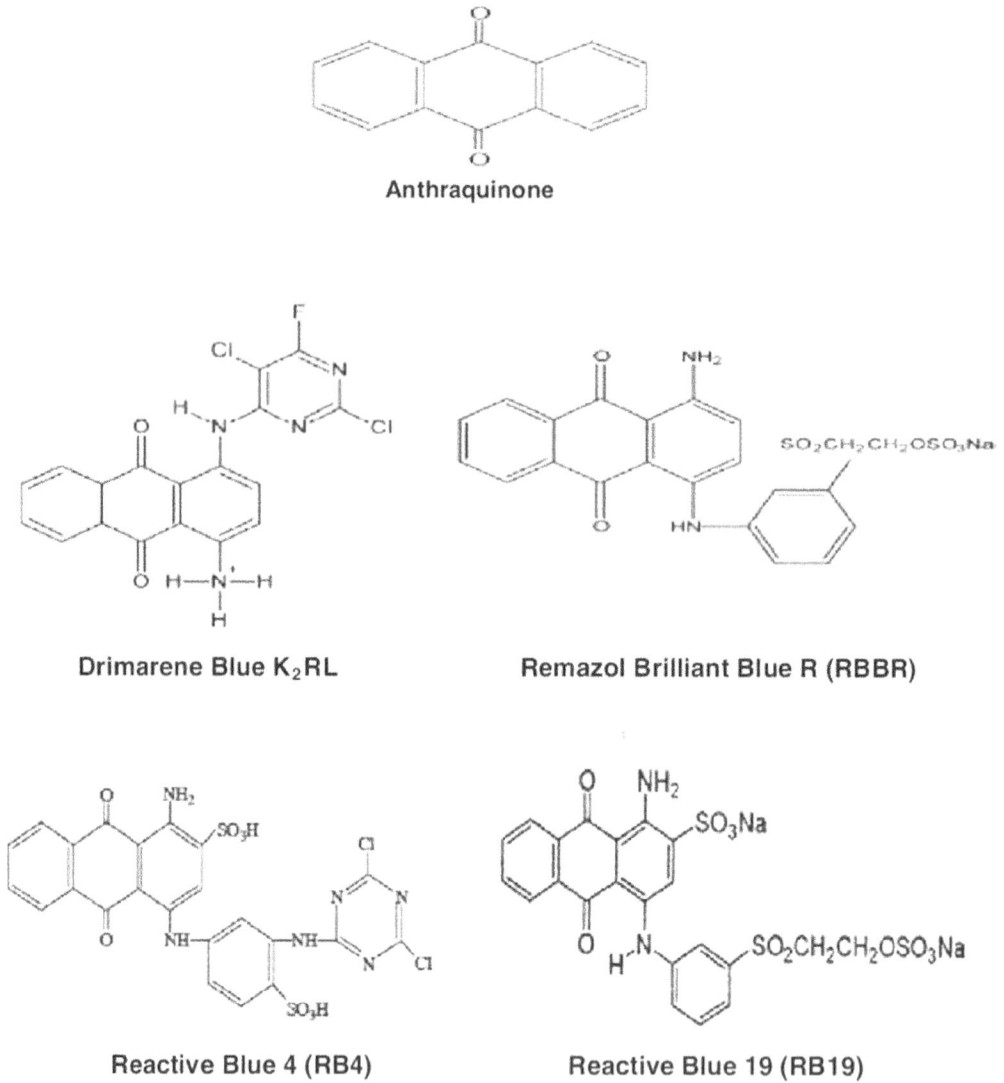

FIGURE 15.2 Anthraquinone and its derivatives.

Source: Singh *et al.* (2014).

The toxic textile dye effluent is not aesthetically pleasing, reduces sunlight penetration, photo-synthetic activity for organisms to survive, O_2 concentration, pH, and water quality, which causes severe and acute toxic effects on aquatic flora and fauna. Furthermore, synthetic dyes and their derivatives have an unfavorable impact on the environment as total organic carbon (TOC), biological oxygen demand and chemical oxygen demand (COD) (Saratale *et al.*, 2009). Researchers have reported that various ecological functions that include the germination rates and the biomass of several important plants and crops have been heavily affected due to dye toxicity. The soil fertility is affected, which is causing heavy erosions in certain areas, degrading the organic matter, and eradication of wildlife habitats.

15.4 TYPES OF BIOREMEDIATION

15.4.1 *In Situ* Bioremediation

In situ bioremediation treats the soil and groundwater at the site of contamination. Therefore, *in situ* techniques are used as a cheaper and lower maintenance option because they require less equipment and there is a minimal amount of soil displaced by exhumation, or by transfer of contaminants to another site. *In situ* bioremediation is a process where organic contaminants are biologically degraded into CO_2 and water in addition to other altered or transformation products. Bioaugmentation and biostimulation are one type of *in situ* bioremediation technique that is used as an eco-friendly, sustainable approach.

15.4.1.1 Bioaugmentation

Bioaugmentation is a process in which highly specific microbial cultures are used (e.g., indigenous and known as allochthonous or autochthonous microorganisms) and added to the contamination site (Figure 15.3). This method of bioremediation is known as autochthonous bioaugmentation (ABA) and was used because non-indigenous microorganisms do not compete well with the other population of microorganisms in the soil, and therefore, do not survive (Mahjoubi *et al.*, 2018). A major application for bioaugmentation is its use as an alternative strategy for the bioremediation of oil-contaminated areas. The major principle of this approach is to increase the rate of degradation of contaminants with the help of special strains of microbes or genetically engineered strains. This increase in the microbial flora helps to increase the genetic pool and gene diversification in the area of contamination. In soil or water several factors affect the successful application of bioaugmentation:

1. Temperature
2. pH
3. Moisture content of the soil
4. Organic matter content
5. Aeration
6. Nutrient content of the soil
7. Type of soil where decontamination is required

Apart from these, any successful bioaugmentation mainly relies on how well the microbial pool adapts to the pollution site and competes with other native microorganisms. In addition, this method is inexpensive and is a sustainable approach to bioremediation (Goswami *et al.*, 2018).

15.4.1.2 Biostimulation

Biostimulation is a remediation technique that incorporates the use of a variety of rate-limiting components, such as phosphorus (P) (ortho-phosphate or organic phosphorous), nitrogen (N) (e.g., ammonic, nitrate, or organic N), O_2, or carbon (C), and electron acceptors, which mostly act as stimulants for the growth of indigenous microbes in the area of contamination (Figure 15.4), such that these microbes can carry out bioremediation by themselves. These nutrients act as the basic building blocks of life and allow the microorganisms to produce the essential enzymes to break down the contaminants, which makes it a favorable option for bioremediation (Pandey *et al.*, 2012).

The optimum levels for the C:N:P ratio was perfected by Wolicka *et al.* (2009), at 100:9:2, 100:10:1, or 250:10:3. This experiment demonstrated how the rate of nutrient supply affects the structure and function of microbial communities on-site.

However, this technique has some disadvantages. The primary challenge for biostimulation is that the distribution of the rate-limiting factors allow for all the components to be readily available to the microorganisms in the soil. Dense and compressed, impregnable surface morphology (e.g., clay or other fine-grained material) make it difficult to spread the components throughout the polluted area.

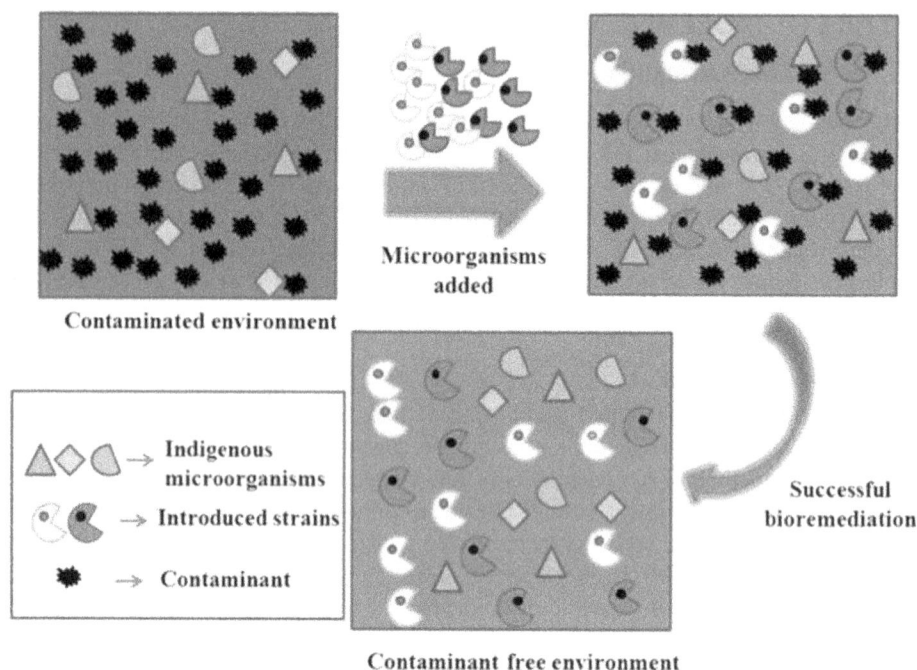

FIGURE 15.3   Pictorial diagram of bioaugmentation.

Source: Goswami *et al.* (2018).

15.4.2   *Ex Situ* Bioremediation

Ex situ remediation is an extensive remediation technique, which involves the use of large equipment for the removal and transportation of the soil, because it is not an on-site remediation technique. Therefore, this method is an expensive remediation process and does not attract many people. Another drawback of this method is the amount of pollution that is caused due to the displacement of contaminants from one site to another. The different types of *ex situ* bioremediation techniques include: landfarming, biopiles, and composting.

15.4.2.1   Landfarming

Landfarming is an attractive bioremediation technique that is composed of tilling the contaminated soil and spreading it over the surface to instigate the natural degradation process, with the help of autochthonous aerobic microbial flora in the area. This technique is successful only with the correct aeration of the soil and the addition of nutrients, moisture, and minerals and is widely used to reduce the levels of petroleum and other components of spillage on the soil. In Iturbe *et al.* (2015), the nutrients were added according to a C:N:P ratio of 100:10:1, to start the biodegradation of the contaminants. The major advantage of landfarming is that it requires low surveillance and maintenance costs and does not require any extra clean-up procedures. However, the major setback of this technique according to the United States Environmental Protection Agency is that it requires a large area of land for the treatment. The dust generated due to the tilling of land and continuous aeration of soil might cause a decrease in the air quality and pose threats to the public living in the vicinity. In addition, the bioaugmentation of oil-contaminated (hydrocarbon) soil from the Mediterranean, which when treated by *ex situ* landfarming gave excellent and effective results.

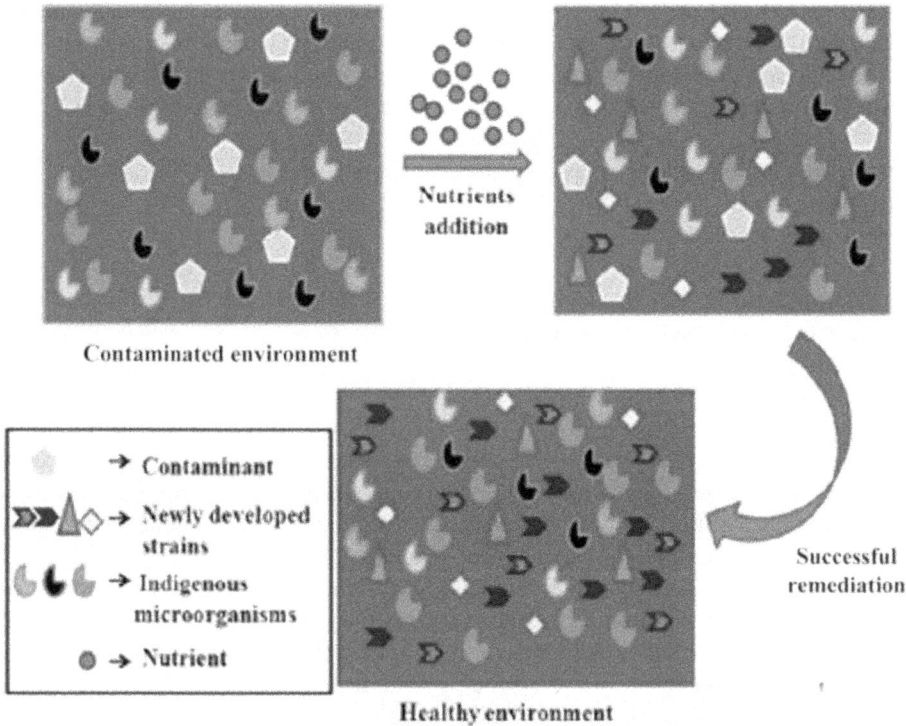

FIGURE 15.4 Pictorial diagram of biostimulation.

Source: Goswami *et al.* (2018).

15.4.2.2 Biopiles

Biopile technology is composed of piling up the contaminated soil (which is usually caused by petroleum) above the ground at approximately 2.13 on an impregnable base. This is continuously aerated, and the microbial population of the soil is stimulated. This triggered microbial activity can be amplified by boosting the soil with moisture and nutrients, such as P and N. This helps to reduce the concentration levels of pollutants by degrading each constituent of the petroleum that has been adsorbed by the soil. Biopiles have irrigation and ventilation systems.

Iturbe *et al.* (2015) used a high-density polyurethane layer (1 mm) as an impermeable base to prevent any leachate or lixiviates from migrating into the soil and entering the water table. Several parameters help to make successful biopiles. These parameters include:

1. Soil characteristics
2. Characteristics of the contaminants
3. Weather conditions

The major benefit of biopile technology is that it is easy to design and construct and the remediation requires very little time (approximately 3–6 months). However, if the site of pollution has high levels of heavy metals (>2,500 ppm), then it might be difficult for the microorganisms to grow.

15.4.2.3 Composting

Remediation by composting is quite similar to natural (biological) degradation that occurs naturally in soil due to the presence of native microorganisms. However, the natural process is quite slow. In composting, the rate of degradation is usually accelerated by applying higher temperatures to the composts, which eventually helps to dissolve the pollutants and enhance the microbial activity in the soil. The polluted soil can be mixed with non-toxic organic additives, such as manure, agricultural, or farm wastes. According to Hsu *et al*. (1999) composting consists of three basic phases:

1. Fast decomposition phase that lasts for approximately 30 days
2. Maturation phase
3. Stabilization phase

The first step for successful composting includes mechanical treatment by grinding, mixing, and then disposing of the non-biodegradable materials, such as plastics, stones, metals, Styrofoam and glass hamper the degradation.

15.4.3 Bioreactors

Bioreactors are large vessels in which contaminated solid materials (soil) and water are processed for decontamination. In these vessels, raw materials can be transformed into substrate-specific products by following a series of biochemical reactions. Most bioreactors are differentiated based on their modes of operation, such as batch, fed-batch, sequencing batch, continuous, and multistage. In this bioremediation technique, the contaminated soil sample is usually fed into the reactor as dry matter or slurry. A bioreactor must mimic the exact optimal growth conditions of a microbial population; therefore, the process parameters, such as temperature, pH, agitation, aeration rates, and substrate and inoculum concentrations must be controlled (Azubuike *et al*., 2016)

15.5 MECHANISMS OF BIOREMEDIATION

Due to the increase in population and urbanization, the increase in various textile industries has had a major impact on the way people live and on the environment. As development increases due to modernization, the release of potentially toxic compounds into the environment at various stages of materials production has become a major threat.

According to statistics, the textile industry accounts for 14% of India's industrial production and 27% of its export revenues and it is the second-largest producer of cotton and silk. Due to the use of commercial dyes and chemical dying agents in the textile industries, tanneries, ink industries, paint and cosmetics industries, almost all of the chemical effluents from the factories are disposed of in the waters or land in the surrounding regions.

Textile industries categorize the coloring dyes as azo, diazo, cationic, basic, anthraquinone, and metal complex based, which depends on their chemical structure. The effluents from textile industries are mostly identified by their highly noticeable color (3,000–4,500 ADMI units), COD (800–1,600 mg/L), alkaline pH (9–11) and total solids (TS) (6000–7000 mg/L) (Manu *et al*., 2002). The dyes used, are non-biodegradable and cause serious environmental and health hazards. The color of wastewater is unpleasant and interferes with the oxygenation ability of water, disrupting the aquatic diversity and food chain.

Because these dyes exist in the water as waste and toxic effluents, they restrict the entry of sunlight; therefore, reducing the photosynthetic activity of aquatic plants. They decrease the levels of O_2 in the water and the chemicals present mean that it is too toxic for consumption. Therefore, recycling this water for other purposes is not viable. However, due to the lack of water worldwide, the only way in which this water can be reused is by removing all the chemicals, either by using dye-degrading microorganisms, physicochemical methods, or using enzymatic methods. The methods to remove dyes or coloring agents are discussed in the following sections of this chapter.

15.5.1   Physicochemical Treatments

This method for the bioremediation of dyes from polluted wastewaters is easy to use; however, it is not very cost-effective and environmentally friendly because it requires high electricity consumption and there are multiple stages in the treatment with a long retention time (Bhatia *et al.*, 2017). Physicochemical treatment generally transfers the pollutants that are present in the wastewaters from one phase to another without completely removing them.

15.5.1.1   Adsorption

This technique is the most widely used bioremediation process for dyes because it is quite simple to use. The only challenge for this technique is the selection of the correct adsorbent material to produce high quality treated water. In certain cases, adsorption is performed with sorption. This combined process is known as biosorption (Gupta *et al.*, 2015). The adsorption uses a variety of adsorbent materials that depends on the type of dyes present in the effluent. Previous research has shown that cetyltrimethylammonium bromide (CTAB)-modified titanium dioxide (TiO_2) nanoparticles could be used as adsorbents for the removal of a mixture of anionic dyes in single and binary systems. The use of nano-γ-alumina as an efficient adsorbent for the removal of a mixture of two textile dyes, for example, Alizarin Red and Alizarin yellow and rubber tires for adsorption of Acid Blue 113 have been documented by researchers.

The porous adsorbents are usually activated so that they stimulate the interaction between the sorbent and the target dye molecule. Other types of adsorbents include activated C, peat, wood chips, silica gel, coir pith, and zeolites.

Vikrant *et al.* (2017) demonstrated that the invention of advanced materials, such as metal-organic frameworks, which are tailored adsorbents and made specifically for the sorption of specific pollutants has made adsorption much easier. However, the major drawback of using these advanced materials are the expense, a longer retention period, the existence of side reactions (i.e., observed in silica gel particularly), and the ineffectiveness of some specific kinds of dyes.

15.5.1.2   Membrane Separation

This method uses membranes of various sizes to separate and purify wastewaters from textile industries. The partial pressure difference on both sides (e.g., permeate and feed) of the membrane is the driving force to separate the dye molecules from effluents (Vikrant *et al.*, 2017). Pressure driven membrane processes, such as microfiltration (MF), ultrafiltration (UF), nanofiltration (NF) and reverse osmosis are some of the techniques used in water treatment and wastewater purification.

MF removes contaminants of 0.1–10.0 μm from fluids, and a UF membrane retains particles of 1,000–1,000,000 molecular weight. Moreover, UF helps in reducing colloidal properties, viruses, endotoxins, pyrogens, volatile organic compounds (VOC), and TOC to typically between 0.1 to 0.05 μm. NF is quite similar to MF but is used where high sodium (Na) removal is not needed. Reverse osmosis is used to remove dissolved salts and other ionic solutes that are present in the contaminant molecules. Despite helping the easy permeation of dye molecules, clogging and fouling of the membrane, disposal of the residue that accumulates after separation of the contaminants, high capital investment in equipment accessories and maintenance, and the inability to treat large amounts of effluents are some of the major drawbacks.

15.5.1.3   Ion-exchange

In ion exchange process, the cation and anion resins are used to react with the functional groups and dye molecules present in concentrated effluent which passes through cation and anion exchange resins. This is performed until all the available exchange sites in the resin are saturated (Robinson *et al.*, 2001). The backbone of the technology is the ion-exchange adsorbents or resins that are usually polymeric granules or beads with several functional groups attached that enable the contaminants of opposite charges (in this case dye molecules) to bind to it. Since most dyes are positive, cation exchange resins are not suitable for the removal of reactive dyes from effluents due to the same charges on them. There are a variety of non-commercial anion exchange resins that have been examined for the removal of reactive effluents, such as poly (acrylic acid -N-isopropylacr ylamidetrimethylolpropane triacrylate) cross-linked with sodium alginate, partial diethylamino-ethylated cotton dust waste, quaternized wood, microcrystalline cellulose gel, and porous chitosan-polyaniline/ZnO hybrid composite. However, this technology is not attractive due to the high maintenance costs and because it cannot separate a wide range of dyes that are present in textile industry effluents.

15.5.1.4 Coagulation

Coagulation–flocculation is one of the easiest to use technologies that is employed to remove dye molecules from factory effluents. This method helps in the effective removal of direct dyes using ferrous sulfate and ferric chloride but cannot remove acidic dyes (Ahmad *et al.*, 2015). However, this is not the only limitation. The chemicals used for coagulation are expensive. In addition, the application of this technology does not give the expected results due to the generation of large amounts of sludge, which is due to the use of, such as lime, alum, polyelectrolyte, and ferrous salts (Manu *et al.*, 2002), and the difficulty in determining the optimum coagulant dose. Based on previous experimental procedures, coagulation–flocculation involves 2 min of rapid mixing at 100 rpm, followed by 30 min of slow mixing at 40 rpm, and 30 min of settling. If required, add-itional centrifugation can be performed at 5000 rpm for approximately 5 min to obtain a much clearer liquid sample before analysis. After complete removal, the leftover dye concentration was measured in a UV-Vis spectrophotometer at 526nm. The percentage removal of dye was calculated by the following equation:

$$\text{Dye removal } (\%) = (C_r - C_t)/\, C_r \times 100$$

Where C_r and C_t = dye concentration in raw and treated solution, respectively.

15.5.1.5 Oxidative Remediation of Dyes

Oxidative bioremediation has been widely used for the decolorization of textile dye effluent and the removal of recalcitrant organic components (Gupta *et al.*, 2015). Oxidation involves the oxidative cleavage of the aromatic ring of the dye molecules. This process can be carried out using a solution of hydrogen peroxide (H_2O_2) and an iron (Fe) catalyst (Fenton's reagent) by activating it and then using it to treat contaminated effluents that are resistant to biological treatment or that are too toxic for treatment with live biomass. This method has good results and is cheaper than ozonation, but the major setback of this technology is the generation of large amounts of sludge, which is extremely difficult to handle and dispose of, the formation of toxic, carcinogenic by-products, and the high costs of energy and chemicals (Robinson *et al.*, 2001). Ozonation is another approach for advanced oxidative bioremediation and is discussed in the following section.

15.5.1.6 Ozonation

This method of chemical oxidation comes is used when dyes in the wastewater are very difficult to treat with the physical methods or by biodegradation (Figure 15.5). This process commonly uses ozone (O_3) as an oxidizing agent to treat textile effluents. The reactive dyes consist of a chromo-phore system groups which is a ring like structure and contains either single or double bonds. The O_3 in the reaction usually oxidizes the double bond of the chromophore group in the dye. The tem-perature for the reaction must be maintained at 35°C. The concentration of the dye can be detected by a UV-Vis spectrophotometer where the absorbance can be measured at 596 nm (Wijannarong *et al.*, 2013).

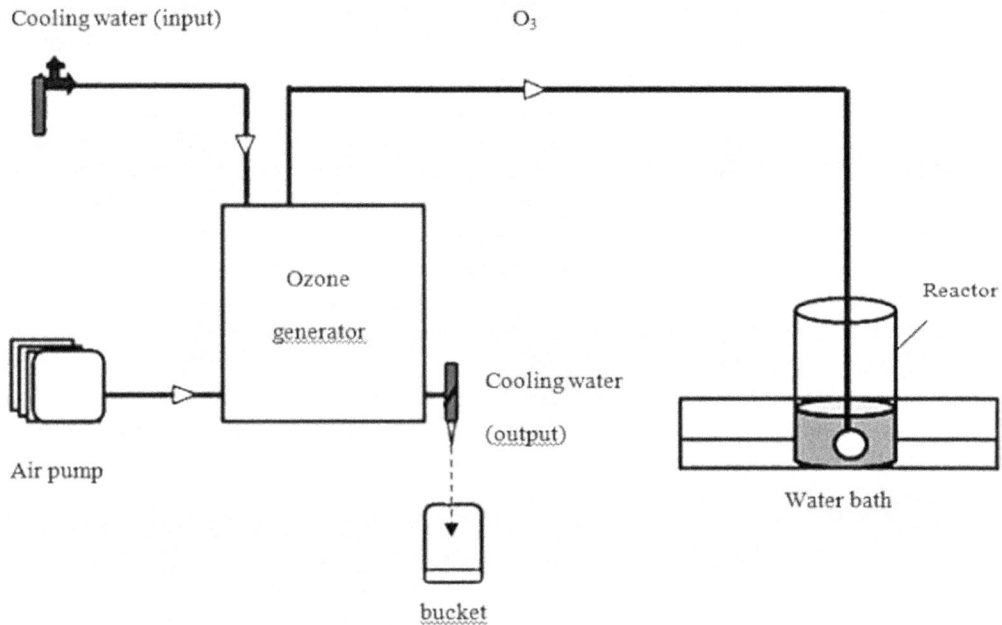

FIGURE 15.5 Schematic diagram showing the experimental system of ozonation.

Source: Wijannarong *et al.* (2013).

The efficiency of decolorization according to Wijannarong *et al.* (2013) was given by:

$$Decolorization\ (\%) = A_0 - A_t / A_0 \times 100$$

where A_0 = initial absorbance; and A_t = absorbance at time t.

Ozonation follows pseudo-first-order kinetics, and simple by-products, such as chlorides, sulfates, and nitrates form due to the oxidation of various substituent groups in the dye particle.

15.6 BIOLOGICAL TREATMENT

Biological methods of treatment have been favored in the scientific community because they have advantages over other conventional methods for cost-effectiveness, eco-friendly treatment processes, easy to operate and safe, with lower sludge generation. Researchers have used various microorganisms, which are natural or genetically modified organisms (GMOs) and have been successful in removing a wide variety of dyes from wastewaters by exploiting either the aerobic or anaerobic pathways of microbes.

The biological approach can be carried out *in situ* or *ex situ* using a mixture of various microbes or by using pure strains. The chemical dyes from textile industries are degraded by breaking the bonds that are present in the chromophore groups in the dye molecules. However, the biological processes for bioremediation can only reduce the COD and turbidity cannot remove the color.

15.6.1 BIODEGRADATION OF DYES USING BACTERIAL STRAINS

Many species of bacteria have been utilized for the decolourization of dyes from effluents and these are given in Chen *et al.* (2003). In the initial stages of bacterial degradation of azo dyes, enzymes, such as azoreductase or laccase (i.e., synthesized in the stationary phase of microbial culture), cause the azo bonds (–N=N–) to cleave reductively under conditions with low O_2. This reaction eventually produces a colorless aromatic amine solution (Bhatia *et al.*, 2017).

The amine products formed can be further catabolized by exploiting either the aerobic or anaerobic pathways of bacterial strains. However, according to Elisangela *et al.*, (2009), it can also be reduced by applying bacterial synthesized enzymes, such as hydroxylase and oxygenase. Other enzymes, such as tyrosinase, NADH-DCIP reductase, and Malachite Green (MG) reductase also participate. Dye decolorization is the most efficient at an optimal pH of 6–10. According to previous research, exploiting bacterial strains sometimes can be limited and cannot be applied to certain complex azo dyes. Naphthylamine sulfonic acids present in the dyes are not affected by bacterial degradation because the strong negatively charged sulfonyl group in the dye molecule cannot pass through the bacterial cell membrane.

Bacillus subtilis, Clostridium perfringens, Proteus sp., Pseudomonas aeruginosa, and *Pseudomonas putida* are some examples of Gram-positive bacteria that are used for decolorizing azo dye from textile wastewaters. Similarly, *Klebsiella pneumonia, Enterococcus sp.,* and *Escherichia coli* are Gram-negative bacteria (Bhatia *et al.*, 2017).

Xiao *et al.* (2018) attained an efficiency of 91.4% for the anaerobic decolourization of amaranth and 71.2% for Methyl red followed by complete rapid decolourization with the help of *Shewanella oneidensis* MR-1 (wild type strain). The first stage in dye decolourization is the cleavage of the (–N= N–) bond in azo dyes, which eventually leads to the formation of aromatic amines with the help of azoreductase that is secreted by various bacterial strains. Azoreductases act when reducing agents, such as NADH, NADPH, and $FADH_2$ are present. However, methanogens require a constant source of C. This can be supplied as simple substrates, such as glucose, starch, acetate, and ethanol. The anaerobic decolorization of azo dyes is a comparatively easy and non-specific process. Bromley-Challenor *et al.* (2004) suggested that the effective decolourization of dye molecules was due to the very low redox potential (≤50 mV) with no O_2. They suggested that azo dye molecules acted as acceptors of the electrons that were supplied by the electron transport chain. However, the major difficulty under anaerobic conditions is the impermeability of the dye molecule through the bacterial cell membranes. This is because most azo dye molecules consist of sulfonate groups, which increase the molecular weight of the dye. The cleavage of the azo bond intracellularly is not possible (Robinson *et al.*, 2001).

15.6.2 BIODEGRADATION OF DYES VIA FUNGAL STRAINS

Many strains of fungi have proved to be useful in dye degradation due to their distinct metabolic capacity to synthesize ligninolytic enzymes. These ligninolytic enzymes (i.e., lignin-degrading molecules), such as manganese peroxidase, manganese independent peroxidase, laccase, tyrosinase, and lignin peroxidase are not specific, and therefore, they can degrade a wide variety of organic compounds. Fungal cultures that produce these enzymes or free enzymes could be used for bioremediation, but the latter has certain cost limitations associated with upstream and downstream processing of the enzymes. This enzymatic method for biodegradation could be used in various types of bioreactors, such as fluidized bed bioreactors and rotating biological contactors. However, the uncontrolled growth of the fungus can clog the continuous reactor. In addition, bacterial contamination of the clogging could restrict the enzymatic activity of the fungus, which could cause longer times textile discoloration. Another method of discoloration for textile dyes is fungal biosorption that uses living or dead fungal cells. Biosorption is a much faster process and allows the fungal cells to attain positive and negative charges on their cell wall. These charges are due to the presence of chitosan and glucuronic acid, which cause positive and negatively charged functional groups, such as amino groups and carboxyl groups respectively. The dye molecules are adsorbed by the functional groups on the surface of the fungal cell wall.

Yesiladalı *et al.* (2005) demonstrated the use of *Trichophyton rubrum*, a wood-degrading fungus for the removal of a mixture of dyes by biodegradation and biosorption. The utilization of white-rot fungi was proved to be successful and was first reported in 1983.

Since most fungi grow at acidic pH, the optimum conditions for fungal dye decolorization are acidic conditions, for example, decolorization by *Aspergillus fumigates* was performed at pH 3–8. This concept proved to be valuable when experiments showed the decolorization of a dye solution using *Aspergillus niger* and *Penicillium sp.* at pH 4.5 and 4, respectively.

15.6.3 BIODEGRADATION OF DYES VIA ALGAL STRAINS

Algae are unusual plant-like organisms that are mostly aquatic and are found in oceans, rivers, lakes, ponds, and even on snow. They are single-celled phytoplankton that floats on the water and most are harmful to humans. Various algal species are abundantly used as biological adsorbents due to their larger surface area and binding capacity. Various reports have suggested that algae can biologically adsorb various functional groups, such as –OH, $RCOO^-$, $-NH_2$, and PO_4^{3-} that are present in wastewater effluent.

There are three mechanisms by which algal decolorization occurs:

1. Harvesting the algal biomass, CO_2, and H_2O, by utilization of chromophores
2. Converting the chromophore material to non-chromophore material
3. The adsorption of the resulting chromophore onto the algal biomass (Alvarez *et al.*, 2015)

According to previous research, algae are responsible for secreting an enzyme called azoreductase that converts azo dyes into aromatic amines for decolorization. Since most algal species utilize azo dyes as a source of C and N for their growth, extensive studies have been conducted using species, such as *Chlorella pyrenoidosa, Spirogyra rhizopus, Cosmarium sp., Pithophora sp., Nostoc muscorum, Ulva lactuca, Sargassum,* and *Desmodesmus sp.*

15.7 MICROORGANISM USED IN BIOREMEDIATION

15.7.1 BACTERIA

The secretion of oxidoreductive enzymes means that bacteria are the best candidates for the degradation of synthetic dyes into less complex products. Various studies were conducted using pure bacterial cultures of *Bacillus cereus, Bacillus subtilis,* and *Aeromonas hydrophila.* According to reports by Coughlin *et al.*, (1999), *Sphingomonas sp.* 1CX can grow on an azo dye and solely use it as C, energy, and N source. Other bacterial strains, such as *Bacillus sp.* OY1-2, *Xanthomonas sp.* NR25-2, and *Pseudomonas sp.* PR41 were used azo dyes (Acid Orange 7 (Orange II) Azodye or Acid Red 88) as a sole C source. It is difficult to achieve complete decolourization using pure bacterial cultures; therefore, mixed cultures are used to provide better results. A variety of bacterial strains have been recognized for decolorizing textile dyes (Table 15.1).

15.7.2 ACTINOMYCETES

Actinomycetes are a group of Gram-positive filamentous bacteria that are found in various habitats and have various metabolic capacities. They can use the dye molecules as their sole source of C and N. In recent studies, most phthalocyanine, azo, and anthraquinone dyes, such as Remazol Brilliant Blue R, Poly B-411, and Poly R-478 were decolorized by *Streptomyces sp.* and *Thermomonospora* (Ball *et al.*, 1989). Actinomycetes are effective against a formazan-copper (Cu) complex dye, which is a Cu base azo dye. Another actinobacterium *Streptomyces ipomoea* produces salt-resistant laccase under versatile pH, to decolorize Orange II. Pillai *et al.* (2014) demonstrated the discolouration of azo blue and azo orange within 48 h using strains isolated from soil. Raja *et al.* (2016) carried

TABLE 15.1
Decolourization of Various Azo Dyes by Azoreductase Producing Bacterial Cultures

Pesticide	Examples
Insecticide	
Organophosphorus	Diazinon, dichlorvos, dimethoate, malathion, parathion
Carbamate	Carbaryl, propoxur, aldicarb methiocarb
Organochlorine	DDT, methoxychlor, toxaphene, mirex, kepone
Cyclodienes	Aldrin, chlordane, dieldrin, endrin, endosulfan, heptachlor
Herbicides	Chlorophenoxy acids, hexachlorobenzene (HCB)
Nitrogen-based	Picloram, Atrazine, diquat, paraquat
Organophosphates	Glyphosate (Roundup)
Fungicide	
Nitrogen-containing	Triazines, dicarboximides, phthalimide
Wood preservatives	Creosote, hexachlorobenzene
Botanicals	Perethrin, permethrin
Antimicrobial	Chlorine, quaternary alcohols

out an extensive study of marine actinobacteria of *Miromonospora sp.* (KPMS 1 and KPMS 9), *Streptomyces sp.* (KPMS2, KPMS5, and KPMS7), and *Micropolyspora sp.* (KPMS4) for the discolouration of Amido Black.

15.7.3 Fungi

Przystas *et al.* (2013) performed fungal degradation using a mixture of three fungal strains: *Pleurotuso streatus* (BWPH), *Gloeophyllum odoratum* (DCa), and *Fusarium oxysporum* (G1), which proved that mixed strain cultures were more efficient than single strain cultures. Another comparative study between pure and mixed fungal strains was conducted by Machado *et al.* (2006) using *Trametes villosa* and *Pycnoporus sanguineus* to decolorize a mixture of 10 dyes. The results showed dye decolourization was faster using mixed strains and the color reduction was approximately 80% and 90%, respectively.

Saroj *et al.* (2014) confirmed the successful utilization of *Penicillium oxalicum* SAR-3 to degrade commercial-grade Acid red 183, Direct Blue 15, and Direct red 75. Harazono *et al.* (2005) indicated the necessity of manganese (Mn^{2+}) and Tween 80/Polysorbate 80, chemically known as polyoxyethylene sorbitan monooleate in the culture media for *Phanerochaete sordida* to grow. This fungus reduced the color of a dye mixture by 90% within 48 h.

Phanerochaete chrysosporium, Trametes (Coriolus) *versicolor, Bjerkandera adusta, Aspergillus ochraceous*, species of *Pleurotus*, and *Phlebia* are isolates of white-rot fungi that have gained widespread recognition in the scientific world. The use of thermophilic fungus *Thermomucorindicae seudaticae* against an anthraquinone dye mixture was successful and it was found that bioremediation by adsorption could be performed at 55°C only at optimum pH.

15.7.4 Algae

A variety of algal strains have been engaged successfully for the effective degradation of various dyes, including *Anabaena flos-aquae* UTCC64, *Cosmarium, Chlorella vulgaris, Chlorella pyrenoidosa, Lyngbya lagerlerimi, Nostoc linckia, Oscillatoria rubescens, Elkatothrix viridis, Volvox aureus, Phormidium autumnale* UTEX1580, and *Synechococcus sp.* PCC7942. The fungal lignin-degrading system is responsible for the degradation of various textile dyes. One ligninolytic enzyme is manganese peroxidases, which is most commonly found in *P. chrysosporium* and is known to degrade high molecular weight chlorolignin in bleach plants and industry effluents.

15.8 ENZYMATIC BIODEGRADATION

The use of enzymes for their biodegradative activity has advantages over the use of microorganisms (e.g., pure culture or consortium). Shorter time for treatment, ability to treat the larger amount of substrates, generation of very low sludge volume, and an easy process to control are the major advantages. However, enzymes are much more expensive than anticipated due to their extended isolation and down streaming (purification) processes. Immobilized enzymes could be used to decrease the costs because they can be used repeatedly for textile dye degradation in a continuously operated reactor.

Oxidoreductive enzymes, also known as lignin modifying enzymes are widely used as dye decolourization enzymes. The ligninase activity of the enzymes and their ability to degrade by oxidation of various xenobiotics, such as aromatic and non-aromatic amines, phenols, chlorophenols, dyes, and polycyclic aromatic hydrocarbons (PAHs) have made them an important set of proteins with immense industrial significance (Wesenberg *et al.*, 2003). They include laccase, manganese peroxidase (MnP), manganese independent peroxidase, lignin peroxidase (LiP), NADH-DCIP reductase, tyrosinase (fungal), and azoreductase (bacterial), some of which are discussed in the following sections.

15.8.1 LACCASE

Laccase (e.g., benzendiol: oxygen oxido reductases Ec. No.1.10.3.2) is the most common biodegradation enzyme this is mostly secreted by white-rot fungi (Figure 15.7). Laccase can degrade a wide range of compounds, such as diphenols, polyphenols, diamines, and aromatic amines. It primarily consists of Cu atoms in its reactive center, which is the reason behind the restricted formation of hazardous substances after interaction with the dye molecule (Chivukula *et al.*, 1996). Laccases remove one or more electrons from the substrate and convert them into less toxic, less complex compounds.

15.8.2 CATALYSIS OF LACCASES

All fungal laccases typically contain four Cu atoms in three different sites and have their absorption wavelength in different ranges (Figure 15.6). Type I Cu produces intense blue color and has a higher absorption wavelength at approximately 610 nm. Unlike Type I, Type II Cu is weakly absorbing, colorless, and can accept one electron. However, Type III consists of a pair of Cu ions with an absorbance at 330 nm and mainly functions as electron acceptors. Type II and III Cu together form a tri-nuclear cluster in the interior of the enzyme and Type I Cu is usually located near the active site. Type I Cu extracts electrons from the substrate (i.e., textile dye effluents) and passes them on to the coupled Type III Cu, which is then handed over to Type II Cu, where O_2 is formed due to reduction of the O_2 molecule, therefore, it is oxidoreductive (Agrawal *et al.*, 2020; Singh *et al.*, 2014).

Yesiladali *et al.* (2006) used *T. rubrum* LSK-27 and reported the detection of laccase activity in the biotic controls at 18 h and a higher concentration of laccase after 26 h for the degradation of Remazol Blue RR, but negligible laccase activity was observed for the degradation of Supranol Turquoise. This proved that the source of the enzyme (different strains) and the chemical structure of the dye molecule were very important when determining the mode of action of the enzyme.

Laccase is secreted by a wide range of fungi, such as *Phanaerochaete chrysoporium, Aspergillus sp., Trichoderma sp, Trametes polyzona*, and *Penicillium sp.*, which have been used to isolate fungal laccase for bioremediation. Other reported white-rot fungi that have been exploited for their

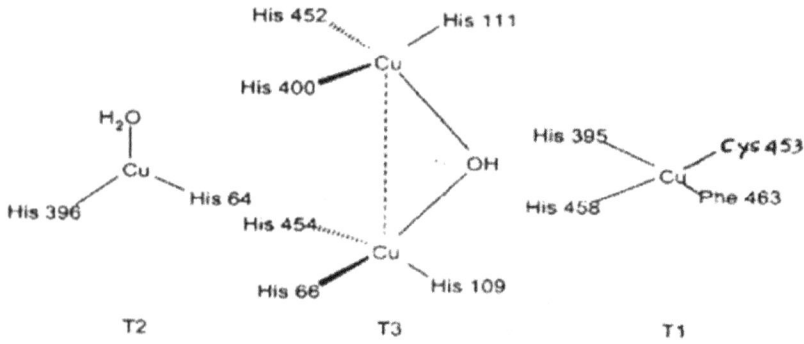

FIGURE 15.6   Three different sites of copper at the catalytic center of laccase.

Source: Chaurasia *et al.* (2013).

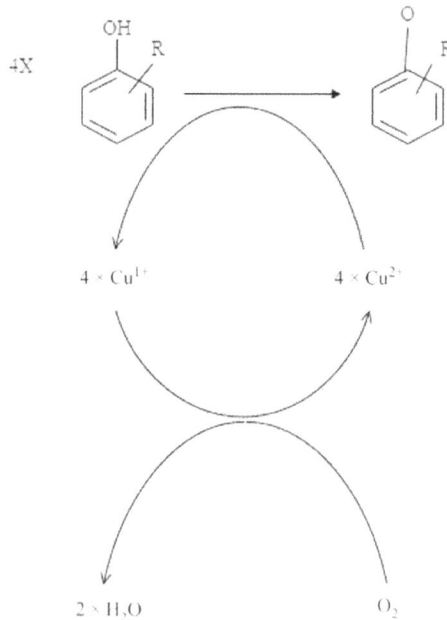

FIGURE 15.7   Laccase mechanism of action.

significant laccase activity are *Hirschioporus larincinus, Inonotus hispidus, Phlebia tremellosa,* and *Coriolus versicolor.*

Campos *et al.* (2001) utilized the laccase isolated from *Trametes hirsuta* and *Sclerotium rolfsii* to determine the indigo dye degradation process. Similarly, *T. versicolor* showed the best biodegradation performance when used for the degradation of syringol derivatives of azo dyes that contained either carboxylic or sulphonic groups. Many such organisms have been used by researchers and most of them have demonstrated positive results for the degradation of dye molecules.

15.8.3 LiP

Similar to laccase, LiP (EC: 1.11.1.14), which belongs to class II peroxidases, and is another lignin-degrading enzyme that is specific to certain bacteria and are mostly extracellular. LiP is a monomeric protein of N- and O-glycosylated protein, which contains heme. The basic nature of the enzyme means it has an isoelectric point between 3 and 5 (i.e., depends on the isomer). H_2O_2 is an essential requirement for LiP to catalyze the oxidation reaction. Lip can oxidize multiple non-phenolic lignin units due to its high redox potential (Sarkar *et al.*, 1997).

LiP was first discovered in *P. chrysosporium* as an extracellular enzyme. However, later various isomers of this enzyme were detected in organisms, such as *T. versicolor, Phlebia radiata,* and *P. sordida.* Further studies suggested that all isomers of LiP had different isoelectric points, amounts of sugar, substrate specificity, and stability.

15.8.4 CATALYSIS OF LiP

LiP can oxidize a wide range of substrates by transferring electrons via many steps. The three-step mechanism is:

1. The resting state of the native enzyme-containing ferric molecule
2. Formation of cation ferryl intermediate (oxoferryl) Compound I (Cpd I)
3. Formation of neutral oxoferryl intermediate Compound II (Cpd II). (Castro *et al.*, 2016)

The four N atoms in the previous diagram represent the porphyrin ring and A is the electron losing substrate. LiP is dependent on H_2O_2 for the oxidative process, where the heme (ferric) carrying native enzyme is oxidized by H_2O_2 with two electrons for the intermediate Compound I as shown in Figure 15.8. The H_2O_2 catalysis enables the formation of ferryl [Fe(IV)] by removing one electron from ferric [Fe(III)] heme. Then, the formation of a porphyrin cation radical is promoted by removing a second electron from the porphyrin ring (Aust, 1995). Then, H_2O_2 is reduced to water at this stage. Eventually, the substrate loses an electron. This means the substrate (i.e., dye enriched effluent) is oxidized by Compound I. Compound I is reduced to Compound II. This step is pH-dependent, as the rate of the reaction increases when the pH is low.

At this step, the lost electron is gained by the porphyrin ring of LiP. This entire catalytic cycle suggests that Compound I has a higher oxidative ability and can oxidize substrates with higher

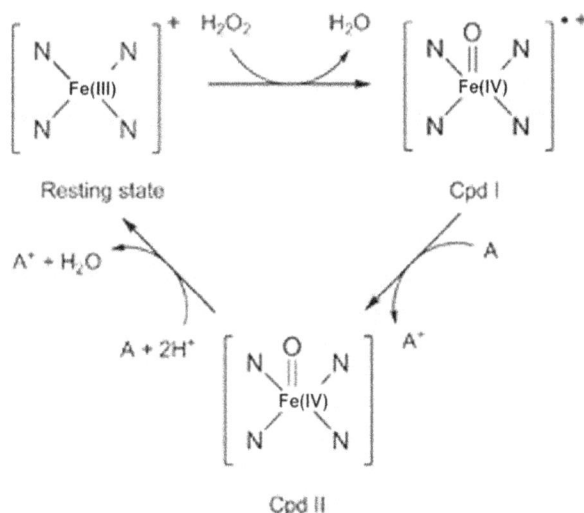

FIGURE 15.8 Catalytic mechanism of lignin peroxidase.

Source: Castro *et al.* (2016).

redox potential than Compound II (Mester *et al.*, 2000). Then, Compound II must return to its resting native enzyme state, which it does by gaining one more electron by interacting with the reducing substrate. LiPs have better oxidation abilities and have higher redox potential than traditional enzymes in the peroxidase family because the heme molecule within the active site of the porphyrin ring in LiPs is more electron deficient than other peroxidases.

Sometimes, certain malfunctions in the cycle might cause the formation of Compound III, which might lead to the inactivation of LiP. However, this can be overcome by the addition of veratryl alcohol, which completes the catalytic cycle of LiP by reducing Compound II to the resting enzyme (Singh *et al.*, 2014).

15.8.5 MnP

Similar to LiP, MnP (EC 1.11.1.13), is a very common lignin-degrading peroxidase, mostly found in white-rot fungi (i.e., basidiomycetes) as multiple isozymes. MnP is a heme-containing glycoprotein that can oxidize a wide range of phenolic compounds (substrates) through intermediary redox reactions with the help of Mn^{2+}/Mn^{3+} ions. The mechanism is similar to other peroxidases, such as LiP and Horse Radish Peroxidase (Figure 15.9).

The major difference is that that MnP utilizes Mn^{2+} as the electron donor to form the oxidized intermediates, Compounds I and II (Zapanta *et al.*, 1997).

Similar to classical peroxidases, MnP utilizes H_2O_2 (the same as LiP) to kick-start its catalysis reaction to produce Fe^{4+}-oxo-porphyrin-radical complex (Compound I). During degradation of substrates (i.e., dye effluents), the monochelated Mn^{2+} is oxidized to Mn^{3+} by donating one electron to the porphyrin intermediate, which leads to the formation of Compound II. The chelation is due to the action of oxalate, which is an organic acid chelator. The synthesis of oxalate and MnP in white-rot fungi, such as *Phanerochaete chrysosporium* is necessary for enhanced MnP activity.

The next major step is the conversion of Compound II into the native state of the enzyme MnP, which is achieved similarly to the previous step, for instance, Mn^{2+} is oxidized to form Mn^{3+} by losing one electron.

Eventually, the chelated Mn^{3+} ions oxidize various substrates, such as phenols, lignin substrates, and amine dyes (Hofrichter *et al.*, 2010). However, comparative studies showed that LiP has better oxidizing properties than MnP and can oxidize non-phenolic compounds easily. Biodegradative studies by Yesiladali *et al.* (2006) using *T. rubrum* LSK-27 showed that this white-rot fungus was an efficient producer of MnP and laccase but not LiP during the degradation of azo dyes. Experiments showed high levels of MnP after 62 h. This suggested that many strains of white-rot fungi might produce specific mediators that help MnP to oxidatively cleave recalcitrant non-phenolic substrates.

15.8.6 Azoreductase

Azoreductases (EC 1.7.1.6) are an important group of oxidoreductive enzymes found in most bacteria (e.g., aerobic and anaerobic) and fungi and is mostly exploited for dye degradation and decolourization. Their first step of catalysis involves reductively cleaving the azo bond (–N=N–) present in the azo dyes and converting the dyes into a colorless aromatic amine. However, the catalytic reaction under anaerobic or aerobic conditions can only occur in the presence of certain agents or cofactors that can donate electrons and are reduced, such as NADH, NADPH, and $FADH_2$ (Figure 15.10). Significant amounts of azoreductase have been found in a wide range of bacteria, *such as Caulobacter subvibrioides* C7-D, *Xenophilus azovorans* KF46F, *Pigmentiphaga kullae* K24, *Staphylococcus aureus*, *E. coli*, *Bacillus sp. OY1-2*, *Enterobacter agglomerans,* and *Enterococcus faecalis.*

Most of the reduction and cleaving activities are extracellular and do not depend on the intracellular uptake of the azo dye molecules, because of their high molecular weight, high

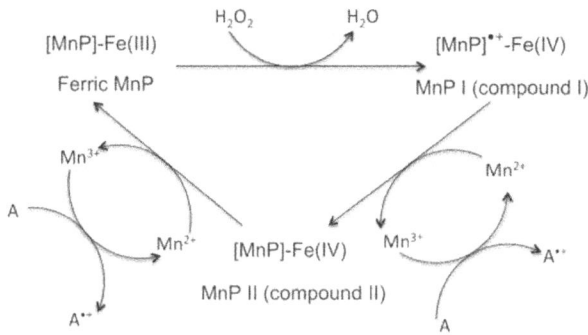

FIGURE 15.9 Catalytic mechanism of manganese peroxidase.

Source: Abdel-Hamid *et al.* (2013).

polarities, and complex structures, which does not allow them to cross the cell membrane. The basic mode of catalysis involves the transfer of electrons from azoreductases to the coenzyme, which then directly or indirectly relay them to the dye molecules (i.e., substrate), which initiates the cleavage of the azo bonds. The coenzymes/cofactors, such as NADH, NADPH, and $FADH_2$ act as mediators of the redox reactions or catalysts, which enhance the reaction rate and transfer of electrons (Rather *et al.*, 2018). However, the origin of these cofactors or mediators remains unknown and are of interest to researchers. Russ *et al.* (2000) suggested, that they could either be metabolic products of certain substrates used by the bacteria, or they could be manually added to the reaction.

Most of the cofactors are present in their oxidized states, such as NAD^+, $NADP^+$, and FAD. The cofactors oxidize the azoreductase; therefore, the azoreductase loses its electrons to the cofactors, which are reduced to NADH, NADPH, and FADH. The reduced cofactors then relay the same electrons to the dye molecules and return to their original states. The lack of O_2 plays a pivotal role in this indirect catalytic mechanism. The participation of O_2 could impede the electron transfer process and the cofactors themselves could become oxidized (Lellis *et al.*, 2019). Based on various experiments and reports, azoreductases can be classified based on their functionality and are dependent or not on flavin.

15.9 GMOS IN BIOREMEDIATION

Recent approaches for degrading the environmental pollutants has shifted from traditional remediation methods to a much more environment-friendly bioremediation method that uses enzymes and a wide variety of microorganisms (e.g., pure culture or microbial consortium). The major goal of bioremediation is to convert all the toxic pollutants into less harmful compounds or, to entirely remove them from the system.

Previously, multiple experiments have proven that using microbes produces better results (Ajaz *et al.*, 2019) than any other physical or chemical method, which is because enzymatic degradation using enzymes is cost-effective, low maintenance, there are no by-products (which is the case in chemical remediation processes) and microorganisms can be applied on-site (i.e., *in situ* bioremediation) and there is no displacement of the sample.

However, research has suggested that each microorganism in the environment has different functionality, and they have specific mechanisms for the degradation and decolourization of dye molecules. Recalcitrant synthetic dyes are difficult to degrade using standard conventional methods, and therefore, use more time, effort, and manpower. However, with the help of enhanced engineering techniques and advanced scientific research, genetic engineering might

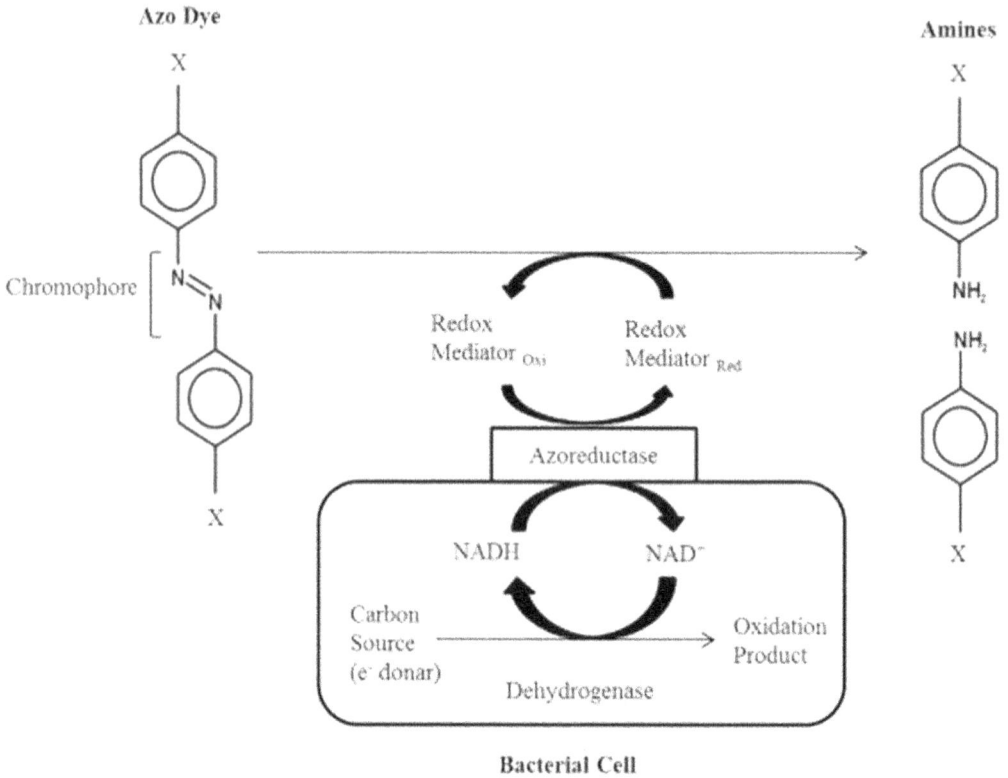

FIGURE 15.10 Catalytic mechanism of azoreductase in degrading azo dyes.

Source: Keck *et al*. (1997); Singh *et al*. (2015).

provide new treatments for bioremediation (Mishra *et al*., 2020). The manipulation of various genetic elements includes the combined knowledge of microbiological abilities and biochemical pathways to influence the biodegradation and bioprocessing of manmade pollutants for environmental clean-up. GMOs have groundbreaking applications in the bioremediation of soil and groundwater and the degradation of organic components in activated sludge. Numerous strains of *Sphingomonas desiccabilis, E. coli, Bacillus idriensis, Pseudomonas putida, Mycobacterium marinum*, and *Ralstonia eutropha* have been used to artificially construct genetically engineered microbes (GEMs).

The addition of a gene of interest into an organism for enhanced microbial or enzymatic degradation of dye molecules produces results that are 10 fold better than traditional bioremediation techniques. Any foreign gene, which is not part of the host system, when inserted into the host secretes the enzyme of interest for the degradation and decolourization of dye. This produces GMOs or designer biocatalysts that are integrated with a manmade metabolic pathway for enzymatic reactions (Paul *et al*., 2005). The modifications can direct the transfer of the segment of the gene that produces the enzyme of interest or by modifying the gene at a molecular level (i.e., site-directed mutagenesis). A similar kind of research was performed by Sandhya *et al*. (2008), which aimed to degrade a highly sulfonated textile dye Remazol Red. This was carried out by transferring the gene that synthesizes azoreductase from *Bacillus latrosporus* RRK1 into *E. coli* DH5a and plasmid pAZR-SS125 to produce the genetically devised strain *E. coli* SS125.

Similar to bacterial gene engineering or manipulation, various recombinant fungi have been exploited for the removal of environmental pollutants. Sakaki *et al.* (2002) proposed the theory for the degradation of dioxins using recombinant *Saccharomyces cerevisiae* that expressed mammalian CYP1A (family of cytochrome P450 complexes) to create recombinant strains. However, most of the strains could not degrade 2,3,7,8-Tetrachlorodibenzo-p-dioxin [i.e, the most toxic polychlorinated dibenzo-p-dioxins (PCDD)]. A similar experiment on genetic engineering failed, which used *Fusarium solani*, a filamentous fungus mostly found naturally in the soil, could not degrade dichloro-diphenyl-trichloroethane (DDT), a synthetic insecticide. A hybrid strain was created using *E. coli* that carried the azoreductase secreting gene from the wild type strain of *Pseudomonas luteola* for the remediation of recalcitrant azo dyes.

Several genetic modifications and genetic pathways can be exploited to generate products for the effective degradation of pollutants. Timmis *et al.* (1999) summarized certain conditions that could help to improve the degradative efficiency using GMOs, such as genetically manipulating the slowest step (i.e., rate-limiting) of known enzymatic reaction systems to increase the rate of degradation of pollutants or integrating a completely new enzymatic pathway into a specific bacterial strain to make degradation more effective as a coherent whole. They suggested that process control, toxicity, stress response assessment, and endpoint analysis for bioremediation could be performed by various strategies using GMOs.

Similarly, Kimura *et al.* (1997) used genetic engineering to treat areas polluted with polychlorinated biphenyls. *Pseudomonas sp.* LB400 and *Pseudomonas alcaligenes* KF707 were used and genetically altered to enhance their substrate specificity. Another method for designing engineered microbial strains is by combining the major catabolic portions from different strains of microorganisms. Therefore, a complete enzymatic (metabolic) route can be constructed. These routes produce products that help in the elimination of xenobiotics. This technique has been applied to degrade highly toxic trihalopropanes. However, attention must be paid to the end products from the artificially engineered routes, and therefore, no toxin should be produced that could be subject to further transformations by other microorganisms which might lead to the formation of reactive molecules.

However, the literature states that GMOs are slow and inefficient in removing any form of pollutants. Lack of motivation and public resistance is the main reason that GMOs have seen very little development in the last few decades for bioremediation. Public safety concerns, legal restrictions, and public health are additional risks related to the release of GEMs into the environment. Various agencies should note all concerns related to the field release of GEMs, determine the extent of the release of GEMs, and identify the fate of the mutated organism and how they could affect the evolution of the existing wild type strains. The fate of the genetically modified strain must be determined because they can either complete the bioremediation and become eliminated from the environment naturally or they can persist and proliferate in the natural environment. Industries are not currently using bioremediation, which is expensive and cannot bear the additional costs of genetically modifying existing strains of microbes. Therefore, genetic and metabolic engineering could present groundbreaking possibilities for bioremediation in the near future; however, they are not currently having an impact.

15.10   NANO TECHNOLOGICAL INTERVENTIONS IN BIOREMEDIATION

The biomimicry aspect of nanotechnology has been the major driving force to expedite the process of bioremediation and to reduce the amount of materials and manpower that are used in traditional remediation techniques. However, nanotechnology could embrace the era of sustainable development by using materials in the nano form (1–100 nm) and could prove to be an effective and efficient form of remediation. The maintenance and improvement in the air, soil, and water quality could be directly attributed to the use of nano materials either *in situ* or *ex situ*.

15.10.1 Nanoremediation

Nano bioremediation can be accomplished in three different stages:

1. Remediation/treatment
2. Sensing and detection
3. Pollution control

Nanotechnology promotes the use of different types of nano materials, such as carbonaceous nanomaterials, C nanotubes, bimetallic particles, zeolites, dendrimers and nanoparticles that adhere to various metals, such as silver (Ag), gold (Au), and Cu Their high surface reaction, surface energy, and interaction abilities that is due to their larger of surface area, higher catalytic efficiency, better mass transfer, spatial confinements, magnetic, electronic and optical properties means that nanoparticles are optimal candidates for wastewater treatment, and mainly water that is enriched by a lot of pollutants (Figure 15.11). Nanoparticles have a higher adsorption capacity, which makes the removal of organic and inorganic pollutants comparatively easy (Tripathi *et al.*, 2018).

Metal-based nanoparticles that use old Ag, platinum (Pt), and palladium (Pd) have been widely explored, mostly for synthesizing magnetic nano particles (MNPs) for bioremediation. However, the major disadvantage of using nanoparticles lies in their production. The manufacture of nanoparticles is mainly carried out by physical or by chemical methods, and it has a large environmental impact due to the toxicity it generates. This major disadvantage led researchers to explore an eco-friendlier, non-toxic, and greener route for manufacturing nanoparticles (Sunkar *et al.*, 2013). Because plants are distributed over a large area, are easily available, and do not require large handling effort, they were chosen as a preferential and reliable source of important metabolites. Nanoparticles preparation using plants has been identified to be a quicker and simplified technique and the manufacturing process can be easily increased to large-scale production. Asthana *et al.* (2016) demonstrated the preparation of zinc (Zn) nanobeads using the seed extracts from *Cuminum cyminum* to treat effluents enriched in Alizarin Red. In addition, evidence of the degradation of dyes was reported by Hong *et al.* (2005), by synthesizing Ag nanoparticles using *Padina tetrastomatica*, which is a type of seaweed.

Most of the production processes are complex and require intricate attention to details, such as process control and monitoring various factors including pH, agitation, illumination, bioagents, metal ion concentration, and the size and shape control of the nanoparticles. Furthermore, the use of autochthonous strains of microbes, combined with nano technological advancements is currently receiving increased attention. The role of naturally occurring microbial flora (e.g., bacteria, fungi, algae, and yeast) plays a pivotal role as nano factories in research because their genes encode for potentially powerful enzymes that act as biotools in the field of nanoremediation (Sharma *et al.*, 2017). The microbial production of nanoparticles can be intracellular or extracellular and depends upon various factors, such as the temperature of microbial growth, synthesis time, the method of nanoparticles extraction, and production ratio of samples. Examples of various bacteria that have been used in nanoremediation for the production of Ag nanoparticles are *Pseudomonas antartica*, *Pseudomonas proteolytica*, *Pseudomonas meridiana*, *Arthrobacter kerguelensis*, *Arthrobacter gangotriensis*, *Bacillus indicus*, and *Bacillus cecembensis*.

Similar to bacteria, fungi have a high metal uptake capacity and require simple nutrients for growth. The use of various types of endophytic fungi, such as *Garcinia xanthochyumus* and *Aravae lanata* for the production of Ag nanoparticles were cited in Ahmad *et al.* (2003) for their participation in textile dye degradation. In addition, microalgae have provided significant input in nano bioremediation and have been able to transform themselves into metal nanoparticle producing factories via the bioaccumulation of metals. The marine algae *Sargassum wightii* and *Gracilaria corticator* have been described to synthesize Ag and Au nanoparticles (Sharma *et al.*, 2017).

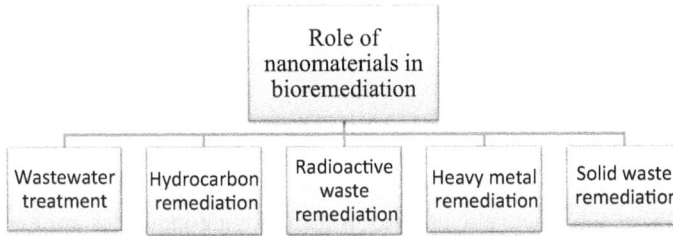

FIGURE 15.11   Application of nanomaterials for remediation purposes.

15.10.2   Nanoscale Tools Used in Bioremediation

15.10.2.1   C Nanotubes and Nanocrystals

The distinctive properties of carbon nanotubes (CNTs), such as high thermal and electrical conductivities, high solidity and stiffness, and special adsorptive properties make CNTs the most favorable candidate to solve a wide range of remediation applications including sorbents, high-flux membranes, depth filters, antimicrobial agents, environmental sensors, renewable energy technologies, and pollution prevention strategies (Mauter *et al.*, 2008). The rate of adsorption depends on the size and the structure of the pores, which enables interaction between the pollutant molecules and the CNTs. Large volumes of the substrate (i.e., dye molecules) can interact with the pores in the CNTs, which are mostly spherical or cylindrical. Stronger adsorption capacities are attributed to the fact that CNTs are rich in activated C, which can adsorb more molecules. The use of single-walled carbon nanotubes (SWCNTs), multi-walled carbon nanotubes (MWCNTs), and hybrid carbon nanotubes (HCNTs) have been suggested for the rapid adsorption of ethylbenzene from industrial effluents. In addition to dyes, CNTs are involved in the removal of pesticides, pharmaceuticals, drugs, metalloids, heavy metals such as chromium (Cr^{3+}), lead (Pb^{2+}), and Zn^{2+}, arsenic (As) and mercuric (Hg) compounds, VOCs, and dioxins (Yu *et al.*, 2014).

15.10.2.2   Nano Iron

Research has been invested in the effectiveness of nanoscale metallic iron (Fe)/zero-valent iron (ZVI). ZVI has been reported as a product for the removal of contaminants, such as halogenated methanes (Cl/Br), trihalomethanes, chlorinated ethenes and benzenes, other polychlorinated hydrocarbons, pesticides, and dyes. The principal mechanism behind ZVI involves dechlorinating the aqueous contaminants in a reduced environment (i.e., if chlorine contaminants are present), or reducing the contaminant to an insoluble form (i.e., if ionic metal contaminants are present). Heavy metals, such as As and Cr, pesticides, and chlorinated solvents have been transformed into less toxic levels by ZVI (Karn *et al.*, 2009).

15.10.2.3   Dendrimers

Dendrimers are highly branched and monodisperse macromolecules that have a cascade-like shape, branching outwards like the roots of a tree. Dendrimers are composed of three integral sections: a central core, interior branches, and terminal functionalized branches (Undre *et al.*, 2013). Dendrimers have profound applications in environmental management and control mainly due to their large number of encapsulating properties. Metal ions and zero-valent metals used in remediation processes can be enveloped or encased within the dendrimer molecule and can easily dissolve in the substrate (i.e, effluent). Since dendrimers have a larger surface area, they can react with the contaminants easily; therefore, increasing their catalytic activity. The most popular class of dendrimers are polyamidoamine (PAMAM) dendrimers, which are manufactured by the repetitive incorporation of amidoamine monomers radially. Dendrimers are mostly used in water treatment facilities for simple filtration (Prasad *et al.*, 2018).

15.11 BIOREMEDIATION OF ORGANIC POLLUTANTS

15.11.1 TEXTILE DYES

Approximately 80% of the reactive dyes used in textile dying industries are hazardous and have major environmental and ecological concerns. Their intense color and high visibility even at very low concentrations are not aesthetic. They form a layer that is impenetrable to light and O_2, making the survival of marine biota impossible due to respiratory problems. From the numerous dyes, such as carbonyl dyes and pigments (i.e., anthraquinone, indigo derivatives), cyanine dyes, Di and triphenylmethane dyes, Azo dyes and pigments (e.g., mono, azo, and diazo) and phthalocyanine dyes and pigments, azo dyes remain the most recalcitrant and hazardous of all dyes mainly due to the azo functional group and they are carcinogens (Benigni *et al.*, 2000).

However, there are limited global resources for land and water. If the textile industries use fresh water each time when dyeing each ton of textile, there would be no water left. Therefore, to reduce this enormous water consumption, there is a strong demand for new recyclable technologies, which would enable us to reuse the water by removing any potential hazards from it. One such potential technology for the removal of dye molecules from effluents is bioremediation. Remediation techniques, such as microbial degradation use microorganisms, such as bacteria and fungi; phytoremediation by plants, which involves several biological mechanisms; and enzyme remediation that uses specific enzymes to degrade pollutants are the most cost-effective, minimalistic, and exciting approaches for the removal of dye molecules today

However, enzymes are preferably exploited for bioremediation, most of which are active redox molecules and have a common mechanism to deal with dye molecules that have a wide range of substrate specificity. Bacteria including *Proteus sp.*, *Enterococcus sp.*, *Streptococcus faecalis*, *Bacillus subtilis*, *Bacillus cereus*, *Pseudomonas sp.*, (Bumpus., 1995) cleave azo bonds. Similarly, fungi, such as *Phanerochaete chrysosporium*, *Irpex lacteus*, *Coriolus versicolor*, *Pleurotus ostreatus*, *Pycnoporus sanguineus*, *Trametes hirsute*, *Neurospora crassa*, and *Geotrichumcandidum*. have been explored widely.

15.11.2 AROMATIC COMPOUNDS

Aromatic amines (AA) are highly toxic, carcinogenic, organic compounds due to the generation of cytotoxic metabolites. They constitute a major part of all chemical dyes and sometimes are used as intermediates for the production of synthetic dyes. Since most AAs are highly water-soluble, it means that they can easily penetrate the soil and enter the natural water table where they can further break down into more toxic components. The most efficient way discovered to render aromatic compounds inert is by fungal bioremediation. *Aspergillus spp*, *Trichoderma spp*, and *Fusarium spp* are the most researched genera. Although the relative growth of the fungus is lower than bacteria, they are powerful tools for bioremediation, because of their efficient uptake of heavy metals.

The use of fungal (e.g., pure culture or consortium) cultures is highly efficient in biocleansing of xenobiotics from the environment. This is because most fungi secrete exoenzymes to supply the necessary nutrients to the growing hyphae. These xenobiotic degrading enzymes are not very specific, which makes them tolerate a wide range of anthropogenic chemicals, including AAs and their derivatives (Zhao *et al.*, 2016). The first step in the biodegradation of AAs according to Bugg *et al.* (1998) begins with the hydroxylation of the aromatic ring with the help of an enzyme that can hydroxylate a wide range of annilinic compounds. The enzyme used for this process is aniline dioxygenase (AtdA) that is derived from *Acinetobacter sp*. YAA. Later, it was shown that this enzyme performed degradation via multiple stages by the simultaneous deamination and oxygenation of aniline and o-toluidine from effluents and converted them to produce catechol and 3-methylcatechol.

15.11.3 HEAVY METALS

Continuous weathering of rocks and volcanic eruptions along with industrial activities, such as tanneries, electroplating, dyeing, and mining have allowed toxic metal species to enter and accumulate in every stage of the food chain, which causes ecological and health hazards. Cr, nickel (Ni), Cu, Fe, Mn, and Zn are some heavy metals present in wastewaters from tanneries and electroplating industries.

The existence of heavy metals even in small quantities in nature is a serious concern. This is because heavy metals are non-biodegradable, and accumulate in humans, animals, and plants and are biomagnified along with the trophic levels. The accumulation of heavy metals in effluents is mainly beyond permissible limits, which are reused to irrigate fields. This causes bioaccumulation in crops, which are then consumed. Some waters are discharged into the rivers; however, some are absorbed by the land, where the metallic contaminants reach the water table. Therefore, the chances of exposure to heavy metals are unlimited due to the increase in their use in the technology, domestic, industrial, and agricultural sectors.

Recently, awareness regarding the nature, effects and hazards that are caused by heavy metals has increased. Therefore, researchers are developing new technologies to eliminate these contaminants. Experiments have shown that traditional techniques, such as adsorption, electrodialysis, precipitation and ion-exchange methods have very little effect on heavy metals because these processes are not natural and cannot remove heavy metals at very low concentrations. Some of the previously mentioned techniques are pH sensitive and almost all of them are expensive methods and require high maintenance. In addition to all the disadvantages, the traditional methods are very slow and have inefficient removal, the sludge generated after cleaning remains contaminated, which then needs to be disposed of and does not serve the purpose of remediation (Gunatilake, 2015).

As an alternative, an eco-friendly method was developed that utilized microorganisms that were already present in nature. The major advantage of using microbial colonies was that remediation was dependent on other environmental factors, such as pH, moisture, temperature, and presence of other ions, In addition, the microbes could convert harmful heavy metals into less toxic intermediate substances, or sometimes eliminate the contaminants from the site. The contaminants could sometimes be degraded or demineralized into organic substances, such as CO_2, water, or N_2 (Kapahi *et al.*, 2019).

Microbes facilitate bioremediation by leaching heavy metals from soil. The process is enhanced by the increased solubility of ions, such as Fe(III) and As(V), and then reducing them to Fe(II) and As(III) (Bachate *et al.*, 2012). Methylation of heavy metals and As can be employed because methylated products are mostly volatiles and are conveniently lost from the soil by evaporation, for example, dimethylmercury and alkyl arsines of Hg and As.

Schizophyllum commune, *Methanothermobacter thermautotrophicus*, *Bacillus cereus* and *Shewanella sp* have been reported to participate actively in the removal of heavy metals from waste waters. In addition, bacterial species, such as *Pseudomonas*, *Desulfovibrio*, *Bacillus*, and *Geobacter*, algal species including *Asparagopsis sp.*, *Codium sp.*, *Spirogyra sp.*, *Chondrus sp.*, *Fucus sp.*, and *Ascophyllum sp.* have been reported. A variety of fungi, such as *S. cerevisiae*, *Aspergillus sp.*, and species of white-rot fungi, including *Pleurotus ostreatus* and *Termitomyces clypeatus* have been reported to degrade persistent pollutants.

15.11.4 Petroleum Hydrocarbons

Petroleum and its derivatives have been posing as a threat to the environment for a long time due to their adverse and harmful effects on human, plant, and animal health. There are four classes of petroleum hydrocarbon: saturates, aromatics, asphaltenes (e.g., phenols, fatty acids, and ketones), and resins (e.g., pyridines, quinolines, and carbazoles). Because of the constituents of petroleum, for example, thousands of aliphatic branched, and aromatic hydrocarbons and other organic compounds, petroleum has long term environmental hazards and is difficult to remove permanently from nature.

Many physicochemical treatments, such as incineration, thermal desorption, cokers, cement kilns, solvent extraction, and land fill have been implemented; however, every technique has some disadvantages. Physicochemical methods are extremely expensive because of mobilizing large quantities of contaminated soil or water samples. An *in situ* green technology is required, which has increased the amount of research on bioremediation. This method does not require moving any samples from the site of contamination; therefore, preventing any spread of pollutants. Bioremediation mitigates the use of microbial cultures for degrading the major constituent of petroleum products. The microbes, in turn, reduce the components into basic organic compounds, such as CO_2, CH_4, water, and biomass, which are comparatively less toxic (Ron *et al.*, 2014).

The most efficient hydrocarbon degradation results have been observed under aerobic conditions. The initial degradative process includes the activation of the microbial strain and the incorporation of O_2 for the oxidation of the pollutant molecules and metabolic reaction, which in most cases is catalyzed by enzymes secreted by the microbe that is being used for degradation (e.g., oxygenases and peroxidases). The pathway involves a steady degradation of the pollutants into intermediates and then finally into less toxic, colorless products (Yuniati *et al.*, 2018).

15.11.5 Pesticides

Pesticide consumption has rapidly increased globally due to the increase in population and urbanization. In addition, the accumulation of a variety of these chemicals has not decreased the quality of the soil and has seeped into the soil to reach the water bed below. Since earlier techniques, such as landfills, recycling, and pyrolysis could not provide a solution for these quantities of chemicals; newer, faster, and cheaper technologies needed to be implemented to eliminate the pollutants from the environment (Debarati *et al.*, 2005).

There are various types of pesticides, which depend on their use: herbicides, insecticides, fungicides, and rodenticides. The structure of the chemicals is based on the classification given in Table 15.2.

Most of the previously mentioned pesticides are carcinogenic and cause tumors, irritation, and convulsions. Similar to most other degradation processes, the remediation of pesticides involves oxidizing the principal component of the pesticide and converting them into CO_2 and water, which are the major sources of nourishment to microbes. The oxidation process is catalyzed by the presence of various enzymes metabolized by the microbes such as, arly aclyamidase, organophosphorus hydrolase, and organophosphorus acid anhydrolase (OPAA) that are obtained from *Bacillus sphaericus*, *Bacillus diminuta*, and *Alteromonas undina*, respectively (Uqab *et al.*, 2016). However, the efficiency of any degradation process is dependent on various external factors, such as temperature, pH, concentration or availability of nutrients, and the presence or absence of O_2 in the system.

15.12 FACTORS THAT AFFECT BIOREMEDIATION

Bacterial metabolism, bacterial growth and sustenance, and their interactions with the contaminants are determined by various external factors. Various researchers have shown that various physical, chemical, and biological factors can inhibit bioremediation. Factors, such as pH, temperature, nutrient status, dissolved O_2 content, presence of electron donors and acceptors, and contaminant load all

TABLE 15.2
Types of Pesticides and Examples

Bacterial Strain	Azo Dye	References
Sphingomonas xenophaga strain BN6	Mordant Yellow 3	Chen *et al.*, 2003
P. aeruginosa	Navitan Fast blue S5R	Nachiyar *et al.*, 2005
Mutant *Bacillus sp.* ACT2	Congo Red	Gopinath *et al.*, 2009
Acinetobacter radioresistens	Acid Red	Ramya *et al.*, 2010
Rhizobium radiobacter MTCC 8161	Reactive Red 141	Telke *et al.*, 2008
Shewanella decolorationis S12	Acid Red GR	Xu *et al.*, 2007
E. coli JM109 (pGEX-AZR)	Direct Blue 71;	Jin *et al.*, 2009
Desulfovibrio desulfuricans	Reactive Orange 96 and Reactive Red 120	Yoo *et al.*, 2000
Alcaligenes sp. AA09	Reactive Red BL	Pandey *et. al.*, 2012

Source: Vaccari *et al.* (2005).

determine the efficiency of remediation and the types of products formed when the process is finished. Some of the factors for the optimal removal of pollutants are discussed in the following sections.

15.12.1 NUTRIENT AVAILABILITY

Microorganisms are already present in the soil and water; however, they might not be present in large quantities, or in quantities that are required for the removal of pollutants contamination sites. This is mainly due to the lack of optimum concentrations of O_2 and limited nutritional supplements, which do not allow the growth and stimulation of the required microbial flora. Nutrients such as C, N, P, K, and Ca are the most vital nutrients that must be readily available for microbial maturation. However, the availability and the concentration of the nutrients could harm degradation. Excessive amounts of any of the previously mentioned elements might have harmful effects on the environment (Van *et al.*, 2004). The nutrients from the building blocks of all microbial life and are required by them to produce the necessary enzymes to break down complex contaminants into their simpler forms. The availability of nutrients ensures that the microbial flora can grow and access the organic pollutants, which in turn affects the rate of biodegradation. Sometimes, even the presence of high numbers of microbes is not sufficient for biodegradation, if the mass transfer is limited. In that case, vigorous mixing and tilling of the soil are required to stimulate and activate the process.

15.12.2 AVAILABILITY OF O_2

The aerobic or anaerobic condition of the system is attributed to the amount of available O_2. Tilling or sparging the soil with O_2 increases the total O_2 levels; however, the introduction of H_2O_2 or MgO_2 can carry out the work. Aerobic conditions are the optimal environments for the degradation of PAHs and hydrocarbons; however, effluents enriched in chlorine can be degraded well in an anaerobic atmosphere. In certain cases, the morphology of the soil particles and the porosity affect the availability of O_2, water, and nutrients. Low soil porosity does not allow the movement of nutrients and O_2 which interferes with the growth of microbes within the soil (Naik *et al.*, 2012).

15.12.3 TEMPERATURE

Bioremediation, *in situ* and *ex situ*, is highly affected by the temperature at which the microbes grow. Research showed that temperatures of 30°C–40°C (e.g., in soil and marine microbes) increased the rate of biodegradation. The biochemical reaction rates and the metabolic reactions of microbes are affected most by any change in temperature (increase or decrease) Even an increase of 10°C can have a significant effect on the rate of enzyme synthesis, and therefore, inhibiting degradation. A temperature range of 15°C–20°C might affect degradation (Naik *et al.*, 2012).

15.12.4 pH

Various types of microbes have been isolated from extreme conditions. This suggests that certain strains of microbes can thrive and reproduce under harsh conditions. However, most of them are known to grow optimally over a range of environmental conditions, such as temperature, salinity, and heat tolerance of which, pH is the most important factor. Usually, the pH is different between sites. This fluctuation in pH is because all sites are contaminated with different types of contaminating molecules, which in turn have different pH. According to the International Center for Soil and Contaminated Sites (2006), the optimal pH of any site is from 6 to 8. Recently, emphasis was placed on the fact that microbes at a high pH could metabolize polyaromatic hydrocarbons. In contrast, the mineralization of petroleum hydrocarbons is favored under neutral pHs (Mishra *et al.*, 2021).

15.13 ECONOMICS OF BIOREMEDIATION

The application of microbes or enzyme-based bioremediation was initiated to offer a low-budget cleaning strategy for soils, sludges, surface, and subsurface water that was contaminated with organic compounds, such as dyes, pesticides, heavy metals, and petroleum by-products. In addition to the cost-effectiveness, bioremediation is a more practical approach to environmental clean-up rather than the conventional procedures, such as adsorption, filtration, or incineration techniques, which are expensive to maintain and handle, require more manpower, and create more pollution as toxic gases and fumes. In addition, it offers a solution to end pollution in an eco-friendly manner, unlike other physicochemical treatments, which mainly focus on converting the contaminants from one form to another. Therefore, this technique does not generate any overheads or liability costs, which might drive the industry to pay less attention to treating the effluents before discharging them (Alvarez *et al.*, 2005).

The biological treatment of pollutants remains vital today and soon mainly due to the advantages it has over classical treatment options including:

1. Bioremediation process does not require any complex or expensive reagents for its source of nutrients, typically because most microbes require small quantities of nourishment for their growth and metabolic reactions
2. Biological treatment does not require heat treatments >10°C–20°C for activation; therefore, minimal energy is consumed when removing pollutants
3. The use of any large equipment, conveyors, or machinery is not required; therefore, no large capital investment is required
4. It is an environmentally friendly process, and any extra pollutants are not released into the environment

However, even with the previous economic advantages, investments must be made into treatability studies, pilot tests, and other studies before setting up a blueprint for biotreatment, especially if the technology has not been proven, and there are no experimental studies or evidence. The total initial investment required by the industry must be known to determine various related aspects, such as what type of remediation process is chosen, what type of equipment is required for the clean-up, whether the clean-up is carried out by contractors or by the industry, and the direct operating costs, overheads, and manpower required for site clean-up. The scale of the clean-up project must be determined based on the overall capital assigned to the project. Therefore, the technology that costs the least but provides the most efficient results must be considered feasible.

Similar to other processes that might pose a risk to humans, animals, the environment, it requires multiple approvals and permits at multiple stages. The approvals by state and local authorities are the most expensive and laborious tasks. However, the most taxing work is to instigate positive public perception. The public that live near or in the vicinity of the site of a clean-up must know any of the harms or risks. This can be achieved by organizing public gatherings and citing experimental documents and expert advice that is related to the project. Any valuable feedback or opposition must be taken into consideration.

However, the clean-up parameters must be identified and alternative strategies for bioremediation must be explored before embarking on long and expensive R&D and on-site clean-up. The technical and economic feasibility of the clean-up project must be examined before implementation (Paul *et al.*, 1993).

15.14   CONCLUSIONS

Recently, bioremediation has become a fast-growing technology with a wide range of applications. Research has meant that bioremediation has emerged as the most sustainable approach to environmental pollution management. However, continuous efforts are being made to directly implement genetic engineering with bioremediation for enhanced microbial action on pollutants. Bioremediation offers several advantages that could potentially eliminate any chances of future hazards following remediation and removes the requirement of disposing of the contaminated material because no waste is left. The advantages outweigh the disadvantages because it is a cheaper option for environmental restoration and it does not use any chemical additives, which could further decrease the soil and water quality. As an ultimate approach, bioremediation could be paired with any other treatment option, such as physical or chemical treatment methods to create a combined process for wastewater and soil treatment. As discussed in this chapter, a broad range of bacterial, fungal, algal, and other physicochemical treatment options have been implicated to improve the efficacy when cleaning up textile dyes from the environment. However, further experiments and knowledge on the understanding of the functionality and responses of the microbial cultures toward the natural environment and the contaminants need to be developed in the near future. In addition, knowledge on the genetics of microbes, their ability to degrade pollutants, enzyme metabolic pathways, and field testing of various techniques need to be strategized more efficiently to release the full potential of bioremediation.

REFERENCES

Abdel-Hamid, A.M., Solbiati, J.O., Cann, I.K.O. (2013) 'Insights into lignin degradation and its potential industrial applications', *Advances in Applied Microbiology*, 82, pp. 1–28.

Agrawal, K., Verma, P. (2020) 'Multicopper oxidase laccases with distinguished spectral properties: A new outlook', *Heliyon, 6.*

Ahmad, P. *et al.* (2003) 'Extracellular biosynthesis of silver nanoparticles using the fungus Fusarium oxysporum', *Colloids Surface. B: Biointerface*, 27, pp. 313–319.

Ahmad, A., Mohd-Setapar, S.H., Chuong, C.S., Khatoon, A., Wani, W.A., Kumar, R. and Rafatullah, M. (2015) 'Recent advances in new generation dye removal technologies: Novel search for approaches to reprocess wastewater', *RSC Advances*, 5(39), pp. 30801–30818.

Ajaz, M. *et al.* (2019) 'Degradation of azo dyes by *Alcaligenes aquatilis* 3c and its potential use in the wastewater treatment', *AMB Express* 9, pp. 64.

Alvarez, M.S., Rodriguez, A., Sanroman, M.A., and Deive, F.J. (2015) 'Microbial adaptation to ionic liquids', *RSC-Advances*, 5, pp. 17379–17382.

Alvarez, P., Illman, W. (2005) *Bioremediation and natural attenuation: Process fundamentals and mathematical models*. John Wiley & Sons. doi: 10.1002/047173862X.ch6.

Asthana, S., Sirisha, D., and Mary, A. (2016) 'Green synthesis of nanoparticle of Zinc and treatment of nanobeads for waste water of Alizarin red dye', *International Journal of Environmental Research and Development*, 6(1), pp. 11–16.

Aust, S.D. (1995) 'Mechanisms of degradation by white rot fungi', *Environmental Health Perspectives*, 103, pp. 59–61.

Azubuike, C.C., Chikere, C.B., and Okpokwasili, G.C. (2016) 'Bioremediation techniques–classification based on site of application: principles, advantages, limitations and prospects', *World Journal of Microbiology and Biotechnology*, 32(11), p. 180.

Bachate, S.P., Khapare, R.M., and Kodam, K.M. (2012) Oxidation of arsenite by two β-proteobacteria isolated from soil', *Applied Microbiology &l Biotechnol*ogy, 93(5), pp. 2135–2145.

Ball, A.S., Betts, W.B., and McCarthy, A.J. (1989) 'Degradation of lignin-related compounds by Actinomycetes', *Applied Environmental Microbio*logy, 55(6), pp. 1642–1644.

Benigni, R., Giuliani, A., Franke, R., and Gruska, A. (2000) 'Quantitative structure-activity relationships of mutagenic and carcinogenic aromatic amines', *Chemical Review*, 100, pp. 3697–3714.

Bhatia, D., Sharma, N.R., Singh, J., and Kanwar, R.S (2017) 'Biological methods for textile dye removal from wastewater: A review', *Critical Reviews in Environmental Science and Technology*, 47, pp. 1–41. doi: 10.1080/10643389.2017.1393263.

Bromley-Challenor, K.C. *et al.* (2004) 'Decolorization of an azo dye by unacclimated activated sludge under anaerobic conditions', *Water Research*, 34, p. 4410.

Bugg, T.D.H., and Winfield, C.J. (1998) 'Enzymatic cleavage of aromatic rings: mechanistic aspects of the catechol dioxygenases and later enzymes of bacterial oxidative cleavage pathways', *Natural Product Reports*, 15, pp. 513–530.

Bumpus, J.A. (1995) 'Microbial degradation of azo dyes. *Progress in Industrial Microbiology*, 32, pp. 157–176.

Campos R. *et al.* (2001) 'Indigo degradation with purified laccases from *Trametes hirsute* and *Sclerotium rolfsii'*, *Journal of Biotechnology*, 89, pp. 131–139.

Castro, L. *et al.* (2016). Insights into structure and redox potential of lignin peroxidase from QM/MM calculations', *Organic & Biomolecular Chemistry*, 14(8), pp. 2385–2389.

Chaurasia, P., Yadav, R., and Yadava, S. (2013) 'A review on mechanism of laccase action', *Research and Reviews in Biosciences*, 7(2), 66–71.

Chen, K.-C., Wu, J.-Y., Liou, D.J., and Hwang, S.-C.J. (2003) 'Decolorization of the textile dyes by newly isolated bacterial strains', *Journal of Biotechnology*, 101(1), pp. 57–68. doi:10.1016/s0168.

Chivukula, M., and Enganathan, V. (1996) 'Phenolic azo dye oxidation by laccase from *Pyricularia oryzae'*, *Applied and Environmental Microbiology*, 61, pp. 4374–4377.

Clark, M. (2011) 'Handbook of Textile and Industrial Dyeing' In M. Clark (ed.) *Principles, Processes and Types of Dyes*. 1, Cambridge, UK: Woodhead publishing, pp. 1–652.

Coughlin, M.F., Kinkle, B.K., and Bishop, P.L. (1999) 'Degradation of azodyes containing amino naphthol by *Sphingomonas sp.* strain ICX', *Journal of Industrial Microbiology and Biotechnology*, 23, pp. 341–346.

Debarati, P., Gunjan, P., Janmejay, P., and Rakesh, V.J.K. (2005) 'Accessing microbial diversity for bioremediation and environmental restoration', *Trends in Biotechnology*, 23, pp. 135–142.

Elisangela, F. *et al.* (2009) 'Biodegradation of textile azo dyes by a facultative *Staphylococcus arlettae* strain VN-11 using a sequential microaerophilic/aerobic process', *International Biodeterioration & Biodegradation*, 63, pp. 280–288.

Ghaly, A.E., Ananthashankar, R., Alhattab, M.V.V.R., and Ramakrishnan, V.V. (2014) 'Production, characterization and treatment of textile effluents: A critical review. *Journal of Chemical Engineering & Process Technology*, 5, pp. 1–18.

Ghodake, G.S., Telke, A A., Jadhav, J.P., and Govindwar, S.P. (2009) Potential of *Brssica juncea* in order to treat textile effluent contaminated sites', *International Journal of Phytoremediation*, 11, p. 1.

Gopinath, K.P., Murugesan, S., Abraham, J., and Muthukumar, K. (2009) '*Bacillus sp.* mutant for improved biodegradation of Congo Red: Random mutagenesis approach', *Bioresource Technol*ogy, 100, pp. 6295–6300.

Goswami, M. et al. (2018) 'Bioaugmentation and biostimulation: a potential strategy for environmental remediation. *Journal of Microbiology & Experimentation*. 6. doi:10.15406/jmen.2018.06.00219.

Gunatilake, S.K. (2015) 'Methods of removing heavy metals from industrial wastewater', *Journal of Multidisciplinary Engineering Science Studies*, 1(1), pp. 12–8.

Gupta, V. *et al.* (2015) 'Decolorization of mixture of dyes: A critical review', *Global Journal of Environmental Science and Management,* 1(1), pp. 71–94.

Harazono, K. and Nakamura, K. (2005) 'Decolorization of mixtures of different reactive textile dyes by the white rot basidiomycete *Phanerochaete sordida* and inhibitory effect of polyvinyl alcohol', *Chemosphere,* 59, pp. 63–68.

Hofrichter, M. *et al.* (2010) 'New and classic families of secreted fungal heme peroxidases', *Applied Microbiology and Biotechnology,* 87(3), pp. 871–897.

Hong, F.S. *et al.* (2005) 'Influences of nano-TiO$_2$ on the chloroplast ageing of spinach under light', *Biological Trace Elements Research,* 104(3), pp. 249–260.

Hsu, J.-H. and Lo, S.L. (1999) 'Chemical and spectroscopic analysis of organic matter transformations during composting of pig manure', *Environmental Pollution* 104, pp. 189–196.

International Centre for Soil and Contaminated Sites (2006) *Manual for biological remediation techniques.* 81, Germany: ICSS at the German Federal Environmental Agency, pp. 5–77.

Iturbe R. and López J. (2015) 'Bioremediation for a soil contaminated with hydrocarbons', *Journal of Petroleum & Environmental Biotechnology,* 6(208). doi:10.4172/2157- 7463.1000208.

Jin, R. *et al.* (2009) Bioaugmentation on decolorization of C.I. Direct Blue 71 by using genetically engineered strain *Escherichia coli* JM109 (pGEX-AZR)', *Journal of Hazardous* Materials, 163, p. 1123

Kapahi, M., and Sachdeva, S. (2019) 'Bioremediation options for heavy metal pollution', *Journal of Health and Pollution,* 9, pp. 191–203.

Karn, B., Kuiken, T., and Otto, M. (2009) 'Nanotechnology and in situ remediation: A review of the benefits and potential risks', *Environmental Health Perspectives,* 117, pp. 1823–1831

Keck, A. *et al.* (1997) 'Reduction of azo dyes by redox mediators originating in the naphthalene sulfonic acid degradation pathway of *Sphingomonas sp.* strain BN6. *Applied Environmental Microbiol*ogy, 63, pp. 3684–3690.

Kimura, N., Nishi, A., Goto, M., and Furukawa, K. (1997) 'Functional analyses of a variety of chimeric dioxygenases constructed from two biphenyl dioxygenases that are similar structurally but different functionally', *Journal of Bacteriology* 179, pp. 3936–3943.

Lellis, B., Fávaro-Polonio, C., Pamphile, J., and Polonio, J. (2019) 'Effects of textile dyes on health and the environment and bioremediation potential of living organisms', *Biotechnology Research and Innovation,* 3, pp. 275–290.

Machado, K.M. *et al.* (2006) 'Biodegradation of reactive textile dyes by *Basidiomycetous* fungi from Brazilian ecosystems', *Brazilian Journal of Microbiology,* 37, pp. 481–487.

Mahjoubi, M. *et al.* (2018) 'Microbial bioremediation of petroleum hydrocarbon– contaminated marine envir-onments. doi:10.5772/intechopen.72207.

Manu, B., and Chaudhari, S. (2002) 'Anaeobic decolourization of simulated textile wastewater containing azo dyes', *Bioresource Technology,* 82, pp. 225–231.

Mauter, M.S., and Elimelech, M. (2008) 'Environmental applications of carbon-based nanomaterials', *Environmental Sciences & Technology,* 42, pp. 5843–5859.

Mester, T., and Tien, M. (2000) 'Oxidation mechanism of ligninolytic enzymes involved in the degradation of environmental pollutants', *International Biodeterioration and Biodegradation,* 46, pp. 51–59.

Mishra, B. *et al.* (2020) 'Microbial approaches for remediation of 672 pollutants: Innovations, future outlook, and challenges', *Energy Environment,* 32, pp. 1–30.

Mishra, M. (2021) *Microbe mediated remediation of environmental contaminants, environmental factors affecting the bioremediation potential of microbes,* Woodhead Publishing Series in Food Science, Technology and Nutrition, pp. 47–58.

Nachiyar, C.V., and Rajkumar, G.S. (2005) 'Purification and characterization of an oxygen insensitive azoreductase from *Pseudomonas aeruginosa*', *Enzyme and Microbial Technology,* 36, pp. 503–509.

Naik, M.G., and Duraphe, M.D. (2012) 'Parameters affecting bioremediation', *International Journal of Life Science and Pharma Research,* 2(3), pp. 77–80.

Pandey, A.K., and Dubey, V. (2012) Biodegradation of azo dye Reactive Red BL by *Alcaligenes sp.* AA09. *International Journal of Engineering Sciences.* 1.

Paul, D., Pandey, G., Pandey, J., and Jain, R.K. (2005) 'Accessing microbial diversity for bioremediation and environmental restoration', *Trends in Biotechnology,* 23, pp. 135–142.

Paul, R.A., and Gayle, S.K. (1993) Technical and economic analyses in the development of bioremedi-ation processes', *The Journal of Environmental Cleanup Costs, Technologies and Techniques,* 4(1), pp. 115–128.

Pillai, H.P.J.S., Girish, K., and Agsar, D. (2014) 'Isolation, characterization and screening of actinomycetes from textile industry effluent for dye degradation', *International Journal of Current Microbiology & Applied Science,* 3(11), pp. 105–115.

Prasad, R., and Aranda, E. (2018) 'Nanotechnology in the life sciences: Approaches in Bioremediation (The New Era of Environmental Microbiology and Nanobiotechnology)' in Ram Prasad and Elisabet Aranda (ed.) *Environmental Nanotechnology: Applications of Nanoparticles for Bioremediation,* Switzerland: Springer, Chama, pp. 301–315.

Przystas, W., Godlewska, E.Z., and Grabinska, E.S. (2013) 'Effectiveness of dyes removal by mixed fungal cultures and toxicity of their metabolites', *Water Air Soil Pollution,* 224, pp. 1–9.

Raja, M.M.M., Raja, A., Salique, S.M., and Gajalakshmi, P. (2016) 'Studies on effect of marine actinomycetes on amido black (azo dye) decolorization', *Journal of Chemical and Pharmaceutical Research,* 8(8), pp. 640–644.

Ramya, M., Iyappan, S., Manju, A., and Jiffe, J.S. (2010) 'Biodegradation and decolorization of Acid Red by *Acinetobacter radioresistens',* *Applied Environmental Microbiology,* 70, pp. 1–105.

Rather, L.J., Akhter, S., and Hassan, Q.P. (2018) 'Bioremediation: Green and sustainable technology for textile effluent treatment', *Sustainable Innovations in Textile Chemistry and Dyes,* pp. 75–91.

Robinson, T., Chandran, B., and Nigam, P. (2001) 'Studies on the decolorization of an artificial textile effluent by white-rot fungi in N-rich and N-limited media', *Applied Microbiology and Biotechnology,* 57, pp. 810–813.

Ron, E. Z., and Rosenberg, E. (2014) 'Enhanced bioremediation of oil spills in the sea', *Current Opinions in Biotechnology,* 27, pp. 191–194.

Russ, R., Rau, J. and Stolz, A., (2000) 'The function of cytoplasmic flavin reductases in the reduction of azo dyes by bacteria', *Applied and Environmental Microbiology,* 66(4), pp. 1429–1434.

Sakaki, T. *et al.* (2002) 'Biodegradation of polychlorinated dibenzo-p-dioxins by recombinant yeast expressing rat CYP1A subfamily', *Archives of Biochemistry and Biophysics,* 401, pp. 91–98.

Sandhya, S., Sarayu, K., Uma, B., and Swaminathan, K. (2008) 'Decolorizing kinetics of a recombinant *Escherichia coli* SS125 strain harboring azoreductase gene from *Bacillus latrosporus* RRK1', *Bioresource Technology,* 99 (7), pp. 2187–2191.

Saratale, R.G. *et al.* (2009) 'Enhanced decolorization and biodegradation of textile azo dye Scarlet R by using developed microbial consortium-GR', *Bioresource Technology,* 100, pp. 2493–2500.

Sarkar, S., Martinez, A.T, and Martinez, M.J. (1997) Biochemical and molecular characterization of a manganese peroxidase isoenzyme from *Pleurotus ostreatus',* *Biochimica et Biophysica Acta,* 1339, pp. 23–30.

Saroj, S. *et al.* (2014) 'Biodegradation of azo dyes acid red 183, direct blue 15 and direct red 75 by the isolate *Penicillium oxalicum* SAR-3', *Chemosphere* 107, pp. 240–248.

Sharma, N. *et al.* (2017) 'Bio nanotechnological intervention: A sustainable alternative to treat dye bearing waste waters', *Indian Journal of Pharmaceutical and Biological Research,* 5, pp. 17–24.

Singh, S., Mishra, S., and Jauhari, N. (2014) 'Degradation of anthroquinone dyes stimulated by fungi', *Environmental Science and Engineering,* pp. 333–356.

Srinivasan, A. and Viraraghavan, T., 2010. 'Decolorization of dye wastewaters by biosorbents: A review', *Journal of environmental management,* 91(10), pp.1915–1929.

Sunkar S., and Nachiyar V. (2013) 'Endophytic fungi mediated extracellular silver nanoparticles as effective antibacterial agents', *International Journal of Pharmacy and Pharmceutical Sciences,* 5, pp. 95–100.

Telke, A., Kalyani, D., Jadhav, J., and Govindwar, S. (2008) 'Kinetics and mechanism of Reactive Red 141 degradation by a bacterial isolate *Rhizobium radiobacter* MTCC 8161', *Acta Chimica Slovenica,* 55, p. 320.

Timmis, K.N., and Pieper, D.H. (1999) 'Bacteria designed for bioremediation. *Trends in Biotechnology,* 17, pp. 201–204.

Tripathi, S. *et al.* (2018) 'Nano-Bioremediation: Nanotechnology and Bioremediation', In Information Resources Management Association (ed.) *Research Anthology on Emerging Techniques in Environmental Remediation,* IGI Global, pp. 202–219.

Undre, S.B, Singh, M., and Kale, R.K. (2013) 'Interaction behaviour of trimesoyl chloride derived 1st tier dendrimers determined with structural and physicochemical properties required for drug designing', *Journal of Molecular Liquids,* 182, pp. 106–120.

Uqab, B., Mudasir, S., and Nazir, R. (2016) 'Review on bioremediation of pesticides', *Journal of Bioremediation & Biodegradation,* 7, p. 343.

Vaccari, D.A., Strom, P.F., Alleman, and J.E. (ed.) (2005). *Environmental Biology for Engineers and Scientists.* Hoboken, New Jersey: John Wiley & Sons.

Van, H. *et al.* (2004) 'Recent advances in petroleum microbiology', *Microbiology and Molecular Biology Reviews,* 67, pp. 503–549.

Vikrant, K., Kumar, V., Kim, K.-H., and Kukkar, D. (2017) 'Metal organic frameworks (MOFs): potential and challenges for capture and abatement of ammonia', *Journal of Materials Chemistry A,* 5, pp. 22877–22896.

Wesenberg, D., Kyriakides, I., and Agathos, S.N. (2003) 'White-rot fungi and their enzymes for the treatment of industrial dye effluents', *Biotechnology Advances,* 22, pp. 161–187.

Wijannarong, S. *et al.* (2013) 'Removal of reactive dyes from textile dyeing industrial effluent by ozonation process', *APCBEE Procedia,* 5, pp. 279–282.

Wolicka, D., Suszek, A., Borkowski, A., and Bielecka, A. (2009) 'Application of aerobic microorganisms in bioremediation in situ of soil contaminated by petroleum products', *Bioresource Technology,* 100, pp. 3221–3227.

Xiao, X. *et al.* (2018) 'A simple method for assaying anaerobic biodegradation of dyes', *Bioresource Technology,* 251, pp. 204–209.

Xu, M., Guo, J., and Sun, G. (2007) 'Biodegradation of textile azo dye by *Shewanella decoloration* S12 under microaerophilic conditions', *Applied Microbiology & Biotechnology,* 76, p. 719.

Yesiladalı, S.K., Pekin, G., Bermek, H., Arslan-Alaton, I., Orhon, D. and Tamerler, C. (2005) 'Bioremediation of textile azo dyes by Trichophyton rubrum LSK-27', *World Journal of Microbiology and Biotechnology,* 22(10), pp.1027–1031.

Yesiladali, S.K., Pekin, G., and Bermek, H. (2006) 'Bioremediation of textile azo dyes by *Trichophyton rubrum* LSK-27', *World Journal of Microbiology & Biotechnology,* 22, pp. 1027–1031.

Yoo, E.S., Libra, J., and Adrian, L. (2000) 'Mechanism of decolorization of azo dyes in anaerobic mixed culture', *Journal of Environmental Engineering,* 127, pp. 844.

Yu, J.G. et al. (2014) 'Aqueous adsorption and removal of organic contaminants by carbon nanotubes', *Science for the Total Environment,* 483, pp. 241–251.

Yuniati, M. (2018) 'Bioremediation of petroleum-contaminated soil: A Review', *IOP Conference Series: Earth and Environmental Science.* 118. doi: 012063. 10.1088/1755-1315/118/1/012063.

Zapanta, L.S., and Tien, M. (1997) 'The roles of veratryl alcohol and oxalate in fungal lignin degradation', *Journal of Biotechnology,* 53, pp. 93–102.

Zhao, L.B., Liu, X.X., Zhang, M., Liu, Z.F., Wu, D.Y. and Tian, Z.Q. (2016) 'Surface plasmon catalytic aerobic oxidation of aromatic amines in metal/molecule/metal junctions', *The Journal of Physical Chemistry C,* 120(2), pp. 944–955.

Zollinger, H. (1991) *Colour chemistry: Synthesis, properties and applications of organic dyes and pigments,* Zollinger, H. (ed.) 5th edn, Weinheim: VCH Publishers, p. 187.

16 Phytoremediation

A Novel and Promising Approach for the Clean-Up of Heavy Metal-Contaminated Soils Associated with Microbes

Hiren K. Patel[1]*, Priyanka H. Jokhakar[2],
Rishee K. Kalaria[3], Divyesh K. Vasava[4], and
Rutu R. Kachhadiya[5]

[1]School of Science, P. P. Savani University, Surat (Gujarat) 394125, India

[2]Department of Biosciences, Veer Narmad South Gujarat University, Surat (Gujarat) 395007, India

[3]ASPEE Shakilam Biotechnology Institute, Navsari Agricultural University, Surat (Gujarat) 395007, India

[4]College of Agriculture, Junagadh Agricultural University, Junagadh (Gujarat) 36200, India

[5]School of Science, P. P. Savani University, Surat (Gujarat) 394125, India

*Corresponding author: drhkpatel1@gmail.com

CONTENTS

DOI: 10.1201/9781003130932-16

16.1   INTRODUCTION

The natural environment consists of biotic and abiotic factors. The balance between abiotic and biotic factors ensures that nutrient recycling, food, and all the basic requirements for life are met. However, interruptions cause damage and imbalances in the environment. The ecosystem is composed of all the elements required by living beings for their growth, development, reproduction, protection, survival, and adaptation. Rapid development, to fulfill the needs of an increased human population, had led to industrialization, urbanization, and deforestation. These anthropogenic activities, the discharge of waste material, agricultural and pharmaceutical wastes, and effluent released from industries without treatment are sources of compounds in the environment that are natural or synthetic. These compounds might be xenobiotics, calcitrant, recalcitrant, or nonrecalcitrant. These wastes contain pesticides, dyes, heavy metals, chemically synthesized fertilizers, radioactive materials, antibiotics, and metals. Currently, heavy metals are at high concentration in the environment because they are nondegradable. Some heavy metals are needed by plants for their growth and metabolism. In addition, heavy metals have adverse effects on humans (Figure 16.1). For example, zinc (Zn) is a cofactor for many enzymes, such as alkaline phosphatase, DNA, and RNA polymerases (Wani et al., 2017).

Heavy metals that have a vital role in cell functions at low concentrations are toxic at their higher concentration. Some heavy metals are toxic at low concentrations when transformed from one form to another. Heavy metals damage living organisms in aquatic and terrestrial habitats. The production

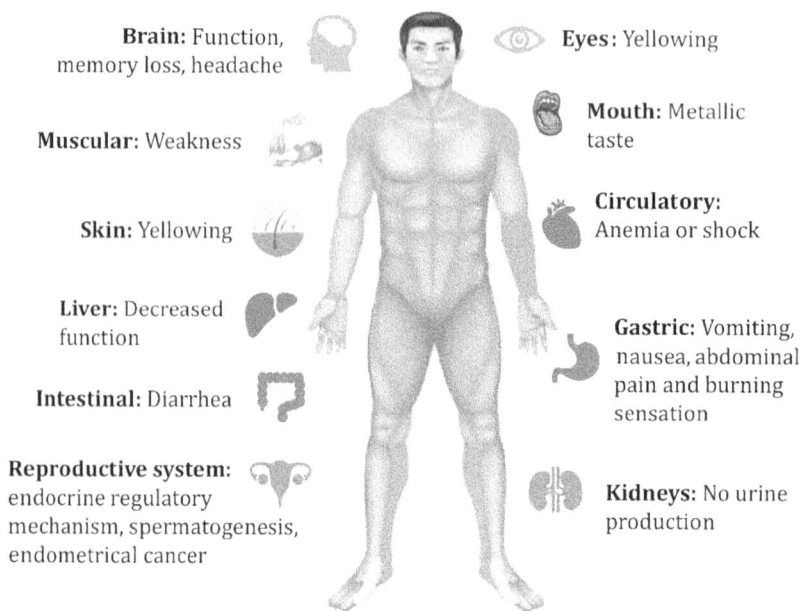

Brain: Function, memory loss, headache

Eyes: Yellowing

Mouth: Metallic taste

Muscular: Weakness

Circulatory: Anemia or shock

Skin: Yellowing

Liver: Decreased function

Gastric: Vomiting, nausea, abdominal pain and burning sensation

Intestinal: Diarrhea

Reproductive system: endocrine regulatory mechanism, spermatogenesis, endometrical cancer

Kidneys: No urine production

FIGURE 16.1   Effect of heavy metals on the human body.

of reactive oxygen (O_2) species, an imbalance in respiration, and damage to photosynthesis are negative effects of heavy metals toxicity in plants (Ansari et al., 2020). From plants, heavy metals enter humans and animals via bioaccumulation and biomagnifications in fish, and other organisms and cause toxicity. For example, mercury (Hg) in humans accumulates in the kidneys and interferes with the regulation of the central nervous system. Chromium(VI) (Cr) is classified as a carcinogen for humans in group I (heavy metal carcinogens) (Muthusaravanan et al., 2018).

Electroplating, mining, and smelting are sources of Zn, copper (Cu) and Cr metals. Phosphate (P) containing fertilizers, geological sources, and sewage are sources of cadmium (Cd) (Lone et al., 2008). Heavy metals released by various mechanisms are transmitted to the soil and water. Heavy metals removal from the soil and water is required because they cause damage and pollution. Various mechanisms can be used for heavy metal remediation. Landfill and soil digging or covering are used for heavy metal remediation. However, could leak into the groundwater or be transmitted to other aquatic sites. Chemical methods involve the use of chemicals, which are not suitable for long term use, because chemicals alter the normal microbial flora of soil and water, affect the pH, elements, and osmolarity of ecosystems (Parmar & Singh, 2015).

Plants are the preferred option for heavy metal remediation because it is based on green technology. There are many advantages of using plants, crops, and aquatic macrophytes for heavy metals detoxification. Planting selected crops at contaminates sites provides *in situ* applications, which is an advantage. Since heavy metal remediation is carried out with solar energy, there is no need to provide electricity. In addition, planted crops help to remove carbon dioxide (CO_2) and heavy metals. Detoxification of sites contaminated with heavy metals via plant mediation does not require any hazardous chemicals or acids. Microbial communities are diverged and functionally active, which means that they can adapt under stress conditions. Microbes associated with plants helps in the phytoremediation of heavy metals. These microbes provide heavy metals resistance and detoxification systems to the plant. Therefore, the plant can accumulate stabilized heavy metals at higher concentrations. In addition, some microbes are plant growth-promoting rhizobacteria (PGPR), which induce plant growth and could be applied as biofertilizers. Microbes associated with plants might be endophytic or root-associated. There are some limitations with heavy metals phytoremediation. For example, plant extracted or accumulated heavy metals enter the food chain and cause toxicity. This results in nonedible plant transformation. Phytovolatilization of heavy metals might create air pollution. Phytoremediation provides many advantages over other methods, and it helps to reduce soil erosion. Phytoremediation is less expensive and is commonly used to manage soil and water supplies that are contaminated with heavy metals. (Muthusaravanan et al., 2018).

This chapter will discuss the mechanisms involved in the remediation and removal of heavy metals by plants and plants associated with microbes. An understanding of microbial diversity for heavy metal phytoremediation can be achieved using metagenomics and transcriptomic studies. In addition, this chapter will discuss the role of genetic modification and its effect on phytoremediation. The analysis of different aquatic and terrestrial plants and microbes, studies of hyperaccumulators and nonhyperaccumulators could help to select plants or microbial consortia in heavy metal detoxification.

16.2 MECHANISMS FOR PHYTOREMEDIATION

Plants use various mechanisms to accumulate, absorb, extract, stabilize, and degrade heavy metals that are present in soil and water. Plants have various and diverse mechanisms for the absorption of nutrients and water from the soil. These mechanisms are utilized by plants to remove, degrade, transform, and remediate xenobiotics and heavy metals. The roots of plants absorb and transport heavy metals to the shoots via proteins of plasma membranes in the cells. Under normal environmental conditions, plants do not uptake compounds that they do not need for metabolism. However, some plants absorb heavy metals and avoid or bypass their toxic effects by accumulating them into vacuoles (Tangahu et al., 2011). Some plants can detoxify the harmful effects of the absorbed

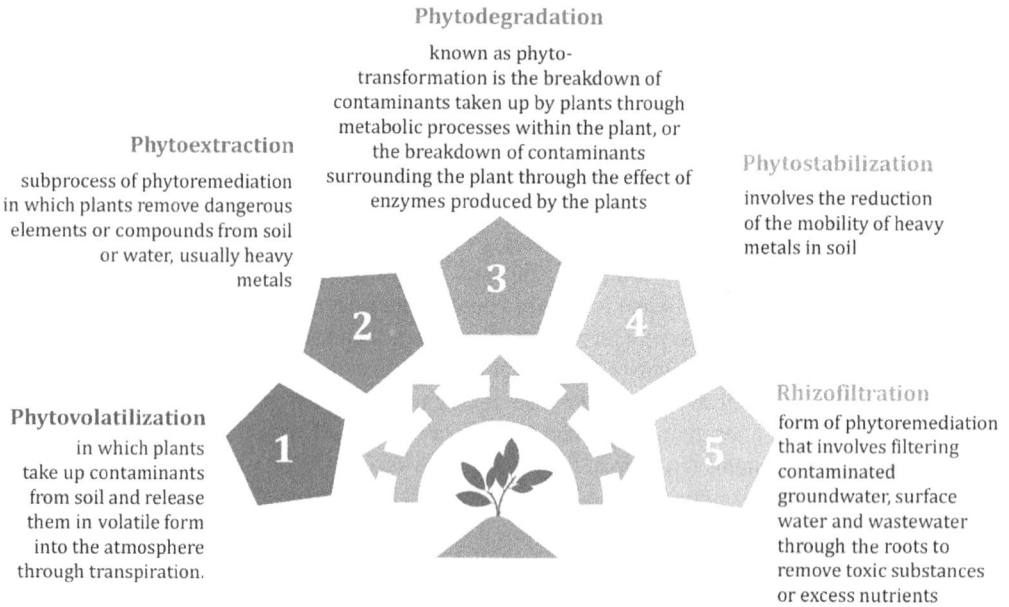

Phytodegradation

known as phyto-transformation is the breakdown of contaminants taken up by plants through metabolic processes within the plant, or the breakdown of contaminants surrounding the plant through the effect of enzymes produced by the plants

Phytoextraction

subprocess of phytoremediation in which plants remove dangerous elements or compounds from soil or water, usually heavy metals

Phytostabilization

involves the reduction of the mobility of heavy metals in soil

Phytovolatilization

in which plants take up contaminants from soil and release them in volatile form into the atmosphere through transpiration.

Rhizofiltration

form of phytoremediation that involves filtering contaminated groundwater, surface water and wastewater through the roots to remove toxic substances or excess nutrients

FIGURE 16.2 Different strategies used by plants for the bioremediation of heavy metals.

and accumulated heavy metals by restricting them to their roots, where they cannot translocate to the shoots. These plants are known as excluders; however, some more adapted plants species that can tolerate and detoxify heavy metals at a higher concentration above their root section and can remediate them are known as hyperaccumulators. Most of the plants of the *Brassicaceae*, *Rubiaceae*, *Asteraceae* and *Proteaceae* families are hyperaccumulators (Suman et al., 2018). One of the driving forces for the absorption of heavy metals into the roots and translocation to the shoots is the transpiration of water from the surface of leaves. Plants achieve heavy metal remediation and degradation through: (1) phytostabilization; (2) phytoextraction or phytoaccumulation; (3) phytofiltration or rhizofiltration; (4) phytodegradation or phytotransformation; and (5) phytovolatilization (Tangahu et al., 2011) (Figure 16.2).

16.2.1 PHYTOEXTRACTION OR PHYTOACCUMULATION

Phytoextraction is the removal from the soil of heavy metals for accumulation and transport in the harvesting sections of plants situated above ground level to eliminate the concentration of heavy metals in the soil. Phytoextraction can be assisted via the application of chemicals or the natural method, which is plant assisted. Chemical mediated phytoextraction has disadvantages of low heavy metal tolerance and the potential of leaching into the soil. Plant mediated phytoextraction has advantages, such as the successful transfer of heavy metals from roots to shoots with environmental safety and a high level of extraction from soil (Nascimento et al., 2006). Phytoextraction involves the cultivation of plants that have heavy metal tolerance at the contaminated site, which can reduce the concentration of heavy metals. Phytoextraction depends on several factors, such as the plant species used for the hyperaccumulation of heavy metals, biomass production in plants, various biotic and abiotic environmental parameters, the bioavailability of heavy metals, the effect of heavy metals on photosynthetic activity, and the nature of the soil for remediation (Nascimento et al., 2006; Suman et al., 2018). The ratio of heavy metal concentration in the plant to the concentration in the soil affects phytoextraction economically. In total, 10 t/ha of plant biomass is required for the effective phytoextraction of heavy metals from soil. In non-hyperaccumulator species of plants, higher concentrations of heavy metals harm photosynthetic activity, which results in a low plant biomass yield (Suman et al., 2018). Some hyperaccumulators, such as barley and rice have a natural ability for the accumulation of heavy metals at higher concentrations. In addition, hyperaccumulation in Indian mustard can be increased (Nascimento et al., 2006).

Phytoextraction can be chemically enhanced using organic acids and chelators. Some compounds, such as low molecular weight containing organic acids are generated and released in the rhizosphere area to stimulate heavy metal uptake by complex formation. The release of organic acids creates an acidic environment in the soil and alters the microbial community for heavy metals removal. Metal reducing chelators are mainly specific and are released by plants to solubilize trace amounts of metals from the soil. Citrate and histidine are chelators for nickel (Ni) in *Thlaspi arvense*. Another chelator is ethylene diamine tetra acetate (EDTA), which can chelate several heavy metals. EDTA, for example, helps lead (Pb) hyperaccumulation in the pea shoots. Zn and Cd are hyperaccumulated in *Brassica juncea* with EDTA. However, the use of EDTA for heavy metal extraction has serious problems with environmental toxicity. Therefore, some artificial and biodegradable chelators, such as methylglycine diacetate and ethylenediamine dissuccinate could be used as an alternative to avoid environmental toxicity. At the cellular level, chelation in the vacuole and cytoplasm is an important mechanism for the hyperaccumulators of heavy metals. A Ni and Zn tolerant and hyperaccumulator vacuole was isolated from *Thlaspi goesingense* and *Thlaspi caerulescens*, respectively. Phytochelatins and metallothioneins play a major role in the chelation of heavy metals (Nascimento et al., 2006).

Hyperaccumulators of heavy metals for phytoextraction belong to; (1) natural hyperaccumulators; (2) high biomass producing nonhyperaccumulators; and (3) genetically engineered and modified hyperaccumulators. All three methods have some advantages and limitations for the phytoextraction of heavy metals. Natural hyperaccumulators have the advantages of a higher phytoextraction rate and can avoid invasive species. However, it has the limitations of low biomass production and low specificity toward heavy metals. High biomass producing but nonaccumulators have the advantages of survival under different climatic conditions, low specificity for heavy metals, and high biomass production; however, they have a low level of extraction. A combination of more than one phenotypic trait with host plant selection and tissue-specific expression with gene modification are the main advantages of genetically engineered hyperaccumulators. Some environmental risks are associated with genetically modified plants and regulatory policies should be followed to avoid these risks. *Salix spp.* produce high biomass in a short time and via deep root systems are high biomass producing nonaccumulators. In this system, harvesting leaves that accumulate heavy metals is easy. For example, the recovery of Cd and Zn is simple, because they accumulate in the leaves of accumulators. Vegetation growth in the target polluted site helps to prevent the spread and transmission of heavy metals into the surrounding soil (Suman et al., 2018). *Alyssum heldreichii* and *Brassica juncea* can accumulate Ni at 1,441 mg/kg and 3,916 mg/kg, respectively in the aboveground section of plants. *Eleocharis acicularis* can accumulate ≤1,470 mg/kg in the shoots, which was extracted from water (Muthusaravanan et al., 2018).

Low phytoextraction and low biomass production have led to the use of genetically engineered plants for heavy metals extraction. Cd tolerance can be induced in *B. juncea* using transgenesis of glutathione synthetase and γ- glutamylcysteine synthetase (Suman et al., 2018). The transgenesis of 1-aminocyclo-propane-1-carboxylic acid deaminase from *Enterobacter cloacae* into *Lycopersicon aesculentum* induced the accumulation of Ni, Pb, Zn, and cobalt (Co) (Zhang et al., 2008). Zn accumulation can be induced in *Nicotiana tabacum* using *Neurospora crassa* origin *tnz1* transporter transgenesis (Dixit et al., 2010). Transgenesis of cysteine synthase (GS) from *Spinacia oleracea* to *N. tabacum* increased the accumulation efficiency of selenium (Se), Ni, and Cd in the shoots, which is easy when harvesting selenium. Similarly, transgenesis of *MRP7*, an ATP binding cassette transporter into *N. tabacum* from *Arabidopsis thaliana* increased Cd tolerance (Suman et al., 2018).

16.2.2 PHYTOSTABILIZATION

Phytostabilization involves the absorption and immobilization of heavy metals that are present in the that surrounds a plant or at contaminated sites to stabilize them. Absorbed and immobilized heavy metals are mainly restricted to the rhizospheric area of plants. The stabilization of heavy metals is mainly carried out using root hairs to prevent them from spreading into the surrounding area. The phytostabilization of heavy metals is accompanied by precipitation, absorption, and reduction in metals valency. In addition, phytostabilization limits the movement of heavy metals from the roots to shoots. Phytostabilization binds heavy metals to the rhizosphere, and therefore, help to prevent their transmission by weathering. The removal of heavy metals by phytostabilization is most suitable for *in situ* applications for soil reuse and it decreases the concentration of heavy metals that enters the food chain (Muthusaravanan et al., 2018; Ojuederie & Babalola, 2017). Modifications in pH and organic matter concentration induce the phytostabilization of heavy metals (Ojuederie & Babalola, 2017). *Wolffia globosa*, *Azolla pinnata,* and *Pteris vittata* were used to stabilize Cr. Some plants can phytostabilize multiple heavy metals, which has a range of remediation applications. For example, *Lemna minor* L. can phytostabilize Cd, selenium (Se), and Cu. Similarly, *Solanum tuberosum* L. can remediate arsenic (As), silver (Ag), antimony (Sb), and Cr. *N. tabacum* L., *Desmostachya bipinnata,* and *Dichanthium annulatum* can remediate Cd via phytostabilization. Phytostabilization of Zn is carried out by *Lemna gibba* L., *Mentha aquatic*, *Ludwigina palustris,* and *Scirpus mucronatus*. *Solenum nigrum* L., *S. oleracea* L., and *Vetiveria zizanioides* can successfully remediate Pb via phytoremediation (Muthusaravanan et al., 2018).

Heavy metals accumulation in plant roots or shoots can prevent them from spreading to parts of plants that are located above ground level and it indicates the phytostabilization ability of a particular plant for respective heavy metals. Phytostabilization capacity or efficiency for heavy metals can be expressed as (Parmar & Singh, 2015):

$$\text{Bioconcentraion factor (BCF)} = \frac{[\text{Concentration of heavy metals}]_{root}}{[\text{Concentration of heavy metals}]_{soil}}$$

$$\text{Translocation factor (TF)} = \frac{[\text{Concentration of heavy metals}]_{shoot}}{[\text{Concentration of heavy metals}]_{root}}$$

Microorganisms that are associated with plants in the rhizosphere help in phytostabilization and heavy metals tolerance development. For example, Pb can be phytostabilized using *Gliricidia sepium* inoculated in maize. Microorganisms associated with plant roots carry out different mechanisms to chelate and reduce heavy metals. The production of organic acids, extracellular polymers, and interactions with heavy metals via carboxyl or amine groups are some phytostabilization mechanisms carried out by microbes. A permeability barrier and the modification of heavy metals via chemicals are mechanisms that microbes use to help to build up resistance to heavy metal toxicity (Muthusaravanan et al., 2018). Radziemska et al. (2017) studied the phytostabilization of a Cu contaminated site by *Festuca rubra* L. Cu had accumulated in the roots, which indicate that the movement of Cu to other parts of the plant had been restricted. Therefore, phytostabilization prevented heavy metals toxicity in the plant biomass. Zgorelec et al. (2020) reported the phytostabilization of Cd and Hg using bioenergy producing *Miscanthus* X *giganteus*. *M.* X *giganteus* can accumulate ≤293.8 µg/year of Cd. In addition, this plant can accumulate Hg at 4.7 µg/year. This study concluded that a lower concentration of Hg above ground limited heavy metal toxicity on the biomass of the plant. The retention of heavy metals in the soil that surrounds the roots is the main disadvantage of phytostabilization. However, it has the has advantage of restricting heavy metals in the surrounding water and soil. (Muthusaravanan et al., 2018).

16.2.3 PHYTOVOLATILIZATION

Phytovolatilization is the extraction of heavy metals by plants that can be transformed into a volatile form and released as a vapor from the stomata, to detoxify soil (Ojuederie & Babalola, 2017; Ahmadpour et al., 2012). Plants and associated microbes can metabolically convert Se into the volatile form dimethyl selenide and Hg into mercuric oxide (Muthusaravanan et al., 2018). Phytovolatilization has the disadvantage that it cannot eliminate heavy metals, because the heavy metals are converted from one form to another, and can spread from the soil to air, which can precipitate back in the rain and be deposited in the soil. The converted volatile form of heavy metals might be more toxic or toxic than the original form. For example, Selenomethionine is converted into dimethylselenide via methylation, which is less toxic (da Conceiçao Gomes et al., 2016).

Gene modification can enhance phytovolatilization to help decontaminate polluted soil by the elimination of volatile substances. For example, modified *merA* (mercuric reductase) helps to develop mercury resistance in transgenic *A. thaliana* (Kumar et al., 2017). Genes, such as mercuric reductase and organomercurial lyase of microbial origin were modified for *A. thaliana* L. and *N. tabacum* L. to eliminate Hg to conserve a polluted site (Muthusaravanan et al., 2018). A high concentration of salts and boron limit phytovolatilization, because their presence can kill plants (Kumar et al., 2017).

Some plants are involved in the phytovolatilization of heavy metals. For example, *Athyrium wardii* and *Thysanolaena maxima* phytovolatilized Pb. Cd can be phytovolatilized using *Acanthus ilicifolius* L., *Lupinus uncinatus* Schld1, and *Silphium perfoliatum*. Phytovolatilization of Ni via *Africallagma elongatum* and *Picea abies* has been reported. Some plants can efficiently volatilize multiple heavy metals, for example, *Typha latifolia* can remove Zn, Mn, Co, Cd, Cr, Ni, and As. *Sorghum bicolor* L. efficiently eliminated Cd and Zn from a polluted site. Phytovolatilization plants could be routinely rotated; therefore, more biomass could be generated and used as livestock feed. (Muthusaravanan et al., 2018).

16.2.4 PHYTODEGRADATION OR PHYTOTRANSFORMATION

Phytodegradation is the uptake and breakdown or conversion of toxic pollutants that are present in soil or water into a nontoxic form. In addition, phytodegradation is known as phytotransformation (Muthusaravanan et al., 2018; Ahmadpour et al., 2012). Heavy metals that are taken up by plants can be degraded using metabolic enzymes present in the plant and their associated microorganisms. Degraded pollutants are converted into simpler forms, which are later utilized as nutrients by the plant to produce biomass. Some environmental and internal factors, such as pH, temperature, nutrient level, concentration of pollutants, and heavy metals tolerance ability affect the rate of phytodegradation. Nitroreductases, which are flavoproteins and dehalogenases, are enzymes that are present in plant cells and are responsible for the removal of halo groups and organic degradation (Ojuederie & Babalola, 2017). Other enzymes involved in phytodegradation are laccases, which are Cu based oxidation carrying enzymes, and peroxidase that carry out redox reactions (Muthusaravanan et al., 2018). Phytodegradation is often limited to organic contaminants because heavy metals are mainly resistant to degradation. Therefore, heavy metals removal is referred to as phytotransformation because they are converted into another more stable and less toxic form by plants and their associated microbes. For example, mobile Cr^{6+} is converted into a less mobile form of Cr^{3+} (da Conceiçao Gomes et al., 2016). *Pseudomonas maltophilia* helps in the transformation of heavy metals. *Aspergillus niger* has been screened for the transformation of an insoluble form of Zn into a soluble form (Parmar & Singh, 2015). *Canna glauca* L., *Cyperus papyrus* L., and *Arundo donax* L photodegrade As. Methylated Hg was photodegraded using *Liriodendron tulipifera*. Hg polluted mine could be cleared using *B. juncea* and *Lupinus sp.* (Muthusaravanan et al., 2018).

16.2.5 Phytofiltration or Rhizofiltration

Rhizofiltration is the absorbance or removal of pollutants present in wastewater or contaminated water at surface level (i.e., heavy metals) by the roots of plants. Absorbed heavy metals accumulate into the biomass of the roots. Rhizofiltration is known as phytofiltration (Kumar et al., 2017; Parmar & Singh, 2015). When heavy metals filtration is carried out using seedlings it is known as blastofiltration. Caulofiltration uses the shoots heavy metals filtration (Da Conceiçao Gomes et al., 2016). Rhizofiltration of heavy metals was less efficient in aquatic plants than in terrestrial plants. Terrestrial hyperaccumulators have a larger root surface area than aquatic hyperaccumulators (Parmar & Singh, 2015). Inoculation of PGPR at heavy metal polluted regions can enhance rhizofiltration. Hyperaccumulators are mostly suited for the phytofiltration of highly concentrated heavy metals (Ojuederie & Babalola, 2017). From the aquatic plants, *Eichhornia crassipes* can efficiently eliminate heavy metals from polluted water. Other aquatic plants, such as *Callitriche stagnalis* S. (star-wort) in pond water and *Potamogeton natans* L. could rhizofilter uranium (Ur) from a water system effectively. In addition, As could be phytofiltered using *Micranthemum umbrosum*. *Helianthus annus* L., *S. oleracea* and *N. tabacum* could remove Pb (Parmar & Singh, 2015). Phytofiltration using seedlings has the advantage of a larger surface area after germination (da Conceiçao Gomes et al., 2016).

16.3 ROLE OF DIFFERENT PLANTS IN HEAVY METAL PHYTOREMEDIATION

In aquatic ecosystems, stress on biotic components increases daily due to the biomagnification of various organic or inorganic pollutants. These stressed and uncontrolled environmental conditions could be remediated and could be balanced by macrophytes that live in water. Based on the habitat of plants, they are aquatic (water) and terrestrial (soil). Some aquatic plants have special applications in the remediation of pollutants and xenobiotics because they have a higher bioaccumulation ability compared with other plants. Aquatic plants are further divided into; (1) submerged; and (2) free-floating in water. *B. juncea* could rhizofilter Cd, Pb, Zn, and As from groundwater. In a hydroponics experiment, *Utricularia gibba* was efficient at the removal of Cr from a 50 μm Cr(VI) solution. Similarly, *Myriophyllum intermedium* accumulates more N, from water and soil than *S. mucronatus* and *Rotala rotundifolia*. *Ceratophyllum demerusum*, *Eichornia crassipes*, *Phragmites australis,* and *T. domingenesis* are native water inhabitants, which were efficient at the removal of heavy metals, such as Cd, Cu, Co, Zn, and Ni from El Temsah Lake, Al Isma'iliyahm Egypt. Because plants have various mechanisms for the removal of heavy metals, aquatic plants are less efficient at rhizofiltration than terrestrial plants, because aquatic plants have weaker roots and less surface area (Wani et al., 2017).

Cr and Hg were removed from contaminated water using *Azolla caroliniana*, which is a water fern. According to the study, Cr and Cd are accumulated by *W. globosa*, which can extract As ≤400 mg/kg. Some antioxidant compounds and thiols are responsible for a plant's resistance to heavy metals. The pigment anthocyanin provides plants with the capacity to remove heavy metal toxicity. Duckweeds recover quickly from a high concentration of heavy metals. *Crinum* and *Typha sp.,* which are mono-cotyledonous flowering plants remediated a higher quantity of heavy metals. *Crysopogon* and *Alternanthera sp.* can remove heavy metals in an aquatic ecosystem (Ansari et al., 2020).

L. minor which belongs to *Lemnaceae* and Lilipsida can accumulate Ur, As, Ni, Cd, and manganese (Mn) from polluted water. Because phytofiltration is carried out by *Spirodela polyrhiza*, commonly known as Giant duck weed. A polyculture of *Lemma aequinoctialis, Landoltia punctata,* and *S. polyrhiza* were effective in heavy metal cleaning in a contaminated site. Using *L. minor* and *S. polyrhiza*, 40%–78% and 52%–75% of Cd was removed. *Potamogeton pectinatus,* which is found in rhizomes in marshes, can remediate Mn, Cu, Zn, and Pb (Soni & Kaur, 2015). *E. crassipes* could remediate Pb and Cd. *Potamogeton lucens, Salvinia herzegoi,* and *E. crassipes* are biosorbents for the remediation of Zn, Ni, Cu, and Cd (Ali et al., 2020).

Some medicinal and aromatic plants can remediate heavy metals. Angelova, (2012) studied heavy metal remediation using selected plants. *Ocimum basilicum* and *Marticaria chamomile* L. accumulate 28.2 mg/kg and 137.3 mg/kg of Pb, respectively; 7.8 mg/kg and 5.5 mg/kg of Cd, respectively; and 543.7 mg/kg and 279.0 mg/kg of Zn, respectively. Heavy metal extraction and remediation capacity can be calculated using the bioconcentration factor (BCF) and TF. In addition, in this study, *O. basilicum* L. had a BCR of 1.140 for Pb and 1.604 for Zn. From selected plants, *Centanthus ruber* L. and *Datura stramonium* L. had a TF of 41.37, 28.54, 11.95, and 29.77, 17.25, 17.85 for Pb, Cd, and Zn. The phytoremediation ability of plants for heavy metals can be calculated as follows (Angelova, 2012):

$$\text{Bioconcentraion factor (BCF)} = \frac{[\text{Concentration of heavy metals}]_{\text{root}}}{[\text{Concentration of heavy metals}]_{\text{soil}}}$$

$$\text{Enrichment factor (EF)} = \frac{[\text{Concentration of heavy metals}]_{\text{shoot}}}{[\text{Concentration of heavy metals}]_{\text{soil}}}$$

$$\text{Translocation factor (TF)} = \frac{[\text{Concentration of heavy metals}]_{\text{shoot}}}{[\text{Concentration of heavy metals}]_{\text{root}}}$$

The value of BCF and TF should be >1 for effective heavy metal remediation and removal.

Barthwal et al. (2008) studied the accumulation of Pb, Cd, Cr, and Ni in the medicinal plants *Abutilon indicum*, *Calotropis procera*, *Euphorbia hirta*, *Peristrophe bycaliculata* and *T. cordifolia* from three different sites. From this study, Pb accumulation was highest in *C. procera* roots and Ni in *A. indicum* from a heavy traffic area (HTA).

Woody plants from the legumes have heavy metal remediation ability. For these plants, increased heavy metal concentration decreases the nitrogenize enzyme activity and affects nodule formation. Some terrestrial and aquatic plants, *Erythrina speciosa* and *Sesbania drummondii* can remediate Pb. From a contaminated site, Cd metal was removed using *Ricinus communis*. *Raphanus sativus* L. and *S. oleracea* L. can remediate Cd, Cu, Ni, Cr, and Zn. Similarly, *Cardaminopsis halleri* L. can remove Zn, Sn, aluminum (Al), and iron (Fe). *Arabis gemmifera* Adans and *S. nigrum* L. can remove Cd and Zn. *Phytolacca acinosa* Roxb can remove Mn, and Co remediation can be carried out by *Tamarix sp.* L. and *Berkheya coddii* L (Sharma et al., 2015). *Schima superba* and *Maytenus bureaviana* can accumulate Mn at 62,412.3 mg/kg and 33,750 mg/kg, respectively. Ni accumulation was studied in different plants with accumulation abilities ≤47,500 mg/kg, 19,100 mg/kg, and 13,500 mg/kg in *Psychotria douarrei*, *Alyssum markgraffi*, and *Alyssum pterocarpum*. *T. caerulescens* could accumulate 51,600 mg/kg of Zn (Yan et al., 2020) (Figure 16.3).

16.4   ROLE OF MICROBES IN HEAVY METAL PHYTOREMEDIATION

Industrialization, anthropogenic activities, and urban area development have led to the introduction, accumulation, magnification, and transmission of various xenobiotic compounds, which include heavy metals into the earth's biosphere. These harm humans, animals, the normal micro flora and plants. Plants are damaged the most because their roots are in direct contact with the soil, which is the accumulation site for heavy metals (Mosa et al., 2016; Mishra et al., 2017). Microbes are present in many habitats, and in the soil that surrounds the roots. These microbes help plants to overcome toxicity and protect against the negative effects of heavy metals because they have various strategies for the remediation of and tolerance to heavy metals. Microorganisms present in the soil are metabolically diverse, have extra and intracellular enzymes, gene regulation systems, and other stress-responsive regulatory systems. Microbes that have PGPR activity establish and provide nutrients to

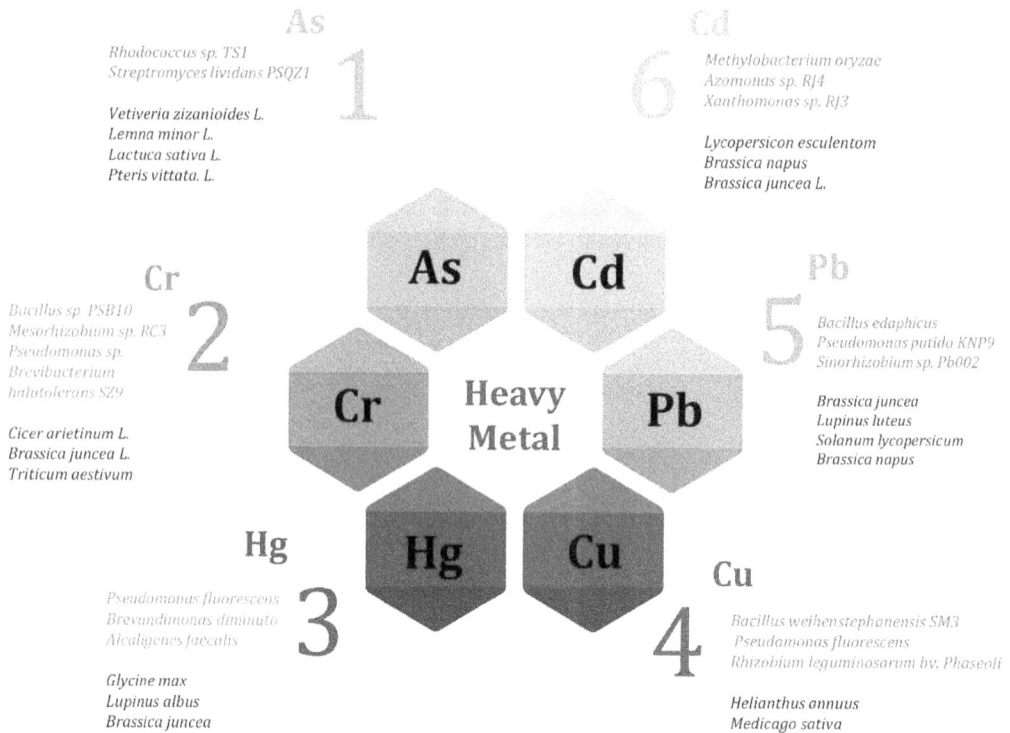

FIGURE 16.3 Plant–microbe interactions for heavy metal bioremediation.

plants and help to combat adverse environmental stress situations (He & Yang, 2007). By detoxification, the negative effects or by transforming heavy metals, microbes help the plant to improve its biomass and lifespan. Biosorption, bioaccumulation, redox reactions, production of biosurfactants, organic acids, and various chelating agents are strategies of microbes for adaptation, accumulation, transformation, and removal of heavy metals. Heavy metal transporter modification can enhance heavy metal tolerance in microorganisms (Mosa et al., 2016; Mishra et al., 2017).

Biosorption of heavy metals is the absorption of heavy metals onto the surface of microorganisms by various chemical groups and it is an energy-independent reversible process. The presence of teichoic acids in bacterial cell walls enhances the absorption of heavy metals. Bacteria absorb more heavy metals than other microbes, because they have a larger surface area. Dead *Bacillus sphaericus* cells could adsorb Cr. Heavy metal binding peptides of polyhistidyl from *Staphylococcus xylosus* can bond with heavy metals with higher affinity. These polypeptides are His_3-Glu-His_3 and His_6, which were genetically engineered. *Escherichia coli,* which was engineered with metallothionein genes from mice, has the potentials for heavy metal adsorption (Mosa et al., 2016). At acidic pH, *Enterobacter sp.* J1 and *Arthrobacter sp.* can biosorb 32.5 mg/g and 17.87 mg/g of Cu, respectively. *Pseudomonas fluorescence* can recover 40.8 mg/g of Ni at pH 2. The nutritional requirements of algae are very low, and they can be cultivated by light and water; therefore, they are cost-effective for heavy metal removal from polluted water and environments. By ion exchange and interactions with different units, brown algae can absorb heavy metals. For example, *Spirogyra sp.*, *Sargassum muticum*, *Chlorella miniata,* and *Spirulina platensis* can biosorb Pb, Zn, Cr, and Co at 140 mg/g, 34.10 mg/g, 34.60 mg/g, and 67.63 mg/g, respectively.

Low numbers of nucleic acids, the ease at biomass purification, secretion of extracellular enzymes, and a high tolerance capacity for heavy metals make fungi suitable microbial candidates for the removal of heavy metals from contaminated sources. *Penicillium chrysogenum* can biosorb

260 mg/g, 204 mg/g, and 92.0 mg/g of Ni, Pb, and Cu, respectively. *Penicillium simpliccium* can adsorb Zn, and Cu can be adsorbed by *A. niger* (Mustapha & Halimoon, 2015). Oves et al. (2013) studied the biosorption of Cd, Cr, Ni, Pb, and Co with an isolated *Bacillus thuringiensis* OSM29. At pH 7, optimum biosorption was observed with 94 % for Ni.

Biosorption of heavy metals can be calculated using the following equations (Oves et al., 2013):

$$Q = V (C_i - C_f)/M$$

where Q=uptake ability of heavy metals/ions (mg/g); V=volume of solution used studied for biosorption (L); C_i=concentration of heavy metals before sorption (mg/g); C_f=concentration of heavy metals before sorption (mg/g); and M=dry weight of biosorbent (g).

A linearized Langmuir equation can be used to calculate maximum adsorption on a monolayer surface of microbial biomass (Oves et al., 2013):

$$1/Q = 1/Q_{max} + 1/b*Q_{max}$$

where Q_{max} and b=Langmuir constants.

A linearized Freundlich equation can be used to calculate heterogenous adsorption on a single-layered surface of microbial biomass (Oves et al., 2013):

$$logQ = logK + (1/n) logC_f$$

where Q=equilibrium constant for the amount of heavy metals absorbed on absorbent material; and K and n=calculated from graph of logQ versus $logC_f$. The Separation factor (S_f), which is a without dimension Langmuir isotherm constant and can be calculated as follows (Oves et al., 2013):

$$S_f = 1/(1+bC_i)$$

The adsorption of heavy metals on microbial biomass can be determined using the following equation (Oves et al., 2013):

$$\Phi \text{ (Surface coverage)} = bC_i/1 + bC_i$$

Bioaccumulation requires energy and mainly is not a reversible process for heavy metals removal. When the rate of heavy metals removal is less than its absorption then the accumulation of heavy metals occurs inside the microbial cell. To be a suitable bioaccumulator of heavy metals, microbes should be able to tolerate higher concentrations of heavy metals. Heavy metal accumulation inside the cell produces toxicity and can kill the microbial cells. To resist the toxicity of heavy metals at higher concentrations, microbial cells must be able to transform heavy metals into their nontoxic form. Nonrecombinant *E. coli* can secrete the (Cys-Gly-Cys-Cys-Gly)$_3$ repeated motif that contains metallothioneins that increase the heavy metal binding affinity of microbial cells. Cd attachment can be induced via recombination in *E. coli*. Recombinants have been produced for the enhanced production of phytochelatins, which have (Glu-Cys)$_n$Gly repeated motifs. DNA microarrays, mass spectrometry, whole-genome sequencing, and transcriptome screening can be used to screen heavy metal accumulating microbes, which involves genes and proteins (Mosa et al., 2016).

Biosurfactants are surfactants produced by living cells to reduce tension between two surfaces. In addition, they are surface-active molecules and are divided into high and low molecular weight containing biosurfactants (Mosa et al., 2016). Vijayanand & Divyashree (2015) studied heavy metals remediation using microbes isolated from the effluent of wastewater that produced biosurfactants. From some isolates, KDM3 and KDM4 proteins could remediate Pb and Zn at 84.13 % and 93.18 %, respectively. Ravindran et al. (2020) isolated a *Bacillus sp.* from a marine sponge, which produced lipopeptide surfactant. It could precipitate heavy metals by micelle formation. An isolated strain could remove 99.93 % of Cd. Other heavy metals, such as Pb, Mn, and Hg were removed with the lipopeptide produced. *Pseudomonas aeruginosa* produces rhamnolipids, which can enhance the solubility of heavy metals. Computational and metagenomics based knowledge of species that cannot be cultivated and novel biosurfactant producers could have advantages to discover diverse microbes (Mosa et al., 2016).

Siderophores are low molecular weight chelators that contain Fe (Fe^{3+}). Carboxylate, hydroxamate (the most common type has C(=O) N-(OH) R), and catecholate/phenolates are types of microbial siderophores. *Rhizobium sp.* can produce carboxylate type siderophores. In Cyanobacteria, siderophore can decrease Cu mediated toxicity. Siderophores that have *DHB* and *Thr* were identified from *Streptomyces lilacinus* via proteomics screening (Mosa et al., 2016).

Redox reactions, which involve oxidation and reduction, can oxidize heavy metals. The reactions are carried out by cytochrome c (cyt c) (present in the outer membrane) and porin associated cyt (present in the trans section of the outer membrane). For example, the reduction of Cr is carried out by chromate reductase. Cu efflux is carried out using Cu oxidases (e.g., CueO, CupD, and CopR). Redox reaction carrying enzymes are generally produced under stressed conditions with heavy metals, which produce enzymes or proteins that remediate heavy metals to reduce their toxic effects on plants. *Geobacillus sp.* could convert the more toxic form of As to a less toxic one due to its oxidization ability.

PGPR are microbes that provide nutrition, growth hormones, help to solubilize insoluble macro or microelements, produce siderophores for Fe chelation, and help to combat biotic and abiotic stress conditions. For example, *Pseudomonas putida*, *Mesorhizobium sp.*, *Klebsiella sp.*, and *Enterobacter asburiae* provide indole acetic acid, which is an auxin hormone, and produce siderophores. *Burkholderia sp.*, *Pseudomonas jessenii* and *Acinetobacter sp.* solubilize P, which is required for nucleic acid and protein production. *Bradyrhizobium sp.* 750, *Ochrobactrum cytisi* and *P. jessenii* help to solubilize and mobilize heavy metals. Some heavy metals are essential for microbes to conduct their normal biometabolic cellular reactions. However, some heavy metals might be toxic above their required concentration. The gradual exposure to heavy metals at different concentrations, for long periods at a site, could enhance the selection and adaptation of heavy metals resistant microbes in soil or water. To overcome the toxicity of heavy metals, microbes adapt various mechanisms, such as immobilization, reduction of bioavailability, and transformation. Extracellular polysaccharides and proteins facilitate the binding of heavy metals to prevent heavy metals mobilization into microbial cells. Siderophore production by microbes can chelate heavy metals, and therefore, they reduce heavy metal bioavailability to microbial cells. If heavy metals enter the microbial cells they can be expelled by proton efflux and the ATP efflux pump. For example, As, Cd, and Cr resistance were reported to have developed in bacteria via an ATPase or proton pump. In addition, precipitation prevents heavy metals from entering microbes. Detoxification of accumulated heavy metals is carried out by microbes using metallothioneins. Ni detoxification in *R. communis* and *H. annus* was carried out by a one-pot assay by inoculation of *Psychrobacter sp.* SRS8, which induced biomass and chlorophyll production. In addition, Cr toxicity could decrease in Mung bean seedlings after treatment with *Bacillus cereus* and *Ochrobactrum*. Uptake of Cd was reduced by *P. aeruginosa* in pumpkin and Indian mustard. Pots assay of *Trifolium repens* inoculated with *Brevibacillus sp.* indicated the provision of nutrients and a reduced concentration of Zn in the tissues of plants (Ahemad, 2019).

16.4.1 ROLE OF ENDOPHYTIC MICROBES

Endophytes are fungi or bacteria that live between living plant cells. Endophytes have a broad host range. Facultative endophytes or obligatory endophytes utilize plant tissues as a host for transmission, growth, and development and sometimes for protection. In plant tissue, endophytes receive carbohydrates and provide tolerance and protection to the plant from various biotic and abiotic factors. In addition, endophytes diversity and the habitation pattern of plants (Aishwarya et al., 2014). Govarthanan et al. (2016) studied the bioremediation of Cu, Zn, Pb, and As by an endophytic isolate from the root of *Tridax procumbens*. From all five isolates, *Paneibacillus* RM could tolerate 750 mg/L, 500 mg/L, 450 mg/L, and 400 mg/L of Cu, Zn, Pb, and As, respectively. In addition, an isolated endophyte could produce indole acetic acid (IAA) with 17.2±1.2 mg/L. This bacterium could produce Fe a chelating siderophore, 1-aminocyclopropane-1-carboxylate (ACC) deaminase enzyme, and biosurfactants for the removal of heavy metals.

High surface area, mycelia, and hyphal biomass production mean that fungi can remediate different xenobiotics and environmental pollutants. Heavy metal resistant and remediating fungi were isolated from hyperaccumulators and nonhyperaccumulators. *A. niger* and *Penicillium spp.* can remove multiple heavy metals, such as Ni, Cd, and Cr. *Microsphaeropsis sp.* can detoxify Cd and was isolated from *S. nigrum* L. *A. niger* and *Aspergillus flavus* were isolated and could remove 50%–76% of As concentration combined with other heavy metals. An isolated *Lasiodiplodia sp.* MXSF31 could accumulate and extract the heavy metals Pb, Zn, and Cd from the solution. It was isolated from *Portulaca oleracea* (Aishwarya et al., 2014).

16.4.2 ROLE OF EXTREMOPHILES

Extremophiles are microbes or archea, which can survive under abnormal or extreme conditions of pH, temperature, and salinity. Therefore, based on these environmental conditions they can be divided into extremophiles or nonextremophiles. Based on this, in the presence of heavy metals above their required concentration and based on their toxicity, they can be considered one of the parameters of extreme conditions in the environment in which, through adaptive strategies, microbes can survive. Therefore, based on heavy metal tolerance and resistance, an extremophile or nonextremophile could grow in the presence of some toxic metals at high concentrations. These are called metallophilic microbes (Nies et al., 2003).

Ralstonia sp. CH34 is an example of a metallophilic bacteria based on the previous concept, because it can survive in the presence of Zn, and Cd, using a CZC (Co-Zn-Cd resistance system). This ability is due to the presence of pMOL30. This is a complex of CzcC, CzcB, and CzcA (resistance nodulation cell division) which are efflux transporters that are responsible for the survival ability in heavy metals. The genes that produce these proteins are under the regulation of CzcD, CzcS, CzcI, CzcN, CzcR, and RpoX. In addition to the previously described system, another heavy metal resistance system for Co and Ni is present in *Ralstonia sp.* CH34. This is due to the presence of *curCBA* genes on another plasmid (pMOL28). This system is based on an efflux pump in the transmembrane of the cell. Cr, which is in the form of oxyanion-Cr is removed by ChrAB proteins from the cell. Three transposons were reported for Hg resistance in *Ralstonia sp.* CH34 (Nies et al., 2003).

Kalita & Joshi (2017) studied the removal of Pb by metallophilic isolates. These were positive for exopolysaccharide production. *Pseudomonas sp.* W6 was isolated from a hot water spring, which can tolerate 1 mM Pb. This isolate was more effective than *P. aeruginosa* MTCC2474, *Pseudomonas alcaligenes* MJ7, and *Pseudomonas ficuserectae* PKRS11 because it could remove 65 % and 61.2 % of Pb from batch culture and a column system. This Pb removal was studied in a synthetic water system, where the production and release of exopolysaccharides by the isolate enhanced hydrophobicity and Pb removal.

16.5 ENHANCEMENT OF PHYTOREMEDIATION

Phytoremediation can be enhanced by chemicals or by the modification of plants via genetic engineering. Induction or enhancement with external or internal forces can overcome the limitations of phytoremediation and can increase heavy metal remediation.

16.5.1 CHEMICAL ENHANCEMENT

Some heavy metals that are present in soil and water are difficult to accumulate or extract by plants *in situ* due to their very low concentrations and insoluble nature. Plants produce and release various chelators, which efficiently and specifically bind to heavy metals and make them available for plants. These chelators are present at very low concentrations in natural systems. Therefore, augmentation of these chelating agents could determine routes for heavy metal uptake, transloca-tion, and availability to plants. Chelating agents form a complex with heavy metals at a cation exchange site and then release these heavy metals into the soil. Therefore, the heavy metals migrate, and the plants can absorb them from the soil via their roots. Some examples of chelating agents are: (1) EDTA; (2) sulfur (in element form); (3) citric acid (4); ammonium sulfate; (5) ethylenedi-amine disuccininc acid (EDDS); and (6) sodium dodecyl sulfate. Of these chelators, citric acid is nontoxic and could be utilized by microbes in the soil. The selection of a specific chelating agent provides fast and enhanced heavy metal remediation (da Conceiçao Gomes et al., 2016). The use of 2.5 mmol/l of citric acid in *Brassica napus* L., showed the uptake and accumulation of Pb and Cd in roots. In addition, it enhanced antioxidant ability and reduced oxidative stress, with the induction of biomass. The application of 10^{-5} mol/L and 10^{-4} mol/L in *Nasturtium officinale* increased Cr accu-mulation and absorption in the root. In *B. juncea* and *Brassica chinesis* treatment with 0.95 g/kg of citric acid, and enhance concentration by accumulation in shoots (da Conceiçao Gomes et al., 2016).

16.5.2 GENETIC ENGINEERING OR MODIFICATION

Genetic modification or genetic engineering is aimed at the over expression of heavy metal remedi-ating genes, proteins, enzymes, chelators, and heavy metal transporter. Enhanced production of metallothioneins was accomplished with genetic engineering of selected plants (da Conceiçao Gomes et al., 2016). An enzyme that changes the oxidation state encoding gene that was inserted into plants helped to remove the toxicity of heavy metals (Fulekar et al., 2009). For example, *N. tabacum* was genetically modified to enhance the expression of metallothionein and Hg reductase gene. *B. juncea* was genetically engineered using genes that were responsible for the expression of ATP sulfurylase and cystathionine-g-synthase from *A. thaliana*. These genetically modified plants could extract twice as much more Se in their roots and shoots. The transgenic cystathionine-g-synthase resulted in a higher rate of Se volatilization compared with the wild type. *B. juncea,* which was genetically modified for the glutathione synthetase gene from *E. coli,* showed a 25 % higher Cd accumulation than the wild type after exposure of mature plants and seedlings to different concentrations of Cd. According to some reports, research has been carried out on hairy root development for potential heavy metals remediation.

In *N. tabacum* and *A. thaliana,* hairy root development was studied for remediation of Cu through Cu and Cu binding periplasmic protein CopC. Some details on the adaptations of hairy roots in harsh conditions and the interaction with other microbes remain unclear (da Conceiçao Gomes et al., 2016). The nicotianamine synthase enzyme encoding gene transformation helps in Ni hyperaccumulation and tolerance to Ni to serpentine soil. Some amino acids do not participate in protein synthesis and some are low molecular weight containing organic acids (Kozminska et al., 2018).

Plants naturally secrete heavy metal accumulating chelators as small-sized peptides, which are known as phytochelators. Phytochelatins contain amino acids, organic acids, and metallothionein. Isophytchelatins are homologous molecules to phytochelatins. Apart from glycine, the C-terminal of Isophytchelatins contain a variety of amino acids. Nitrate reductase can be reactivated 1,000 times more by peptide phytochelatins than citric acid. In addition, phytochelators play an important role in heavy metals resistance capacity development. For example, mutant *A. thaliana* with *cad1* were sensitive to Cd and noted for the formation of a Cd-phytochelatins complex, because they did not have phytochelatin synthase (PCS) activity. From HPLC analysis, they could not accumulate phytochelatins, and therefore, were sensitive to Cd toxicity (Fulekar et al., 2009). Coexpression of PCS and c-glutamyl cysteine synthase (GCS) increased the synthesis of phytochelatins. Transformation of *A. thaliana* with PCS and *ABC* improved Cd and As coaccumulation and tolerance in the cell vacuoles (Kozminska et al., 2018).

Some proteins are located on the cell membrane, which allows heavy metals to enter the plant cells. These are known as transporters. For example, in *Arabidopsis sp.* IRT1 and MRP1 genes encode transporter proteins that facilitate the entry of Fe and Mg-ATPase into cells. Transporters are mainly divided into: (1) AtNramp1 and OsNramp1; and (2) AtNramp2 to 5 and OsNramp2 classes. The disintegration of *AtNramp3* induces tolerance to Cd. *RAN1* are found on the Golgi complex in plant cells. *CAX2* and *ZAT* were found in vacuole. At the boundary phase of the membrane and cytosol ZIP1 to 4 and COPT 1 were detected (Fulekar et al., 2009). ATP binding cassette transporters (ABC) are located in the tonoplast, which is expressed by *ABC* and help in heavy metals removal. Transformation through *ABC* induces mobilization of Cd, Pb, and Cu. Metal transporters of tolerance proteins are encoded by cation diffusion facilitators (*CDF*). High levels of expression of these genes result in the synthesis of thiols. These help in the hyperaccumulation of heavy metals. For example, the hyperaccumulation of Cd and As was accomplished by a transgenic *Nicotiana sp.* by the enhanced expression of OsMTP1 that was obtained from *Oryza sativa* L. cv. IR64. Zn and Fe metals transportation into plant cells is regulated by proteins related to the ZRT/IRT (ZIP) family. Highly expressed AtIRT1, enables *A. thaliana to* hyperaccumulate Zn and Cd. The ZIP proteins do not always increase heavy metal transportation, some specific ZIP proteins might downregulate this process. To overcome this problem, some other ZIP regulatory genes have been transferred with these genes to control their expression (Kozminska et al., 2018).

Some transgenic plants produce proteins that bind with heavy metals for their accumulation and tolerance capacity, which are known as metallothioneins. Introduction of these metallothionein expressing genes into selected plants confers heavy metal remediation. For example, by introducing *CUP1* into *Nicotiana sp.* Cd and Cu phytoextraction was induced (Fulekar et al., 2009). Insertion of *ThMT3* from *Tamarix hispida* into *Salix matsudana* and *Saccharomyces cerevisiae* resulted in the induction of tolerance to Cu, Cd, and Zn, respectively. Introduction of *SpMT1* into *Sedum plumbizincicola* contributed to Cd hyperaccumulation by increasing the transcription of *SpMTL*. The expression of αMT1a in *Linum usitatissimum* L. increased Cd accumulation by the overexpression of Cd binding peptides. Homeostatic levels of transgenics are maintained by reactive O_2 species (Kozminska et al., 2018).

16.5.3 TRANSPLASTOMICS APPROACH

Transplastomics can be generated or produced by the genetically modified genomes of the chloroplast. Homologous site-specific recombination is possible between the genes of interest and plastid. Therefore, the selected or genes of interest can be inserted into the plastid genome with the advantages of maternal type inheritance and no requirement for gene codon optimization. When multiple gene expression is required, for example, in genetically modified plants for heavy metals remediation, this is one of the most suitable methods. Transplastomic *Nicotiana sp.* can tolerate 400 mM of Hg by transformation with *merB* and *merA* of bacterial origin. Independent from the Hg contained in the soil, the transformed *Nicotiana sp.* could volatilize Hg, which indicated the efficient expression of the mercuric reductase enzyme. In another experiment, metallothionein gene expression was observed by transformed tobacco plants, which could tolerate 20 μM of Hg with increased chlorophyll concentration in the plant. Some antioxidant systems can be stimulated by transplastomic production under heavy metals treatment, to overcome the stressful condition. For example, treatment with Zn and Cd enhanced glutathione S-transferase in transplastomic *Nicotiana sp.* Genes utilized for screening tolerance, for example, *BrMT1,* were transformed into *Arabidopsis sp.* This gene encodes metallothionein, which is specific to *Brassica rapa* type-1. This allows the remediation of Cd into a nontoxic form (Kozminska et al., 2018).

16.5.4 CRISPR Technique

Genome editing for heavy metals tolerance is possible with manipulation of a plant's genome, which is already used for heavy metal remediation when the genome sequence information is available. For example, the genome sequence for *T. caerulescens* and *Arabidopsis halleri* is available. Both of these are good for the remediation of Cd. The remediating sequence for Pb by *Hirschfeldia incana* is available. The genetic modification of these plant's sequences can induce various processes, which includes those that are involved in heavy metal remediation, such as phytovolatilization, phytoextraction, and phytoaccumulation (Zarrin & Azra, 2018).

Specific information or instructions can be inserted into selected plants using high throughput software in bioinformatics to optimize and enhance heavy metal remediation. Heavy metal detoxification can be improved by increasing the productions of siderophores, which act as roots that secrets and metallothioneins, which are heavy metal binding ligands. In addition, the induced expression of phytochelatins, which are and secreted by plants and have an important role in the removal of heavy metals, could enhance phytoremediation. CRISPR-mediated expression control of these heavy metal remediating molecule-producing genes can be beneficial in a variety of ways. Cd phytoaccumulation in *Oryza sp.* can be decreased by the knockout of *OsNramp*, which is responsible for the expression of proteins that are involved in heavy metal transportation (Zarrin & Azra, 2018). Point mutations have been created in *Lemna,* which is an aquatic plant. This mutation was carried out using CRISPR/ *Cas9* to enhance the As(V)/P transporter. In addition, it induced the As(III) phytochelatin (PC)transporter in vacuoles (Mateo et al., 2019).

16.6 PHYTOREMEDIATION OF HEAVY METALS

Details about the phytoremediation of selected and most widely used heavy metals are described in the following sections

16.6.1 Cadmium

Cd can cause cancer and damage enzymes in the respiratory system (Wani et al., 2017). Electronic waste, painting chemicals, and the combustion of fuel are the main sources of Cd waste (Aishwarya et al., 2014). Cd induces chlorosis and slow growth in plants. In addition, it interferes with photosynthesis. In soil, Cd is present at 0.53 mg/kg and in rock at 0.3 mg/kg. Cd causes harm to the male reproductive system and the motility of sperm and female reproductive organs. In wheat seedlings, elevated reactive O_2 species are responsible for hydrogen peroxide (H_2O_2) concentration and malondialdehyde. This indicates toxicity at the gene level. In general, the root has a higher concentration of Cd than stems and stems have a higher concentration of Cd than the leaves (Raza et al., 2020). In soil, Cd is present at 0.1–345 mg/kg. However, according to regulations, the permissible limit is 100 mg/kg (Ahemad et al., 2015).

Salix mucronata, N. tabacum, and *Solanum melonaona* can extract Cd efficiently. In addition, *Vigna unguiculata* and *Swietenia macrophylla* are potential phytoextractors of Cd. *Suaeda salsa* in combination with mycorrhizal fungi could effectively extract Cd. Use of EDTA and citric acid enhanced chelator mediated phytoextraction. Plants that are hyperaccumulators are generally specific toward heavy metals. *Virola surinamensis* has the potential for the phytostabilization of Cd. Rhizofiltration of Cd was carried out using *Limnocharis flava* and *M. umbrosum. A. donax* effectively filters Cd with Zn. Caulofiltration of Cd was observed using *Berkheya coddii*. The mobile characteristics of Cd mean that it is easier and faster to extract through plants (Raza et al., 2020). *Phytolacca americana, A. pinnata,* and *Prosopis laevigata* can accumulate ≤10,700, 740, and 8,176 mg/kg of Cd (Yan et al., 2020).

From soil or water sources Cd enters the root via an apoplastic pathway. Chelators present on the epidermis facilitate Cd entry into the root along with proton exchange. These protons are generated by the respiration of the root cells. The diffusion of Cd is enhanced by the high surface area of the root hairs. Zn/Fe transporters, ABC and ATPase, which are P-type transporters, play a vital role in Cd transportation. In addition, MTPI, which is a nonselective transporter channel, and CAX family proteins= help in the transportation of Cd across the plasma membrane. From the root, Cd is transported above ground via the long-distance method. TaLCT1 ZIP, IRT, TcZIP4/1, and AtRT help in the transportation of ionic Cd, from the root to shoots or other plant parts. In rice plants, the Mn^{2+} transporter OsNRAMP can transport Cd. In addition, Cd can o be transported via transporters of other heavy metals. For example, the ZIP and LCT family of transporters, which mainly transport Zn and Fe can transport Cd. After being taken up by the root, homeostasis of Cd is maintained by AtPDR8. This protein is a type of ABC transporter. Metal tolerance proteins (MTPs) are part of Cd homeostasis. Combined with this, the P_{18} type metal transporter of ATPase plays a key role in homeostasis. GSH, OsLCT1, and phytochelators help in Cd transportation to the phloem (Raza et al., 2020).

At the cellular level, transporters, such as HMA, NRAMP, YSL, and MTP transport Cd inside the cell from the external environment. In the cell, Cd forms a complex with PCs. These complexes enter the vacuoles via the uniport of NRAMP. Here Cd is released from the PC complex. The released Cd binds with vacuolar ligands. In the vacuole, Cd can be transported both ways, inside and outside of the vacuole via the symport of MTP with a proton exchange mechanism. At the vacuolar level, removal can be carried out with ABCC transporter chelatins. The mobility of Cd is possible via the AtCAX2 and At CAX4 transporters. This mobility essentially provides heavy metals to the cytosol of the cell from the vacuole chamber. Phytochelatin and Cd conjugators OsHMA3 and SpHMA3 are involved in the sequestration of Cd in the vacuole. In addition, SpHMA3 was highly expressed in *Sednum alfredii*, which is a hyperaccumulator of Cd. AtNRAMP3 helps Cd efflux in the vacuoles of the mesophyll. Plants have some defense mechanisms and responses by which they can tolerate and resist heavy metal toxicity. For example, in *Triticum aestivum* 100 μm of Cd increase SOD, CAT, and α-tocopherol levels with AsA and POD. Similarly, in *Macleaya cordata* Cd decreased CAT, but enhanced MDA and SOD levels. In *Petroselinum Hortense,* different concentrations of $CaCl_2$ administered for 15 days increased the POD level and reduced APX, SOD, and CAT activity. Cd toxicity reduces transpiration and affects the hormonal balance in plants. In addition, it affects the stomatal opening and closing and the photosynthesis rate (Raza et al., 2020).

In the chelation assisted phytoremediation of Cd, EDTA and APCA enhance phytoextraction by increasing its solubility to the plant. However, the use of chelators might restrict Cd to the roots. Metallothioneins helps in the reduction of Cd toxicity by repairing damaged DNA and by scavenging reactive O_2 species. MT expressing genes are regulated by heavy metals. For example, *CsMTL2* in cucumber was regulated through Cd. Under stress conditions, *CsMT*4 was over expressed (Raza et al., 2020). Homeostasis maintaining *AvMT*2 and *AvPCS* had enhanced expression in *Avicennia* in the presence of Cd (Gonzalez-Mendoza et al., 2007).

NRT1.1 and *AtHMA4* modification in *Arabidopsis* enhanced the uptake and translocation of Cd. In *O. sativa* the knockout of *OsMHA*3 induce Cd translocation and sequestration in the vacuoles of the root. Modification of *AtFC*1, *AtPDF*2.5, and *AtPDF*2.6 in *Arabidopsis* increased Cd extraction and tolerance levels. The genetically engineered gene for γ-glutamyl cysteine synthetase in a transgenic plant of Indian mustard increased phytochelatins (Raza et al., 2020).

Plant associated microbes help in heavy metal remediation via various mechanisms and the secretion of chemicals and proteins. In total, 70 % of the Cd uptake ability of *S. nigrum* was enhanced via inoculation with *Serratia nematodiphila* LRE07. Another study indicated that phytostabilization and extraction of Cd were increased via *Bacillus sp.* SLS18 in *S. bicolor. Streptomyces sp.* in *Salix dasyclados* induced siderophore production. In another study, *P. putida* ATCC 39,213 successfully increased Cd extraction by 29% in *Eruca sativa*. Some *Arbuscular mycorrhizae* help plants to remediate heavy metals and provide protection from stress and nutrition elements. For example, *Glomus intraradices* enhanced Cd phytoextraction in *Salix viminalis* and *Tagetes erecta. Glomus constrictum* and *Glomus mosseae* helped in the stabilization of Cd in *Zea mays*. *Rhizophagus intraradices* enhanced the stabilization of Cd in *Lonicera japonica*. In *Phragmites communis*, Cd biosorption was triggered by inoculation with *Simplicillium chinese* (Raza et al., 2020).

Zhia et al. (2020) studied Cd phytoremediation using a pot culture of snapdragon with 1.0 mg/kg and 2.5 mg/kg of Cd. They concluded that exudates of snapdragon roots enhanced Cd absorption, with a pH reduction up to 0.3–0.6 units. Therefore, it helps in the solubilization of Cd. Raman et al. (2012) studied Cd remediation through three different varieties of tuberose, with five different concentrations from 0 mg/kg to 100 mg/kg in soil. When Cd was >50 mg/kg in the soil, some minute reductions in photosynthesis and the dry weight of the plant were observed. This study discovered that all three types efficiently collected Cd, with a >1 value for Cd in the shoot to bulb.

16.6.2 Arsenic

As exist in multiple oxidative forms of –3, 0, +3 and +5 in the natural environment. Arsenate [As(V)] and arsenite [As(III)] are predominantly found in nature. As causes various toxic effects, such as skin cancer, lung, and kidney damage. In plants, it negatively affects mitosis, growth, and enzymes. In addition, As metal is responsible for chlorosis in leaves. In soil, it exists in cobaltite and chloride complex conjugates. As is resent in an enargite form in the soil at 9.36 mg/kg worldwide. Groundwater contains more As compared with surface water. As can be mixed in drinking water by leaching. Anthropogenic activities, the use of chemically synthesized pesticides and fertilizers in agriculture, and industrial effluents are the main source of As contamination in soil and water (Mirza et al., 2018). In the soil, As is between 0.1 mg/kg and 102 mg/kg; however, its permissible limit in the soil is 20 mg/kg (Ahemad, 2015).

As that is present in soil and water can be remediated using plants, which is one of the chemical-free methods. In *A. thaliana* As(V) is taken up through a P transporter that has a high affinity (PHT), for example, PHT1;1 and PHT1;4. Intrinsic type proteins (NIP) are responsible for the entry of As(III) in combination with aquaporin. After entering the cell, As(III) is converted into As(IV) by a reductase enzyme. *At2g21045* is responsible for the expression of As reductase. After reduction to As(III), it is detoxified in a complex with phytochelatins. PC peptides are similar to ArsD in functionality. After conjugating with PCs, the PC-AS(III) complex enters the cell vacuole via ABC transporters. These transporters require energy as ATP (Mateo et al., 2019).

Carboxylic acids and carboxylates, such as citrate, malic and isocitric acids, metallothioneins, and phytochelatins are involved in the detoxification of As via the formation of a complex and of charge. Overproduction of PC2 was observed in *P. vittata* after exposure to As. Binding of As to GSH can detoxify As, because these complexes are sequestered to the vacuole via ABC transporter proteins, and they are similar to multiple drug resistance containing proteins. As(V) is predominantly found in roots (Mirza et al., 2018).

V. zizanioides L., *Helianthus annuus* L., *L. minor* L., *P. vittata* L., *Lepidum sativum* L, and *Lactuca sativa* L. can remediate As from polluted environments (Sharma et al., 2015). The generation of transgenic plants has emerged to enhance As free plants and As free areas. As extraction can be enhanced via the induced expression of $PvPHT_{1;3}$, which has an affinity for As(V) and with the overexpression of PvACR3, which has an affinity for As(III). PvACR3 translocates As in the vacuole. Nonhyperaccumulator *N. tabacum* can extract high quantities of As via the constitutive expression of *AtPCS1*. This gene is responsible for PC production and can be constitutively expressed under the CaMV 355 promoter. This enhances the accumulation of As in vacuoles. The overproduction of GST makes *A. thaliana* more resistant to stress generated by As toxicity (Mateo et al., 2019).

Microbes associated with the plant's rhizosphere help plants combat As toxicity and provide some phytohormones for plant growth. A concentration gradient is a mechanism by which As can be translocated to the plant. For example, in the roots of rice plants knockout of $OsNIP_{2;1}$ or decreased

Lsi1, As entrance is allowed up to a certain point, but at higher levels of exposure, bidirectional transport via a concentration gradient begins, which causes As to accumulate. *Lsi2* is responsible for the symplastic movement of As(III) in the xylem. As volatilization is mainly carried out via As(III) methyl transferase, which is expressed by microbes associated with plants and rhizospheres. In rice plants, the induced expression of this enzyme reduced As concentrations by volatilization. Fe plaques formed on the root surfaces of rice, forming a Fe–As complex, and therefore, reduced the availability of As to the plants. WRKY are transcription factors that regulate the uptake of As(V) in *A. thaliana.* In rice, OsARM1 was enhanced by As(III) and repressed As-associated transporter genes, for example, OsLsi6, OsLsi2 and OsLsi1 (Mateo et al., 2019).

Raj and Singh (2015) studied As remediation from As polluted soil using *P. vittata, A. capillus veneris, C. dentata,* and *Phragmites karka.* This study was carried out for approximately 3 years with sampling at 6-month intervals. From the previously described plants 70 %, 60 %, 55.1 %, and 56.1 % As removal was obtained, respectively. Therefore, these plants could successfully remove As.

Francis et al. (2017) screened for the phytoremediation of As using *Pteria vittata* with the heavy metal chelators EDDS and citric acid *in situ.* From this study, EDDS made As and Cd more soluble compared with citric acid. Without treatment of chelator 0.9 mg/kg As accumulated. The use of only EDDS and citric acid accumulate As in plants at 1.03 mg/kg and 0.12 mg/kg, respectively. When the concentration of EDDS was doubled compared with citric acid, the plant could accumulate 0.98 mg/kg As. From this study, EDDS at 5 mmol/kg could extract As better with the plant used. Jasrotia et al. (2017) studied As phytoremediation in the aquatic system using one macrophyte (*E. crassipes*) and two microphytes (*Chlorodesmis sp.* and *Cladophora sp.*) and screened different parameters. Within 15 days COD reduction was between 50% and 65%. In addition, 2, 4, and 6 mg/L of As were tolerated by *E. crassipes, Chlorodesmis sp.,* and *Cladophora sp,* respectively. *Pteris ryukyuensis* can accumulate 3,647 mg/kg of As (Yan et al., 2020).

16.6.3 COPPER

Electroplating and mining are the main sources of Cu (Aishwarya et al., 2014). Soil has various concentrations of Cu between 0.03 and 1,550 mg/kg. The permissible limit for Cu is 600 mg/kg (Ahemad, 2015). Liestianty and Abdullah (2014) studied the phytoremediation of Cu polluted soil by *Glycine max* L Merril. This study was carried out by adding compost. Because the obtained EF value during this study was >1, it was concluded that *G. max* L Marril is a hyperaccumulator of Cu at 68.509 mg/g. Afrousheh et al. (2015) screened *Calendula officinalis* for Cu removal combined with salicylic acid, to observe its effect on the growth and remediation of Cu. Cu was used at four different concentrations from 0 to 400 mg/kg. In addition, 0, 1, and 2 mM salicylic acid were used. This study resulted in 1,000 mg/kg of Cu tolerance in the shoot. For TF and BCF, a value of >1 was obtained, which indicated the hyperaccumulation ability of *C. officinalis.* The application of salicylic acid enhanced Cu phytoremediation and plant growth. Cu stress was overcome with the help of salicylic acid.

Xu et al. (2019) studied the remediation of Cu and Cd using *Pennisetum sp., E. splendens, S. lutescens.* and *S. plumbizincicola.* With the use of hydroxyapatite, *E. splendens* gave the best results for Cu and Cd remediation from all four species. *Pennisetum sp.* could remove Cd and Cu maximally with biomass production. In addition, hydroxyapatite enhanced the stability of aggregates in soil and water. A pot level study of artificially polluted soil for Cu phytoremediation indicated that *Brassica nigra* and *L. esculentum* Mill were more effective at removing Cu than *H. annuus. B. nigra,* which is known as black mustard, could remove 42 % of Cu metal after 120 days (Lothe et al., 2016).

From *Jatropha curcas, Acacia mangium,* and *Hopea odorata* the roots of *J. curcas* could accumulate 100 % Cu after sewage sludge treatment with the highest BCF value. At 100 % sludge provision, the highest TF was found for *J. curcas* and *Acacia mangium,* and for *H. odorata,* the TF was highest at 60 % soil and 40 % sludge concentration (Maryam et al., 2015). *Bacillus subtilis* S3, *P. alcaligenes* S22, and *B. subtilis* SJ-101 help Cu accumulation to plants. Other reported Cu hyperaccumulator plants include *Clerodendrum infortunatum, Croton bonplandianus,* and *Thordisa vilosa* (Ahmadpour et al., 2012). The aquatic plants, *Leptodictyum ripatium, Elodea Canadensis* and *Potamogeton alpines* detoxify and remediate Cu. Other Cu remediating aquatic plants include *Typha domingensis* and *Myriophyllum spicatum* (Wani et al., 2017). *Aeolanthus biformifolins* and *E. acicularis* can accumulate Cu at 13,700 mg/kg and 20,200 mg/kg, respectively (Yan et al., 2020).

16.6.4 LEAD

The permissible limit of Pb in the soil is 600 mg/kg; however, due to pollution and untreated discharge from various industries, its actual concentration is from 1 to 6,900 mg/kg (Ahemad, 2015).

Discharge from various industrial sewage plants, combustion processes, batteries, and color paints are sources of Pb. In aquatic systems, Pb causes serious damage to phytoplankton and zooplankton. In mammals, it is responsible for the cause of damage to teeth and other tissues. Reverse osmosis, ion exchange, electrodialysis, and precipitation are some conventional methods that are used to remove Pb. However, these methods have some disadvantages, such as high cost, production of sludge, accumulation of toxic compounds, and clogging of the membrane during reverse osmosis and electrodialysis (Singh et al., 2012). Pb is present in two oxidative forms (0 and +2), and Pb(II) is the most common form. In combination with chlorine, sulfate, P, humic acid, and amino acids it is mainly insoluble (Ahemad et al., 2019).

Aransiola et al. (2013) screened phytoextraction of Pb with the seeds of *G. max* L. using a pot assay. In this study, 0 25 ppm Pb was used and studied for 12 weeks of incubation. From this analysis, the seeds had the maximum concentration of Pb (4.2 mg/kg), and the stem had a minimum concentration of Pb (1.37 mg/kg). In addition, the highest BCF value (1.38) was obtained for 20 ppm of Pb. From the stem, leaves, and seeds the highest TF was obtained for the leaves (5.65) at 10 ppm of Pb. Therefore, *G. max* L. can be used to clean up Pb polluted sites. Huang et al. (1997) studied the role of chelators in the extraction of Pb at 2,500 mg/kg. The use of EDTA in *Z. mays* L. cv. Fiesta and *Pisum sativum* L. enhanced Pb accumulation in shoots. In this study, 1 g/kg of EDTA enhanced Pb translocation to the shoots at 99.80±5.52 µg/plant/day.

Alaboudi et al. (2018) studied Pb and Cd removal using *H. annuus* L. at 200 mg/kg. The maximum BCF for Cd was 1.67 for the root. In the shoot and root, 40.1 and 107.7 mg/kg of Pb was found, respectively. An inverse relationship was found with increased concentrations of Pb and root length in this study. For 0.00 mg/kg of Cd a shoot length of 53.11±0.84 cm was found, and at 200 mg/kg Cd the shoot length was minimum (i.e., 8.93±2.72 cm). *Betula occidentalis, Medicago sativa,* and *Thlaspi rotundifolium* could accumulate Pb at 1,000, 43,300, and 8,200 mg/kg, respectively (Yan et al., 2020). PGPR bacteria, such as *B. edaphicus* can remove Pb in *B. juncea*. Other microbes, such as *P. putida* KNP9, *Kluyvera ascorbata* SUD165, and *Sinorhizobium sp.* Pb002 are associated with Pb phytoremediation (Ahemad, 2019).

16.6.5 MERCURY

Currently, the soil has Hg concentrations from 0.001 to 1,800 mg/kg; however, the permitted value is 280 mg/kg (Ahemad, 2015). Oh et al. (2015) studied the effect of *Agrobacterium tumefaciens* for the removal of Hg by sweet sorghum. The results indicated that *A. tumefaciens* enhanced plant growth but did not promote Hg uptake. There was no significant difference observed in Hg accumulation with and without *A. tumefaciens* and the obtained values were 14.0 and 6.2 µg/plant, respectively. Yavar et al. (2014) analyzed the effect of *Brevundimonas diminuta* and *Alcaligens fecalis* on Hg contaminated soil using *S. mucronatus*. From the results, both bacteria had PGPR ability and could effectively remove Hg from polluted soil.

Franchi et al. (2017) carried out a comparative screening on the phytoextraction of Hg and As by supplementing with chemicals and an isolated PGPR bacteria on *B. juncea* and *L. albus*. As was supplemented with ammonium thiosulfate [$(NH_4)_2S_2O_3$] and potassium dihydrogen phosphate (KH_2PO_4) and the PGPR bacteria were isolated *Gordonia alkanivorans* 185.1, *G. alkanivorans* 195.23, *G. alkanivorans* 207.37, *G. necator* 188.14, and *S. luteola* 209.1, which were used in consortia. Here, 0.27 M of ammonium sulfate $(NH4)_2SO_4$ (extracted 12.9 mg/kg of Hg and 7.4 mg/kg of As. The shoot dry weight of *L. albus* was significantly promoted by treatment with $(NH_4)_2S_2O_3$ and PGPR. A mixture of $(NH_4)_2S_2O_3$ and PGPR had 576±21, 679±32, and 658±29, compared with the control, which was 632±33 mg. The concentration of Hg and As were higher in the root. $(NH_4)_2S_2O_3$ enhanced the extraction of Hg in the root of *B. juncea* and *L. albus* from 30 to 1,000 mg/kg and 25 to 800 mg/kg, respectively. With the addition of $(NH_4)_2S_2O_3$ to *L. albus*, As was five times higher in the shoot compared with the control. For PGPR application, Hg accumulation increased in *L. albus* and *B. juncea* by 35.8% and 44.7%, respectively. Therefore, from this study, an isolated PGPR consortium and a thiosulfate source enhanced the phytoextraction efficiency of Hg and As. Other reported plants, such as *Achillea millefolium*, *Marrubium vulgare,* and *Rumex induratus* successfully accumulated 18.275, 13.8, and 6.45 mg/kg of Hg (Yan et al., 2020).

16.6.6 CHROMIUM

The main sources of Cr are aircraft, electronic material, glass, colors, artificial dyes, Cr containing waste, and anthropogenic activity. The concentration of Cr in the soil is generally 0.005 to 3,950 mg/kg, which is higher than the permissible limit of 100 mg/kg. Cr(III) is mainly used by microbes as a nutrient and Cr(VI) is toxic and is produced by the oxidation of Cr(III). Cr(VI) is highly toxic, due to its diffusion ability across cell membranes via anionic channels, which are not specific. Cr(VI) enters the cell via a sulfate transporter, which has structural similarity to sulfate. After entering the cell, it is transformed into Cr(III) with free radicals production. Cr (VI) is reduced to Cr(III) via an NADPH-dependent reductase, which is present in the electron transport chain with cofactors cytb and cytc. Reduced Cr(III) cannot pass through the membrane, because it is insoluble. Some Cr(VI) was reduced to Cr(V) in the absence of O_2 via chromate reductase, which is embedded in the cell membrane. The reduced Cr(V) is responsible for the generation of reactive O_2 species. Microbes counteract these reactive O_2 species via the production of catalase and GSH. Inside microbial cells, Cr(III) can easily bind to DNA and cause errors or inhibition in replication. Bacterial cells can resist the toxicity of Cr(III) via DNA damage repair pathways that have a chrA transporter and RecG, RuvB, UvrD (Ahemad, 2015).

Revathi et al. (2011) studied Cr remediation using *S. bicolur* L. with garden soil and polluted soil with the addition of vermicompost. For garden soil, 112% and 101% Cr accumulation occurred in the roots and shoots, respectively. At 50 % garden soil and 50 % vermicompost Cr accumulation was 98% and 78% for the roots and shoots of *S. bicolor* L. Therefore, vermicompost decreased Cr accumulation. Ehsan et al. (2016) studied Cr removal using *Vinca rosea* L. using a pot assay. In this assay, 10, 20, 30, 40, 50, and 60 ppm were used with a control of 0 ppm. After 6 weeks of incubation, Cr was screened using atomic absorption spectroscopy. The height of the plant was measured at 43 cm for 40 ppm of Cr, which was higher than the control. For 10 ppm of Cr, the TF value was 1.3 and it increase at 30p pm and was constant for the other concentrations of Cr. The BCF and TF in this experiment indicated the potential applications of selected plants for Cr removal from polluted sites.

Some PGPR help plants to resist the toxicity of Cr and increase tolerance levels. For example, *Pseudomonas* sp. VRK3, *Bacillus sp.*, *Pseudomonas sp.* RNP4, and *Pseudomonas* sp. NBRI 4014 are IAA generating siderophore secreting and phosphate solubilizing bacteria *Delfia sp.* JD2 carries out nitrogen fixation with Cr resistance and Cr reducing ability. *Ochrobactrum intermedium* C32413 and *Rhodococcus erythropolis* MTCC 7905 have Cr resistance. In *Z. mays, Mixcrobacterium sp.*

SUCR 140 and *A. tumefaciens* improved the plant growth with a reduction in Cr(VI) toxicity and enhanced Cr(VI) uptake. In *Lens esculenta, Brevibacterium sp., B. cereus,* and *O. intermedium* significantly enhanced the weight of grain and increased the length of the shoot and root. *R. erythropolis* MTCC 7905 induced plant growth even at low temperatures in the presence of Cr and *P. sativum* (Ahemad, 2015).

16.7   ROLE OF METAGENOMICS AND TRANSCRIPTOMICS

Salam et al. (2020) studied changes in microbial communities in agricultural soil that was treated with Cd. Illumina sequencing of Cd treated soil samples revealed the presence of Proteobacteria (50.50% at phyla level) and at species level *Methylobacterium radiotolerans* (12.80%) were detected. In a control sample of soil (not treated with Cd) Proteobacteria and *Conexibacter woesei* were dominant with 37.38% and 8.93% at phyla and species level, respectively. The loss of some bacterial members of *Proteobacteria* and *Actinobacteria* highlighted Cd oxidative stress. The functional screening revealed the presence of genes for carbohydrate and amino acid metabolism. Some genes were only found in Cd treated soil samples. The metagenome of the control soil sample indicated the presence of *copA, copB, copC, copP, cueO, cutC* and *cutE* for Cu efflux, ATPase for Cd translocation, *furA* for Fe resistance and Fe^{2+} or Mn^{2+} transporter system for Mn resistance. *czcA, czcD, czrA, czrB, MnSOD, SodA,* and *SodB* were screened in Cd treated samples, which are present in microbes for the removal of Cd toxicity. *czcA, czcD, czrA,* and *czrB* were Cd resistant through an efflux system. The detected *sodA, sodB,* and *ahpC* were responsible for the protection of microbes from Cd stress. Sequences for alkyl hydroperoxide reductase were detected in Cd treated soil samples.

Shi et al. (2019) screened the potential of methanotroph *Methylocystis* for the remediation of Hg and As using metagenomic tools. Hg(II) reductase, which is responsible for Hg reduction, was expressed from *merA*, which is similar to the genes of *P. Halophilus* and *Bradyrhizobium sp.* CCH5-F6. From four As(V) reductase, two were included in the glutaredoxin of *E.coli* and r two reductases were included in the thioredoxin plasmid in *Staphylococcus aureus.* As(V) reducing genes were found on the arsRCCB operon. In addition, this study reveals that horizontal gene transfer was responsible for the successful inclusion of *ArsC* in isolates.

Hur and Park (2019) studied the taxonomical and functional profiling of microbial communities in a heavy metal polluted site. Using PacBioRSII, dominant Proteobacteria were screened out in all samples. *Leptospirillum, Rhodoplanes,* and *Thiobacillus* were found at the genera level in all samples. Fe oxidizing and sulfur oxidizing bacterium were identified from samples BF and BB. The presence of habitat-specific *Gallionella, Halothiobacillus,* and *Leptolyngbya* could be used as an indicator. This study revealed that *Halothiobacillus,* which was detected in the BB sample, could be used as an indicator in the detection of drainage from acid mining. *Rhodoplanes,* a purple nonsulfur light dependent containing genus bacterium was discovered in a sample called Hwaseong and Daegu.

Feng et al. (2018) studied the metagenomics of Cd contaminated soil. *Proteobacteria* and *Acidobacteria* at the phylum level were the most abundant in the collected samples. *Candidatus koribacter* and *Nitrososphaera* were the most abundant at the genus level were samples S1 and S2, respectively. At the functional level, ABC transporters, which help in the transportation of metal ions, were predominant. From the total detected pathways, 90 % were lower in sample S2, which indicated the toxic effects of Cd on this selected site.

Wang et al. (2020) studied microbial communities associated with the rhizosphere of *S. alfredii* for a Pb accumulating and nonaccumulating ecotypes. In the Pb lead accumulating ecotype of *S. alfredii,* stable and diverse microbes with *Flavobacterium* were detected with siderophore secretion ability. These microbes could produce IAA, which is responsible for plant growth. In the Pb nonaccumulating ecotype of *S. alfredii, Pseudomonas* was predominantly found with other less diverse microbial communities. The detected *Pseudomonas* could solubilize P. Therefore, Pb accumulated *S. alfredii* could extract Pb, and the nonaccumulated type could precipitate Pb. In addition, Pb enhanced PGPR-like ACC deaminase producing *Sphingomonadaceae*. The accumulating ecotype could adapt to and tolerate higher stress from Pb than the nonaccumulated type.

RNA sequencing for transcriptome analysis indicated that As and Cd responding genes belong to five categories: (1) redox reaction carrying; (2) glutathione metabolomics; (3) biogenesis of cell wall; (4) regulation of expression; and (5) gene for transmembrane transporter. In total, 27 genes were common for As and Cd resistance and remediation in rice plants and to *A. thaliana*. These are homologous genes. For example, redox related genes found in rice were *Os07g0418500* and *Os03g0227700* and in *A. thaliana* they were $AT_2G_4 6_{950}$ and $AT_3G_{50}660$. In *A. thaliana*, $AT_2G_3 6_{910}$, AT_1G6_{5730}, and AT_2G_{41560} and in rice plants *Os01g0695800, Os04g0524500,* and *Os01g0939100* were detected for transmembrane transporter related genes (Mateo et al., 2019).

16.8 CONCLUSIONS

From this chapter, heavy metal pollution occurs due discharge of untreated waste into the environment. The cultivation of hyperaccumulator and nonhyperaccumulator plants could effectively extract various heavy metals from soil and water. Phytovolatilization from leaves helps in the removal of heavy metals from plants. Phytochelatins, metallothioneins, and biosurfactants make heavy metals available to plants that have phytostabilization and aid in the removal of heavy metals. The application of genetic engineering, modifications, the study of CRISPER, and research into transplastomics could help to optimize heavy metals remediation. The inoculation of extremophilic and nonextremophilic PGPRs could act as a prokaryotic living catalyst, which could help plants make heavy metals removal faster and provide protection from toxicity *in situ*. With omics studies, data on microbes that cannot be cultivated for heavy metals remediation could be extracted. Therefore, the selection of suitable microbes and plants for specific heavy metals could help the viability of the environment.

REFERENCES

Afrousheh, M., Shoor, M., Tehranifar, A., & Safari, V. R. (2015). Phytoremediation potential of copper contaminated soils in *Calendula officinalis* and effect of salicylic acid on the growth and copper toxicity. *International Letters of Chemistry, Physics and Astronomy, 50*, 159–168.

Ahemad, M. (2015). Enhancing phytoremediation of chromium-stressed soils through plant-growth-promoting bacteria. *Journal of Genetic Engineering and Biotechnology, 13*(1), 51–58.

Ahemad, M. (2019). Remediation of metalliferous soils through the heavy metal resistant plant growth promoting bacteria: paradigms and prospects. *Arabian Journal of Chemistry, 12*(7), 1365–1377.

Ahmadpour, P., Ahmadpour, F., Mahmud, T. M. M., Abdu, A., Soleimani, M., & Tayefeh, F. H. (2012). Phytoremediation of heavy metals: A green technology. *African Journal of Biotechnology, 11*(76), 14036–14043.

Aishwarya, S., Venkateswarlu, N., Chandramouli, K., &Vijaya, T. (2014). Role of endophytic fungi in the restoration of heavy metal contaminated soils. *Indo American Journal of Pharmaceutical Sciences, 4*, 5427–5436.

Alaboudi, K. A., Ahmed, B., & Brodie, G. (2018). Phytoremediation of Pb and Cd contaminated soils by using sunflower (*Helianthus annuus*) plant. *Annals of Agricultural Sciences, 63*(1), 123–127.

Ali, S., Abbas, Z., Rizwan, M., Zaheer, I. E., Yavaş, İ., Ünay, A., & Kalderis, D. (2020). Application of floating aquatic plants in phytoremediation of heavy metals polluted water: A review. Sustainability, *12*(5), 1927.

Angelova, V. (2012). Potential of some medicinal and aromatic plants for phytoremedation of soils contaminated with heavy metals. *Agrarni Nauki, 4*(11), 61–66.

Ansari, A. A., Naeem, M., Gill, S. S., & AlZuaibr, F. M. (2020). Phytoremediation of contaminated waters: An eco-friendly technology based on aquatic macrophytes application. *The Egyptian Journal of Aquatic Research, 12*, 1927–1960.

Aransiola, S. A., Ijah, U. J. J., & Abioye, O. P. (2013). Phytoremediation of Lead Polluted Soil by *Glycine max* L. *Applied and Environmental Soil Science,1–7*.

Barthwal, J., Smitha, N. A. I. R., & Kakkar, P. (2008). Heavy metal accumulation in medicinal plants collected from environmentally different sites. *Biomedical and Environmental Sciences*, *21*(4), 319–324.

Da Conceiçao Gomes, M. A., Hauser-Davis, R. A., de Souza, A. N., & Vitoria, A. P. (2016). Metal phytoremediation: General strategies, genetically modified plants and applications in metal nanoparticle contamination. *Ecotoxicology and Environmental Safety*, *134*, 133–147.

Dixit, P., Singh, S., Vancheeswaran, R., Patnala, K., & Eapen, S. (2010). Expression of a Neurospora crassa zinc transporter gene in transgenic *Nicotiana tabacum* enhances plant zinc accumulation without co-transport of cadmium. *Plant, Cell & Environment*, *33*(10), 1697–1707.

Ehsan, N., Nawaz, R., Ahmad, S., Khan, M. M., & Hayat, J. (2016). Phytoremediation of chromium contaminated soil by an ornamental plant Vinca (*Vinca rosea L.*). *Journal of Agriculture and Environmental Sciences*, *7*, 29–34.

Feng, G., Xie, T., Wang, X., Bai, J., Tang, L., Zhao, H., &Zhao, Y. (2018). Metagenomic analysis of microbial community and function involved in cd-contaminated soil. *BMC Microbiology*, *18*(1), 11.

Franchi, E., Rolli, E., Marasco, R., Agazzi, G., Borin, S., Cosmina, P., & Petruzzelli, G. (2017). Phytoremediation of a multi contaminated soil: mercury and arsenic phytoextraction assisted by mobilizing agent and plant growth promoting bacteria. *Journal of Soils and Sediments*, *17*(5), 1224–1236.

Francis, O. O., Xiang, L., & Alepu, O. E. (2017). In-situ phytoremediation of arsenic from contaminated soil. *International Journal of Waste Resources*, *7*(1), 1–5.

Fulekar, M. H., Singh, A., & Bhaduri, A. M. (2009). Genetic engineering strategies for enhancing phytoremediation of heavy metals. *African Journal of Biotechnology*, *8*(4), 529–535

Gonzalez-Mendoza, D., Moreno, A. Q., & Zapata-Perez, O. (2007). Coordinated responses of phytochelatin synthase and metallothionein genes in black mangrove, *Avicennia germinans*, exposed to cadmium and copper. *Aquatic Toxicology*, *83*(4), 306–314.

Govarthanan, M., Mythili, R., Selvankumar, T., Kamala-Kannan, S., Rajasekar, A., & Chang, Y. C. (2016). Bioremediation of heavy metals using an endophytic bacterium *Paenibacillus sp.* RM isolated from the roots of *Tridax procumbens*, 3. *Biotechnology*, *6*, 242.

He, Z. L., & Yang, X. E. (2007). Role of soil rhizobacteria in phytoremediation of heavy metal contaminated soils. *Journal of Zhejiang University Science B*, *8*(3), 192–207.

Huang, J. W., Chen, J., Berti, W. R., & Cunningham, S. D. (1997). Phytoremediation of lead-contaminated soils: role of synthetic chelates in lead phytoextraction. *Environmental Science & Technology*, *31*(3), 800–805.

Hur, M., & Park, S. J. (2019). Identification of Microbial Profiles in Heavy-Metal-Contaminated Soil from Full-Length 16SrRNA Reads Sequenced by a PacBio System. *Microorganisms*, *7*(9), 357.

Jasrotia, S., Kansal, A., & Mehra, A. (2017). Performance of aquatic plant species for phytoremediation of arsenic-contaminated water. *Applied Water Science*, *7*(2), 889–896.

Kalita, D., & Joshi, S. R. (2017). Study on bioremediation of lead by exopolysaccharide producing metallophilic bacterium isolated from extreme habitat. *Biotechnology Reports*, *16*, 48–57.

Kozminska, A., Wiszniewska, A., Hanus-Fajerska, E., & Muszynska, E. (2018). Recent strategies of increasing metal tolerance and phytoremediation potential using genetic transformation of plants. *Plant Biotechnology Reports*, *12*(1), 1–14.

Kumar, B., Smita, K., & Flores, L. C. (2017). Plant mediated detoxification of mercury and lead. *Arabian Journal of Chemistry*, *10*, S2335–S2342.

Liestianty, D., & Abdullah, M. (2014). Phytoremediation study of copper-contaminated soil using soybean (*Glycine max* (L) Merril) with compost addition. *International Journal of Innovation and Applied Studies*, *9*(4), 1938–1943.

Lone, M. I., He, Z. L., Stoffella, P. J., & Yang, X. E. (2008). Phytoremediation of heavy metal polluted soils and water: progresses and perspectives. *Journal of Zhejiang University Science B*, *9*(3), 210–220.

Lothe, A. G., Hansda, A., & Kumar, V. (2016). Phytoremediation of copper-contaminated soil using *Helianthus annuus*, *Brassica nigra*, and *Lycopersicon esculentum* Mill.: A pot scale study. *Environmental Quality Management*, *25*(4), 63–70.

Maryam, G., Majid, N. M., Islam, M. M., Ahmed, O. H., & Abdu, A. (2015). Phytoremediation of copper-contaminated sewage sludge by tropical plants. *Journal of Tropical Forest Science*, *27*(4), 535–547.

Mateo, C., Navarro, M., Navarro, C., & Leyva, A. (2019). Arsenic Phytoremediation: Finally a Feasible Approach in the Near Future. *Bioremediation*,1–14.

Mirza, N., Mahmood, Q., Maroof Shah, M., Pervez, A., & Sultan, S. (2018). corrigendum to "Plants as Useful Vectors to Reduce Environmental Toxic Arsenic Content". *The Scientific World Journal*. doi:10.1155/2014/921581,2014.

Mishra, J., Singh, R., & Arora, N. K. (2017). Alleviation of heavy metal stress in plants and remediation of soil by rhizosphere microorganisms. *Frontiers in Microbiology, 8,* 1706.

Mosa, K. A., Saadoun, I., Kumar, K., Helmy, M., & Dhankher, O. P. (2016). Potential biotechnological strategies for the cleanup of heavy metals and metalloids. *Frontiers in Plant Science*, *7*, 303.

Mustapha, M. U., & Halimoon, N. (2015). Microorganisms and biosorption of heavy metals in the environment: a review paper. *Journal of Microbial &Biochemical Technology, 7*(5), 253–256.

Muthusaravanan, S., Sivarajasekar, N., Vivek, J. S., Paramasivan, T., Naushad, M., Prakashmaran, J., & Al-Duaij, O. K. (2018). Phytoremediation of heavy metals: mechanisms, methods and enhancements. *Environmental Chemistry Letters*, *16*(4), 1339–1359.

Nascimento, C. W. A. D., & Xing, B. (2006). Phytoextraction: a review on enhanced metal availability and plant accumulation. *Scientia Agricola*, *63*(3), 299–311.

Nies, D. H. (2003). Efflux-mediated heavy metal resistance in prokaryotes. *FEMS Microbiology Reviews*, *27*(2–3), 313–339.

Oh, K., Takahi, S., Wedhastri, S., Sudarmawan, H. L., Rosariastuti, R., & Prijambada, I. D. (2015). Phytoremediation of mercury contaminated soils in a small scale artisanal gold mining region of Indonesia. *International Journal of Bioscience and Biotechnology*, *3*(1), 14–21.

Ojuederie, O. B., and Babalola, O. O. (2017). Microbial and plant-assisted bioremediation of heavy metal polluted environments: a review. *International journal of environmental research and public health*, *14*(12), 1504.

Oves, M., Khan, M. S., & Zaidi, A. (2013). Biosorption of heavy metals by *Bacillus thuringiensis* strain OSM29 originating from industrial effluent contaminated north Indian soil. *Saudi Journal of Biological Sciences*, *20*(2), 121–129.

Parmar, S., & Singh, V. (2015). Phytoremediation approaches for heavy metal pollution: A review. *Journal of Plant Science Research*, *2*(2), 135.

Radziemska, M., Vaverkova, M. D., & Baryła, A. (2017). Phytostabilization: management strategy for stabilizing trace elements in contaminated soils. *International Journal of Environmental Research and Public Health*, *14*(9), 958.

Raj, A., & Singh, N. (2015). Phytoremediation of arsenic contaminated soil by arsenic accumulators: a three year study. *Bulletin of Environmental Contamination and Toxicology*, *94*(3), 308–313.

Ramana, S., Biswas, A. K., Singh, A. B., Kumar, P. N., Ahirwar, N. K., Behera, S. K., & Rao, A. S. (2012). Phytoremediation of cadmium contaminated soils by tuberose. *Indian Journal of Plant Physiology*, *17*(1), 61–64.

Ravindran, A., Sajayan, A., Priyadharshini, G. B., Selvin, J., & Kiran, G. S. (2020). Revealing the efficacy of thermostable biosurfactant in heavy metal bioremediation and surface treatment in vegetables. *Frontiers in Microbiology*, *11*, 222.

Raza, A., Habib, M., Kakavand, S. N., Zahid, Z., Zahra, N., Sharif, R., & Hasanuzzaman, M. (2020). Phytoremediation of cadmium: Physiological, biochemical, and molecular mechanisms. *Biology*, *9*(7), 177.

Revathi, K., Haribabu, T. E., & Sudha, P. N. (2011). Phytoremediation of chromium contaminated soil using sorghum plant. *International Journal of Environmental Sciences*, *2*(2), 417–428.

Salam, L. B., Obayori, O. S., Ilori, M. O., & Amund, O. O. (2020). Effects of cadmium perturbation on the microbial community structure and heavy metal resistome of a tropical agricultural soil. *Bioresources and Bioprocessing*, *7*(1), 1–19.

Sharma, S., Singh, B., & Manchanda, V. K. (2015). Phytoremediation: role of terrestrial plants and aquatic macrophytes in the remediation of radionuclides and heavy metal contaminated soil and water. *Environmental Science and Pollution Research*, *22*(2), 946–962.

Shi, L. D., Chen, Y. S., Du, J. J., Hu, Y. Q., Shapleigh, J. P., & Zhao, H. P. (2019). Metagenomic evidence for a Methylocystis species capable of bioremediation of diverse heavy metals. *Frontiers in Microbiology*, *9*, 3297.

Singh, D., Tiwari, A., & Gupta, R. (2012). Phytoremediation of lead from wastewater using aquatic plants. *Journal of Agricultural Technology*, 8(1), 1–11.

Soni, V., & Kaur, P. (2015). Efficacy of aquatic plants for removal of heavy metals from wastewater.

Suman, J., Uhlik, O., Viktorova, J., & Macek, T. (2018). Phytoextraction of heavy metals: a promising tool for clean-up of polluted environment?. *Frontiers in Plant Science*, 9, 1476.

Tangahu, B. V., Sheikh Abdullah, S. R., Basri, H., Idris, M., Anuar, N., & Mukhlisin, M. (2011). A review on heavy metals (As, Pb, and Hg) uptake by plants through phytoremediation. *International Journal of Chemical Engineering*. doi:10.1155/2011/939161

Vijayanand, S., & Divyashree, M. (2015). Bioremediation of heavy metals using biosurfactant producing microorganisms. *International Journal of Pharmaceutical Sciences and Research*, 6(5), 840–847.

Wang, R., Hou, D., Chen, J., Li, J., Fu, Y., Wang, S., & Tian, S. (2020). Distinct rhizobacterial functional assemblies assist two *Sedum alfredii* ecotypes to adopt different survival strategies under lead stress. *Environment International*, 143, 105912.

Wani, R. A., Ganai, B. A., Shah, M. A., & Uqab, B. (2017). Heavy metal uptake potential of aquatic plants through phytoremediation technique—a review. *Journal of Bioremediation and Biodegradation*, 8(404), 2.

Xu, L., Xing, X., Liang, J., Peng, J., & Zhou, J. (2019). In situ phytoremediation of copper and cadmium in a co-contaminated soil and its biological and physical effects. *RSC Advances*, 9(2), 993–1003.

Yan, A., Wang, Y., Tan, S. N., Yusof, M. L. M., Ghosh, S., & Chen, Z. (2020). Phytoremediation: a promising approach for revegetation of heavy metal-polluted land. *Frontiers in Plant Science*. doi: 10.3390/ijerph18052435

Yavar, A., Sarmani, S., Hamzah, A., & Khoo, K. S. (2014). Phytoremediation of mercury contaminated soil using *Scirpus mucronatus* L. exposed to bacteria. Paper presented at International Conference on Agricultural, Ecological and Medical Sciences. Kuta, Indonesia.

Zarrin, B., & Azra, Y. (2018). Genome editing weds CRISPR: what is in it for phytoremediation?. *Plants*, 7(3).

Zgorelec, Z., Bilandzija, N., Knez, K., Galic, M., & Zuzul, S. (2020). Cadmium and mercury phytostabilization from soil using *Miscanthus X giganteus*. *Scientific Reports*, 10(1), 1–10.

Zhang, Y., Zhao, L., Wang, Y., Yang, B., & Chen, S. (2008). Enhancement of heavy metal accumulation by tissue specific co-expression of *iaaM* and *ACC* deaminase genes in plants. *Chemosphere*, 72(4), 564–571.

Zhi, Y., Zhou, Q., Leng, X., & Zhao, C. (2020). Mechanism of remediation of cadmium-contaminated soil with low-energy plant Snapdragon. *Frontiers in Chemistry*, 8, 222.

17 A Combination of Biosorption and Enzymatic Degradation of Azo Dyes

Nelson Libardi[1]*, Juliana Barden Schallemberger[1], Maria Eliza Nagel Hassemer[1], Rejane Helena Ribeiro da Costa[1], Carlos Ricardo Soccol[2], and Luciana Porto de Souza Vandenberghe[2]

[1]Department of Sanitary and Environmental Engineering, Federal University of Santa Catarina, UFSC, 88040-970, Florianópolis, Brazil

[2]Department of Bioprocess Engineering and Biotechnology, Federal University of Paraná, UFPR, 81531-980, Curitiba, Brazil

*Corresponding author: nelson.libardi@ufsc.br

CONTENTS

17.1 INTRODUCTION

Since the 1800s, industrial development has improved living standards, which has resulted in the release of toxic pollutants that impact water, soil and air. Harmful pollutants, such as pesticides, petroleum-based hydrocarbons, heavy metals, polychlorinated biphenyls, and synthetic dyestuff shave high toxicity and might bioaccumulate, which causes toxic, carcinogenic, teratogenic, and mutagenic effects on humans and other organisms (Bilal and Iqbal, 2020).

The textile industry is characterized by its high energy consumption and the use of natural resources, which contributes to the rapid generation of post-consumption wastewater. Therefore, this is the most significant source of surface water pollution due to the discharge of dyes, which are typically recalcitrant and toxic. In this industrial sector, 70% of all used dyes are mainly azoic, which are easily recognized by the nitrogen–nitrogen double bonds (N=N), bright colors, simple application techniques, and low energy consumption. Due to the molecular structure and synthetic origin of the dyes, these compounds are stable and difficult to degrade, which makes the discoloration of effluents a complex process. In general, conventional effluent treatment processes cannot degrade

DOI: 10.1201/9781003130932-17

or remove dyes, which requires the excessive use of chemicals to achieve the levels of discoloration that are required by environmental legislation. This generates large amounts of contaminated sludge, which is an additional cost for the industries (Shirvanimoghaddam *et al.*, 2020).

Several remediation techniques have been developed to degrade environmentally-related pollutants. Despite the technological efforts to efficiently degrade anthropogenic pollutants, these techniques are expensive, involve the use of excessive use of harsh chemicals, have a low-cost-effective ratio, and release toxic by-products. Conventional methods used to treat textile effluents are oxidation, coagulation and flocculation, biological treatment, and membrane filtration. However, a single conventional treatment does not ensure the satisfactory removal of recalcitrant azo dyes in textile wastewater due to their colorfastness, stability, and resistance to degradation (Liu *et al.*, 2019; Bilal and Iqbal, 2020).

Despite the previously mentioned disadvantages to wastewater treatment techniques, adsorption remains a popular method, due to its ease of operation, process control, high efficiency, and economic feasibility. Low-cost adsorbents, such as bentonite clay are of interest; however, since they are a single-use material, a post-sorption material is generated that needs to be discharged (Morsy *et al.*, 2020). Biosorption has been designated as an adsorption subclass, and biological waste or nonwaste materials are used as adsorbents for pollutants removal. These renewable and environmentally friendly materials have good removal efficiency and are widely available in nature, including agro–industrial wastes, microbial biomass, and wastewater sludge. In addition, an advantage of them is the possibility of converting waste materials into high value-added biomass. Adsorption does not allow any kind of pollutant biodegradation, neither mineralization nor transformation. The pollutant is trapped on the adsorbent surface via chemisorption of physisorption mechanisms, which results in mass transfer from the liquid to the solid phase (Ray, Gusain, and Kumar, 2020).

Bioremediation is microbial action via the secretion of catalytic enzymes that can transform harmful pollutant into less toxic by-products, which is designated as enzymatic bioremediation. The enzymatic reactions are the key factor in any bioremediation process and can involve microbial strains or enzymes previously produced or isolated from living cells. The enzymes maintain their catalytic activity outside the living cells. Therefore, bioremediation does not depend on the use of living microbial cells to catalyze biochemical reactions to degrade pollutants.

Enzymatic catalysis has several advantages over whole microbial cells. Enzymes accelerate the chemical reactions and are not consumed by the reaction and do not generate toxic by-products, as in some microbial processes. The enzymes are digested *in situ* by indigenous microorganisms. They can work under extreme temperatures, pH, and ionic strength conditions, in contrast to microbial cells that require mild environmental conditions. In addition, the enzyme concentration and cofactors availability can be controlled, which means that the results of the bioremediation can be predicted. Enzymes could be produced at large scales, with enhanced stability, or activity, or both and at lower costs with higher yields using recombinant-DNA technology. Finally, enzymes cannot replicate, as microbes or genetically modified organisms can, which are restricted by several regulatory policies (Alcalde *et al.*, 2006).

Despite the advantages offered by enzymes, the high enzyme production costs, low enzyme activity, low substrate specificity, and low temperature and pH stability might prevent enzymatic bioremediation applications (Alcalde *et al.*, 2006). Enzyme immobilization is an effective approach to improve enzyme activity and stability, which allows the enzymes to be reused, without wash-out in enzymatic reactors. Immobilization insolubilizes water-soluble enzymes. Some natural biopolymers have been used for enzyme immobilization strategies, such as alginate, agarose, carrageenan, and chitosan, due to their nontoxicity, biocompatibility, physiological inertness, cost, and biodegradability (Bilal and Iqbal, 2020). The spent mushroom substrate acts as a natural enzyme immobilization material, which includes extracellular oxidative enzymes that are generated from the fungal metabolism. The extracellular polymeric substances produced during fungal mycelia development in the solid substrate act as an adhesion compound, which prevents fungal dehydration, storage of excess nutrients, and extracellular enzyme immobilization. Immobilized laccases could be an effective, eco-friendly, and commercial alternative to the physical, chemical, and oxidative dye decolorization methods. The use of enzymatic bioremediation of dyes in real wastewater could be reduced due to the presence of salts, chelating agents, by-products, or surfactants. The combination of bioremediation with a chemical method, such as adsorption might increase the efficiency of degradation and removal from the waste stream (Morsy *et al.*, 2020).

The combination of adsorption and enzyme bioremediation display a synergic effect since adsorption can concentrate the diluted dyes from the water phase, which are then degraded by enzymes.

17.2 ADSORPTION OF AZO DYES

Due to their low cost and effectiveness, azo dyes are the largest class of dyes used in industrial applications. They are predominantly used in the textile, printing, food, paper, leather, cosmetics, and pharmaceutical industries. The global production of these dyes was estimated to be approximately 460,000 t/year, which represents approximately 60%–70% of the total dyes produced (Shah, 2019).

Azo dyes are characterized by the presence of ≥1 azo groups (–N=N–) in their chemical structure, which can be classified according to the number of these groups (e.g., monoazo, diazo, triazo, and polyazo), that are responsible for the color, denominated chromophores (Benkhaya, M'rabet, and Harfi, 2020). The azo group is associated with auxochromic hydroxyl (–OH) or amino (–NH$_2$) groups that intensify the performance of the chromophores.

One of the main disadvantages associated with azo dyes is their ability to induce mutagenicity or carcinogenicity, or both due to the dye or its metabolites. Functional groups and their position in the dye molecule determine the toxicity. After the degradation of azo dyes, xenobiotic compounds can be generated, such as β-naphthylamine, aniline, triazine, p-phenylenediamine, and β-amino-α-naphthol. On average, 20%–50% of the dyes used in textile dyeing are eliminated in the effluents. In total, ≤50% of the annual production of dyes directly affects the environment, mainly due to the inefficiency or lack of industrial effluent treatment plants. In addition to the toxic characteristics of azo dyes, they cause several environmental impacts, such as impairing aesthetics, altering biological cycles, and causing damage to aquatic communities (Khatri *et al.*, 2015).

In general, conventional effluent treatment processes cannot degrade or remove dyes. Therefore, new complementary methods for the discoloration of effluents must be developed, which consider sustainability, efficiency, viability, ease of operation, and cost. Therefore, enzymatic degradation and adsorption are suitable methods for the removal of azo dyes from effluents, which will be presented and discussed in the following section.

17.2.1 ADSORPTION MECHANISMS

Adsorption is a mass transfer mechanism where certain substances, which are present in the liquid or gaseous fluids, are concentrated on the surface of solids, which allows the components of these fluids to be separated (Worch, 2012). Adsorption mechanisms can be classified as physisorption, chemisorption, and ion exchange. Understanding these mechanisms is essential to determine the favorable conditions for adsorption, which enables the regeneration and recycling of the adsorbent in a sustainable and economic process (Ray, Gusain, and Kumar, 2020).

The most described adsorption mechanisms are physisorption and chemisorption. Physisorption is a weak and reversible intramolecular interaction between the adsorbent and the adsorbate. This mechanism is an exothermic phenomenon that promotes the formation of multilayer adsorbates on the adsorbent surface. The functional groups present on the surface of the adsorbent and in the adsorbate are critical for physisorption. Chemisorption generates irreversible chemical bonds, which results in a stronger bond compared with physisorption. Chemisorption is endothermic and changes the electronic state of the adsorbent. For the process to be energetically favorable, there must be efficient interactions with the active sites of the adsorbent. Ion exchange, which is considered a chemisorption mechanism by some researchers, is an irreversible process in which the ions on the surface of the adsorbent are replaced by the ions of the adsorbate (Ray, Gusain, and Kumar, 2020).

More than one of these mechanisms can occur simultaneously during adsorption and depends on the composition, structure, and properties of the adsorbent and the contaminant, and some solution conditions, such as pH, ionic strength, and temperature. Several interactions are involved in each mechanism: Physisorption involves surface adsorption, Van der Waals forces, hydrogen bonds, hydrophobic interactions (π–π interactions, Yoshida interactions) and diffusion in the material network. Chemisorption involves electrostatic interactions, complexation, chelation, formation of inclusion complexes, proton displacement, covalent bonding and oxidation or reduction. Ion exchange involves reversible ion exchange (Crini *et al.*, 2019). The adsorption of Congo red azo nanocrystalline hydroxyapatite can be explained by three different mechanisms: (1) hydrogen bonds in the azo or $-NH_2$ groups of the dye with hydroxyl (OH^-) and phosphate (PO_4^{3-}) ions of the adsorbent; (2) chelation of calcium (Ca^{2+}) ions on the surface of the adsorbent by the azo, $-NH_2$, or sulfonate (SO_3) groups of the dye; (3) ion exchange between PO_4^{3-} of the adsorbent and the SO_3 of Congo red. Liu et al. (2020), when evaluating the effect of pH on the efficiency of metallic structures based on Co(II) in adsorbing Congo red and Orange IV, found that electrostatic interaction was one of the main adsorption mechanisms. They concluded that in addition to electrostatic interaction and pore filling, adsorption was due to the interaction between π-π stacking and H bonds, which meant that adsorption was a result of the synergistic effect of different mechanisms.

17.2.2 Adsorption of Azo Dyes

The characteristics of the adsorbents are mainly responsible for the efficiency during the separation of the contaminants. An adsorbent commonly used to remove azo dyes is activated carbon (C), which is characterized by a porous structure formed from carbonaceous materials after chemical or physical activation (Worch, 2012). New materials and activation techniques for the production of activated C were evaluated and developed by several researchers, which aimed at the removal of azo dyes.

Kittappa *et al.* (2020) synthesized a magnetized activated C from palm bark coated with silica. According to the Langmuir model, the activated C showed an adsorption capacity of 528.98 and 257.25 mg/g of the Methyl orange and Congo red azo dyes, respectively. This adsorbent was efficient in removing the mixture of these dyes. The applied kinetic models revealed that the mechanism was physical with endothermic and spontaneous thermodynamic parameters. A benefit of using this material is that it can be separated by a magnet, regenerated, and reused four times without significant loss of adsorption capacity.

An activated C was developed from biological sludge from a beverage effluent treatment plant. For Allura red AC, this adsorbent had a maximum removal capacity of 287.1 mg/g at pH 2. The pseudo-second-order kinetic model represented the experimental data satisfactorily and indicated that adsorption was favored at the highest initial concentrations of the dye. The increase in temperature favored adsorption, which suggested endothermic adsorption, because the increase in temperature reduced the interaction of the solvent with the adsorbent and exposing a higher number of active sites (Streit *et al.*, 2019).

Despite its efficiency at removing azo dyes, adsorption with commercially available activated C (e.g., wood and coal) is relatively expensive due to the cost of the material and its preparation. The use of cheap and abundant natural materials, such as waste and different types of biomass, in the synthesis of activated C, contributes to a reduction in cost and the sustainability of the process. However, regeneration of the exhausted activated C is expensive and complex and results in the loss of adsorbent. Therefore, to replace the use of activated C, new and low-cost adsorbents have been tested for their efficiency and application for the removal of dyes. Among the wide variety of materials proposed in the literature, Crini *et al.*, (2019) reported that low-cost adsorbents used in the removal of dyes could be classified in the following categories: solid urban; agricultural and forest residues; industrial by-products; natural materials; biosorbents; and others.

Agricultural and forest residues are abundant, cheap, and readily available. The variety of functional groups on their surface, such as –OH, carbonyl, –NH$_2$, carboxyl, methyl, and others, allows these to be used as adsorbents for a variety of pollutants, such as azo dyes (Zhou *et al.*, 2019).

The adsorption of cationic and anionic dyes by banana, cucumber, and potato peels without any modification was studied by Stavrinou, Aggelopoulos, and Tsakiroglou (2018). The Orange G anionic dye showed greater removal under acidic conditions (pH<4), because the adsorbent surface was positively charged, which promoted the adsorption of the negatively charged dye. Based on the analysis performed, chemisorption was the predominant adsorption mechanism and the kinetics were governed by the Elovich and pseudo-second-order models. The Langmuir isotherm revealed maximum adsorption capacities of 20.9, 23.6, and 40.5 mg/g for banana, cucumber, and potato peels, respectively.

Chicken eggshell was evaluated as an adsorbent for Remazol Brilliant Violet 5R by Rápó et al. (2020). They obtained a 95% removal with ideal parameters (20 mg/L of dye, 1.5 g of biomass, adsorbent size 160 μm, pH 6, and 20°C). The adsorption capacity and removal efficiency increased with decreasing particle size. The thermodynamic parameters indicated that the process was physical and spontaneous, and from the Langmuir model, the adsorption occurred in a monolayer. For the kinetic study, intraparticle diffusion did not affect adsorption and the pseudo-second-order model governed the process.

The fly ash produced by coal-fired power stations has been tested as an azo dye adsorbent. Kumar *et al.* (2018) incorporated magnesium oxide (MgO) into fly ash to improve the removal efficiency of Reactive Blue 5. In this study, an adsorption capacity of 48.78 mg/g was obtained by the Langmuir model. However, when reusing the compound, the adsorption capacity decreased by 50% in each successive 90 min cycle, and therefore, required an alternative regeneration process.

Among the natural materials used as adsorbents, clays, zeolites, and their compounds are of interest. The limitations of using clay as an adsorbent are the negative natural charge that prevents the adsorption of acid dyes and the difficulty in reuse due to high. Therefore, new methods of clay modification and crosslinking have been developed, which are aimed at the better adsorption of different dyes. Gamoudi and Srasra (2019) analyzed the removal of Methyl orange in a modified clay with the hexadecyl pyridinium (HDPy$^+$) cationic surfactant. They obtained a maximum adsorption capacity of 227.27 mg/g by the Langmuir model and the process was exothermic and independent of pH.

The adsorption capacity of zeolites is mainly due to the potential of ion exchange (which is similar in clays), in addition to the high porosity and surface area. Zeolites have a lower adsorption capacity than clay, with unsatisfactory anion and organic adsorption performance, which require modifications. However, these materials are inexpensive and abundant (Zhou *et al.*, 2019).

Mahmoodi and Saffar-Dastgerdi (2019) synthesized zeolite nanoparticles and modified them with (3-aminopropyl) triethoxy silane. In this study, the maximum adsorption capacities of 4842 and 2415 mg/g of Direct Red 23 and 80, respectively. The increase in temperature caused an improvement in adsorption, which was attributed to the greater energy of the dye molecules that interacted with the active sites of the adsorbent. The results indicated that the adsorption was physical, exothermic, and spontaneous.

Pillar[5]arene-modified zeolite was developed and evaluated for its adsorption capacity of Orange dye. At 100 mg/L dye, the removal efficiency of 81.51% was achieved after 960 mins. The pseudo-second-order kinetic model was efficient in describing the kinetic behavior of adsorption. For the adsorption isotherms, the Langmuir model represented the results and indicated a q$_{max}$ of 18.33 mg/g (Yang *et al.*, 2019).

Low-cost biological adsorbent materials include chitosan and its compounds, peat, yeast, fungi, or bacterial biomass. Biosorbents are inexpensive materials that have a variety of functional groups that can adsorb dyes (Crini *et al.*, 2019). Currently, chitosan is one of the most explored biosorbents for the adsorption of dyes. They are biopolymers that are commercially extracted from the crustacean exoskeleton and are generated as a by-product during food processing. The $-NH_2$ and OH functional groups present in large quantities on the chitosan surface are responsible for the high potential for the adsorption of dyes. The protonation of the $-NH_2$ group, which adsorbs cationic dyes by electrostatic interaction, is the main mechanism of adsorption (Zhou *et al.*, 2019). An advantage of chitosan is its versatility, because it can be used in different forms, such as flakes, gels, spheres, or fibers. However, the use of chitosan to remove pollutants is restricted due to its solubility in acidic solutions, low mechanical resistance, and deformation after drying, which requires modifications, such as crosslinking, impregnation, and grafting for it to be an effective adsorbent (Crini *et al.*, 2019).

When developing an ecological and less toxic cross-linked chitosan, Santos and Zaritzky (2020) used oxalic acid as a crosslinking agent for chitosan hydrogels. This study proved that the crosslinking employed reduced swelling and improved the chemical stability of chitosan at a pH<2.5. In addition, the adsorption of Direct Red increased, which resulted in maximum removal of 90.6% at pH 4. Exposure of the adsorbent to continuous adsorption/desorption cycles revealed that the synthesized material could be reused without losing its adsorption capacity.

Jawad, Mubarak, and Abdulhameed (2020) modified the cross-linked chitosan with glutaraldehyde and added titanium dioxide (TiO_2) nanoparticles, for the adsorption of Reactive Red 120. They verified that the equilibrium between the content of the $-NH_2$ functional group and the surface area was the factor that determined the adsorption capacity of the nanocomposite. The best adsorption results were achieved using 25% TiO_2 nanoparticles, with a q_{max} of 103.1 mg/g (obtained by the Langmuir model).

Cyclodextrins, which are cyclic derivatives of starch, have been evaluated as adsorbent materials. These polymers have a hydrophilic outer cavity that swells in water and allows for the quick diffusion of dyes, and the highly hydrophobic inner cavity of cyclodextrin captures nonpolar dyes (Crini *et al.*, 2019). The main disadvantage of using cyclodextrin is its solubility, which prevents its direct use in solutions. To improve the adsorption potential, Jiang *et al.* (2020) cross-linked β-cyclodextrin with tetrafluoroterephthalonitrile, which has a highly porous surface area and modified it with ethanolamine to introduce a positively charged group, which attempted to remove anionic dyes. The resulting polymer had a maximum adsorption capacity of 290 and 1,250 mg/g of Methyl orange and Congo red anionic dyes. The high adsorption was attributed to the effect of three mechanisms involved in the process: electrostatic interaction, inclusion complexation, and pore capture. For the simultaneous removal of organic dyes and heavy metal ions, Qin *et al.* (2019) developed a polymeric β-cyclodextrin adsorbent with an expanded pocket structure. In the simultaneous adsorption of Congo red and cadmium [Cd(II)], the adsorption of the dye, when present at ≤400 mg/L, was not affected by Cd(II). In concentrations ≥ 600 mg/L, inhibition by Cd(II) was verified. The removal of Cd(II) was favored in the presence of the dye, especially at Cd(II) ≥50 mg/L. The results inferred that the $-NH_2-$, $-N=N-$ and sulfonic acid groups of the adsorbed dye molecule acted as extra binding sites to increase Cd(II) adsorption. This study proved that the developed adsorbent did not promote competitive adsorption and had a high adsorption capacity for Congo red azo dye and Cd(II) in binary systems. However, the recovery of the adsorbent was not achieved and the cost of this adsorbent is higher compared with other adsorbents, such as polymer containing β-cyclodextrin and activated C (Qin *et al.*, 2019).

Nanomaterials and metal–organic frameworks are promising emerging adsorbents for dye adsorption. Nanosponge β-cyclodextrin (β-CD) polyurethane was modified with phosphorylated multiwalled C nanotubes and coated with silver (Ag) and TiO_2 by Taka *et al.* (2020). The nanobiosorbent showed a high surface area (352.5 m²/g) and high adsorption capacity for Congo red in synthetic samples (q_{max}= 146.96 mg/g) and in low concentrations real effluent. In addition, the nanobiosorbent was effectively reused after desorption and regeneration, without significant loss of efficiency in ≤3 cycles.

Hafdi *et al.* (2020) proposed a simple method to synthesize a natural phosphate (P) adsorbent doped with nickel oxide (NiO) nanoparticles. The adsorbent promoted a maximum removal of 96% of Reactive ed 141 under ideal conditions (20 mg/L, pH 6, and 0.1 g/L of adsorbent). The estimated cost of the adsorbent in this study indicated the commercial viability compared with other adsorbents. Despite the efficiency of nanomaterials in the removal of dyes, the possible toxicity of these adsorbents must be considered and evaluated for their effective application.

Metal–organic frameworks based on cobalt [Co(II)] were efficient in adsorbing Congo red and Orange IV, with a maximum adsorption capacity of 762.3 and 739.9 mg/g according to the Langmuir isotherm (Liu *et al.*, 2020). Liu *et al.* (2020) found that metal–organic frameworks with different ions [zinc (Zn), copper (Cu) and Ni] showed an adsorption capacity for Alizarin Yellow GG >500 mg/g. The type of metal ion significantly influenced adsorption and the use of Zn as a binder resulted in a greater adsorption capacity compared with other metals. Despite the promising results of metal–organic frameworks for the adsorption of dyes, they are high-cost materials that require a complex synthesis process, and new methods need to be studied and developed to enable the application of these adsorbents on a large scale (Zhou *et al.*, 2019).

Biochars are an ecological and inexpensive alternative to activated C, which are defined as carbonaceous materials produced by pyrolysis or gasification of biomass under a controlled atmosphere at a temperature from 350°C to 800°C. Kelm *et al.*, (2019) produced biochar by gasifying solid wood residues for the adsorption of Indosol Black NF1200. At pH 2, granulometry of the biochar of 40 and 100 mesh, and initial concentration of 50 mg/L, 99% removal were obtained. However, the maximum adsorption capacity at pH 12 (14 mg/g) was 15 times lower than that obtained at pH 2 (185 mg/g).

Biochars were produced from agricultural waste by rapid pyrolysis (Khan *et al.*, 2020). Maximum removal efficiencies were 96.9% and 98.8% of Congo red using rice husk and cow manure biochars in a compact bed bioreactor, respectively. This study produced two bio oils that resulted from the biochar pyrolysis, which are products of commercial interest, mainly for the cosmetics and pharmaceutical industries.

Table 17.1 presents a synthesis of recent studies that developed and evaluated different adsorbents and their ability to adsorb azo dyes. The reported optimum conditions of pH, temperature, and adsorbent concentration are the result of optimization tests. The exact or approximate equilibrium times obtained by the researchers are listed.

17.3 ENZYMATIC CATALYSIS OF AZO DYES

The enzymatic treatment of textile effluents to remove dyes involves the use of enzymes that are produced by different microorganisms, such as fungi, bacteria, and yeasts. An essential requirement for the application of enzymes in the removal of azo dyes is their resistance to the toxic effects of the dye and the substances present in the textile effluent as well as to pH and temperature conditions. When employing enzymatic systems in the treatment of textile effluents, special attention must be given to the formation of toxic metabolites by cleavage of the azo bond, mainly amines and their derivatives. Therefore, the identification of the compounds formed and toxicity or mutagenicity tests are essential for the effectiveness of the treatment (Chacko and Kalidass, 2011).

A wide variety of azo dyes with different chemical structures are used in the textile industries and only a few enzymes can degrade the dyes. Promising results for the degradation of azo dyes have been obtained using the enzymes azoreductase, laccase, manganese peroxidase (MnP), lignin peroxidase (LiP), polyphenol oxidase, and versatile peroxidase (Singh, Singh, and Singh, 2015). Due to their ability to degrade lignin, laccase, LiP, and MnP are efficient at the degradation of azo dyes

TABLE 17.1
Azo Dyes Adsorption Studies

Adsorbent	Azo Dye	Maximum Adsorption Capacity[a]	pH	Temperature (°C)	Dose of Adsorbent	Equilibrium Time	References
Cross-linked chitosan/oxalic acid hydrogels	Reactive Red 195	110.7 mg/g	4	45	4 g/L	16 h	Pérez-calderón, Santos, and Zaritzky (2020)
3D Superhydrophilic polypyrrole nanofiber	Metanil Yellow, Acid Orange 7, Acid Yellow 9	84.70 mg/g, 81.15 mg/g, 68.77 mg/g	1	45	1 g/L	3 h	Qi et al. (2020)
Hydroxyapatite nanorods synthesized from phosphogypsum waste	Congo red	122.4 mg/g	5.5	25	2 g/L	20 min	Bensalah et al. (2020)
Magnetized and chemically modified chitosan beads	Cibacron Brilliant Yellow 3G-P, Cibacron Brilliant Red 3B-A	243.90mg/g, 666.66 mg/g	2	25	1 g/L	30 min, 45 min	Muedas-Taipe et al. (2020)
Zr-based MOF nanoparticles	Methyl orange, Methyl Red	454.54, 384.62	4	25	–	–	Ahmadijokani et al. (2020)
Spent mushroom waste	Direct Black 22, Direct Red 5B, Reactive Black 5, Direct Blue 71	15.46 mg/g, 18.08 mg/g, 14.62 mg/g, 20.19 mg/g	2	50	30 g/L	240 min	Alhujaily et al. (2020)

with various structures. In the following section, the degradation of azo dyes by these enzymes will be discussed.

17.3.1 AZO DYE DEGRADATION PATHWAYS BY LACCASES

Laccases are glycoproteins phenoloxidase that belongs to the multi-Cu group and has low substrate specificity. This enzyme directly catalyzes the oxidation of phenolic compounds with the simultaneous reduction of molecular O_2 in water, without the need for hydrogen peroxide (H_2O_2). In the presence of suitable mediators, such as 2,2'-azino-bis-(3-ethylbenzo-thiazol-6-sulfonic acid) (ABTS), 1-hydroxybenzotriazole, 3-hydroxyanthranilic acid, and syringaldehyde, laccase oxidation capacity increases, and it can degrade nonphenolic structures and other compounds with high redox potential. The laccase molecule has four Cu atoms of three ionic types, which have different catalytic functions and spectroscopic and paramagnetic characteristics: Type 1 (one Cu atom); Type 2 (two Cu atom) and Type 3 (three atoms Cu). The oxidation of substrates is related to the reduction of molecular O_2 and the generation of two water molecules (H_2O); therefore, for each reduced O_2, four substrate molecules are oxidized. Therefore, laccase is considered a green catalyst, because it uses O_2 as a cosubstrate and generates H_2O as a by-product (Riva, 2006). The degradation of azo dyes by laccase occurs by symmetrical or asymmetric cycling with a mechanism of highly nonspecific free radicals that form phenolic compounds. The degradation of azo dyes by laccase starts with the cleavage of the azo bond followed by oxidative cycling, desulfonation, deamination, demethylation, and dihydroxylation, which depend on the structure of the dye and the laccase enzyme.

Adnan *et al.* (2015) determined that laccase plays an important role in the biodegradation of Reactive Black 5 by *Trichoderma atroviride* F0, and the presence of LiP and MnP was not detected. In this study, the degradation of the dye was initiated by the cleavage of the azo bond, which resulted in naphthalene-1,2,8-triol and sulfuric acid mono-[2- (toluene-4-sulfonyl)-ethyl] ester. Then, the sulfuric acid mono-[2- (toluene-4-sulfonyl)-ethyl] ester was desulfonated in 1,2,4-trimethylbenzene. The oxygenated ring of naphthalene-1,2,8-triol was cleaved into 2- (2-carboxy-ethyl)-6-hydroxy-benzoic acid, which might have been degraded by two methods, decarboxylation and methylation with the formation of 2,4-di-tertbutyl phenol, and decarboxylation that resulted in benzoic acid-trimethylsilyl ester. Because toxic aromatic amines were not formed during degradation, it was proven that this laccase has the potential to be applied in the treatment of textile effluents.

The degradation of Congo red azo by a laccase produced by the white-rot fungus *Oudemansiella canarii* was investigated by Iark *et al.* (2019). They verified the asymmetric cycling of the azo bond and the generation of several products. The results indicated that Congo red was oxidized and resulted in a species with NH_2 nitrification and SO_3 loss. The release of naphthalene, which is a nitrified or hydroxylated species, was due to the oxidation of the $-NH_2$ groups present in the dye molecule and its intermediates. In the final stage of degradation, the benzene ring was cleaved, forming a fully oxygenated compound. This study concluded that the degradation of the compounds resulting from the disintegration of the Congo red molecule was easier and that they were less toxic compared with the dye, which indicated the viability of the process.

Pandi, Marichetti, and Numbi (2019) evaluated the discoloration of azo dyes by laccase from a new strain of *Peroneutypa scoparia* and reported that the degradation pathway for Acid Red 97 could be described in three phases. First, the laccase oxidized the phenolic group of the dye with an electron, which produced a phenoxy radical. Second, the oxidation of the carbon ion occurred. Finally, the carbon in the phenolic ring with the azo bond underwent a nucleophilic attack by water and produced the metabolites naphthalene 1,2-dione and 3-(2-hydroxy-1-naphthylazo) benzenesulfonic acid.

Chhabra, Mishra, and Sreekrishnan (2015) determined the degradation pathways of Acid Red 27 by immobilized laccase of *Cyathus bulleri* in the presence of the 1-hydroxybenzotriazole mono-hydrate (HOBT) mediator. Here, 4-((2-oxo-3,6-disulfo-2,3-dihydronaphthalen-1-yl)diazenyl) naphthalene-1-sulfonate (Compound I), which was identified as a product of degradation, was formed by the oxidation of the phenolic group in quinine. The laccase-mediator system aided the oxidation of two electrons to form the phenoxy radical and the dye molecule undergoes additional conversion. Then, a water molecule promoted a nucleophilic attack, which created a site for the subsequent action of the laccase-mediator system. Compound I underwent oxidation that resulted in 4-diazenylnaphthalene-1-sulfonate (Compound II), which is oxidized again and results in naphthalene-1-sulfonate (Compound III). The N from the azo bond was released as molecular N. In addition, this study revealed that the laccase-mediator system did not add toxicity and reduced mutagenicity according to an Ames test.

17.3.2 AZO DYE DEGRADATION PATHWAYS BY LIP AND MNP

LiP is an oligomannose-type glycoprotein that contains protoporphyrinic iron [Fe(IX)] (heme) as a prosthetic group and requires H_2O_2 for catalysis. This is one of the most important enzymes in lignin degradation, because it has a high redox potential compared with other enzymes. For the substrate, this enzyme is relatively nonspecific and can oxidize both the usual substrates of peroxidases (phenols and aromatic amines) and a variety of other nonphenolic aromatic structures and organic compounds (Valli, Wariishi, and Gold, 1999).

In the LiP catalytic cycle, the Fe present in the heme group undergoes different redox states. The preferred reducing agent of LiP is veratryl alcohol, a metabolite of the white-rot fungus that prevents the inactivation of the enzyme by excess H_2O_2. Veratryl alcohol is oxidized to veratraldehyde by LiP and is used in tests to determine the activity of these enzymes.

MnP is similar to LiP and is a glycoprotein that presents heme Fe as a prosthetic group and needs H_2O_2 for its activity. The main characteristic that distinguishes MnP from other peroxidases is the requirement for the presence of manganese (Mn^{2+}) for the catalytic cycle to be completed. In addition, the redox potential is lower compared with LiP and, therefore, its enzymatic mechanisms cannot oxidize nonphenolic compounds (Kirk and Cullen, 1998).

MnP catalyzes the oxidation of Mn^{2+} to Mn^{3+}, which is stabilized by organic acids produced by fungi (e.g., oxalate, malonate, tartrate, or lactate), becoming a low molecular mass agent that removes electrons from nonspecific organic compounds, which promotes the formation of highly reactive radicals that act on the oxidation of various recalcitrant pollutants (Perminova, Hatfield, and Hertkorn, 2002).

Liu *et al.* (2020) evaluated the degradation pathways and detoxification of Congo red with recombinant ligninolytic enzymes from *Aspergillus sp.* expressed in *Pichia pastoris*. Dye degradation by LiP started by the asymmetric cleavage of N=N bonds, which formed pyrrolo [1,2-a] pyrazine-1, 4-dione, hexahydro-3-(phenylmethyl) and 1-proline-N-allyloxycarbonyl-tetradecyl ester. Then, the asymmetric cycling of the C–N bond occurred, removing N from the aromatic ring and generating various products, such as 9-octadecenamide, (Z)- and phenylethyl alcohol and *p*-xylene. Finally, the breakdown of benzene resulted in stable low molecular weight products, such as butanoic acid, and 3-methyl and 4-methyl-2-oxovaleric acid. The degradation of Congo red by MnP was different from that of LiP. After the asymmetric cycling of the N=N group by MnP, dibutyl phthalate was detected, which, with the breakdown of benzene, was converted into 1,9-dioxacyclohexadeca-4,13-diene-2-10-dione, 7,8,15,16- tetramethyl, and then transformed into triethyl citrate. Other degradation products were detected, such as tryptophol, diethyltrisulfide (from the SO_3 group of the dye) and 2-methyl-butanoic acid. Dye degradation by LiP and MnP promoted the reduction of phytotoxicity in pepper seeds (Liu *et al.*, 2020)

Gomare, Jadhav, and Govindwar (2008) studied the degradation pathways of diazo Blu-2B by the LiP from *Brevibacillus laterosporuc* MTCC 2298. With the removal of two N_2 molecules, 1'-dicyclohex-2, 5-ene-4-one (Compound 1) which underwent α-cleavage was transformed into 4-(2-hexenoic acid)-2, 5-cyclohexadieneone. Redox reactions and the subsequent cleavage of the benzene ring of α-naphthol derivatives resulted in dehydroacetic acid. The results indicated that in addition to the cleavage of the azo bond, LiP can remove sulfate groups from the dye.

The discoloration of the Congo red dye by *Aspergillus niger* was studied by Asses *et al.* (2018). Dye degradation occurred due to deamination and the action of LiP and MnP. Different degradation stages generated different intermediates: (1) simultaneous total deamination and oxygenation of the dye; (2) partial deamination and asymmetric cleavage of the C–N bond between the aromatic ring and the azo group with the loss of a sodium (Na) atom; (3) asymmetric cleavage of the C–N bond followed by deprotonation; (4) opening of the benzene ring and dehydrogenation; and (5) cleavage by peroxidases resulted in the metabolites sodium naphthalene sulfonate and cycloheptadienylium.

17.3.3 STUDIES ON THE ENZYMATIC DEGRADATION OF AZO DYES

The efficiency of degradation of azo dyes by enzymes depends on different factors that involve the culture conditions, the microorganism, and the nature of the extract (raw or purified), and the degradation conditions (e.g., temperature, pH, type, and concentration of the dye, and reaction time) (Martínez *et al.*, 2020). Several studies have been attempted to improve the production of enzymes by different microorganisms, effective applications in the treatment plants for the removal of dyes, the reduction of costs, and the sustainability of the process.

The use of agro–industrial waste as a fermentation medium for microbial growth and the production of enzymes is an attractive alternative due to the cost benefits and circular bioeconomy. Omeje *et al.*(2020) evaluated the degradation of different azo dyes by laccase produced by *Aspergillus sp.* Omeje. This microorganism was isolated from a site polluted by oil spills and grown in a medium formulated with different agro–industrial residues (e.g., rice husks, mango husks, orange peel, peanuts husks, and sawdust) as the only C source. The partially purified laccase had a high affinity for the studied azo dyes, which produced >60% discoloration.

Yehia and Rodriguez-Couto (2017) evaluated the degradation of Congo red by MnP produced by under fermentation in a medium made up of agro–industrial residues. Among the residues evaluated, the banana peel provided the greatest laccase activity and promoted a maximum discoloration of 95% at pH 4 and 35°C after one hour of contact. The phytotoxicity of Congo red was reduced after treatment with MnP, according to a test carried out with the bioindicators of world beans (*Phaseolus radiatus*) and sorghum (*Sorghum vulgare*).

There is a difficulty in enzymatic discoloration of textile effluents as they do not only contain dyes but also salts, extreme pH values, chelating agents, by-products, precursors, surfactants, metals and others, which can negatively influence the catalytic activity of enzymes. Dye mixing can also result in compounds of different structures that significantly affect discoloration. Vantamuri and Kaliwal (2016) found that the purified laccase of the new white-rot fungus *Marasmius sp.* BBKAV79 can degrade 88% to 93% of Navy blue HER, Green HE4BD, and Orange HE2R azo dyes in 96 h. However, the metal ions mercury [Hg^{2+} (96.5%)], Ag^+ (96.6%), and Fe^{3+} (41.7%) and the agents H_2O_2 (98.51%), sodium chloride [NaCl (76.75%)], and sodium dodecyl sulfate (SDS) (66.83) at a concentration of 20 mM and 1%, respectively, showed a high percentage of laccase inhibition. Yang *et al.* (2018) cloned a laccase from samples of marine sediments and found that in addition to promoting an efficient degradation of azo dyes in the presence of ABTS as a mediator (77.6% of Acid Violet and 95.4% of Congo red), this enzyme is thermostable and tolerant to organic solvents and salt, being suitable for the treatment of textile effluents.

In a study, the application of a commercial laccase was more efficient in the degradation of Direct Red 31 and Acid Blue 92 in a simple system than in a binary system. The addition of 0.02 M NaCl caused a significant reduction in the discoloration of the dyes due to enzyme inhibition, and the addition of sodium carbonate, sodium bicarbonate, and sodium sulfate had no significant effect on degradation. The optimum conditions determined were the dye concentration (500 mg/L), reaction time (10 min) and pH 5 (Mohajershojaei, Khosravi, and Mahmoodi, 2013).

The use of mediators in the enzymatic reaction has enabled the action of enzymes in a variety of compounds and the achievement of better degradation efficiencies. In the enzyme-mediator system, the mediator is oxidized by the enzyme and subsequently, the oxidized mediator oxidizes the dye molecule, forming the product and the mediator is regenerated. Wang *et al.* (2018) analyzed a new purified laccase from *Trametes sp.* F1635 and its ability to degrade azo dyes in the presence of different mediators. The discoloration rate s increased from 89%, 20%, and 61% without a mediator to 95.2%, 88.9%, and 75.7% for Evans Blue, Methyl Orange with violuric acid, and Eriochrome Black with acetosyringone, respectively.

Forootanfar *et al.* (2016) found that the discoloration of the studied azo dyes (e.g., Acid Orange 67, Disperse Yellow 79, Basic Yellow 28, Basic Red 18, Direct Yellow 107, and Direct Black 166) by a purified laccase from *Paraconiothyrium variabile* increased in the presence of the hydroxybenzotriazole mediator in a concentration of ≤5 mM and decreased by 10 mM. Bilal *et al.* (2016) studied a purified MnP produced by *Ganoderma lucidum* IBL-05 and obtained Reactive red 195A removal of 57% and 69.5% by free and immobilized MnP, and in the presence of the $MnSO_4$ mediator (1 Mm) of 59.7% and 78.6%, respectively.

To improve enzyme production that might be low in native hosts, the development of recombinant proteins has been studied, which involves homologous and heterologous expression in prokaryotic and eukaryotic systems. Several studies have evaluated the ability of recombinant enzymes to degrade azo dyes. A mutant composed of CotA-laccase produced by *Bacillus pumilus* and expressed in *P. pastoris* could remove 80.35%, 71.44%, and 64.72% of Evans blue, Reactive brilliant orange K-7R, and Reactive Black KN-B in 24 h and with ABTS 0.3 mM as a mediator, respectively (Xia *et al.*, 2019). Chen *et al.* (2015) found that a recombinant MnP from *Irpex lacteus* F17 expressed in *Escherichia coli* promoted a maximum degradation of 80.2% of the Reactive Black 5 at pH 3.5 and 35 °C after a 90 min reaction. The results indicated that the degradation capacity of recombinant MnP was the same as that of native MnP.

To eliminate costs with the enzyme purification process, the use of crude extract for the degradation of dyes has been a studied alternative. Sosa-Martínez *et al.* (2020) evaluated the potential for discoloration of different dyes by the crude enzymatic extract produced by *Phanerochaete chrysosporium* CDBB 686, containing LiP and MnP as main enzymes. This study found that the extract provided a greater or similar and faster degradation with discoloration in vivo. Under the optimized conditions, the crude extract was able to discolor 41.84% of Congo red. However, degradation products showed greater toxicity compared with Congo red.

Vats and Mishra (2017) obtained the removal of 84%, 81%, and 76.6% for Orange 16, Reactive Black 5, and Reactive Violet 5 with a crude extract containing laccase from *C. bulleri* in the presence of ABTS (100 μM). Without the presence of a mediator, Tavares *et al.* (2020) found a removal greater than 90% of Reactive Black 5 and Reactive Blue 220 using the crude enzyme laccase extract produced by *Lentinus crinitus*. Fernandes *et al.* (2020) found that a crude enzymatic extract produced by *Pleurotus sajor-caju* produced 86% removal of Reactive Black 5 without a mediator.

Due to the cost and time required for the production of enzymes by microorganisms in suitable culture media and controlled conditions, the extraction of readily available enzymes could be a promising alternative, mainly in solid residues from the production of mushrooms, which are fungi that can produce ligninolytic enzymes. Singh *et al.* (2011) evaluated the potential of crude concentrated enzymes extracted from the residual substrate from the production of *P. sajor-caju* for the discoloration of dyes. The removal efficiency was 84.0% for Black 5 and 80.9% for Orange 16 after 4 h of incubation with 45 U of LiP at pH 4.5, 1 mM veratryl alcohol and 0.2 mM H_2O_2 at 30°C. In another study, Singh *et al.* (2010) obtained discoloration of 87.8%, 87.2%, and 63.8% for Tryphan blue, Starch black, and Congo red with 25 U of LiP extracted from the residual substrate of the production of *P. sajor-caju* after 24 h under the same conditions used in the study by Singh *et al.* (2011).

Some studies on the enzymatic degradation of azo dyes by laccase, MnP, and LiP, purified or as crude extracts, are presented in Table 17.2.

TABLE 17.2
Studies of the Degradation of Azo Dyes by LiP, MnP and Laccase

Organism	Main enzyme	Assay Conditions	Azo Dye	Removal (%)	Amount of enzyme	Dye (mg/L)	pH	T^a (°C)	Time	References
Aspergillus sp. Omeje	Laccase	Partially purified laccase with H_2O_2	Yellow 6, Acid Red 337, Basic Blue 22, Purple, Yellow, Brilliant black	41 39 58 73 65 62	–	–	5	37	120 min	Omeje et al. (2020)
Alcaligenes faecalis XFI	Laccase	Partial purified and immobilized laccase on chitosan-clay composite beads	Methyl Red, Reactive Black 5	82 69	–	100	9	50	24h	Mehandia, Sharma, and Arya (2020)
Anthracophyllum discolor	MnP	Ultra-purified and immobilized MnP on Fe oxide/chitosan magnetic nanocomposite	Orange 16	98	–	50	7	37	50 min	Siddeeg et al. (2020)
Phanerochaete chrysosporium CDBB 686	MnP LiP	Crude enzymatic extract and H_2O_2 1.83 mM	Congo red	41.84 (LiP) 35.13 (MnP)	8.60 U/L (LiP), 5.87 U/L (MnP), 1.57 U/L (Laccase)	50	-	48.36	39,47 h 44.07 h	Sosa-martinez et al. (2020)
Pleurotus sajor-caju CCIBt 020	Laccase MnP	Crude enzymatic extract	Reactive Black 5	86	–	50	5	30	600 min	Fernandes et al. (2020)
Lentinus crinitus	Laccase	Crude enzymatic extract	Reactive Black 5 Reactive blue 220	95 90	–	1000	5	28	24 h	Tavares et al. (2020)

[a] Temperature.

17.4 COMBINED BIOSORPTION AND ENZYME CATALYSIS

Due to the heterogeneity of real effluents that contain azo dyes and their chemical complexity, single technologies for dye removal are limited. A combination of biological and physico-chemical methods might overcome the limitations of single removal techniques, which could result in the effective and desired discoloration and detoxification of azo dyes (Morsy *et al.*, 2020).

The enzyme immobilization support could adsorb the dyes diluted in the liquid phase. Some of the enzyme immobilization materials are similar to those used for dye adsorption, especially biobased materials, such as biopolymers or residues from agro–industrial processes. These materials include various organic (e.g., cellulose, nanocellulose, chitin, chitosan, carrageenan, alginate, gelatin, and collagen) and inorganic matrices [e.g., silica-based supports, zeolite, activated C, celite, magnetic-based supports, acrylic resins, and supported cross-linked enzyme aggregates (CLEAs)], and agro–industrial residues (Wahab *et al.*, 2020).

The combined mechanism of adsorption and immobilized enzyme catalysis is presented in Figure 17.1. The lack of dye degradation during adsorption could be overcome by using enzymes immobilized in the adsorbents. The adsorbed dyes are removed from the liquid phase, and they are degraded by the enzyme catalyst inside the solid matrix of the adsorbent. Then, the residual products are released from adsorbent material (Morsy *et al.*, 2020; Wahab *et al.*, 2020).

The diffusion of the substrates through the solution and the immobilization support of the active enzyme and the release of products and their diffusion into the bulk liquid are limitations for the catalysis of immobilized enzymes. Enzymes located inside the pores of the supporting material would receive less substrate than those located on the support surface, which results in reduced enzymatic catalysis (Wahab *et al.*, 2020).

The diffusion of the dye from the bulk liquid into the liquid film on the adsorbent surface is the first step of adsorption. This is followed by internal diffusion and dye adherence to the active site of the adsorbent (Wang and Guo, 2020). Several diffusion models have been developed to explain the mechanisms of adsorption, which are considered external and internal diffusion. The interaction between the enzyme and the support matrix might cause conformational changes that affect the reaction kinetics. The mass transfer through the stagnant layer around the enzyme and inside the support's pores and the liquid bulk might result in substrate and product gradient concentrations and reduced enzyme catalysis (Valencia and Ibanez, 2019).

The immobilization of enzymes in the supporting materials might rigidify their structure via the formation of multiple covalent linkages, which results in the retention of high activity, even in the presence of agents that are responsible for distorting the enzyme structure, such as organic solvents, urea, acids or bases. Free enzymes are more susceptible to structural modifications caused by these compounds. Enzyme immobilization might provide a stabilization effect and prevent protein unfolding, via intra and intermolecular crosslinking (Mohan *et al.*, 2005; Bilal and Iqbal, 2020; Wahab *et al.*, 2020).

Coimmobilization has been proposed as a solution for pollutants catalyzed by enzyme cascade reactions, and multiple enzymes could be attached to the immobilization matrix to act synergistically, and the immobilization of the mediator for the enzyme reactions that require mediators (Wahab *et al.*, 2020). Enzymes could be released from the immobilized matrix into the liquid phase and act as free enzymes due to their leakage from the carrier.

Dye removal mediated by fungal biomass is an example of a combined technique, which can be categorized as biosorption, bioaccumulation, and biodegradation. The adsorption of dyes in fungal cells is a physicochemical, metabolically independent process that involves the binding of dye molecules to the fungal biomass, in living or dead fungal cells. Many classes of dyes have an affinity to the chitin and chitosan present in many fungal species. Bioaccumulation is related to the biological accumulation of dyes inside the cell cytoplasm. Biodegradation is the enzymatic breakdown

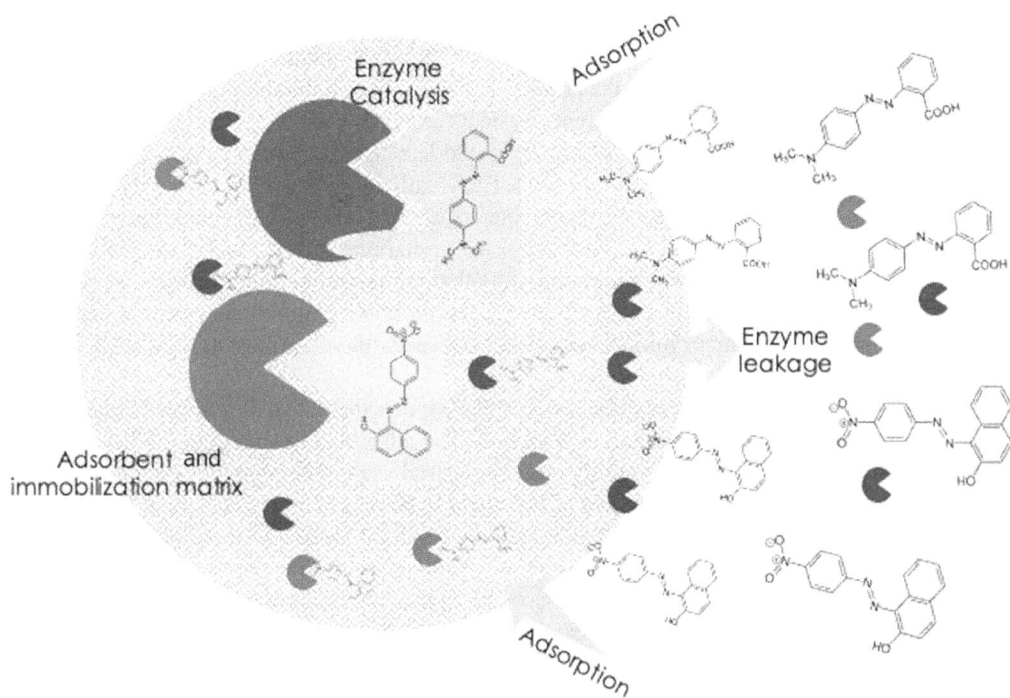

FIGURE 17.1 Mechanistic representation of combined azo dyes adsorption and immobilized enzymatic biodegradation in a solid matrix in a liquid medium.

of azo dyes into simpler molecules. The use of fungal biomass could synergistically combine these three mechanisms to improve dye removal (Singh, 2017).

Bilal *et al.*, (2017) identified an enhanced biocatalytic performance for dye degradation that used horseradish peroxidase immobilized in chitosan beads. They found that immobilized enzymes were more stable at higher temperatures and in the presence of inhibitors and achieved a Remazol Black 5 removal efficiency of 97.82%.

Most research data are related to tests that used dyes with simple, low molecular weight structures that eventually showed high discoloration results. However, real industrial effluents that contain dyes have complex and different chromophore structures in mixed dye wastes. In addition, salts and other compounds are present that might inhibit enzymatic performance. Industrial-scale removal requires cost-effective processes, the use of enzymes (especially immobilized enzymes), which could be reused (Morsy *et al.*, 2020).

17.5 CONCLUSIONS

Increased manufacturing and the use of synthetic compounds in industrial processes are related to the presence of undesirable xenobiotic compounds in environmental matrices, such as water and soil. Azo dyes are used in the textile industry due to their effective ink effects. However, inefficient processes to treat them and their discharge into water bodies is a significant concern, since azo dyes have carcinogenic, genotoxic, and mutagenic effects on humans and animals. Adsorption and enzymatic degradation have been highlighted as efficient azo dyes removal techniques due to their efficiency, viability, ease of operation, and low cost. Adsorption for azo dye removal has been extensively explored; however, pilot and real-scale experiments are scarce, and most of the cited literature describes laboratory-scale experiments. Although it is a cost-effective and robust technique developed, adsorption results in a post-sorption material, which requires further treatment and disposal. The regeneration of adsorbents allows their reuse; however, the desorption phase is generally a liquid phase that contains the contaminant, such as the azo dye. It involves moving the toxic pollutant from one place to another or a different physical form.

Bioremediation can transform harmful azo dyes into less toxic compounds and is mediated by microbes, or enzyme-catalyzed reactions, or both. Enzymatic bioremediation of azo dyes has been tested using azoreductase, laccase, MnP, LiP, polyphenol oxidase, and versatile peroxidase. Although they are efficient at degrading azo dyes, there are some limitations with the full-scale application of this technology. Enzymes are water-soluble proteins that are lost in the liquid medium, especially in continuous reactors, where wash-out is observed. The immobilization of enzymes has been used to overcome some of their main limitations, which improves their stability against toxic pH and inhibitors, achieving better efficiency.

Adsorption and enzyme immobilization have similar aspects that mean they are potential technological options for azo dye removal and degradation. The adsorbent regeneration ability is one of the most important aspects for their viability at an industrial scale. The main advantage of enzyme immobilization is their regeneration, which allows them to be reused and makes them insoluble.

A combined adsorption and biodegradation technique is a future trend for the effective removal and degradation of azo dyes, which generates a low quantity (or no) of toxic by-products. Research is required into a sustainable technique that uses biological or residual materials as adsorbents, which contains enzymes that can degrade azo dyes.

REFERENCES

Adnan, L. A. *et al.* (2015) 'Metabolites characterisation of laccase mediated Reactive Black 5 biodegradation by fast growing ascomycete fungus *Trichoderma atroviride* F03', *International Biodeterioration & Biodegradation*. 104, pp. 274–282. doi:10.1016/j.ibiod.2015.05.019.

Ahmadijokani, F. *et al.* (2020) 'Superior chemical stability of UiO-66 metal-organic frameworks (MOFs) for selective dye adsorption', *Chemical Engineering Journal*. 399, p. 13. doi: 10.1016/j.cej.2020.125346.

Alcalde, M. *et al.* (2006) 'Environmental biocatalysis: from remediation with enzymes to novel green processes', *Trends in Biotechnology*, 24(6), pp. 281–287. doi: 10.1016/j.tibtech.2006.04.002.

Alhujaily, A. *et al.* (2020) 'Adsorptive removal of anionic dyes from aqueous solutions using spent mushroom waste', *Applied Water Science*. 10, pp. 1–12. doi: 10.1007/s13201-020-01268-2.

Asses, N. *et al.* (2018) 'Congo red decolorization and detoxification by *Aspergillus niger*: Removal mechanisms and dye degradation pathway', *BioMed Research International*, 2018, p. 9.

Benkhaya, S., M'rabet, S. and El Harfi, A. (2020) 'Classifications, properties, recent synthesis and applications of azo dyes', *Heliyon*, 6, p. 26. doi: 10.1016/j.heliyon.2020.e03271.

Bilal, M. *et al.* (2016) 'Characteristic features and dye degrading capability of agar–agar gel immobilized manganese peroxidase', *International Journal of Biological Macromolecules*, 86, pp. 728–740. doi: 10.1016/j.ijbiomac.2016.02.014.

Bilal, M. *et al.* (2017) 'Enhanced bio-catalytic performance and dye degradation potential of chitosan-encapsulated horseradish peroxidase in a packed bed reactor system', *Science of the Total Environment*, 575, pp. 1352–1360. doi: 10.1016/j.scitotenv.2016.09.215.

Bilal, M. and Iqbal, H.M.N. (2020) 'Microbial bioremediation as a robust process to mitigate pollutants of environmental concern', *Case Studies in Chemical and Environmental Engineering*. 2, p. 100011. doi: 10.1016/j.cscee.2020.100011.

Chacko, J. and Kalidass, S. (2011) 'Enzymatic degradation of azo dyes: A Review', *International journal of environmental sciences*, 1(6), pp. 1250–1260.

Chen, W. *et al.* (2015) 'Cloning and expression of a new manganese peroxidase from *Irpex lacteus* F17 and its application in decolorization of reactive black 5', *Process Biochemistry*. 50(11), pp. 1748–1759. doi: 10.1016/j.procbio.2015.07.009.

Chhabra, M., Mishra, S. and Sreekrishnan, T.R. (2015) 'Immobilized laccase mediated dye decolorization and transformation pathway of azo dye acid red 27', *Journal of Environmental Health Science & Engineering*. 13, pp. 1–9. doi: 10.1186/s40201-015-0192-0.

Crini, G. *et al.* (2019) 'Conventional and non-conventional adsorbents for wastewater treatment', *Environmental Chemistry Letters.* 17(1), pp. 195–213. doi: 10.1007/s10311-018-0786-8.

Fernandes, C.D. *et al.* (2020) 'Fungal biosynthesis of lignin-modifying enzymes from pulp wash and *Luffa cylindrica* for azo dye RB5 biodecolorization using modeling by response surface methodology and artificial neural network', *Journal of Hazardous Materials*, 399, p. 12. doi: 10.1016/j.jhazmat.2020.123094.

Forootanfar, H. *et al.* (2016) 'Studies on the laccase-mediated decolorization, kinetic, and microtoxicity of some synthetic azo dyes', *Journal of Environmental Health Science and Engineering,* 14, pp. 1–9. doi: 10.1186/s40201-016-0248-9.

Gamoudi, S. and Srasra, E. (2019) 'Adsorption of organic dyes by HDPy þ -modified clay: Effect of molecular structure on the adsorption', *Journal of Molecular Structure*, 1193, pp. 522–531. doi: 10.1016/j.molstruc.2019.05.055.

Gomare, S.S., Jadhav, J.P. and Govindwar, S.P. (2008) 'Degradation of sulfonated azo dyes by the purified lignin peroxidase from *Brevibacillus laterosporus* MTCC 2298', *Biotechnology and Bioprocess Engineering*, 13, pp. 136–143. doi: 10.1007/s12257-008-0008-5.

Hafdi, H. *et al.* (2020) 'Design of a new low cost natural phosphate doped by nickel oxide nanoparticles for capacitive adsorption of reactive red 141 azo dye', *Environmental Research*, 184, p. 14. doi: 10.1016/j.envres.2020.109322.

Iark, D. *et al.* (2019) 'Enzymatic degradation and detoxification of azo dye Congo red by a new laccase from *Oudemansiella canarii*', *Bioresource Technology,* 289, p. 7. doi: 10.1016/j.biortech.2019.121655.

Jawad, A.H., Mubarak, N.S.A.M. and Abdulhameed, A.S. (2020) 'Tunable Schiff's base-cross-linked chitosan composite for the removal of reactive red 120 dye: Adsorption and mechanism study', *International Journal of Biological Macromolecules*, 142, pp. 732–741. doi: 10.1016/j.ijbiomac.2019.10.014.

Jiang, H. *et al.* (2020) 'Selective adsorption of anionic dyes from aqueous solution by a novel b-cyclodextrin-based polymer', *Journal of Molecular Structure*, 1203, p. 10. doi: 10.1016/j.molstruc.2019.127373.

Kelm, M. A. P. *et al.* (2019) 'Removal of azo dye from water via adsorption on biochar produced by the gasification of wood wastes', *Environmental Science and Pollution Research.* 26, pp. 28558–28573.

Khan, N. *et al.* (2020) 'Hydrothermal liquefaction of rice husk and cow dung in Mixed-Bed-Rotating pyrolyzer and application of biochar for dye removal', *Bioresource Technology.* Elsevier, 309, p. 9. doi: 10.1016/j.biortech.2020.123294.

Khatri, A. *et al.* (2015) 'A review on developments in dyeing cotton fabrics with reactive dyes for reducing effluent pollution', *Journal of Cleaner Production*, 87(1), pp. 50–57. doi: 10.1016/j.jclepro.2014.09.017.

Kirk, T.K. and Cullen, D. (1998) 'Enzymology and molecular genetics of wood degradation by white-rot fungi', In Young, R.A. and Akhtar, M M. (ed.) *Environmentally friendly technologies for the pulp and paper industry.* John Wiley & Sons, pp. 273–307.

Kittappa, S. *et al.* (2020) 'Engineering Functionalized magnetic mesoporous palm shell activated carbon for enhanced removal of azo dyes', *Journal of Environmental Chemical Engineering,* 8, p. 10. doi: 10.1016/j.jece.2020.104081.

Kumar, T.H.V. *et al.* (2018) 'Synthesis and characterization of coral-like hierarchical MgO incorporated fly ash composite for the effective adsorption of azo dye from aqueous solution', *Applied Surface Science,* 449, pp. 719–728. doi: 10.1016/j.apsusc.2018.01.060.

Liu, L. *et al.* (2019) 'Mitigation of environmental pollution by genetically engineered bacteria: Current challenges and future perspectives', *Science of the Total Environment*, 667, pp. 444–454. doi: 10.1016/j.scitotenv.2019.02.390.

Liu, S. *et al.* (2020) 'Degradation and detoxification of azo dyes with recombinant ligninolytic enzymes from *Aspergillus sp.* with secretory overexpression in *Pichia pastoris*', *Royal Society Open Science*, 7, p. 15.

Mahmoodi, N.M. and Saffar-dastgerdi, M H. (2019) 'Zeolite nanoparticle as a superior adsorbent with high capacity: Synthesis, surface modification and pollutant adsorption ability from wastewater', *Microchemical Journal*, 145, pp. 74–83. doi: 10.1016/j.microc.2018.10.018.

Mehandia, S., Sharma, S.C. and Arya, S.K. (2020) 'Immobilization of laccase on chitosan-clay composite beads to improve its catalytic efficiency to degrade industrial dyes', *Materials Today Communications*, 25, p. 10. doi: 10.1016/j.mtcomm.2020.101513.

Mohajershojaei, K., Khosravi, A. and Mahmoodi, N.M. (2013) 'Decolorization of dyes using laccase enzyme from single and binary systems', *Desalination and Water Treatment*, (52, pp. 37–41. doi: 10.1080/19443994.2013.792010.

Mohan, S.V. *et al.* (2005) 'Acid azo dye degradation by free and immobilized horseradish peroxidase (HRP) catalyzed process', *Chemosphere*, 58(8), pp. 1097–1105. doi: 10.1016/j.chemosphere.2004.09.070.

Morsy, S.A.G.Z. *et al.* (2020) 'Current development in decolorization of synthetic dyes by immobilized laccases', *Frontiers in Microbiology*, 11, pp. 1–8. doi: 10.3389/fmicb.2020.572309.

Muedas-Taipe, G. *et al.* (2020) 'Removal of azo dyes in aqueous solutions using magnetized and chemically modified chitosan beads', *Materials Chemistry and Physics*, 256, p. 13. doi: 10.1016/j.matchemphys.2020.123595.

Omeje, K.O. *et al.* (2020) 'Synthetic dyes decolorization potential of agroindustrial waste-derived thermoactive laccase from *Aspergillus species*', *Biocatalysis and Agricultural Biotechnology*, 29, p. 7. doi: 10.1016/j.bcab.2020.101800.

Pandi, A., Marichetti, G. and Numbi, K. (2019) 'A sustainable approach for degradation of leather dyes by a new fungal laccase', *Journal of Cleaner Production*, 211, pp. 590–597. doi: 10.1016/j.jclepro.2018.11.048.

Pérez-calderón, J., Santos, M.V. and Zaritzky, N. (2020) 'Synthesis, characterization and application of cross-linked chitosan/oxalic acid hydrogels to improve azo dye (Reactive Red 195) adsorption', *Reactive and Functional Polymers*, 155, p. 14. doi: 10.1016/j.reactfunctpolym.2020.104699.

Perminova, I.V, Hatfield, K. and Hertkorn, N. (2002) *Use of Humic Substances to Remediate Polluted Environments: From Theory to Practice*. Dordrecht: Springer.

Qi, F. *et al.* (2020) '3D superhydrophilic polypyrrole nanofiber mat for highly efficient adsorption of anionic azo dyes', *Microchemical Journal*, 15, p. 9. doi: 10.1016/j.microc.2020.105389.

Qin, X. *et al.* (2019) 'β -Cyclodextrin-crosslinked polymeric adsorbent for simultaneous removal and stepwise recovery of organic dyes and heavy metal ions: Fabrication, performance and mechanisms', *Chemical Engineering Journal*, 372, pp. 1007–1018. doi: 10.1016/j.cej.2019.05.006.

Rápó, E. *et al.* (2020) 'Adsorption of Remazol Brilliant Violet-5R textile dye from aqueous solutions by using eggshell waste biosorbent', *Scientific Reports*, (1), pp. 1–12. doi: 10.1038/s41598-020-65334-0.

Ray, S.S., Gusain, R. and Kumar, N. (2020) 'Adsorption in the context of water purification', in *Carbon Nanomaterial-Based Adsorbents for Water Purification*. Elsevier Inc., Netherlands. pp. 67–100. doi: 10.1016/B978-0-12-821959-1.00004-0.

Riva, S. (2006) 'Laccases: blue enzymes for green chemistry', *Trends in Biotechnology*, 24(5), pp. 219–226. doi: 10.1016/j.tibtech.2006.03.006.

Shah, M.P. (2019) 'Bioremediation of azo dye', in *Microbial Wastewater Treatment*. Elsevier Inc., Netherlands. pp. 103–126. doi: 10.1016/B978-0-12-816809-7.00006-3.

Shirvanimoghaddam, K. *et al.* (2020) 'Death by waste: Fashion and textile circular economy case', *Science of the Total Environment*, 718, p. 137317. doi: 10.1016/j.scitotenv.2020.137317.

Siddeeg, S.M. *et al.* (2020) 'Iron Oxide/Chitosan magnetic nanocomposite immobilized manganese peroxidase for decolorization of textile wastewater', *Processes*, 8(5), p. 12.

Singh, A.D. *et al.* (2010) 'Decolourisation of chemically different dyes by enzymes from spent compost of *Pleurotus sajor-caju* and their kinetics', *African Journal of Biotechnology*, 9(1), pp. 41–54.

Singh, A.D. *et al.* (2011) 'Enzymes from spent mushroom substrate of *Pleurotus sajor-caju* for the decolourisation and detoxification of textile dyes', *World Journal of Microbiology & Biotechnology*, 27, pp. 535–545. doi: 10.1007/s11274-010-0487-3.

Singh, L. (2017) 'Biodegradation of synthetic dyes: A mycoremediation approach for degradation/decolourization of textile dyes and effluents', *Journal of Applied Biotechnology & Bioengineering*, 3(5), pp. 430–435. doi: 10.15406/jabb.2017.03.00081.

Singh, R.L., Singh, P.K. and Singh, R.P. (2015) 'Enzymatic decolorization and degradation of azo dyes: A review', *International Biodeterioration & Biodegradation*, 104, pp. 21–31. doi: 10.1016/j.ibiod.2015.04.027.

Sosa-martínez, J. D. *et al.* (2020) 'Synthetic dyes biodegradation by fungal ligninolytic enzymes: Process optimization, metabolites evaluation and toxicity assessment', *Journal of Hazardous Materials*, 400, p. 12. doi: 10.1016/j.jhazmat.2020.123254.

Stavrinou, A., Aggelopoulos, C.A. and Tsakiroglou, C.D. (2018) 'Exploring the adsorption mechanisms of cationic and anionic dyes onto agricultural waste peels of banana, cucumber and potato: Adsorption kinetics and equilibrium isotherms as a tool', *Journal of Environmental Chemical Engineering*, 6, pp. 6958–6970. doi: 10.1016/j.jece.2018.10.063.

Streit, A.F.M. *et al.* (2019) 'Development of high quality activated carbon from biological sludge and its application for dyes removal from aqueous solutions', *Science of the Total Environment*, 660, pp. 277–287. doi: 10.1016/j.scitotenv.2019.01.027.

Taka, A.L. *et al.* (2020) 'Metal nanoparticles decorated phosphorylated carbon nanotube/cyclodextrin nanosponge for trichloroethylene and Congo red dye adsorption from wastewater', *Journal of Environmental Chemical Engineering*, 8, p. 11. doi: 10.1016/j.jece.2019.103602.

Tavares, M.F. *et al.* (2020) 'Decolorization of azo and anthraquinone dyes by crude laccase produced by *Lentinus crinitus* in solid state cultivation', *Brazilian Journal of Microbiology,* 51, pp. 99–106.

Valencia, P. and Ibañez, F. (2019) 'Estimation of the effectiveness factor for immobilized enzyme catalysts through a simple conversion assay', *Catalysts*, 9(11). doi: 10.3390/catal9110930.

Valli, K., Wariishi, H. and Gold, M.H. (1999) 'Oxidation of monomethoxylated aromatic compounds by lignin peroxidase: Role of veratryl alcohol in lignin biodegradation', *Biochemistry*, 29(37), pp. 8535–8539. doi: 10.1021/bi00489a005.

Vantamuri, A.B. and Kaliwal, B.B. (2016) 'Purification and characterization of laccase from *Marasmius species* BBKAV79 and effective decolorization of selected textile dyes', *3 Biotech*, 6(2), pp. 1–10. doi: 10.1007/s13205-016-0504-9.

Vats, A. and Mishra, S. (2017) 'Decolorization of complex dyes and textile effluent by extracellular enzymes of *Cyathus bulleri* cultivated on agro-residues / domestic wastes and proposed pathway of degradation of Kiton blue A and reactive orange 16', Environmental Science and Pollution Research, 24, pp. 11650–11662. doi: 10.1007/s11356-017-8802-2.

Wahab, R.A. *et al.* (2020) 'On the taught new tricks of enzymes immobilization: An all-inclusive overview', *Reactive and Functional Polymers*, 152, p. 104613. doi: 10.1016/j.reactfunctpolym.2020.104613.

Wang, J. and Guo, X. (2020) 'Adsorption kinetic models: Physical meanings, applications, and solving methods', *Journal of Hazardous Materials*, 390, p. 18. doi: 10.1016/j.jhazmat.2020.122156.

Wang, S. *et al.* (2018) 'An extracellular yellow laccase from white rot fungus *Trametes sp.* F1635 and its mediator systems for dye decolorization', *Biochimie*, 148, pp. 46–54. doi: 10.1016/j.biochi.2018.02.015.

Worch, E. (2012) *Adsorption Technology in Water Treatment: Fundamentals, Processes and Modeling.* Berlin: Walter de Gruyter GmbH & Co.

Xia, J. *et al.* (2019) 'Secretory expression and optimization of *Bacillus pumilus* CotA-laccase mutant GWLF in *Pichia pastoris* and its mechanism on Evans blue degradation', *Process Biochemistry*, 78, pp. 33–41. doi: 10.1016/j.procbio.2018.12.034.

Yang, Q. *et al.* (2018) 'Characterization of a novel, cold-adapted, and thermostable laccase-like enzyme with high tolerance for organic solvents and salt and potent dye decolorization ability, derived from a marine metagenomic library', *Frontiers in Microbiology*, 9, pp. 1–9. doi: 10.3389/fmicb.2018.02998.

Yang, Y. *et al.* (2019) 'Preparation and characterization of cationic water-soluble pillar [5] arene-modified zeolite for adsorption of Methyl orange', *ACS Omega*, 4, pp. 17741–17751. doi: 10.1021/acsomega.9b02180.

Yehia, R.S. and Rodriguez-couto, S. (2017) 'Discoloration of the azo dye Congo red by manganese-dependent peroxidase from *Pleurotus sajor-caju* 1', *Applied Biochemistry and Microbiology*, 53(2), pp. 222–229. doi: 10.1134/S0003683817020181.

Zhou, Y. *et al.* (2019) 'Recent advances for dyes removal using novel adsorbents: A review', *Environmental Pollution*, 252, pp. 352–365. doi: 10.1016/j.envpol.2019.05.072.

18 A Combination of Physicochemical and Biological Methods for Azo Dye Degradation

Ambreen Ashar[1], Muhammad Shoaib[2],
Zeeshan Ahmad Bhutta[3], Huma Munir[4],
Iqra Muzammil[5], and Moazam Ali[6]*

[1]Department of Chemistry, University of Agriculture, Faisalabad, 38000, Pakistan

[2]Institute of Microbiology, Faculty of Veterinary Science, University of Agriculture Faisalabad, 38000, Pakistan and Key Laboratory of New Animal Drug Project, Gansu Province, Key Laboratory of Veterinary Pharmaceutical Development, Ministry of Agriculture and Rural Affairs, Lanzhou Institute of Husbandry and Pharmaceutical Sciences of Chinese Academy of Agriculture Sciences, PR, China

[3]Laboratory of Biochemistry and Immunology, College of Veterinary Medicine, Chungbuk National University, Cheongju, Chungbuk, 28644, Republic of Korea

[4]Department of Medical Laboratory Technologist, College of Allied Health Professionals, Faculty of Medical Sciences, Government College University Faisalabad, 38000, Pakistan

[5]Department of Veterinary Medicine, University of Veterinary and Animal Sciences, Lahore, 54000, Pakistan

[6]Department of Clinical Medicine and Surgery, University of Agriculture Faisalabad, Pakistan

*Corresponding author: ambreenashar2013@gmail.com

CONTENTS

18.1   INTRODUCTION

Dyes are colored materials that are often used in aqueous solutions due to their strong attraction to water. The dye is colored because of the presence of color groups in the chemical structure. Synthetic dyes are usually made from oil by-products and soil minerals. Azo dyes are the major type of artificial aromatic dyes used in the textile industry for dyeing. They are composed of ≥ 1 azo ($-N=N-$) and sulfonic acid (SO_3^-) groups, because of its high commercial value[1]. Azo dyes usually contain 1, 2, 3, or more azo bonds, and the naphthalene ring that connects the phenyl groups is usually replaced with a specific functional group, such as triazine amine, chlorine, hydroxyl (OH), methyl, nitro, or sulfonate acid salt[2]. There are two main types of azo dyes based on their hydrophobicity: (1) azohydrophobic dye, which is absorbed by bacterial cells and is reduced intracellularly; and (2) hydrophilic dyes that are degraded outside the bacterial cells. In addition, various azo dyes can be widely used in industry. Since azo dyes do not fluoresce, a fluorescent probe is used to track their binding to azo dyes through alkyl bonds.

Azo, synthetic, reactive, sulfur, oxidative, anthraquinone, pyridine, and many other dyes are used in textiles. The main reason for using specific types of azo dye in the dyeing process is due to the production of cellulose fibers, synthetic fiber, and protein fiber. Not all dyes adhere to the fibers during dyeing; therefore, a certain percentage of the unfixed dyes are released in the wastewater and cause environmental pollution[3]. The use of synthetic dyes started after the industrial revolution and became an important part of the textile industry. Of the 900,000 tons of dyes produced annually, 70% belongs to the azo group[4]. Previously, there were no methods to decompose the azo dyes. Without proper treatment, the textile industry discharges large volume of colored wastewater into the environment that causes serious environmental pollution. Azo dyes have long-term effects on living life due to their heterogeneous and recalcitrant properties. Wastewater from the textile industry contains a large number of dyes, which include many toxic metals that increase the pH, biological oxygen demand (BOD), and chemical oxygen demand (COD) of the wastewater discharged from industrial source[5]. In addition, it disproportionately increases the amount of organic and inorganic chemicals in the environment, which affects the biological life in the water. Mixing dyes with water reduces the

FIGURE 18.1 Flow diagram of various physicochemical and biological methods for the degradation of azo dyes.

efficiency of light transmission, which affects the aquatic ecosystem. Toxic azo dye compounds that are mixed with water are absorbed by fish and other aquatic animals and then absorbed by humans, which causes problems like high blood pressure, sporadic illness, and seizures. Benzidine-based azo dyes are thought to be carcinogenic because they cause liver cancer, spleen sarcoma, nuclear abnormalities in laboratory animals, and chromosomal abnormalities in mammalian cells[6].

Since azo dyes are easily inhaled or dissolved in water, they are quickly absorbed by the skin and can cause allergic reactions, cancer, and eye irritations. After the reduction of azo dyes, aromatic amines are formed. These aromatic amines are metabolized and oxidized into reactive electrophilic particles, which irreversibly bind to DNA[1]. The existing methods for removing dyes can be divided into the following three processes: physical, chemical, and biological treatment (Figure 18.1). Today, attention is focused on the development of inexpensive wastewater treatment methods for the textile industry, which could protect aquatic ecosystems. Therefore, these processes are physicochemical, biochemical, or a combination of both, and are effective methods for removing pollutants from the water.

18.2 METHODS FOR AZO DYE DEGRADATION

In industrialized countries, coagulation is the most common method used in textile wastewater treatment plants. It can be used as a pre, post, and main treatment system[7]. This process effectively removes the basic sulfur and disperses the colorant; however, the acid, direct, reactive, and reducing dyes are not effectively removed by coagulation and agglutination[8]. Filtration methods, such as like ultrafiltration, nanofiltration, and repenetration can be used for water reuse and chemical recovery. The main disadvantages of filtration technologies are the high investment costs, potential of membrane contamination, and secondary waste that requires further processing. An anaerobic pretreatment should be followed by aerobic and membrane posttreatment for water reuse[9].

18.2.1 Physical Methods

18.2.1.1 Adsorption

Adsorption is an efficient wastewater treatment for the textile industry, because it is a cost-effective method to remove dyes from fibers and wastewater. Soluble organic dyes transfer from the wastewater to the surface of the adsorbent. Adsorbents are solid, very porous materials[10]. It is the most efficient wastewater treatment method in the textile industry, the adsorbents adsorb and remove most of the compounds, which is a cost-effective method to remove dyes from fibers and changes the color of the wastewater. The adsorbent should absorb a large amount of the compound and be replaced it with new material when it is used up. The used adsorbent can be recycled or incinerated. The main factors that affect the adsorption of a dye are the interactions between the dye and the adsorbent, adsorbent surface, particle size, and contact time. Activated carbon (C) has been formulated for the best adsorption of large negatively charged or polar dye molecules[11]. Activated C in powder or granules with a specific surface area of 500–1,500 m^2/g, a pore volume of 0.3–1.0 cm^3/g and a bulk density of 300–550 g/L. Decomposition is seen when another filtration step is used. The removal rate for cationic mediators and acid dyes is very high, but the rate of removal for dispersed, reducing, direct, and reactive dyes are moderate[12]. A variety of cheap agricultural wastes (e.g., rice husks, sugarcane, tortillas, and cone corn), industrial wastes (e.g., coal ash, peat, clay, bentonite, red soil, bokisite, rice husks, and sheet powder), saw, peanut hulls, rice hulls, bags, and other cellulose wastes can absorb and accumulate dyes and other organic compounds in textile wastewaters, 40%–90% of which are basic dyes. Agricultural wastes can remove 40% of direct dyes[13].

The advantages of using these materials are their wide availability and low cost without the need for recycling. The waste materials can be burned after solid fermentation, which can enrich proteins. Although it is an advantage to use cheap adsorbents to remove dyes from textiles, they are still not as effective as activated C and require a large amount of adsorbents[14].

18.2.1.2 Irradiation

Radiation is used from monochromatic ultraviolet (UV) lamps, which operate at 253.7 nm. This process requires a constant supply of oxygen (O_2) for the effective decomposition of organic dyes[15]. Irradiation of secondary wastewater from wastewater treatment plants with 15,000 Gy gamma rays reduces COD, total organic carbon (TOC), and color by 64%, 34%, and 88%, respectively. Titanium dioxide (TiO_2) catalysts can be used to improve the efficiency of the irradiation process[16].

18.2.1.3 Filtration Processes

Commonly used membrane filtration methods will be discussed in the following sections.

18.2.1.3.1 Microfiltration

Microfiltration (MF) is mainly used for the treatment colorants, such as pigment colorants and for the treatment of rinsing baths[17]. It can be used as a pretreatment for nanofiltration or refiltration because it can remove suspended solids, particles in wastewater (0.1–1.0 μm), or high molecular weight colloids. MF can effectively remove approximately 90% of haze or sludge. The microfilter membrane is made from a polymer, such as a polyethersulfone, polytetrafluoroethylene (PTFE), polyphenylidene fluoride, polyvinylidene fluoride, polysulfone, polypropylene, or polycarbonate[18].

If very high chemical resistance or high temperature is required, C foil coated with ceramic, glass, zirconium oxide, alumina, and sintered metal film can be used. Typically, MF and ultrafiltration (UF) operate at a transmembrane pressure (tmp) of 20–100 psi and a velocity of 20–100 cm/s.

18.2.1.3.2 Ultrafiltration
This removes approximately 31.76% of contaminants, such as dyes and removes polymers and particles. Treated wastewater cannot be used for fabric dyeing. Extra filtration is used as a pretreatment to reintroduce or remove metal hydroxides[19].

18.2.1.3.3 Nanofiltration
Nanofiltration (NF) is used for the treatment of colored wastewaters that are very sensitive to colloid and polymer deposition[20]. NF films are composed of cellulose acetate and aramid. Inorganic materials, such as carbon-based membranes, ceramics, and zirconia can be used to create NF and reverse osmosis (RO) membranes. NF systems operate at 8 bar and18°C report approximately 70% ink removal. The molecular weight limits of the four polyether sulfonic acid films are 40, 10, 5, and 3 kDa, which are used on three different types of wastewater in industrial coating cycles[21]. NF can be used as an alternative for the discoloration of wastewater fibers.

18.2.1.4 Reverse Osmosis (RO)
This removes hydrolyzed reactive dyes and chemical additives in one step[17]. Repenetration is very sensitive to deposits; therefore, very contaminated water requires pretreatment. An RO membrane is made from cellulose acetate, aramid, and some inorganic materials. The combination of membrane technology and physical and chemical treatments is superior to other traditional treatments. The major drawback of the physical methods is the production of sludge. Therefore, the disposal of the adsorbent is a challenge. In contrast, if the amount of wastewater is small, a physical method could help.

18.2.2 Chemical Methods

18.2.2.1 Oxidative Process
Chemical oxidation is the use of chemical oxidizing agents [e.g., chlorine, ozone, Fenton's reagents, UV/peroxide, and UV/ozone (O_3)] other than O_2/air or bacteria that convert pollutants into easily degraded organic compounds.

18.2.2.1.1 Oxidation with Sodium Hypochlorite
The chloride cation (Cl^+) attacks the amino dye molecule and accelerates the breakdown of the azo bond. Increasing the chlorine (Cl) concentration increases the dye removal and fading process and lowers the pH. This method is useful for dyes that contain an amino group or a substituted amino group on the naphthalene ring, which are based on amino naphthol and naphthyl amino-sulfonic acid. This treatment is not suitable for dispersed dyes. This method is rarely used due to the adverse effects of aromatic amines that enter the water bodies.

18.2.2.1.2 Oxidation with Hydrogen Peroxide
Oxidation with hydrogen peroxide (H_2O_2) can be used in both wastewater treatment systems. Hydroxylation (OH) catalysts in visible light or UV-soluble (i.e., Fenton's reagent) is based on a homogeneous system that uses oxidase [Hydrogen (H) is activated by some iron (Fe) salts without UV radiation], H_2O_2, and other chemicals, such as active oxidants (e.g., O_3 and peroxide). Wastewater treatment using this method removes 31.5% Remazol Arancio 3R and Remazole Rose RB dye at pH 4 by utilizing 0.18–0.35 M H_2O_2 and 1.45 mM Fe^{+2} by oxidation for 30 min[13]. It is based on

a heterogeneous system that uses clay with or without UV light, semiconductors, and zeolite. For example, it uses 20 mM H_2O_2 and FeY 11.5 (1 g/L) for heterogeneous catalytic oxidation at pH 3–5 for 10 mins and contains TiO_2 zeolite with aluminum (Al) and Fe for 53%–83% zeolite removal, 68%–76% COD removal, and 32%–37% TOC removal[22]. The benefits of this oxidative treatment include reduced COD, wastewater color, toxicity, and removal of soluble and insoluble colorants, such as the dispersing dyes. The complete discoloration is usually achieved after the completion of the Fenton reagent step (i.e., after 24 h). The main disadvantage is the separation of the solid photocatalyst at the end of the process[23].

18.2.2.1.3   Oxidation by Ozonation

Ozone (O_3) is a powerful oxidizing agent that can oxidize aromatic rings and break down certain fiber dyes. O_3 can break the conjugated double bonds in the colored groups of organic dyes and cause discoloration. However, this process can generate toxic products. Therefore, O_3 treatment could be combined with physical methods. In the direct method, O_3 act as an electron acceptor, hydroxide ions (at high pH) catalyze the automatic decomposition of O_3 and react with organic and inorganic pollutants in wastewater to form OH radicals. At low pH, O_3 reacts efficiently via a direct reaction and binds to unsaturated chromospheres in the dye molecule[24]. The main advantage of O_3 oxidation is that O_3 can be used in the gaseous form, without adding wastewater or sludge. The disadvantage of O_3 treatment is that it has a very short half-life of approximately 20 min, which makes it unstable to salt, pH, and temperature and increases the cost of installing an O_3 generator. A combination of radiation or membrane filtration techniques could improve the efficiency of O_3 treatment[25].

18.2.2.1.4   Photochemical Oxidation

When wastewater is treated with UV in the presence of H_2O_2, the dye molecules split into smaller organic molecules, which are the end products of carbon dioxide (CO_2), H_2O, and other inorganic oxides. Dye decomposition is caused by the formation of OH groups[26]. The effectiveness of this method depends on the UV intensity, pH, dye and bath composition. For example, when 400–500 mg/L H_2O_2 at pH 3–7 is used, color removal of ≥60%– 90% was observed in wastewater that contained M5B red dye, H acid, and MR blue dye[11].

18.2.2.2   Condensation and Precipitation

Aluminium (Al) Fe rods, organic polymers, and coagulants are used for the coagulation of dyes and other additives in textile wastes. Colloidal particles in fibrous wastewater cannot be separated using a simple gravity method. However, they are affected by certain chemicals (e.g., lime, ferrous sulfate, ferric chloride, aluminum sulfate, aluminum chloride, and cationic organic polymers). This means that repellents and other auxiliary substances enter textile wastewater. These chemicals primarily neutralize charges and disrupt the stability of microparticles and emulsions. Therefore, these particles form clusters. These clusters are large enough to settle (condense) by gravity or filtration[27]. The main disadvantages of this method are, it is a controlled process, residual impurities in wastewater (i.e., nonionic detergents), and the resulting sludge must be separated and dried for further processing. A mixture of activated C and bentonite is used as a coagulant with aluminum chloride to remove direct and alkaline dyes such as CI Direct Orange 26, CI Moldant Black 11, and CI Basic Green 4. In addition, it can be used to absorb paint. Bentonite and high molecular weight electrolytes can be added to ferrous sulfate to improve the bleaching efficiency and sludge quality. The combination of sodium alginate and alum and a naturally occurring high molecular weight electrolyte as the primary coagulant significantly improved the color removal of a highly colored Remazol dye solution[28].

18.2.2.3   Electrocoagulation

This is an advanced electrochemical treatment method for the removal of dyes and colors. These processes include electrolysis using electrodes, condensation in wastewater, and the adsorption of soluble contaminants by coagulants and removal by precipitation. Even at high pH, this process can effectively remove color and COD, and the effectiveness depends on current density and response time. According to the available data, most of the dyes Orange II and Acid Red 14 had >98% removal during electrocoagulation (EC) processing[29]. In general, under the best conditions, EC treatment has a bleaching efficiency of 90%–95% and a COD removal rate of 30%–36%. The main disadvantage of chemical processes is their high cost, which makes them unsuitable for industrial use.

18.2.2.4   Fenton Process

The reaction between Fe and H_2O_2 is called the Fenton reaction. It is often used to decompose reactive dyes[30]. The Fenton process is very economical, easy to use, and can effectively break down organic contaminants. The main advantage of the Fenton method is that it does not require energy to activate H_2O_2. The Fenton process is a very efficient process when chemical oxidation needs to be reduced[31].

18.2.2.5   Photolytic Chemical Processes

18.2.2.5.1   Homogeneous Photolytic Chemical Processes

18.2.2.5.1.1   UV Lamp   The degradation of oxidants and dyes (i.e., H_2O_2) is primarily influenced by pH, dye structure, dye composition, and UV intensity during UV processes[32]. The UV process can covert the dyes into CO_2, H_2O, and salt molecules. The UV process is mainly performed at low pressures at 254 mm. Under the influence of UV, a powerful oxidant forms free OH radicals and decomposes dyes in the wastewater. The main advantage of this method is that no sludge remains after treatment[33].

18.2.2.5.1.2 $O_3/H_2O_2/UV$ The addition of H_2O_2 to O_2/H_2O_2 significantly increases the formation of OH radicals. The formation of these OH radicals is very efficient and fast during mineralization of pollutants. A complete color change was observed during the $O_3/H_2O_2/UV$ treatment of wastewater[34].

18.2.2.5.1.3   Simple Fenton Method   Using the photo-Fenton process, UV lamps can significantly promote the formation of OH radicals. The efficiency of the Fenton process, which changes the color of the wastewater, is similar to that of the photo-Fenton process, but the photo-Fenton process significantly improved the mineralization[35].

18.2.2.5.2   Heterogeneous Photolytic Chemical Process

Semiconductors are composed of two energy bands: a high energy band and a low energy valence band. The chemical oxidation during the decay of semiconductors in hypersensitivity reactions is used to form OH radicals in a heterogeneous process. Zinc oxide (ZnO), strontium titanate and TiO_2 are used commercially. The valence and conduction bands of semiconductors are separated by band gaps[36]. In a photocatalytic process, photoinduced reactions increase efficiency. The absorption of photons by energy (exceeding the ban on catalysts) is very important for photocatalytic reactions[37]. Photocatalytic processes are often used to decompose dyes from textile wastewaters because they can be completely mineralized. TiO_2 and ZnO are commonly used as photocatalysts to remove organic pollutants and dyes from wastewater[38].

18.3   BIOLOGICAL METHODS FOR THE DEGRADATION OF AZO DYES

Decomposition of dyes using biological methods, such as biological phenomena (i.e., bioremediation), are environmentally friendly methods that allow the dye to be removed from the textile wastewater with minimal costs and optimal run time. In most countries or regions, biological methods are widely used to remove dyes from wastewater[39].

A biological method only removes substances that are dissolved in the wastewater. The degradation is affected by the ratio of material, colorants, microbes, the O_2 concentration, and the temperature of the system. Depending on the O_2 requirement, biological processes can be categorized into aerobic, anaerobic, hypoxic, permeable, or a combination of these. The aerobic process uses microorganisms to remove dyes from industrial wastewater in the presence of O_2, and the anaerobic process uses microorganisms to remove dyes from industrial wastewater in the absence of O_2. In practice, combinations of treatments, such as aerobic and anaerobic, produce better results. This uses an anaerobic process to remove COD from the wastewater, followed by an aerobic polishing process to purify the wastewater with low COD[40]. When the demand for O_2 in the wastewater is high and >3 g/L, the anaerobic process can only produce methane biogas. It contains organic compounds, for example, polyvinyl alcohol (PVA) and starch that are more degradable[41]. Therefore, the anaerobic process produces biogas (methane) that has a specific calorific value. However, the colored water is visible in the environment, because this treatment is not enough to completely remove harmful particles from the textile wastewater. Traditional methods can meet the COD of the wastewater, but not the COD of water. Water must be dye-free or nontoxic. Therefore, a variety of microorganisms and enzymes have been isolated, and attempts have been made to degrade certain dyes. The use for efficient separation and decomposition by microbes is an interesting biological aspect of wastewater treatment. In addition to this, other traditional methods for the removal of biological pigments include adsorption by microbial biomass, decomposition by algae, enzymatic decomposition, and cultivation of fungi and microbes. This method should be used with caution.

The use of enzymes to remove pigments is very popular because people believe that biological pigment removal is the cheapest and safest decolorization method. The efficiency of biodegradation depends on the adaptability of the selected microorganism and the activity of the enzymes[42].

Biological processes for complete decomposition of fibrous wastewater have the following advantages: (1) environmental friendliness; (2) low cost; (3) low sludge formation; (4) harmless metabolites or ensur4full mineralization; and (e) low water use (e.g., increased concentration or low dilution) over physical or oxidative methods[43]. Biological processes involve living organisms and the main disadvantage is the growth rate. System instability often occurs when biological dyes are removed because growth and reaction rates are difficult to predict.

18.3.1 Decolorization and Degradation of Azo Dyes by Bacteria

Using bacteria for azo dye discoloration and mineralization is faster than using fungi. Extensive research has been carried out to determine the role of various bacteria during the discoloration of azo dyes under different conditions.

18.3.1.1 Bacterial Degradation of Azo Dyes by Pure Cultures

Pure cultures of bacteria under anaerobic conditions, such as *Pseudomonas luteola*, *Proteus mirabilis*, and *Pseudomonas spp.*, showed good results for the degradation of azo dyes[44]. In addition, other bacteria, such as *Rhizobium actinobacillus*, *Desulfovibrio desulfuricans*, *Exiguobacterium spp.*, *Comamonas sp.*, and *Sphingomonas spp.* effectively degrade several azo dyes. Among these strains, *Pseudomonas sp.* are commonly used to decolorize azo dyes and commercially available azo dyes in textiles, such as Red HE7B, Red BLI, Reactive Red 22, Reactive Red 2, Orange I and II, and reactive blue 172. Bacterial strains can degrade azo bonds using azoreductase, which is insensitive to aerobic conditions. The bleaching of azo dyes by *Aeromonas hydrophila* under aerobic conditions has been investigated[45]. Under anaerobic conditions [nitrogen gas (N_2) atmosphere], *Bacillus lentus* showed no degradation, which indicated that the bleaching process requires a certain amount of O_2. The use of a pure culture system ensures the reproducibility of the data and facilitates the interpretation of experimental observations. In addition, these studies investigate the mechanisms of biodegradation using biochemical and molecular biological tools. This information could be used to modify strains to achieve increased enzymatic activity for faster pigment degradation[44].

18.3.1.2   Bacterial Degradation of Azo Dyes by Mixed Cultures and Consortia

Due to a large amount of wastewater that contains various dyes, it is impossible to cultivate it once in the field. Therefore, a mixed culture is useful because microbial consortia biodegrade together[46]. The is possible only in the presence of certain enzymes obtained by mixing culture[47].

The first cleavage of the azo bond occurs during the decomposition of the azo dye, which forms aromatic amines. It is difficult to isolate a single bacterial strain from a wastewater sample that contains dyes, and lengthy adaptation procedures are required to effectively decolorize and degrade the azo dye[48]. *Micrococcus luteus*, *Micrococcus spp.*, and *Paenibacillus polymyxa* can discolor nine dyes when used in a mixed culture; however, when used in a pure culture they are ineffective at removing the dyes[46]. A consortium of four bacterial isolates, *Pseudomonas putida*, *Bacillus cereus*, *Pseudomonas fluorescens,* and *Stenotrophomonas acidaminiphila* effectively degraded Acid Red 88. Consortium dynamics might contribute to the faster biotransformation of methyl red and *Klebsiella spp. Bacillus spp*, and *Buttiauxella spp.*, were identified in mixed culture and *Klebsiella spp*, *Clostridium spp. Escherichia spp.*, and *Bacillus spp.* were mixed consortia under aerobic and anaerobic conditions, respectively [44].

The microorganisms in consortia work together to improve bleaching activity.[49] In addition, a consortium decolorizes wastewater dyes more efficiently than individual strains. Compared with individual strains, mixed cultures increased the degradation rate and the proportion of azo dyes, decreased degradation time and improved the mineralization of azo dyes[50]. When forming a consortium, the concentration of each microorganism is important to achieve an effective azo dye processing system.[47] Four isolates in the same proportions in a consortium increased the efficiency of dye discoloration three-fold compared with other combinations of the same isolates.

The discoloration rate of Reactive Black 5 and Reactive Black 2 was more proficient due to *Bjerkandera spp.* and natural microflora at a 2: 3 ratio compared with a 2:4 ratio[51]. Therefore, the portion of the culture associated with the process changes. The effective bleaching of azo dyes depends on a combination of aerobic and anaerobic processes. Consortia have been reported for the continuous bleaching of azo dyes. Under microaerobic conditions, the azo dye partially decomposes and forms some intermediate products. Then, under aerobic conditions, these intermediates undergo further metabolism, which results in complete mineralization. The anaerobic–aerobic system mineralizes azo dyes. The use of consortia improves the decomposition of Golden Yellow when aerobic conditions are followed by microaerobic conditions[52]. In many cases, aromatic amines are obtained as a result of the anaerobic degradation of azo dyes. These are partially oxidized under aerobic conditions, and the metabolites of azo dyes are automatically oxidized by O_2, and the products cannot decompose further. The density of individual species in the consortium depends on the amount of O_2[50]. Under microaerobic conditions, *Enterococcus casseliflavus* was prominent, and *Enterobacter cloacae* and *Citrobacter freundii* prevailed under aerobic conditions [53].

18.3.1.3   Bacterial Degradation of Azo Dyes by Immobilized Cells

Immobilized cells can be used to break down azo dyes. Cell immobilization is carried by adhesion and wrapping. In adhesion, cells attach to the surface of inert substances or other organisms. In wrapping, the bacterial cells are enclosed in cavities of fibrous or porous material. Recently, various reactor designs have been suggested for the efficient treatment of azo dyes. Compared with free cells, fixed bacterial cells are more resistant to environmental damage (e.g., changes in pH and exposure to high concentrations of dye). Immobilization is a good technique because it can be easily performed *in situ*, under sterile conditions, prevents cell washout, maintains a high cell density in a continuous reactor, and can be easily expanded. In general, immobilization improves catalyst stability and substrate absorption due to high nutrient utilization. Many carriers can be used to immobilize bacterial cells, such as sintered glass and nylon mesh. In a reactor that used immobilized *A. hydropylla, Streptococci* sp, *Comamonas testosterone,* and *Acinetobacter baumannii* on PVA, Red RBN was discolored, and discoloration was effective even at high dye concentrations[54]. Different conditions, such as layer swelling, number of cell beads, density and initial concentration of the dye, hydraulic residence time, and the diameter of the beads used to decolorize the dye were investigated[6]. In addition, large-scale (reactor) immobilization of bacteria on natural or synthetic materials could be used to degrade azo dyes, because it creates an anaerobic environment in which bacterial enzymes can degrade the dye. Bacterial cells that were immobilized in charcoal in a sequential solid membrane bioreactor could accelerate the degradation of industrial wastewater during dye production[55].

18.3.1.4   Microbial Fuel Cell Azo Dye Decomposition

In microbial fuel cells (MCF), microorganisms use electrons to work with electrodes, and electrons are removed or transferred by reactions[56]. MFCs are a basic type of bioelectrochemical system that spontaneously converts biomass into electrical energy using microorganisms. MFCs are a green technology that meets increased energy demands, in particular, via the use of wastewater as a substrate for the simultaneous generation of electricity and complete wastewater treatment. Therefore, the operating costs of the wastewater treatment plant can be offset[57]. Bacterial fuel cells can generate energy and reduce the color in textile wastewater[58]. The discoloration of Reactive Brilliant Red X3 was investigated using glucose as a substrate. An increase in the degrading activity of the MFC was observed during power generation.

18.3.2   Enzymatic Degradation of Azo Dyes

The first step in the degradation of azo dyes is to cleave the azo-bound chromophore group ($-N=N-$). This reduction involves a variety of mechanisms, including enzymes, low molecular weight redox mediators, chemical reduction using biological reducing agents (e.g., sulfides), or combinations of these. For the enzymatic discoloration and degradation of azo dyes, two families of enzymes, azoreductases and laccases have potential. Laccase has potential for the discoloration of many known industrial dyes[59]. Certain enzymes, such as manganese peroxidase (MnP), lignin peroxidase (LiP), and polyphenol oxidase (PPO), could be involved in the discoloration and removal of azo dyes.

18.3.2.1   Azoreductase

Azoreductase is expressed in bacteria and fungi that degrade azo dyes and is the main enzyme that decolorizes or degrades these dyes. The azo dye can be decolorized by reducing the azo bond to form an aromatic amine (a colorless product)[60]. Azoreductase catalyzes the reaction in the presence of NADH, NADPH, and $FADH_2$. These reducing molecules act as electron donors and are involved in breaking the azo bonds inside and outside the bacterial cell membrane. Azoreductase is very important in the development of biological wastewater treatment processes that contain azo dyes because it can catalyze the degradation of azo groups in some microorganisms[61].

Recently, many researchers have reported several bacteria that produce azoreductase. Catalytic proteins with azo reductase activity have been isolated from *Rhodobacter sphaeroides, Escherichia coli, Enterococcus faecalis, Pigmentiphaga kullae, Staphylococcus aureus, Bacillus spp,.* and *Xenophilus azovorans*[62]. However, many azo dyes have a very polar and complex structure and cannot diffuse across the cell membrane, which suggests that azoreductase is involved in the intracellular discoloration by bacteria. Enzymatic degradation of azo dyes occurs in a variety of organisms and includes rat liver enzymes, cytochrome P450, rabbit liver aldehyde oxidase, and azo reductases from intestinal flora, bacteria, ecological fungi, salt requiring and resistant microorganisms[61].

The classification of azoreductases is mainly based on their O_2 demand and structure. Flavin-dependent azoreductases are reclassified according to the required coenzymes, NADH, NADPH, or both. Three systems for classifying the groups have recently been proposed. The first group consists of FMN-dependent enzymes that use NADH, the second group consists of NADPH enzymes, and the third group consists of flavin-free reductases[63]. The structure of azoreductase is primarily a monomer, but there are several reports of dimers and tetramers[64]. The optimum temperature range for bacterial azoreductase is 25°C–45°C, and pH 7.0. Azoreductase activity is independent of intracellular dye uptake because high molecular weight azo dyes are unlikely to penetrate the cell membrane of bacterial cells[63]. Azoreductase is located inside or outside the bacterial cell membrane. These azo reductases require NADH, NADPH, or FADH as electron donors to cleave azo bonds[65]. Azoreductase activity has been observed in cell extracts. $FADH_2$, NADH, NADPH, $FMNH_2$, and their reductases are found in the cytoplasm as cofactors. During cell lysis, these cofactors and enzymes are released into the extracellular environment. If the cells are intact, the membrane transport system might be a prerequisite for the recovery of azo dyes. Riboflavin can penetrate the cell membrane; however, FAD and FMN cannot easily penetrate the cell wall. Similarly, due to their complex structure and high polarity, many azo dyes cannot penetrate the cell membrane, and azoreductase is found in many bacterial cells. Therefore, cell extracts from lysed cells exhibit a higher pigment reductase activity than intact cells. Many researchers have noted that the azoreductase system has no specificity and have demonstrated that the substrate specificity of azoreductase depends on the functional groups near the azo bond[63].

Bacterial azoreductase is a new family and it is very similar to other known reductases. Several researchers have purified and biochemically characterized azoreductase, which reduces azo dyes. *E. coli, E. faecalis, Klebsiella spp.*, and *S. aureus* are primary sources of aerobic FMN-dependent azoreductase. A better understanding of the sequence levels requires DNA screening and probe development for NADPH-dependent azoreductase derived from *Bacillus sp.* (20 kDa). They also isolated genes from *Bacillus subtilis* strains, and *G. stearothermophilus* was identified. Another 30 kDa azoreductase was identified in *Xenophilus azovorans*[53].

18.3.2.2   NADH–2,6-dichloroindophenol Reductase

NADH–2,6-dichloroindophenol (DCIP) reductase uses NADH as an electron donor to reduce DCIP. DCIP is blue in the oxidized form and becomes colorless after reduction with reductase. In 1980, *Bacillus stearothermophilus* NADH–DSIP reductase was reported. Numerous researchers have reported that NADH–DCIP reductase is a labeled enzyme for reducing azo dyes. *Bacillus spp.* significantly induces DCIP reductase activity during the discoloration of various dyes[50].

18.3.2.3   Tyrosinase

Tyrosinase (monophenol monooxygenase) is found in many organisms. These are copper-containing (Cu) enzymes that can catalyze two types of reactions: (1) the orthohydroxylation of certain monophenols (monophenolase, cresolase); and (2) the oxidation of orthodiphenol to orthoquinone (diphenolase and theaphenolase) using O_2[66]. This enzyme acts as a marker for oxidases involved in the degradation of azo dyes. Tyrosinase could be induced during bacterial discoloration of azo dyes. Tyrosinase induction following exposure to azo dyes was reported in a consortium of *Alcaligenes faecalis, Rhodosporidium, Galactomyces geotrichum*, and *Bacillus pp.*[67]

18.3.2.4   Laccases

Laccase is the most abundant member of the Cu oxidase protein family. Laccase is a family of Cu-containing PPOs commonly referred to as multiple copper oxidases (MCOs)[68]. Laccase is a very important oxidoreductase in various bioengineering processes, mainly due to its unique properties, such as nonspecific oxidative capacity, which does not require cofactors, and it accepts electrons. Due to its ability for O_2 free bioremediation, which is readily available to the microorganisms.

In addition, laccase purified from *Hypsizygus ulmarius* decolorized methyl orange without redox mediators. Extensive studies have been carried out on the ability of laccase to degrade azo dyes. The most well-known laccase is obtained from fungi (i.e., white-rot) or plants, but a small amount of laccase has been identified and isolated from bacteria. With the simultaneous use of azoreductase and laccase, a model dye effluent had maximum degrading and detoxifying effect[61]. Two laccase genes were cloned from RNA of mycelium of *Lentinula edodes* L54. Efficient biodegradation of dyes and polycyclic aromatic hydrocarbons is expressed by the methylotrophic yeast *Pichia pastoris* in two allelic forms. These genes are 1,573 bp long and are presented as alleles with a difference of only 21 nucleotides. Laccase oxidizes phenol to form phenoxy groups, which are enzymatically reoxidized to form phenoxy radicals, and a C ion, the charge of which has an azo bond on the C of the phenolic ring. The water-induced nucleophilic attack produces 4-sulfophenyldiazene and benzoquinone. 4-sulfophenyldiazonium becomes unstable in the presence of O_2 and can be oxidized to the corresponding phenyldiazonium radical. The latter loses N_2 with the formation of a sulfophenyl group, which is removed by O_2 and then, four sulfophenyl hydroperoxides (SPH) are formed.

18.3.2.5 Peroxidase

Fungal systems are the most suitable for treating azo dyes. The ability of fungi to break down azo dyes is based on the synthesis of exonucleases, such as peroxidase and phenoloxidase[61]. The main structures and substrates of hemperoxidase can be divided into: (1) peroxidase of nonanimal origin; (2) peroxidase of animal origin; (3) catalase; (4) haloperoxidase; (5) dihemechytochrome C peroxidase; and (6) peroxidase of the DyP family. Peroxidase has three substrate binding sites called edges d and c, and open tryptophan residues. The complex structure of the d-edge substrate in many peroxidases has been identified using X-ray crystallization studies. The process of H_2O_2 oxidation catalyzed by chloroperoxidase (CPO) and the decolorization of azo dyes in buffers, such as Orange G and Sunset yellow was investigated and chloroperoxidase (CPO), which is a heme-containing enzyme produced by *Caldariomyces fumago,* was investigated[69]. The effect of specific enzymes,[70] such as LiP, laccase, tyrosinase, and DCIP reductase on the spontaneous degradation of reactive blue 160 present in the roots of *Tagetespatula spp.* for the discoloration and detoxification of textile dyes were investigated[61]. LiP produced during the fermentation of sewage sludge, could be used to purify textile wastewater to discolor dyes. The optimization procedure was under static conditions and the removal of the methylene blue dye was high (65%) at an initial concentration of 15 mg/L, and the activity of lignin peroxidase was 0.687 U/mL with a reaction time of 60 min[71].

18.3.2.6 PPO

PPO is a tetrameric enzyme, each molecule of which contains four Cu atoms and has two binding sites, an aromatic compound and O_2. This enzyme catalyzes the orthohydroxylation of monophenols to orthodiphenols. It can further catalyze the oxidation of o-diphenol to o-quinone. The rapid polymerization of o-quinone into black, brown, or red pigments leads to the darkening of fruit. The amino acid tyrosine contains one phenolic ring, which can be oxidized to o-quinone by PPO. PPO acts on a wide variety of substrates, removing organic pollutants that are present in very low concentrations[61]. Ammonium sulfate proteins precipitated from potatoes and brinjal have high polyphenolic activity and can degrade fibrous and nonfibrous dyes[72]. The use of pure banana pulp PPO to degrade fiber dyes, direct red 5B, and direct blue GLL has been studied[61]. The purified enzyme can denature 90% of Direct Red Dye 5B in 48 h and ≤85% of GLL Direct blue Dye in 90 h.

18.3.3 Algae (Phycoremediation)

Live and dead algae have been used for the biological recovery of fibrous wastewaters[73]. Several research groups have concluded that algae use different internal mechanisms for the decolorization of azo dyes[74]. This includes the use of chromophores to create algal biomass by assimilation, the production of CO_2 and H_2O by converting colors into unstained molecules, and the adsorption of chromophores by algal biomass[75]. The mechanism of algal discoloration includes enzymatic degradation, adsorption, or both. The most studied algae associated with the degradation of azo dyes are blue and green algae and diatoms. *Chlorella* and *Oscillatoria* are effective species as decoloring agents for azo dyes. Metabolism produces aromatic amines. Similar to bacteria, algal-mediated degradation of azo dyes depends on an induced form of azoreductase that cleaves azo bonds to form aromatic amines[76]. The role of derivatives of *Oscillatoria curviceps* enzymes azoreductase, laccase, and PPO in the degradation of black acid has been determined. Unlike bacteria and fungi, which require C spills and other additives to remove colorants, algae do not require additives.[77] The use of immobilized microalgae, such as *Chara vulgaris* and *Scenedesmus quadricauda* is another method to decolorize and degrade dyes immobilized on alginate[78]. The higher the proportion of the dye in the textile, the better it copes with floating algae.

18.3.4   FUNGAL DEGRADATION OF AZO DYES

18.3.4.1   Degradation of Azo Dyes Using Yeast

Yeast strains have been used to decolorize various azo dyes due to their many bioremediation benefits, which includes their high ability to accumulate dyes and heavy metals [e.g., lead (Pb(II)) and cadmium (Cd (II)][79], they grow faster than filamentous fungi, discolor faster, and can survive under adverse environmental conditions. Sludge produced following wastewater treatment contains a large number of diverse yeast strains compared with other media; however, the yeast constitutes a small fraction of the microorganisms in activated sludge[80]. Yeast has many advantages over bacteria and filamentous fungi. They grow as fast as bacteria, and similar to filamentous fungi, they can withstand adverse environmental conditions. A recent literature review demonstrated that yeast strains, such as *Galactomyces geotrichum*, *Saccharomyces cerevisiae*, *Trichosporon beigelii*, and *Kluyveromyces marxianus* were promising dye adsorbents that could absorb high concentrations of pigments and were involved in the discoloration of Remazole Black B. *T. beigelii* can decolorize various azo dyes, such as HER dark blue (100%), HE7B red (85%), 4BD gold (60%), HE4BD green (70%), and HE2R orange (50%). The decolorization rate of some dyes is not ideal. Several studies have shown that yeast strains are promising pigment adsorbents that can absorb pigments at higher concentrations[5]. After 22 h, the *Candida zeylanoids* had degraded many simple azo dyes with a degradation rate of 46%–67%. *C. zeylanoids* discolored four model azo dyes; however, it took a long time (40–60 h). Similar to microalgae, color change by yeasts is via adsorption[81], enzymatic degradation, or a combination of both. At low pH, adsorption on yeast biomass is more efficient[82]. Different azo dyes are adsorbed to varying degrees. *Candida tropicalis* absorbs 94% Remazole blue and 44% reactive red, and *Trichosporon akiyoshidainum* absorbs 63% reactive blue and 90% reactive red 141. Dye degradation is related to yeast growth and their primary metabolism. Without glucose or easily absorbed C and energy, yeast cells cannot grow. In addition, degradation requires a C source. The azo dyes are the products of oxidases and reductases, such as MnP, Tyr, and yeast NADH. May cause an increase in DCIP reductase[5]. Increasing the production of laccase will decrease the degradation time. For example, in a basal medium the discoloration of azo dyes and non-azo dyes, Sudanese black and crystal violet using, a culture filtrate from *Paraconiothyrium variabile* in the basal medium was 9.4% and 16.8%, respectively, within 12 h. When xylidine and Cu were added to the medium and the filtrate from the optimized culture solution was used, the activity of laccase from *P. variabile* increased from 970 to 16,678 U/L. After 3 h, the discoloration of the dyes was 84%, 94%, 93%, and 87%, respectively[83]. The color change mechanism is pH-dependent. *T. akiyoshidainum* adsorbs 63% of reactive blue 221 and 90% of reactive red 141 at pH 2.0, but these dyes are almost completely degraded at pH 7.0. Decomposition by yeast can be achieved under aerobic or anaerobic conditions[84]. The bioadsorption of textile dyes is achieved by the biomass obtained from the *Kluyveromyces marxianus*, which can discolor Remazole Black B via physical adsorption[85]. Oxygenated *Rhodotorula spp.* and *Rhodotorula rubra* can completely degrade crystal violet within 4 days. *Candida zeylanoides* can degrade various azo dyes. In addition, *S. cerevisiae* can effectively remove colorants from honey media[86].

18.3.4.2 Decoloration of Azo Dyes Using Filamentous Fungi

Filamentous fungi are ubiquitous in the environment, and their rapid metabolic adaptation to different sources of C and N is essential for their survival. Due to the increased surface activity and kinetic energy of the dye, the adsorption capacity of fungal biomass increases with increasing temperature. However, at very high temperatures, discoloration is reduced. This might be due to the inactivation of the adsorbent surface or the destruction of some active sites. Dye adsorption depends on the concentration of the dye. Adsorption decreases at high concentrations[87].

Filamentous fungi oxidize azo dyes using peroxidase and phenoloxidase to avoid the formation of amines during the reduction of azo dyes. Cultivation and nutritional conditions, especially N restriction affect filament growth, enzyme production, and subsequent degradation of pigments. Aerobic degradation is more effective than anaerobic degradation. A higher shaking speed improved the process by transferring O2 more efficiently, for Blue 49, Orange 12, and Brilliant blue by *Pleurotus sanguineus*[5]; Congo Red by *Aspergillus niger, Aspergillus flavus, Alternaria spp.*, and *Penicillium spp.* effectively degraded Acid Red 151 and Orange II. Maintaining a constant pH during this process (i.e., suitable for fungal growth) enhanced the degradation. Sources of C and N, for example, have a significant impact on fading[88] where discoloration is achieved in N-restricted environments. *Trametes spp.* degrade Orange II and Brilliant blue R250 by improving the C:N ratio in growing media. Adding additives to the medium can improve the efficiency of degradation. In the presence of *Phanerochaete chrysosporium*, the production of MnP increased, veratryl alcohol increased the production of LiP, and therefore, degraded Reactive Black 5. Tween 80 and manganese (Mn^{2+}) increased the efficiency of MnP. The presence of veratryl alcohol is dependent on its ability to induce enzyme activity in immobilized *P. chrysosporium*. Reactive Blue 220 can be used as a redox mediator in the enzymatic reactions used by *Pleurotus ostreatus* to degrade azo dye[5].

Various materials were used to eliminate fungi. The immobilization of *P. chrysosporium* could be carried out on zirconium oxychloride ($ZrOCl_2$)-activated pumice, expanded polystyrene, nylon sponge, sunflower skins, polyurethane foam, luffa sponge, calcium alginate beads, lignite xylite and lignite particles, and a packed-bed bioreactor. Lyophilization can be used to test fungal biomass, which is similar to lyophilized petrolatum mycelium, which can decolorize >85% of indigo carmine at 1,000 ppm[5].

18.3.5 Genetically Modified Organisms for the Degradation of Azo Dyes

Biological restoration is an environmentally friendly method for the treatment of textile wastewater. However, the physical and chemical properties of wastewater (e.g., pH, NaCl and other salts, temperature, and the presence of organic compounds) can inactivate enzymes and fungi. Therefore, meeting the requirements of wastewater treatment in the textile industry requires high stability, high yield, low cost, and more active and versatile enzymes and fungi. For the complete biodegradation of contaminants, the process should be iterative[89]. For example, in white-rot spoilage *Trametes spp.* the *lac48424-1* was cloned into *P. pastoris*. Compared with other known laccases, the purified recombinant laccase (rLAC48424-1) can discolor a variety of dyes (e.g., methyl orange and bromophenol blue) more efficiently. In the presence of redox mediators, the expression of laccase *lcc1* from *Trametes strogii* in *P. pastris* resulted in higher *Lac* production and improved azo dye discoloration[5]. A laccase gene from *Ganoderma lucidum* was manufactured using an optimized codon and a two-step PCR-based DNA synthesis method. The resulting recombinant *Lac* was overexpressed in *P. pastoris* and showed a high degree of methyl orange degradation[90]. In the absence of mediators, purified recombinant *Lac* from the heterologous production of *Pleurotus sanguineus*

Lac in *P. pastors* can effectively decolorize synthetic dyes. Compared with wild-type *Lac*, 2 variants from 2,300 randomly mutated variants of *P. ostreatus* OXA1b *Lac* showed higher stability and ability to decolorize azo dyes under various environmental conditions[91]. In contrast, fungal origin laccase has a better pigment reducing effect at acidic pH and with redox mediators[5].

18.3.6 DEGRADATION OF AZO DYES BY PLANTS (PHYTOREMEDIATION)

Recently, some studies have described the use of plants for the biological removal of dyes from wastewater. Researchers have used *Petunia grandiflora* Juss for the degradation of a mixture of dyes and dye-containing wastewater[92]. Three plants that can change the color of azo dyes in textile wastewater: mustard, beans, and sorghum were reviewed[76]. The degree of discoloration of these plants (*Brassica juncea*, *Phaseolus mungo*, and *Sargassum vulgare*) were 79%, 53% and 53%, respectively, for wastewater containing fibers. Similarly, many other plants have been identified for discoloration of dyes, such as *Blumeamalcolmii* and *Typicalhonium flagelliforme*, which can degrade Direct Red 5B and Brilliant blue R, respectively. The potential of three fast-growing plants, *Parthenium hysterophorus*, *Alternanthera sessilis*, and *Jatropha curcas*, to simultaneously degrade two fiber dyes, yellow 5G and brown R have been investigated[93].

More information is required on the unique metabolic pathways that plants use to metabolize toxic substances. The main advantages of plant pigment removal are that it is a self-feeding system that contains large amounts of biomass, has low nutritional requirements, is easy to use, and is widely recognized for its aesthetic demands and environmental sustainability[94]. Extensive research has been carried out to develop effective plant recovery methods to degrade azo dyes; however, large-scale plant recovery applications are not possible, and many challenges remain, which includes plant tolerance to pollutants, the bioavailability of pollutants, evaporation, evaporation of volatile organic pollutants, and the wastewater treatment plants[95].

18.4 CONCLUSIONS AND FUTURE DIRECTIONS

Azo dyes are dyes that are mainly manufactured and used in the fabric industry. They are known as electron-deficient heterologous biological compounds that contain aromatic functional groups with ≥ 1 azo groups (–N=N–). The fabric industry is the biggest producer of wastewater that contains azo dyes. Due to the aesthetic value, toxicity, mutagenicity, and carcinogenicity of wastewater that contains dyes, their discharge into the environment have received great attention. Therefore, wastewater from the textile industry cause serious environmental problems and the colorant must be removed from the wastewater before discharge and disposal. The existing methods for the removal of dyes can be divided into three processes: physical, chemical, and biological treatment. Adsorption, ion exchange, oxidation, and irradiation are various common physical methods used to treat wastewater and have achieved good results. Chemical processes include various oxidation processes, condensation and precipitation, electrocoagulation, Fenton processes, and various photodegradation processes. Chemical processes are commonly used to remove organic contaminants via coagulation and agglomeration processes. Although physicochemical wastewater treatment methods are easy to use, they are not always low cost and environmentally friendly. High energy consumption and low productivity are involved, and large amounts of by-products are produced and the slurry cannot be reused. These methods are not one-step processes and involve multistep processes that require long storage. Biological methods that involve microorganisms (e.g., fungi, enzymes, bacteria, algae, and plants) overcome the limitations of the physical and chemical methods and provide another way to remove azo dyes by conventional methods. These processes are cost-effective, environmentally friendly, and suitable for a variety of colorants. Bacteria are most often used to remove dyes from textile wastewaters because they grow easily, can live under adverse environmental conditions, and can discolor azo dyes faster than other microorganisms. The usefulness of microbial processes for the biological removal of azo dyes is based on the adaptability and functions of the selected microorganism. For the degradation of azo dyes, different microbial populations or a consortium processing system is more efficient due to the synergistic metabolic activity of the microbial population. This review described the importance of oxidases and reductases and their possible mechanisms to understand the biochemical mechanisms of azo dye discoloration and degradation. In addition, the development of innovative nanotechnologies is the next solution to the issue of dye wastewater in the textile, printing, and dyeing industries. The combination of nanotechnology and traditional biological methods could play an important role in improving environmental protection.

REFERENCES

1. Sudha M, Saranya A, Selvakumar G, Sivakumar N. Microbial degradation of azo dyes: A review. Int J Curr Microbiol and Appl Sci. 2014; 3:670–90.
2. Bell J, Plumb JJ, Buckley CA, Stuckey DC. Treatment and decolorization of dyes in an anaerobic baffled reactor. J Environ Eng. 2000; 126:1026–32.
3. Barragan BE, Costa C, Marquez MC. Biodegradation of azo dyes by bacteria inoculated on solid media. Dyes Pigm. 2007; 75:73–81.
4. Kanagaraj J, Senthilvelan T, Panda R. Degradation of azo dyes by laccase: biological method to reduce pollution load in dye wastewater. Clean Technol Environ Pol. 2015; 17:1443–56.
5. Sen SK, Raut S, Bandyopadhyay P, Raut S. Fungal decolouration and degradation of azo dyes: A review. Fungal Biol Rev. 2016; 30:112–33.
6. Puvaneswari N, Muthukrishnan J, Gunasekaran P. Toxicity assessment and microbial degradation of azo dyes 2006.
7. Gahr F, Hermanutz F, Oppermann W. Ozonation: An important technique to comply with new German laws for textile wastewater treatment. Water Sci Technol. 1994; 30:255.
8. Vandevivere PC, Bianchi R,. Verstraete W. Treatment and reuse of wastewater from the textile wet-processing industry: Review of emerging technologies. *Journal of Chemical Technology & Biotechnology: International Research in Process, Environmental and Clean Technology.* 1998; 72:289–302.
9. Robinson T, Chandran B, Nigam P. Studies on the production of enzymes by white-rot fungi for the decolourisation of textile dyes. Enzyme Microb Technol. 2001; 29:575–9.
10. Cooper P. *Colour in Dyehouse Effluent.* Society of Dyers and Colourists. 1995.
11. Anjaneyulu Y, Chary NS, Raj DSS. Decolourization of industrial effluents–available methods and emerging technologies–a review. Rev Environ Sci Biotechnol. 2005; 4:245–73.
12. Nigam P, Armour G, Banat I, Singh D, Marchant R. Physical removal of textile dyes from effluents and solid-state fermentation of dye-adsorbed agricultural residues. Bioresour Technol. 2000; 72:219–26.
13. Zaharia C, Suteu D, Muresan A. Options and solutions for textile effluent decolorization using some specific physico-chemical treatment steps. Environ Eng Manag J. 2012; 11:493–509.
14. Suteu D, Zaharia C, Malutan T. Biosorbents based on lignin used in biosorption processes from wastewater treatment. A review. Lignin: Properties and Applications in Biotechnology and Bioenergy. 2012; 279–306.
15. Borrely S, Cruz A, Del Mastro N, Sampa M, Somessari E. Radiation processing of sewage and sludge. A review. Prog Nucl Energy.1998; 33:3–21.
16. Krapfenbauer K, Robinson M, Getoff N. Development and testing of TiO_2 Catalysts for EDTA-Radiolysis using X-Rays. J Adv Oxod Technol. 1999; 4:213–217.
17. Babu BR, Parande A, Raghu S, Kumar TP. Cotton textile processing: waste generation and effluent treatment. J Cott Sci. 2007.
18. Ghayeni SS, Beatson P, Schneider R, Fane A. Water reclamation from municipal wastewater using combined microfiltration-reverse osmosis (ME-RO): preliminary performance data and microbiological aspects of system operation. Desalination. 1998; 116:65–80.
19. Naveed S, Bhatti I, Ali K. Membrane technology and its suitability for treatment of textile wastewater in Pakistan. J Sci Res. 2006; 17:155–64.
20. Ciardelli G, Ranieri N. The treatment and reuse of wastewater in the textile industry by means of ozonation and electroflocculation. Water Res. 2001; 35:567–72.

21. Alves AMB, de Pinho MN. Ultrafiltration for colour removal of tannery dyeing wastewaters. Desalination. 2000; 130:147–54.
22. Neamtu M, Zaharia C, Catrinescu C, Yediler A, Macoveanu M, Kettrup A. Fe-exchanged Y zeolite as catalyst for wet peroxide oxidation of reactive azo dye Procion Marine H-EXL. Appl Catal B. 2004; 48:287–94.
23. Zaharia C, Suteu D, Muresan A, Muresan R, Popescu A. Textile wastewater treatment by homogenous oxidation with hydrogen peroxide. Environ Eng Manag J. 2009; 8:1359–69.
24. Adams CD, Gorg S. Effect of pH and gas-phase ozone concentration on the decolorization of common textile dyes. J Environ Eng. 2002; 128:93–298.
25. Lopez A, Ricco G, Ciannarella R, Rozzi A, Di Pinto A, Passino R. Textile wastewater reuse: ozonation of membrane concentrated secondary effluent Water Sci Technol. 1999; 40:99–105.
26. Slokar YM, Le Marechal AM. Methods of decoloration of textile wastewaters. Dyes Pigm. 1998; 37:335–56.
27. Zaharia C, Diaconescu R, Surpateanu M. Study of flocculation with Ponilit GT-2 anionic polyelectrolyte applied into a chemical wastewater treatment. Open Chem. 2007; 5:239–56.
28. Singh K, Arora S. Removal of synthetic textile dyes from wastewaters: a critical review on present treatment technologies. Crit Revin Environ Sci Technol. 2011; 41:807–78.
29. Daneshvar N, Ashassi-Sorkhabi H, Tizpar A. Decolorization of orange II by electrocoagulation method. Sep Puri Technol. 2003; 31:153–62.
30. Liu R, Chiu H, Shiau C-S,. Yeh RY-L, Hung Y-T. Degradation and sludge production of textile dyes by Fenton and photo-Fenton processes. Dyes Pigm. 2007; 73:1–6.
31. Yonar T. Treatability studies on traditional hand-printed textile industry wastewaters using Fenton and Fenton-like processes: plant design and cost analysis. Fresenius Environ Bul. 2010; 19:2758–68.
32. Chun H, Yizhong W. Decolorization and biodegradability of photocatalytic treated azo dyes and wool textile wastewater. Chemosphere. 1999; 39: 2107–15.
33. Ledakowicz S, Gonera M. Optimisation of oxidants dose for combined chemical and biological treatment of textile wastewater. Water Res. 1999; 33:2511–16.
34. Perkowski J, Kos L. Decolouration of model dyehouse wastewater with advanced oxidation processes. Fibres Text East Eur. 2003; 11:67–71.
35. Lucas MS, Peres JA. Decolorization of the azo dye Reactive Black 5 by Fenton and photo-Fenton oxidation. Dyes Pigm. 2006; 71:236–44.
36. Tang WZ, An H. Photocatalytic degradation kinetics and mechanism of acid blue 40 by TiO_2/UV in aqueous solution. Chemosphere. 1995: 31:4171–83.
37. Carp O, Huisman CL, Reller A. Photoinduced reactivity of titanium dioxide. Prog Solid State Ch. 2004; 32:33–177.
38. Mirkhani V, Tangestaninejad S, Moghadam M, Habibi M, Rostami-Vartooni A. Photocatalytic degradation of azo dyes catalyzed by Ag doped TiO_2 photocatalyst. J Iran Chem Soc. 2009; 6:578–87.
39. Al-Alwani MA, Ludin NA, Mohamad AB, Kadhum AAH, Mukhlus A. Application of dyes extracted from *Alternanthera dentata* leaves and *Musa acuminata* acts as natural sensitizers for dye-sensitized solar cells. Spectrochim Acta A: Mol Biomol Spectrosc. 2018; 192:487–98.
40. Gurses A, Gunes K, Sahin E. The performance evaluation of hybrid anaerobic baffled reactor for treatment of PVA-containing desizing wastewater. Green Chem Water Rem: Res Appl. 2020, 135.
41. Rongrong L, Xujie L, Qing T, Bo Y, Jihua C. The performance evaluation of hybrid anaerobic baffled reactor for treatment of PVA-containing desizing wastewater. Desalination. 2011; 271:287–94.
42. Solis-Oba MM, Solis A, Perez HI, Manjarrez N, Flores M. Microbial decolouration of azo dyes. Process Biochem. 2012; 47:1723–48.
43. Hayat H, Mahmood A, Pervez ZA, Baig SA. Comparative decolorization of dyes in textile wastewater using biological and chemical treatment. Sep Purif Technol. 2015; 154:149–53.
44. Saratale RG, Saratale GD, Chang J-S, Govindwar SP. Bacterial decolorization and degradation of azo dyes: A review. J Taiwan Inst Chem Eng. 2011; 42:138–57.
45. Chen B-Y, Lin KW, Wang YM, Yen CY. Revealing interactive toxicity of aromatic amines to azo dye decolorizer *Aeromonas hydrophila*. J Hazard Mater. 2009; 166:187–94.
46. Nigam P, Banat IM, Singh D, Marchant R. Microbial process for the decolorization of textile effluent containing azo, diazo and reactive dyes. Process Biochem. 1996; 31:435–42.

47. Khehra MS, Saini HS, Sharma DK, Chadha BS. Chimni SS. Comparative studies on potential of consortium and constituent pure bacterial isolates to decolorize azo dyes. Water Res. 2005; 39:5135–41.

48. Moosvi S, Kher K, Madamwar D. Isolation, characterization and decolorization of textile dyes by a mixed bacterial consortium JW-2. Dyes Pigm. 2007; 74:723–9.

49. Jadhav J, Kalyani D, Telke A. Phugare S, Govindwar SP. Evaluation of the efficacy of a bacterial consortium for the removal of color, reduction of heavy metals, and toxicity from textile dye effluent. Bioresour Technol. 2010; 101:165–73.

50. Solis M, Solis A, Perez HI, Manjarrez N, Flores M. Microbial decolouration of azo dyes: A review. Process Biochem. 2012; 47:1723–48.

51. Forss J, Welander U. Biodegradation of azo and anthraquinone dyes in continuous systems. Int Biodeterior Biodegr. 2011; 65:227–37.

52. Waghmode TR, Kurade MB, Khandare RV, Govindwar SP. A sequential aerobic/microaerophilic decolorization of sulfonated mono azo dye Golden Yellow HER by microbial consortium GG-BL. Int Biodeterior Biodegr. 2011; 65:1024–34.

53. Dave SR, Patel TL, Tipre DR. Bacterial degradation of azo dye containing wastes. Microbial Degradation of Synthetic Dyes in Wastewaters. In *Microbial Degradation of Synthetic Dyes in Wastewaters*. Springer, 2015; 57–83.

54. Chen K-C, Wu J-Y, Liou D-J, Hwang S-CJ. Decolorization of the textile dyes by newly isolated bacterial strains. J Biotechnol. 2003; 101:57–68.

55. Sheth N, Dave S. Enhanced biodegradation of Reactive Violet 5R manufacturing wastewater using down flow fixed film bioreactor. Bioresour Technol. 2010; 101: 8627–31.

56. Rabaey K, Van de Sompel K, Maignien L, Boon N, Aelterman P, Clauwaert, L, et al. Microbial fuel cells for sulfide removal. Environ Sci Technol. 2006; 40:5218–24.

57. Lu L, Zhao M, Liang SC, Zhao LY, Li DB, Zhang BB. Production and synthetic dyes decolourization capacity of a recombinant laccase from *Pichia pastoris*. J Appl Microbiol. 2009; 107:1149–56.

58. Sun J, Hu Y-Y, Bi Z, Cao YQ. Simultaneous decolorization of azo dye and bioelectricity generation using a microfiltration membrane air-cathode single-chamber microbial fuel cell. Bioresour Technol. 2009; 100:3185–92.

59. Reyes P, Pickard MA, Vazquez-Duhalt R. Hydroxybenzotriazole increases the range of textile dyes decolorized by immobilized laccase. Biotechnol Lett. 1999; 21:875–80.

60. Pandey A, Singh P, Iyengar L. Bacterial decolorization and degradation of azo dyes. Int Biodeterior Biodegr. 2007; 59:73–84.

61. Singh RL, Singh PK, Singh RP. Enzymatic decolorization and degradation of azo dyes: A review. Int Biodeterior Biodegr. 2015; 104:21–31.

62. Chen H, Hopper SL, Cerniglia CE. Biochemical and molecular characterization of an azoreductase from *Staphylococcus aureus*, a tetrameric NADPH-dependent flavoprotein. Microbiol. 2005; 151:1433.

63. Chengalroyen M, Dabbs E. The microbial degradation of azo dyes: minireview. World J Microbiol Biotechnol. 2013; 29:389–99.

64. Bafana A, Chakrabarti T. Lateral gene transfer in phylogeny of azoreductase enzyme. Comput Biol Chem. 2008; 32:191–7.

65. Russ R, Rau J, Stolz A. The function of cytoplasmic flavin reductases in the reduction of azo dyes by bacteria. Appl Environ Microbiol. 2000; 66:1429–34.

66. Chen L, Flurkey W. Effect of protease inhibitors on the extraction of Crimini mushroom tyrosinase isoforms. Curr Top Phytochem. 2002; 5:109–20.

67. Shah PD, Dave SR, Rao M. Enzymatic degradation of textile dye Reactive Orange 13 by newly isolated bacterial strain *Alcaligenes faecalis* PMS-1. Int Biodterior Biodegr. 2012; 69:41–50.

68. Arora DS, Sharma RK. Ligninolytic fungal laccases and their biotechnological applications. Appl Biochem Biotechnol. 2010; 160:1760–88.

69. Zhang J, Feng M, Jiang Y, Hu M, Li S, Zhai Q. Efficient decolorization/degradation of aqueous azo dyes using buffered H_2O_2 oxidation catalyzed by a dosage below ppm level of chloroperoxidase Chem Eng J. 2012; 191:236–42.

70. Patil AV, Jadhav JP. Evaluation of phytoremediation potential of *Tagetes patula* L. for the degradation of textile dye Reactive Blue 160 and assessment of the toxicity of degraded metabolites by cytogenotoxicity. Chemosphere. 2013; 92:225–32.

71. Alam MZ, Mansor MF, Jalal K. Optimization of decolorization of methylene blue by lignin peroxidase enzyme produced from sewage sludge with *Phanerocheate chrysosporium*. J Hazard Mat. 2009; 162:708–15.

72. Husain Q, Jan U. Detoxification of phenols and aromatic amines from polluted wastewater by using phenol oxidases. 2000.

73. Lim, DA, Huang Y-C, Alvarez-Buylla A. Neurosurg Clin N Am. 2007; 18:81–92.

74. Parikh A, Madamwar D. Partial characterization of extracellular polysaccharides from cyanobacteria. Bioresour Technol. 2006; 97:1822–27.

75. Dubey SK, Dubey J, Mehra S, Tiwari P, Bishwas A. Potential use of cyanobacterial species in bio-remediation of industrial effluents African J Biotechnol. 2011; 10:1125–32.

76. Singh PK, Singh RL. Bio-removal of azo dyes: A review. Int J Appl Sci Biotechnol. 2017; 5:108–26.

77. Patil KJ, Mahajan R, Lautre HK, Hadda TB. Bioprecipitation and biodegradation of fabric dyes by using *Chara sp.* and *Scenedesmus obliquus*. J Chem Pharmaceut Res. 2015; 7:783–91.

78. Ergene A, Ada K, Tan S, Katircioglu H. Removal of Remazol Brilliant Blue R dye from aqueous solutions by adsorption onto immobilized *Scenedesmus quadricauda*: Equilibrium and kinetic modeling studies Desalination. 2009; 249:1308–14.

79. Fairhead M, Thony-Meyer L. Bacterial tyrosinases: old enzymes with new relevance to biotechnology. New Biotechnol. 2012; 29:183–91.

80. Yang X, Wang J, Zhao X, Wang Q, Xue R. Increasing manganese peroxidase production and biodecolorization of triphenylmethane dyes by novel fungal consortium. Bioresour Technol. 2011; 102:10535–41.

81. Safarik I, Rego LFT, Borovska M, Mosiniewicz-Szablewska E, Weyda F, Safarikova M. New magnetically responsive yeast-based biosorbent for the efficient removal of water-soluble dyes. Enzyme Microb Technol. 2007; 40:1551–6.

82. Charumathi D, Das N. Bioaccumulation of synthetic dyes by *Candida tropicalis* growing in sugarcane bagasse extract medium. Adv Biol Res. 2010; 4:233–40.

83. Aghaie-Khouzani M, Forootanfar H, Moshfegh M, Khoshayand M, Faramarzi M. Decolorization of some synthetic dyes using optimized culture broth of laccase producing ascomycete *Paraconiothyrium variabile*. Biochem Eng J. 2012; 60:9–15.

84. Pajot HJ, Farina JI, de Figueroa LIC. Evidence on manganese peroxidase and tyrosinase expression during decolorization of textile industry dyes by *Trichosporon akiyoshidainum*. Int Biodeterior Biodegr. 2011; 65:1199–207.

85. Meehan C, Banat I, McMullan G, Nigam P, Smyth F, Marchant R. Decolorization of Remazol Black-B using a thermotolerant yeast, *Kluyveromyces marxianus* IMB3. Environ Int. 2000; 26:75–9.

86. Aksu Z. Reactive dye bioaccumulation by *Saccharomyces cerevisiae*. Process Biochem. 2003; 38:1437–44.

87. Erden E, Kaymaz Y, Pazarlioglu NK. Biosorption kinetics of a direct azo dye Sirius Blue K-CFN by *Trametes versicolor*. Electron J Biotech. 2011; 14:3.

88. Arroyo-Figueroa G, Ruiz-Aguilar GM, Lopez-Martinez L, Gonzalez-Sanchez G, Cuevas-Rodriguez G, Rodriguez-Vazquez R. Treatment of a textile effluent from dyeing with cochineal extracts using *Trametes versicolor* fungus. Sci World J. 2011; 11.

89. Desai H, Pathak H, Madamwar D. Advances in molecular and "-omics" technologies to gauge microbial communities and bioremediation at xenobiotic/anthropogenic contaminated sites. Bioresour Technol. 2010; 101:1558–69.

90. Sun J, Peng R-H, Xiong A-S, Tian Y, Zhao W, Xu H, et al. Secretory expression and characterization of a soluble laccase from the *Ganoderma lucidum* strain 7071-9 in *Pichia pastoris*. Mol Biol Rep. 2012; 39:3807–14.

91. Miele A, Giardina P, Sannia G, Faraco V. Random mutants of a *Pleurotus ostreatus* laccase as new biocatalysts for industrial effluents bioremediation. J Appl Microbiol. 2010; 108:998–1006.

92. Watharkar AD, Khandare RV, Kamble AA, Mulla AY, Govindwar SP, Jadhav JP. Phytoremediation potential of *Petunia grandiflora* Juss., an ornamental plant to degrade a disperse, disulfonated triphenylmethane textile dye Brilliant Blue G. Environ Sci Pollut Res. 2013; 20:939–49.

93. Shinde U, Metkar S, Bodkhe R, Khosare G, Harke S. Potential of polyphenol oxidases of *Parthenium hysterophorus*, *Alternanthera sessilis* and *Jotrapha curcas* for simultaneous degradation of two textiles dyes: Yellow 5G and Brown R. Trends Biotechnol. 2012; 1:24–8.

94. Kagalkar AN, Jagtap UB, Jadhav JP, Bapat VA, Govindwar SP. Biotechnological strategies for phytoremediation of the sulfonated azo dye Direct Red 5B using *Blumea malcolmii* Hook. Bioresour Technol. 2009; 100:4104–10.
95. Williams JB. Phytoremediation in wetland ecosystems: progress, problems, and potential. Crit Rev Plant Sci. 2002; 21:607–35.

19 Cyanobacteria Mediated Bioremediation of Hazardous Dyes

Sougata Ghosh,[1,2]* Tanay Bhagwat,[2] and Thomas J. Webster[2]

[1]Department of Microbiology, School of Science, RK. University, Rajkot 360020, Gujarat, India

[2]Department of Chemical Engineering, Northeastern University, Boston, MA02115, USA

*Corresponding author:ghoshsibb@gmail.com

CONTENTS

19.1   INTRODUCTION

Dyes are coloring agents that stain fibers permanently and are resistant to fading on exposure to light, water, sweat, oxidizing chemicals, and even microbial action (1). The first commercial synthetic dye was discovered by William Henry Perkin in 1856, following which >10,000 synthetic dyes had been developed and used in manufacturing by the end of the nineteenth century. Global increases in the textile industries have led to a significant increase in the use of synthetic dyes, and therefore, water pollution. Synthetic dyes are major water pollutants that have potent mutagenic and carcinogenic effects on living organisms. Wastewater contaminated with dyestuffs is a potential threat to the environment and aquatic life. Moreover, when it enters the food chain, it can affect higher animals and human health. The different types of synthetic dyes used in dry processing mills and woven fabric finishing mills determine the variable wastewater characteristics for pH, total organic carbon, biological oxygen demand (BOD), chemical oxygen demand (COD), dissolved oxygen, and organic and inorganic chemical content. It was estimated that >280,000 t of textile dyes are discharged in industrial effluents annually worldwide (2).

A comprehensive overview of the critical toxicological implications of the industrial dyes is shown in Figure 19.1(a) (3). Some of the azo basic, acid, and direct dyes are highly toxic to fish, crustaceans, algae, and bacteria, and reactive azo dyes are toxic at very high concentrations (i.e., effective concentration levels >100 mg/L). Occupational sensitivity to hazardous dyes in the textile industries has been reported since 1930, where disperse dyes, such as monoazo or anthraquinones resulted in allergic reactions. Textile dyes, such as tartrazine and carmoisine can cause severe hepatic and renal damage at higher doses due to inducing oxidative stress due to the formation of free radicals (4). When untreated dyes are released, they can generate DNA adducts that might result in toxic effects to microorganisms that actively participate in the discoloration of dyes. The cytotoxic

DOI: 10.1201/9781003130932-19

A

B

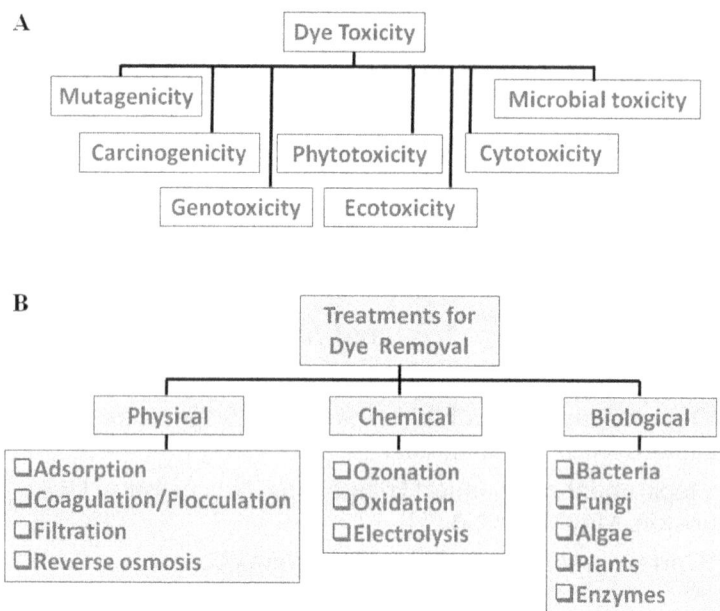

FIGURE 19.1   Showing: (a) Toxic effects of hazardous dyes; and (b) treatment methods for dye removal.

and genotoxic effects of p-dimethyl aminobenzene (p-DAB) were confirmed using chromosome aberration tests, micronucleus tests, and mitotic index in bone marrow cells and spermatozoids of rats (5). The mutagenic effects for chlorotriazine reactive azo red 120 by the induction of micronuclei in erythrocytes of fish were confirmed (6). Similarly, genotoxic and mutagenic effects of Disperse Blue 291dye were attributed to DNA fragmentation, micronuclei formation, and an increase in the index of apoptosis in mammalian cells (HepG2). Considering these potent hazardous effects, various physical, chemical, and biological dye removal methods have been developed, as shown in Figure 19.1(b).

19.2   CONVENTIONAL METHODS FOR DYE REMOVAL

Several physicochemical methods are used for the removal of hazardous dyes from industrial effluents. Sulfur and dispersed dyes can be removed by physical methods, such as coagulation–flocculation. However, acid, direct, reactive, and vat dyes cannot efficiently be removed using this process. In addition, the generation of a large amount of sludge limits the application of these techniques (7). Another method, adsorption, is preferred due to the higher efficiency in the removal of a wide range of dyes. Multiple parameters, such as high affinity, capacity for target compounds, and the potential of adsorbent regeneration play a critical role in the selection of an ideal adsorbent (8).

One of the most effective dye removing entities is activated carbon (C), although its high cost is one of its major drawbacks (9). Therefore, low-cost adsorbent materials, such as peat, bentonite clay, fly ash, polymeric resins, ion exchangers, and many biological materials, such as corn or maize cobs, maize stalks, and wheat straw can be used for dye removal from industrial effluents (10).

Several filtration techniques, such as ultrafiltration, nanofiltration, and reverse osmosis are used for the effective removal of dyes in addition to the membranes that provide potential for the simultaneous separation of hydrolyzed dyestuffs and dyeing auxiliaries (11). However, high investment costs, potential membrane fouling, and the production of secondary waste streams that need further treatment are some of the major disadvantages of the membrane-based dye removal processes (11,9).

Chemical oxidation that uses oxidizing agents, such as ozone (O_3), hydrogen peroxide (H_2O_2), and permanganate (MnO_4) lead to degradation or decomposition of dye molecules (12). Although ozonation is popularly used to treat azo dyes, the major limitations are its short lifetime, ineffectiveness toward dispersed and insoluble dyes, low COD removal capacity, and the high cost of O_3 (13). Other advanced oxidation processes include photochemical and photocatalytic degradation of dyes where O_3, H_2O_2, titanium dioxide (TiO_2), zinc oxide (ZnO_2), manganese (Mn), and iron (Fe), are employed either in the presence or absence of irradiation (13,14).

Most of the physical and chemical techniques for dye removal either concentrate the colored matter into sludge or destroy them (15). Therefore, most of the available physical and chemical methods for dye removal have several disadvantages that include being economically unfeasible, cannot completely remove recalcitrant azo dyes, the generation of a large amount of sludge, and the production of hazardous and toxic by-products (13).

Based on the previous discussion, an alternative cost-effective and environmentally friendly tertiary treatment is recommended for effective dye removal and degradation. In this chapter, various decolorization and degradation techniques that use cyanobacteria will be discussed, which could be effectively used for the treatment of textile effluent that contains hazardous industrial dyes. In addition, the effect of the physicochemical conditions, enzymatic mechanisms of decolorization, and the toxicity of the degradation products are considered.

19.3 CYANOBACTERIA MEDIATED DYE REMOVAL

Cyanobacteria are promising microorganisms that can effectively remove dyes from industrial effluents. They have potential mechanisms, such as biosorption, bioaccumulation, and biodegradation that can remove and detoxify hazardous dyes. Table 19.1 lists various cyanobacteria that can be used for the efficient bioremediation of industrial dyes. Several cyanobacteria with dye removing potential will be discussed in detail in the following section.

Alphanocapsa elachista was isolated from a polluted location in the industrial region of Quisna and Sadat City, Menoufia Governorate, Egypt. It was purified in axenic culture in Allen medium at pH 6.8. After inoculation, *A. elachista* was maintained at 28°C±1°C under continuous illumination (60±4µE/m²s). *A. elachista* can degrade and remove the color from various dyes in wastewater effluents due to different mechanisms for the assimilative utilization of chromophores to produce algal biomass, carbon dioxide (CO_2), and H_2O. Dyes and azo dyes used included Disperse Orange 2RL (4-nitro-4`-[N-ethyl-N-(2-cyanoethyl)-amino[azobenzene), Reactive Yellow 3RN (3-carboxy-4-Hydroxy-4`-nitro azobenzene), Reactive Black NN (1,5-diamino pentane), and Disperse Red BS (Tracid Red BS, $C_{30}H_{37}N_3Na_2O_8S_2$) at 20 ppm and 7 days incubation. *A. elachista* degraded Disperse Orange 2RL by 26.89%, Tracid Red BS by 48.27%, Reactive Black NN by 31.5%, and Reactive Yellow 3RN by 49.16% after 7 days incubation. The effect of these dyes was confirmed on protein content using the Lowry method of estimation. After 7 days of incubation, Disperse Red BS showed the largest protein content at 33.84% followed by Reactive Black NN at 29.47%, 29.43% for Disperse Orange 2RL, and 27.23% for Reactive Yellow 3RN. These were lower than the control (44.00%). Azoreductase activity was determined in which the addition of Reactive Yellow 3RN dye induced azoreductase activity in *A. elachista* by approximately 47.06%, 52.23%, and 52.48% after 3, 5, and 7 days incubation, respectively compared with the control. Fourier-transform infrared spectroscopy (FTIR) analysis showed differences in spectral intensity and the occurrence of stretched vibrations in infra-red (IR) spectra of the algal biomass before and after treatment with dyes and showed evidence of possible biosorption in addition to the algal degradation activities. After treatment by Reactive Yellow 3RN, *A. elachista* showed a reduction in azo bonds within 1,654–1,542 cm^{-1} (16).

In another study, *Anabaena fertilissima, Nostoc muscorum, Phormidium fragile,* and *Wollea* sp. were grown and maintained on BG11 medium under white continuous illumination (3,000 lux) at 25°C±2°C and were tested for the removal of red color and heavy metals from crude effluent discharged from the El-Shafie textile factory at EL- Mahalla EL- Kobra East Delta, Gharbia Governorate, Egypt. The effluent was polluted with a diazo dye. A 200 mL of textile effluent was inoculated with 10 mL of each cyanobacterial strain (10^{12} CFU/mL) and incubated at 25°C±2°C for 7, 14, 21, and 28 days under continuous white light exposure (3,000 lux).

N. muscorum recorded the highest dye removal at100% followed by Wollea sp., P. fragile, and A. fertilissima that exhibited 99.23%, 92.31%, and 63.46% dye removal, respectively after 28 days incubation. All four cyanobacterial strains grew and increased their biomass in dry weight increase from 7 to 28 days under continuous white illumination (3,000 lux). At 28 days, N. muscorum had the highest dry weight of 12.78 mg/L followed by Wollea sp. (10.00 mg/L), P. fragile (6.00 mg/L), and A. fertilissima (5.18 mg/L).

Of note, the pH and electric conductivity (EC) of the crude textile effluent changed over the 28-day incubation. The pH increased to 7.91 for A. fertilissima, followed by pH 7.82, 7.72, and 7.71 for P. fragile, N. muscorum, and Wollea sp., respectively. The EC were 1.05, 1.04, 1.02, and 1.01 for A. fertilissima, P. fragile, N. muscorum, and Wollea sp., respectively. Further, the BOD and COD of the effluent decreased due to the cyanobacterial treatment. The highest percent reduction in COD was achieved by N. muscorum (90.04%) followed by Wollea sp. (85.24%), P. fragile (79.60%), and A. fertilissima (79.40%). Similarly, the highest BOD reduction was observed for N. muscorum (58.78%) followed by A. fertilissima (58.47%), Wollea sp. (57.16%), and P. fragile (48.71%). The four cyanobacterial strains decolorized the industrial effluent and removed the heavy metals from the textile crude discharge (17).

Cyanobacterial potential for dye removal from effluent have been reported (8). Dye-containing liquid effluent and solid waste were collected from the Municipal Treatment Station, Sao Paulo, Brazil and three textile dyes Indigo, Remazol Brilliant Blue R (RBBR), and Sulfur Black were used. The capacity for degradation of dyes was confirmed using cyanobacterial strains Anabaena flos-aquae UTCC64, Phormidium autumnale UTEX1580, and Synechococcus sp. PCC7942 and efficiency were compared with anaerobic and anaerobic–aerobic systems for discoloration and toxicity evaluations. A. flos-aquae UTCC64 was grown on Allen and Arnon (AA) medium and P. autunmnale UTEX1580 and Synechococcus PCC7942 were grown on BG11 medium. Initially, 0.02% (m/v) of each dye was added to 50 mL of AA or BG11 medium and inoculated with 1 mL of cyanobacterium culture. P. autumnale could decolorize Indigo by 91% but could not decolorize Sulfur Black and RBBR. Synechococcus sp. resulted in less discoloration for all three dyes and A. flos-aquae could degrade Indigo ≤71% and it partially degraded Sulfur Black.

Since P. autunmnale UTEX1580 showed the highest percent Indigo discoloration, it was further used for bioreactor based degradation of Indigo. P. autunmnale was grown in 500 mL of BG11 medium for 14 days with Indigo added at 0.02%. Indigo biodegradation by P. autunmnale UTEX1580 was indicated by the complete discoloration after 19 days of incubation. The biofilm that formed on the bioreactor wall adsorbed the Indigo dye. Metabolite analysis showed that the original Indigo molecule was detected until day 18, which demonstrated that the dye was completely degraded by P. autunmnale after that. Anthranilic acid (m/z 399) and isatin or iondole-2,3-dione (m/z 421) appeared after 17 days incubation, which indicated that the degradation of Indigo molecules was not complete, and they were detectable until day 19.

Anaerobic–aerobic degradation systems for dye discoloration were evaluated. Anaerobic conditions were created by filling 150 mL of bottles with the effluent that was then sealed with rubber tops in a CO_2 atmosphere. After incubation for 15 days at 28°C in the dark, the bottles were opened, and 50 mL of effluent was transferred to a sterilized flask and re-enriched with 1 mL of inoculum from a fresh effluent. The flasks were incubated aerobically for 15 days at 28°C, which resulted in 17% and 33% discoloration after anaerobic treatment and combined anaerobic–aerobic treatment, respectively. Although high color removal was obtained by anaerobic degradation, it produced an effluent with higher turbidity. Due to the complexity and being recalcitrant and xenobiotic, a combination of anaerobic and aerobic conditions is required for the complete degradation of dyes. The cyanobacterial strains used could reduce effluent toxicity and showed the presence of degraded compounds in the effluent. The Hydra bioassay confirmed that no toxic effects or toxic by-products were formed after dye treatment by *P. autunmnale*. The presence of the metabolites isatin and anthranilic acid did not show increased toxicity, which confirmed that treatment of Indigo by *P. autunmnale* has potential (18).

Soil and water samples were collected from the vicinity of dye manufacturing units in Gujarat, India (19), which were used for the enrichment and isolation of cyanobacteria identified as *Chroococcus minutus, Gleocapsa pleurocapsoides, Gleothece* sp., *Oscillatoria splendid*, and *Phormidium ceylanicum*. The cyanobacteria were isolated, purified, and identified by growing and maintaining the cultures at 27°C in BG11 medium with a light/dark cycle of 12:12 h under white fluorescent light of at 30 W/m². Textile dyes used to confirm degradation ability were the cyclic azo dyes Acid Red 97, FF Sky Blue, and Amino Black 10B. Initial screening of isolated cultures for dye decolorization was made in BG11 medium amended with 50 mg/L of dye, which indicated 70%–90% decolorization.

All the cultures grew efficiently in the medium that contained dyes. *P. ceylanicum, G. pleurocapsoides,* and *C. minutes* demonstrated rapid decolorization of dye at 100 mg/L. *G. pleurocapsoides* decolorized 90% of FF Sky Blue and 83% of Acid Red 97, and *P. ceylanicum* decolorized 88% of Acid Red 97 and 77% of FF Sky Blue. *C. minutus* decolorized Amido Black 10B by 55% after 26 days. *O. splendida* decolorized FF Sky Blue <58% and *Gleothece* sp. decolorized Acid Red 97 by 40%. A large decrease in chlorophyll a content was observed in the presence of dyes, possibly due to the attenuation of light transmission that reduced the rate of photosynthesis, and therefore, the growth rate. The slower rate of decolorization observed might be attributed to higher molecular weight, structural complexity, and the presence of inhibitory functional groups, such as $-NO_2$ and $-SO_3Na$ in the dyes. The absorption ratio at two distinct wavelengths changed with time, with a sequential reduction in absorbance at the dye's absorption maxima (λ_{max}). This phenomenon was attributed to the cleavage of color imparting chromophores (e.g., $-NO_2$, $-N=N-$, and $-NH_2$) and fused aromatic rings with the simultaneous formation of ultraviolet (UV) absorbing intermediates. FF Sky Blue was converted to highly fluorescent compounds due to photooxidative changes, which was degraded with cyanobacterial treatment (19).

Arthrospira maxima were reported to degrade Congo red from 2 to 20 ppm on days 5, 10, and 15. *A. maxima* was maintained on Zarrouks medium at 24°C±10°C in a thermostatically controlled room, illuminated with fluorescent tubes that provided 30 µEm²/s and 12:12 h light/dark cycles. Pigment analysis showed the presence of chlorophyll and protein in varying amounts when *A. maxima* were allowed to interact with Congo Red. Maximum decolourization was observed (at 2 ppm) of 46% (20).

In another study, *Lyngbya* sp. BDU 9001 with coir pith (in a ratio of 0.1:1.0 g) was used to decolorize 78.04% textile dye effluent by treating it for 15 days under a photoperiod of 1,500 lux, at pH 7 and 29°C. In the mixture of cyanobacterium with coir pith, pH decreased from an initial pH of 8.9±0.3 to 7.12±0.3 due to the secretion of enzymes formed to degrade the coir pith by the cyanobacterium. A large amount of oxidizable organic matter in addition to calcium (Ca), traces of dissolved oxygen (O_2), considerable amounts of nitrate and phosphates in the effluents were factors that favored the growth of cyanobacterium *Lyngbya* sp. BDU 9001. Nitrate and nitrites present in the effluent from the textile and tannery industries could be significantly reduced by 18% and 73.91%, respectively following the cyanobacterial treatment. Of interest, *Lyngbya* sp. BDU 9001 with coir pith showed an increase in chlorophyll a content that might be attributed to the growth of cyanobacterium. This indicated that the presence of textile dye in the effluent did not inhibit the growth of *Lyngbya* sp. BDU 9001. The original protein concentration of *Lyngbya* sp. was 118.0 mg/g, which significantly increased ≤148.0 mg/g following treatment. Although the initial salinity was higher, the reduction in chloride content could have been attributed to the reduction in salinity during decolorization. Dissolved O_2 increased from 1.51 mg/L to 2.21 mg/L due to the decomposition of coir pith, which indicated a higher rate of photosynthesis in the presence of high light intensity (21).

In another study, *Lyngbya lagerlerimi, Nostoc linckia,* and *Oscillatoria rubescens* that were isolated from an industrial region of Quisna, Menoufiya Governorate, Egypt decolorized and removed Methyl Red, Orange II, G-Red (FN-3G), Basic cation and Basic fuchsin. Decolorizing ability was confirmed by adding 20 ppm of dyes to 150mL Allen medium inoculated with 2×10^4 cells/mL of cyanobacteria. Variations in the decolourization ability against azo dyes were evident for *L. lagerlerimi, N. linckia,* and *O. rubescens,* which depended on the type of dye, concentration, and the cyanobacterium used. *L. lagerlerimi* decolourized basic cationic ≤84.1% and 89.4% after 3 and 7 days of incubation, respectively. Decolourization and removal was approximately 83% against Basic Fuchsin at 5 ppm. *L. lagerlerimi* decolourized Methyl Red by 35% after 7 days. Orange II was only slightly decolourized and removed by *L. lagerlerimi* by 47% after7-day incubation.

N. linckia decolourized Basic cationic and Basic Fuchsin by 91% and 83% after 3 days of incubation, respectively. *N. linckia* could decolorize and degrade Methyl Red by 75.40% and 81.97% after 3 and 7 days incubation. This degradation activity was accompanied by a spectral shift. *O. rubescens* displayed the highest ability to rapidly remove Basic fuchsin ≤94.79% after 3 days incubation. Basic cationic was degraded by 70% in 3 days and 85% in 7 days by *O. rubescens*. *O. rubescens* decolorized Methyl Red by 81.30% and 81.16% after 3 and 7 days incubation, respectively. *O. rubescens* decolourized Orange II, which increased from 12.8% in 3 days to 33.63% after7 days incubation.

After 7 days of incubation, *L. lagerlerimi, N. linckia,* and *O. rubescens* could only decolourized and degrade G-Red (FN 3G) by 10%. During the reduction of azo dyes, the azo bond was broken down by azoreductase and aromatic amines were produced. IR spectrum analysis showed differences in the IR peaks that resulted from the biomass of *L. lagerlerimi* before and after treatment with Methyl Red. IR analysis of the biomass of *N. linckia* with Methyl red demonstrated a stretching vibration of the azo bond that diminished within 1,630–1,575 cm^{-1}. *N. linckia* affected the chemical structure of basic cationic with a stretching vibration that diminished at 1,600 cm^{-1}, which was a result of the activity of cyanobacterial species against basic cationic. These variations in IR spectra might be attributed to inducing azoreductase and cleavage of the azo bonds in Methyl Red and the formation of aromatic amines (22).

Cyanobacterial isolates, such as *Leptolyngbya* sp., *Phormidium* sp., and *Synechococcus* sp. from the mangroves in Cardoso Island, Sao Paulo State, Brazil and *Phormidium autumnale* UTEX1580 were cultured in BG11, SWBG11, and ASNIII and maintained at 24°C±1°C under constant illumination by white fluorescent light (40 μmol/photon/m^2/s) for dye removal. The dyes were Indigo BANN 20 (in 30% aqueous solution), Palanil Yellow 3G, Drimarene Yellow CL-R, Indanthrene Yellow 5GF, Drimarene Red CL-5B, Indanthrene Red FBB, Dispersol Red C-4G, Dispersol Blue C-2R, Drimarene Blue CL-R, Indanthrene Blue RCL, and the polymeric RBBR with absorbance spectrum in the visible range between 400 and 900 nm.

For decolourization studies, the cyanobacterial inoculum was prepared from a 15-day growth culture. 10 mL of each cyanobacterial medium was mixed with 0.02% of different dyes and inoculated with the respective cyanobacteria followed by incubation at 23°C under static conditions for 26 days. Similarly, a solid culture medium was used. After 26 days of cyanobacterial growth, *Leptolyngbya* sp. CENA103 discolored three dyes: Indigo (86%), Palanil Yellow (83%), and Indanthrene Yellow (68%). *Leptolyngbya* sp. CENA104 could discolor the dyes Indigo (68%), and Palanil Yellow (78%). *P. autumnale* UTEX1580 removed 6 dyes: Indigo (88%), Indanthrene Blue (60%), Dispersol Blue (65%), Indanthrene Red (88%), Palanil Yellow (62%), and Indanthrene Yellow (79%). *Leptolyngbya* sp. CENA134 removed 3 dyes: Indigo (63%), Palanil Yellow (77%), and Indanthrene Yellow (58%). *Phormidium* sp. CENA135 was able to remove three dyes: Indigo (100%), Palanil Yellow (51%), and Indanthrene Yellow (70%). *Synechococcus* sp. CENA136 discolored Dispersol Red (54%). Drimarene Blue, Red, and Yellow were not removed by any of the tested cyanobacteria, except for *Leptolyngbya* sp. CENA104, which removed 34% of Drimarene Red. *P. autumnale* UTEX1580 only removed Dispersol Blue and Indanthrene Red by >50%. The removal of palanil yellow was >50%

by almost all of the cyanobacteria. Indanthrene Yellow was removed by >50% by *P. autumnale* UTEX1580, *Leptolyngbya* sp. CENA103, *Leptolyngbya* sp. CENA134, and *Phormidium* sp. CENA135 (23).

Degradation and decolourization efficiency of *Nostoc carneum* for azo dyes Methyl Orange (MO) and Yellow Azo (YA) was confirmed at 10–70 mg/L. *N. carneum* was initially grown in BG11 medium at pH 7 for 14 days under static incubation conditions and continuous illumination (3,000 lux) at 28°C±2°C. BG11 medium was mixed with 10mL *N. carneum* culture supplemented with MO and YA dyes. The highest dye removal observed for 30 mg/L of MO was 69.8% and for 50 mg/L YA was 84.9%. The dye removal efficiency was a function of contact time and was maximum at day 12.

A progressive increment in the cyanobacterial biomass with MO and YA dye supplementation was observed with the maximum recorded biomass were 911.20 mg/L and 746.57 mg/L for 30 mg/L of MO and 50 mg/L of YA, respectively. MO induced maximum chlorophyll a content at day 10, which 0.0328 mg/g dry biomass and carotenoid content reached their maximum value at day 12 (6.6972 mg/g dry biomass). Chlorophyll a and carotenoid content by YA supplements exhibited the same pattern of response with maximum chlorophyll a content at day 10 of 0.0382 mg/g dry biomass for 50 mg/L addition and carotenoids content reached its maximum value on day 12 of 6.7142 mg/g dry biomass for 70 m/L supplements.

Cyanobacterial cellular protein content varied with dye supplementation. A 30 mg/L MO supplement enhanced protein production to 306.248 mg/L and a 50 mg/L YA supplement increased protein content to 368.710 mg/L.

Laccase activity was higher in YA supplementation than MO at the start of the stationary phase of culture, which agreed with *N. carneum* growth (biomass production). This indicated an exponential relationship between dye concentration and laccase activity (24).

Marine *Osciallatoria formosa* NTDM02 efficiently decolorized Amido Black G200 and textile dye effluent collected from a textile industry at Karur, Tamil Nadu, India. At the start of incubation, pH was 11± 0.2 which decreased to 7.5±0.2 during decolourization and decreased further and was approximately pH 7.2 at the end of the experiments. The net negative charge on the surface of cyanobacterium, which is due to the functional groups associated with the cell wall, might be responsible for the decrease in pH along with the production of acids and enzymes in the medium for the dye reduction. Effluent treated with *O. Formosa* NTDM02 showed a larger amount of decolourization than Amido Black G200, which was attributed to the production of reductase enzyme for dye removal. Additional functional groups were exposed due to the alkaline conditions that ruptured the cell walls of microbes; therefore, solubilizing the cell constituents, such as lipids and enhancing decolorization by the organism (25).

Similarly, *Oscillateria tenuis* degraded Eriochrome Blue SE and Black T effectively when introduced separately into flasks that contained an azo compound at 20 ppm. Of note, the compounds that contained a hydroxy or amino group were readily decolorized. Some compounds that contained an amino or hydroxy group counteracted the inhibition of the SO_3H group on azo reduction. This indicated that the hydroxy or amino group might have an important contribution to the ability of cyanobacteria to decolorize a particular azo dye. A lower reduction rate was observed for compounds that contained methyl, methoxy, nitro, or sulfo derivatives, which were hardly decolorized. The highest activity was observed for the culture medium that was free from organic carbon (C) and nitrogen (N), which suggested that *O. tenuis* could utilize Eriochrome Blue SE and Black T as a sole source of C and N. An obvious change in the UV-Vis absorption spectrum of azo dyes was observed after treatment with *O. tenuis*. A new peak appeared at 3,500–3,100 cm^{-1} and represented the stretching vibration of aromatic amines on treatment with cyanobacteria and the stretching vibration of azo bonds diminished at 1,630–1,575 cm^{-1} which suggests the cleavage of azo bonds in Eriochrome Blue SE and the formation of aromatic amines. In addition, 100% of aniline was lost and with no detection of any other organic compounds, which indicated that aniline was effectively

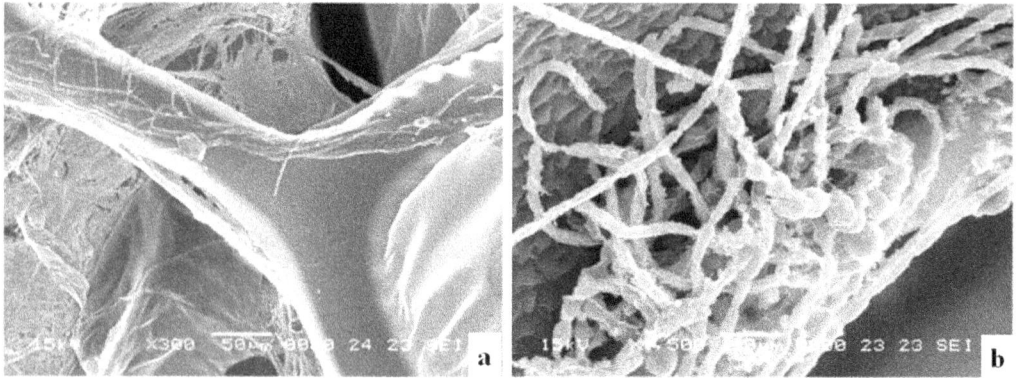

FIGURE 19.2   SEM view of PUF immobilized: (a) *Oscillatoria;* and (b) *Westiellopsis*. (Reprinted with permission from (27). Treatment of dye industry effluent using free and immobilized cyanobacteria. J Bioremed Biodeg. 3:1000165.)

degraded into simple organic intermediates. The reduction rate of azo dyes is related to the molecular structure of the dye and to the utilization capability of the species of blue-green algae used combined with other environmental conditions. The azo reductase of *O. tenuis* is responsible for degrading the azo dyes into aromatic amine by breaking the azo linkage (26).

In another study, dye industrial effluent was treated with cyanobacteria *Oscillatoria brevis* and *Westiellopsis prolifica* to remove color and nutrients, which was based on their dominant presence in the effluent. The microorganisms removed the organic and inorganic content from Remazol and Venyl sulfone and reduced the intensity of the color in the effluent. Reduction of BOD and COD ≤95% was observed. The microorganisms were used in free and immobilized conditions. In addition, the effluent supported the growth of *Oscillatoria* and *Westiellopsis* to a certain extent. *O. brevis* and *W. prolifica* were maintained in BG11 medium and the cyanobacteria were immobilized by entrapment in polyurethane foam (PUF) cubes that allowed high cell loading as shown in Figure 19.2. A higher quantity of oxidizable organic matter, traces of dissolved O_2, a considerable amount of Ca, phosphates, nitrates, and alkaline pH were probably the factors that favored the growth of cyanobacterium although the growth in the control was well pronounced compared with the effluent. *O. brevis* and *W. prolific* significantly reduced the color from the dye effluent under free and immobilized conditions. Overall, >74% of color was removed after 30 days. Color removal with immobilized cyanobacteria was more efficient than with free cells when 100% undiluted effluent was used. *Oscillatoria* sp. was more efficient than *Westiellopsis* sp. Ammonia was completely removed from the effluent followed by nitrite, but nitrate was not completely removed from the effluent. Therefore, the cyanobacteria might utilize ammonia, nitrite, and then nitrate, which might be attributed to the ammonium requiring the least processing to be incorporated into cell constituents. However, nitrate requires a special enzyme for the transport of nitrate into the cell and to then be converted to nitrate and ammonium by the sequential action of nitrate and nitrite reductase. Therefore, the complete removal of other nutrients with only 50% nitrite removal was observed in the dye effluent. A gradual reduction of nitrate level was observed from day 9 onwards and on day 30, the percent removal of nitrate was maximum in effluent immobilized on PUF. Complete removal of total organic and inorganic phosphates was observed on day 30. Gradual increase in dissolved O_2 content from day 5 onwards was seen with maximum level of dissolved O_2 record on day 30. Approximately 50% reduction in chloride and >90% removal of Ca and magnesium were achieved (27).

The removal of Remazol Blue and Reactive Black B was achieved by immobilized thermophilic *Phormidium* sp. that were obtained from a hot spring in Ayas, Turkey. Ca alginate gel immobilized *Phormidium* sp. beads were added to 100 mL BG11 medium and incubated under stationary conditions illuminated by cool white fluorescent lamps at 2,400 lux at 45°C±1°C for 8 days in a plant growth chamber. Different dyes were used by dissolving powdered dyestuff in distilled water at 2% (w/v). Dye removal increased with an increase in pH, the highest removal was at pH 8.5 for both dyes although the highest decolorization was obtained in Reactive Black B. Dye without beads did not show any reaction, unlike the spontaneous dye removal observed between 12% and 18% with cell-free beads. Dye removal of Remazol Blue at different initial dye concentrations varied between 9.1 to 82.1 mg/L at 45°C after 8 days. With the increase in dye concentration, dye uptake by *Phormidium* sp. decreased and the highest dye removal was obtained in the samples that had slower initial dye concentrations. At 9.1 mg/L, 88% of the initial dye was decolorized after 8 days of incubation. At higher dye concentrations, 56.74 and 82.09 mg/L, the removal efficiency was ≤60%. Reactive Black B removal by *Phormidium* sp. was investigated at different initial dye concentrations between 12.1 and79.47 mg/L. No significant difference was observed in the removal of the dyes. However, the Reactive Black B uptake yield of immobilized *Phormidium sp.* was higher than that of Remazol Blue at high dye concentrations; approximately 68% and 60% of Reactive Black B removal occurred at 57.54 and 79.47 mg/L, respectively. At the end of incubation, the highest dye uptake was measured as 87.42% for 12.1 mg/L initial Reactive Black B. Temperature variation had a significant effect on the decolorization of the tested dyes by *Phormidium* sp. for 8 days incubation with the highest removal at 45°C and 50°C. With the decreasing incubation temperatures, the removal of dyes by immobilized *Phormidium* sp. decreased. At 40°C, maximum removal of Remazol Blue (42.66%) and Reactive Black B (56.63%) was achieved by the immobilized *Phormidium* sp. with an initial 75 mg/L dye.

Ca alginate was used to immobilize *Phormidium* sp. at pH 8.5 and under different dye concentrations (9.1 to 82.1 mg/L) at 45°C. High dye degradation was observed with maximum removal from 50% to 88%. Remazol Blue removal was maximum ≤50.3% and Reactive Black B removal was 60.0%. Maximum dye removal was observed at 45°C and 50°C, and a decrease in temperature resulted in a decrease in the dye removal efficiency. At approximately 75 mg/L initial dye, the highest specific dye uptake measured was 41.29–41.17 mg/g for Remazol Blue and 47.69–43.82 mg/g for Reactive Black B at 45°C and 50°C,respectively, after 8 days incubation (28).

The dye removal capacities of *Phormidium* sp. and *Synechocystis* sp. were investigated against Reactive Red, Remazol Blue, and Reactive Black B. Dye removal by *Synechocystis* sp. and *Phormidium* sp. was a function of initial pH and initial dye concentration in media with and without triacontanol (TRIA). *Synechocystis* sp. had a similar dry weight in media with 1and 10 mg/L TRIA and *Phormidium* sp. had the highest dry weight at 10 mg/L TRIA. Different initial pH values in media with approximately 25 mg/L dye and 10 mg/L TRIA was used. *Synechocystis* sp. removed maximum dye ≤34.4% for Reactive Red at pH 9.5, 23.1% for Remazol Blue at pH 8.5, and 13.7% for Remazol Black B at pH 8.5. For *Phormidium* sp. suitable pH values were pH 9.5 for Reactive Red and pH 8.5 for Remazol Blue and Reactive Black B.

Similarly, the dye removal efficiency of *Synechocystis* sp. was determined at different initial dye concentrations between 22.9 and 86.9 mg/L. After incubation, the maximum removal yields were approximately 25 % for Reactive Red. In the media with Reactive Red but no TRIA, the *Synechocystis* sp. removed dye ≤ 24.4%. Dye removal capacity decreased with an increase in dye concentration. Remazol Blue removal was highest ≤ 37.5% with 29.7 mg/L initial dye with TRIA. For Reactive Black B, the highest removal was in media with TRIA, and the lowest Reactive Black B concentration (22.9 mg/L) was 29.2%. Increased dye concentrations decreased the removal of the dye, regardless of the effect of TRIA.

Dye removal by *Phormidium* sp. was investigated at different initial dye concentrations. *Phormidium* sp. removed Reactive Red ≤35.4% in presence of TRIA. *Phormidium* sp. showed overall dye removal capacity between 27.5% and 30.0% in media with increasing Reactive Red concentrations regardless of TRIA. Dye removal decreased with increasing dye concentrations, which was similar to *Synechocystis* sp. (29).

Phormidium sp. and *Synechococcus* sp. isolated from hot springs in Ayas, Turkey can remove textile industry dyes, such as Remazol Blue, Reactive Black, and Reactive Red RB. Bioaccumulation varied between 7.5% and 31.3% for *Synechococcus* sp. and between 8.1% and 44.5% for *Phormidium* sp. after 9 days for all tested pH and dyes. For Reactive Red RB, bioaccumulation increased with the increase in pH, which was highest at pH 9.5 for *Phormidium* sp. and *Synechococcus* sp. Maximum removal was observed at pH 8.5 for Remazol Blue and Reactive Black B. Dye bioaccumulation in *Synechococcus* sp. was between 8.7% and 66.9%; however, it could not grow in high concentrations of Remazol Blue and Reactive Black B. Of interest, at increasing dye concentrations the maximum specific dye uptake by *Synechococcus* sp. increased for all tested dyes. Bioaccumulation of Remazol Blue by *Phormidium* sp. was 80% of the initial dye at the end of 8 days incubation. At high initial dye concentrations of 49.8 and 75.8 mg/L, bioaccumulation reached approximately 35% despite the longer incubation. In samples that contained Reactive Black B, bioaccumulation was >90% within 6 days at low initial dye concentrations. However, at high dye concentrations the bioaccumulation achieved was only ≤ 31% – 41% after 14 days. *Phormidium* sp., bioaccumulated 50% and 20% of Reactive Red RB at lower and higher concentrations, respectively within 14 days in BG11 medium under thermophilic conditions.

Although cyanobacterial growth was observed at 40°C, 45°C, and 50°C, temperature variations had a significant effect on the bioaccumulation of all the tested dyes by cyanobacterial cultures within 9–14 days incubation. *Synechococcus* sp. could not grow when incubated with Remazol Blue and Reactive Black B at 45°C and 50°C. Increased incubation temperatures significantly affected bioaccumulation yield at all concentrations and maximum dye removal was detected at 40°C for all dyes after 9 days. The bioaccumulation yield of *Synechococcus* sp. decreased but uptake was higher at higher initial dye concentrations. The highest specific dye uptake measured was 56.25 mg/g, 80.22 mg/g, and 35.33 mg/g for Reactive Red RB, Remazol Blue, and Reactive Black B at an initial dye concentration of 78.3, 72.4, and 62.0 mg/L, respectively. At increasing dye concentrations, the maximum specific dye uptake of *Phormidium* sp. increased for all tested dyes and temperatures. The highest specific dye uptake measured were 14.49, 16.76, and 16.95 mg/g for Reactive Red RB, Remazol Blue, and Reactive Black B with initial dye concentrations of 78.3, 72.4, and 61.69 mg/L, respectively (30).

A marine cyanobacterium, *Phormidium valderianum* was used for the removal of three textile dyes: Acid Red, Acid Red 119, and Direct Black 155. *P. valderianum* biomass (30 mg dry weight) effectively removed >90% of Acid Red, Acid Red 119, and Direct Black 155 with initial concentrations of 700, 500, and 400 mg/L, respectively. An increase in concentration reduced the efficiency of the cells to remove dyes from the solution. In addition, *P. valderianum* could efficiently remove >95% dye above pH 11 and Acid Red was maximally adsorbed >50% at all pH tested. The presence of phenolic compounds and metal chelators significantly decreased the ability of *P. valderianum* to remove the dye and the removal effect was in the order of phenol>EDTA>tartrate>resorcinol>glucose>glycerol. An increase in phenol concentration with a fixed concentration of dye revealed that the amount of dye removal reduced proportionally with the concentration of phenol. Monovalent metal ions (Na$^+$ and K$^+$) helped to reduce the dye normally; however, divalent metal ions showed marginal or no effect (31).

19.4 CONCLUSIONS

Cyanobacteria are considered ideal candidates for wastewater treatment due to their promising capability to remove toxic dyes. Owing to the diverse functional groups associated with their cell surface, cyanobacteria can entrap dye molecules on binding sites that are located on the filamentous surfaces. The method of dye uptake can be passive, or active, or both, which is known as bioaccumulation. Immobilized cyanobacteria could be a very powerful method for wastewater treatment due to their reusability and recyclability. Several parameters, such as duration, biomass and dye concentration, temperature, pH, aeration, and media components should be developed to optimize and develop

TABLE 19.1
Cyanobacterial Mediated Removal of Hazardous Dyes

Name of Organism	Name of Dye	Degradation (%)	References
A. elachista	Disperse Orange 2RL	26.89	(16)
	Reactive Yellow 3RN	48.27	
	Reactive Black NN	31.5	
	Disp Red BS	49.16	
Anabaena fertilissima	Azo dye-containing effluent	63.46	(17)
A. flos-aquae UTCC64	Indigo	71	(18)
A. maxima	Congo Red	46	(20)
C. minutus	Amido Black 10B	55	(19)
G. pleurocapsoides	FF Sky Blue	90	(19)
	Acid Red 97	83	
Gleothece sp.	Acid Red 97	40	(19)
Leptolyngbya sp. CENA 103	Indigo	86	(23)
	Palanil Yellow	83	
	Indanthrene Yellow	68	
Leptolyngbya sp. CENA104	Indigo	68	(23)
	Palanil Yellow	78	
	Drimarene Red	34	
Leptolyngbya sp. CENA134	Indigo	63	(23)
	Palanil Yellow	77	
	Indanthrene Yellow	58	
Lyngbya sp. BDU 9001	Dye effluent	78.04	(21)
L. lagerlerimi	Basic cationic	89.4	(22)
	Basic Fuchsin	83	
	Methyl Red	35	
	Orange II	47	
	G-Red (FN 3G)	10	
N. carneum	MO	69.8	(24)
	YA	84.9	
N. linckia	Basic cationic	91	(22)
	Basic Fuchsin	83	
	Methyl Red	81.97	
	Orange II	25	
	G-Red (FN 3G)	10	
N. muscorum	Azo dye-containing effluent	100	(17)
O. brevis	Remazol	>74	(27)
	Venyl Sulfone	>74	
O. formosa NTDM02	Amido Black G200		(25)
	Dye effluent		
O. tenuis	Eriochrome BlueSE		(26)
	Eriochrome BlackT		
O. rubescens	Basic Fuchsin	94.79	(22)
	Basic cationic	85	
	Methyl Red	81.30	
	Orange II	33.63	
	G-Red (FN 3G)	10	
O. splendida	FF Sky Blue	58	(19)
P. autumnale UTEX1580	Dispersol Blue	65	(23)
	Indanthrene Red	88	
	Indigo	88	
	Indanthrene Blue	60	

(*continued*)

TABLE 19.1 (Continued)
Cyanobacterial Mediated Removal of Hazardous Dyes

Name of Organism	Name of Dye	Degradation (%)	References
	Palanil Yellow	62	
	Indanthrene Yellow	79	
	Indigo	91	(18)
P. ceylanicum	FF Sky Blue	77	(19)
	Acid Red 97	88	
P. fragile	Azo dye-containing effluent	92.31	(17)
Phormidium sp.	Remazol Blue	88	(28)
	Reactive Black B	87.42	
	Reactive Red	35.4	(29)
	Reactive Black B	28.3	
	Remazol Blue	33.5	(30)
	Reactive Black B	44.5	
	Reactive Red RB	19.2	
Phormidium sp.CENA135	Indigo	100	(23)
	Palanil Yellow	51	
	Indanthrene Yellow	70	
P. valderianum	Acid Red	>90	(31)
	Acid Red 119	>90	
	Direct Black 155	>90	
Synechococcus sp.	Remazol Blue	39.9	(30)
	Reactive Black B	13.7	
	Reactive Red RB	23.0	
Synechococcus sp.CENA136	Dispersol Red	54	(23)
Synechococcus sp.PCC7942	SulfurBlack	–	(18)
Synechocystis sp.	Reactive Red	34.4	(29)
	Remazol Blue	23.1	
	Reactive Black B	13.7	
W. prolifica	Remazol	<74	(27)
	Venyl Sulfone	<74	
Wollea sp.	Azo dye-containing effluent	99.23	(17)

a bioreactor based microbial process for dye removal. Further, scale-up of the process could help to design biological water treatment plants at the community scale. Therefore, cyanobacterial cultures could be ideal candidates for the development of bioprocesses for the treatment of wastewater that is contaminated with hazardous dyes.

ACKNOWLEDGMENTS

Dr. Sougata Ghosh acknowledges the Department of Science and Technology (DST), Ministry of Science and Technology, Government of India and Jawaharlal Nehru Centre for Advanced Scientific Research, India for funding under Post-doctoral Overseas Fellowship in Nano Science and Technology(Ref. JNC/AO/A.0610.1(4) 2019–2260 dated August19, 2019).

REFERENCES

1. Saratale RG, Saratale GD, Chang JS, Govindwar SP. Bacterial decolorization and degradation of azo dyes: A review. J Taiwan Inst Chem Engrs. 2011; 42:138–57.

2. Jin X, Liu G, Xu Z, Tao W. Decolourisation of a dye industry effluent by *Aspergillus fumigatus* XC6. Appl Microbiol Biotechnol. 2007; 74:239–43.

3. Ventura-Camargo BC, Marin-Morales MA. Azo Dyes: Characterization and toxicity– a review. TLIST. 2013; 2(2):85–103.

4. Amin KA, Hameid HA, Abd Elsttar AH. Effect of food azo dyes tartrazine and carmoisine on biochemical parameters related to renal, hepatic function and oxidative stress biomarkers in young male rats. Food Chem Toxicol. 2010; 48:2994–9.

5. Biswas SJ, Khuda-Bukhsh AR. Cytotoxic and genotoxic effects of the azo-dye p-dimethylaminoazobenzene in mice: a time-course study. Mutat Res. 2005; 587:1–8.

6. Al-Sabti K. Chlorotriazine reactive azo red 120 textile dye induces micronuclei in fish. Ecotoxicol Environ Saf. 2000; 47:149–55.

7. Vandevivere PC, Bianchi R, Verstraete W. Treatment and reuse of wastewater from the textile wet-processing industry: Review of emerging technologies. J Chem Technol Biotechnol. 1998; 72(4):289–302.

8. Subramaniam S, Savitha S, Swaminathan K, Lin FH. Metabolically inactive *Trichoderma harzianum* mediated adsorption of synthetic dyes: Equilibrium and kinetic studies. J Taiwan Inst Chem Engrs. 2009; 40(4):394–402.

9. Robinson T, McMullan G, Marchant R, Nigam P. Remediation of dyes in textile effluent: A critical review on current treatment technologies with a proposed alternative. Bioresour Technol. 2001; 77(3):247–55.

10. Ramakrishna KR, Viraraghavan T. Dye removal using low-cost adsorbents. Water Sci Technol. 1997; 36(2–3):189–96.

11. dos Santos AB, Cervantes FJ, van Lier JB. Review paper on current technologies for decolourisation of textile wastewaters: Perspectives for anaerobic biotechnology. Bioresour Technol. 2007; 98 (12):2369–385.

12. Metcalf E. Wastewater Engineering: Treatment and Reuse. 4th ed. New York: McGraw-Hill; 2003.

13. Anjaneyulu Y, Sreedhara Chary N, Raj DSS. Decolourization of industrial effluents-available methods and emerging technologies: A review. Rev Environ Sci Biotechnol. 2005; 4:245–73.

14. Forgacs E, Cserhati T, Oros G. Removal of synthetic dyes from wastewaters: A Review. Environ Int. 2004; 30:953–71.

15. Adav SS, Lee DJ, Lai JY. Treating chemical industries influent using aerobic granular sludge: Recent development. J Taiwan Inst Chem Engrs. 2009; 40(3):333–6.

16. El-Sheekh MM, Abou-El-Souod GW, El Asrag HA. Biodegradation of some dyes by the green alga *Chlorella vulgaris* and the cyanobacterium *Aphanocapsa elachista*. Egypt J Bot. 2018; 58:311–20.

17. Ghazal FM, Battah MG, El-Aal AAA, Eladel HM, Adly SE. Studies on the efficiency of cyanobacteria on textile wastewater treatment. Res J Pharm Biol. Chem Sci. 2016; 7:2925–31.

18. Dellamatrice PM, Silva-Stenico ME, de Moraes LAB, Fiore MF, Monteiro RT R. Degradation of textile dyes by cyanobacteria. Braz J Microbiol. 2017; 48(1):25–31.

19. Parikh A, Madamwar D. Textile dye decolorization using cyanobacteria. Biotechnol Lett. 2005; 27(5):323–6.

20. Mahalakshmi S, Lakshmi D, Menaga U. Biodegradation of different concentrations of dye (congo red dye) by using green and blue-green algae. Int J Environ Res. 2015; 9(2):735–44.

21. Henciya S, Shankar MA, Malliga P. Decolorization of textile dye effluent by marine cyanobacterium *Lyngbya* sp. BDU 9001 with coir pith. Int J Environ Sci. 2013; 3:1909–18.

22. El-Sheekh MM, Gharieb MM, Abou-El-Souod GW. Biodegradation of dyes by some green algae and cyanobacteria. Int Biodeterio Biodegrad. 2009;63(6):699–704.

23. Silva-Stenico ME, Vieira FDP, Genuario DB, Silva CSP, Moraes LAB, Fiore MF. Decolorization of textile dyes by cyanobacteria. J Braz Chem Soc. 2012; 23:1863–70.

24. Hussein MH, El-Wafa GSA, Shaaban-Dessuki SA, Allah AMA, El-Morsy R M. Efficiency of azo dyes biodegradation by *Nostoc carneum*. J Agric Forest Meteorol Res. 2019; 2:160–76.

25. Mubarak Ali D, Suresh A, Praveen Kumar R, Gunasekaran M, Thajuddin N. Efficiency of textile dye decolorization by marine cyanobacterium *Osciallatoria formosa* NTDM02. Afr J Basic Appl Sci. 2011;3:9–13.

26. Jinqi L, Houtian L. Degradation of azo dyes by algae. Environ Pollut. 1992; 75:273–8.

27. Vijaykumar S, Manoharan C. Treatment of dye industry effluent using free and immobilized cyanobacteria. J Bioremed Biodeg. 2012; 3:165.

28. Ertugrul S, Bakir M, Donmez G. Treatment of dye-rich wastewater by an immobilized thermophilic cyanobacterial strain: *Phormidium* sp. Ecol Eng. 2008; 32(3):244–8.
29. Karacakaya P, Kilic NK, Duygu E, Donmez G. Stimulation of reactive dye removal by cyanobacteria in media containing triacontanol hormone. J Hazard Mater. 2009; 172(2–3):1635–9.
30. Sadettin, S., Donmez, G. Bioaccumulation of reactive dyes by thermophilic cyanobacteria. Process Biochem. 2006; 41(4):836–841.
31. Shah V, Garg N, Madamwar D. An integrated process of textile dye removal and hydrogen evolution using cyanobacterium, *Phormidium valderianum*. World J Microbiol Biotechnol. 2001; 17:499–504.

20 Microbial Enzymes and their Role in the Bioremediation of Environmental Pollutants
Prospects and Challenges

Muhilan BM[1] and Indranil Chattopadhyay[1]*

[1]Department of Life Sciences, Central University of Tamil Nadu, Thiruvarur, Tamil Nadu, India

*Corresponding author: indranil@cutn.ac.in; indranil_ch@yahoo.com

CONTENTS

DOI: 10.1201/9781003130932-20

20.1 INTRODUCTION

Environmental pollution continues to increase at an astounding rate and is caused by human activities, such as urbanization, technical development, unhealthy agricultural practices, and fast industrialization that deteriorates the ecosystem. Due to their toxicity, the heavy metals that are released into the environment pose a significant danger to organisms that are exposed to elevated concentrations of these contaminants. Metals are important for plant and animal biological functions but interfere with metabolic processes in organism systems at higher levels (Ojuederie and Babalola, 2017). Currently, pollution, in particular, heavy metal pollution, is potentially one of the most important problems in the world. Of note, a small concentration of certain heavy metals is essential to ensure the correct metabolism of some organisms; however, a high concentration of these heavy metals creates many issues in several animals and plants (Govind, 2014). Because these heavy metals cannot be degraded, they infiltrate the water and soil and eventually enter the food chain (Azimi *et al.,* 2017). These heavy metals and their derivatives are extremely toxic, carcinogenic, mutagenic, and teratogenic, even at very low concentrations. Most of the contamination caused by heavy metals is serious, relatively long term, and irreversible (Tang *et al.,* 2014). There are a wide range of toxins in the environment, such as polychlorinated biphenyls (PCBs), hydrocarbons, dyes, pesticides, esters, heavy metals, petroleum products, and nitrogen (N) containing chemicals that are emitted from various industrial and agricultural sources (Dua *et al.,* 2002). Recently, the reduction and deterioration of toxins have become of interest to environmental scientists. Initially, the waste from different factories and agricultural products was handled by depositing it in a hole, using high-temperature incineration, or using UV rays. However, these approaches are not very successful due to their low efficiency, complicated processes, high costs, and the emergence of other recalcitrant derivatives. Therefore, bioremediation offers a method for the removal of these contaminants (Dzionek *et al.,* 2016). To date, the variety of approaches that could remove these hazardous heavy metals from the polluted system include dialysis, reverse osmosis, chemical precipitation, solvent extraction, and ion exchange (Xu *et al.,* 2017). However, these methods are expensive and are not very successful, or they can have harmful effects on the soil and alter the original composition (Azimi *et al.,* 2017). Other approaches that are environmentally sustainable and do not have harmful consequences are designed to address these restrictions (e.g., low performance and severe effects) (Uqab *et al.,* 2016). These are the bioremediation techniques and could be achieved using microorganisms, such as plants, bacteria, or fungi. By modifying their valance state and rendering them less toxic, they can consume, absorb, or transform heavy metals (Ayangbenro and Babalola, 2017). Bioremediation requires the use of microorganisms and their enzymes to degrade and convert contaminants into less toxic derivatives. Various species of archaea, bacteria, algae, and fungi can bioremediate (Dua *et al.*, 2002). To solve the problem of industrial waste, especially by degrading chemicals and toxic substances, microbial enzymes are required for the development of bioremediation techniques in the environmental bioindustry. An efficient, healthy, and low-cost approach involves the use of microbes and their enzymes for the removal of contaminants (Karigar and Rao, 2011). Compared with the chemical and physical methods for remediation, the environmentally sustainable and cost-saving aspects are the main advantages of bioremediation. To date, there are various forms of bioremediation, which have focused on degradation. This chapter will discuss the environmental impact of heavy metals and how the use of microorganisms and their enzymes could successfully address this. In addition, it will discuss the different methods used to remediate heavy metal toxicity by microorganisms and their enzymes. The future opportunities and challenges of bioremediation for genetically modified organisms (GMOs) will be discussed.

20.2 PRINCIPLES OF BIOREMEDIATION

Bioremediation is a method that uses microbial systems to remove contaminants from a contaminated site. Contaminants are used as food and energy sources by microorganisms (Azubuike *et al.*, 2016). Microorganisms effectively degrade or turn the toxic and hazardous components into less toxic components and by-products (Ayangbenro and Babalola, 2017). Bioremediation aims to enhance the natural microflora in a polluted site by supplying the nutrients food and ideal conditions for growth; therefore, the microflora can achieve their maximum capacity. The generation of secondary metabolites then generates more enzymes. These metabolites effectively break down the complex pollutants into simpler pollutants (Chen and Wang, 2017). Chemical bonds are broken, and energy is emitted during bioremediation of the contaminant, which is further used by microorganisms for their development. The microbial species that convert heavy metals can be isolated from aerobic and anaerobic environments. Aerobic microorganisms are mainly used for bioremediation (Azubuike *et al.*, 2016). Analysis has shown that the overall percentage of transformation by microorganisms of various heavy metals are: cobalt (Co) 20%, lead (Pb) 22%, cadmium (Cd) 31%, and chromium (Cr) 27%, which constitutes approximately 70% of the total material. In addition, there is approximately 18% of other metals, such as nickel (Ni) 7%, zinc (Zn) 5%, arsenic (As), and lead (Hg) (Pratush *et al.*, 2018). Table 20.1 lists the names of microorganisms that belong to four main classes (e.g., archaea, yeast, bacteria, and fungi) with metal-transforming properties (Ayangbenro and Babalola, 2017; Gupta and Singh, 2017).

20.3 TYPES OF BIOREMEDIATION

Developments in bioremediation have been made in the last two decades, which aim to rapidly regenerate contaminated ecosystems in an environmentally friendly and low-cost way. Researchers have created and designed numerous methods for bioremediation; however, there is still no specific bioremediation strategy that acts as a perfect solution to recover degraded ecosystems due to the form of the pollutant. The removal process relies on the type of the contaminant, which might include dyes, hydrocarbons, agrochemicals, heavy metals, chlorinated compounds, plastics, nuclear waste, sewage development, and greenhouse gases (Azubuike *et al.*, 2016).

Bioremediation methods can be categorized as *ex situ* or *in situ*, which considers the application site. The selection criteria that are addressed when selecting the bioremediation technique are pollutant composition, degree and depth of contamination, climate type, expense, location, and policies for the environment (Frutos *et al.*, 2012; Smith *et al.*, 2015). In addition to the selection criteria, before the bioremediation project, the criteria for performance [e.g., temperature, oxygen (O_2),

TABLE 20.1
Heavy Metal Biotransformation Potential of Some Microorganisms

Organisms	Genus or Species
Bacteria	*P. aeruginosa*
	P. putida
	Pseudomonas veronii
	Sporosarcina ginsengisoli
	Streptomyces sp.
	Z. ramigera
	Arthrobacter sp.
	Bacillus cereus
	B. cereus XMCr-6
	Bacillus subtilis
	Citrobacter sp.
	Cupriavidus metallidurans
	Phylum Cyanobacteria sp.
	Enterobacter cloacae
	E. cloacae B2-DHA
	Kocuria flava
Archaea	*P. chrysosporium*
	Phylum Crenarchaeota sp.
Fungi	*Aspergillus fumigatus*
	Aspergillus tereus
	A. versicolor
	Gloeophyllum sepiarium
	P. chrysogenum
	Rhizopus oryzae (MPRO)
Yeast	*Rhodotorula mucilaginosa*
	Rhodotorula rubra GVa5
	S. cerevisiae
	Candida utilis
	Hansenula anomala

pH, nutrient concentration, and other abiotic factors] that decide the effectiveness of the process of bioremediation are important (Azubuike *et al.*, 2016). Bioremediation techniques are diverse (Figure 20.1).

20.3.1 *In Situ*

In situ techniques include the disposal of polluted chemicals at the polluted site. They need no removal. They involve minimal or no disruption to the composition of the earth. In contrast to *ex situ* bioremediation, these techniques should be less expensive because no additional costs are required for excavation; however, the expense of constructing and installing advanced equipment on-site to enhance the activities of the microbes for bioremediation is a significant concern. *In situ* bioremediation techniques, such as biosparging, bioventing, and phytoremediation might be developed; however, other methods can continue without any form of development (e.g., intrinsic bioremediation or natural attenuation) (Azubuike *et al.*, 2016). The key benefit of this process is that the soil does not need to be removed.

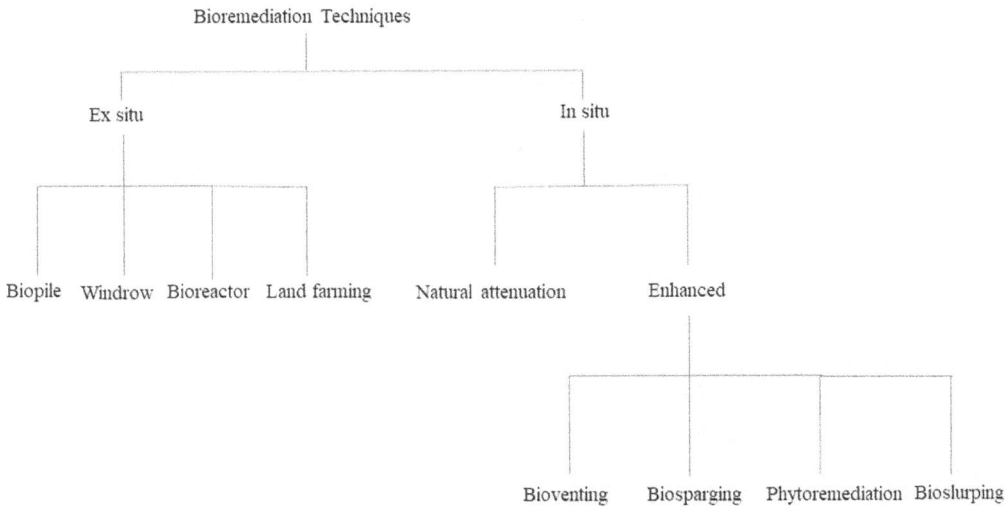

FIGURE 20.1 Bioremediation techniques.

Bioremediation involves the use of non-pathogenic microorganisms (Gomes et al., 2013). *In situ* bioremediation involves an aerobic procedure, via bioventing and adding hydrogen peroxide (H_2O_2), which delivers O_2 to the soil that enhances the growth of microorganisms (Brown *et al.*, 2017). Of note, electron acceptor status, moisture content, supply of nutrients, temperature, and pH and are essential conditions, which must be correct to achieve effective *in situ* bioremediation (Philp and Atlas, 2005).

20.3.2 *Ex Situ*

Ex situ techniques include removing contaminants from contaminated areas and transferring them for disposal to another location. *Ex situ* bioremediation is based on the treatment cost, contamination depth, pollutant type, pollution degree, and location and geology of the contaminated area. The performance parameters must be identified, which define the options for *ex situ* bioremediation (Philp and Atlas, 2005). Two *ex situ* methods are slurry phase bioremediation and solid-phase bioremediation (Aislabie *et al.*, 2006).

During the bioremediation of slurry, water and other reagents are combined with the soil in a bioreactor, which provides the microorganisms and the optimal growth conditions. This method of degradation adds nutrients and O_2. Water is removed and the soil is replaced following the process (Chen *et al.*, 2010).

Solid-phase bioremediation involves an above-ground recovery facility for the treatment of polluted soil. The above-ground treatment method includes the extraction of soil that is deposited and scattered on the treatment bed above-ground. An aeration device is delivered to the treatment bed. To optimize the productivity of bioremediation, nutrients, O_2, pH, moisture, and heat should be regulated. This technique is primarily used to treat different toxins and pesticides. This technique is fast and more effective than other techniques (Coulon *et al.*, 2010).

20.4 EFFECTS OF HEAVY METALS ON THE ENVIRONMENT AND HUMAN HEALTH

Heavy metals are approximately five times denser than water (Sarubbo *et al.*, 2015). Heavy metals exist naturally in small quantities on the earth but are concentrated in specific sites due to unregulated human activities (e.g., industry, mining, and smelting) (Das *et al.*, 2014). Because heavy metals cannot be biodegraded, they cannot be extracted from polluted biological tissues. Due to their toxicity, this is a significant problem for global health (Ayangberno and Babalola, 2017). The common heavy metals that contaminate the water and soil are As, Pb, Hg, and Cr. For organisms to survive, heavy metals, such as copper (Cu), Co, manganese (Mn), iron (Fe), and molybdenum (Mo) are required in low concentrations; however, at high concentrations, they might be harmful (Kushwaha *et al.,* 2015).

20.4.1 ARSENIC

As is a trace element that is present in human diets at very low concentrations. Many foods, such as meat, seafood, grains (particularly rice), cereals, dairy products, mushrooms, and bread contain very low concentrations and levels of As (Cheyns *et al.*, 2017). In adults and children, As is the most prevalent cause of acute heavy metal toxicity and can cause respiratory disorders, such as lung cancer or impaired pulmonary function (Dadzie, 2012).

The typical As biotransforming microorganisms are: *Acinetobacter sp., Brevundimonas sp., Pseudomonas sp., Rhizobium sp., Aeromonas sp.,* and *Penicillium canescens* (Layton *et al.*, 2014).

20.4.2 LEAD

Pb is the second most uncontrolled metal on earth that is present as a contaminant. Exposure to elevated amounts of Pb in the body can have severe health effects, such as loss of balance and paralysis, and serious exposure to Pb destroys the body's internal organs, such as the kidneys, liver, and heart tissue (Flora, 2012).

Bioremediation of Pb is carried out using the immobilization of different microbial strains. Several resistant mechanisms have been developed by a microorganism that enables them to tolerate the toxic effects of Pb. Accumulation, biosorption, and precipitation are the most common Pb resistance process employed by various microbial processes to remove Pb from the medium or soil (Naik and Dubey, 2013). *Aspergillus terrus, Aspergillus versicolor, Aspergillus niger, Penicillium canescens, Neurospora crassa, Penicillium decumbens, Penicillium chrysogenum, Penicillium simplicissimum, and Saccharomyces cerevisiae* are the main biotransforming species for Pb (Jebara *et al.*, 2015; Joshi *et al.*, 2011).

20.4.3 MERCURY

Hg, which is widely known as quick silver, can enter the ecosystem via several pathways. Most of the emissions are in the gaseous form of Pb, which can be transferred a long way from the source of emissions (Driscoll *et al.*, 2013). Excessive exposure to Hg induces disease, such as asthma, speech impairment, nervous, and renal system disorders (Azimi and Moghaddam, 2013). Using mobilization and immobilization methods, it can be bioremediated from polluted soil or water. Changing the metal ion's valance status and making it less toxic by reducing extracellular volatilization or precipitation is the most common process used by microorganisms for Pb bioremediation (Sinha *et al.*, 2009). Hg can be transformed using the following microorganisms: *Rhizopus arrhizus, Penicillium canescens, Geobacter metallireducens, Shewanella oneidensis,* and *Geobacter sulfurreducens* (Sinha *et al.*, 2009).

20.4.4 CHROMIUM

Cr is a heavy metal that is the 17th most plentiful element in the earth's mantle (Bhalerao and Sharma, 2015). Different effects on plant and human health are caused by increased accumulation of Cr in the soil or water. Cr aggregation causes root damage and affects growth in the shoots and photosynthesis in plants (Rodriguez *et al.*, 2012). However, in microorganisms and animals, it stimulates the hyperexpression of various enzyme antioxidants [e.g., catalase (CAT), peroxidase, superoxide dismutase (SOD) and glucose-6-phosphate dehydrogenase], which contributes to some severe disorders (Pratush *et al.*, 2018). Using different mechanisms, Cr can be bioremediated by many microbial species including *Bacillus megaterium, Pantoea sp., Bacillus circulans, Zoogloea ramigera, Bacillus coaglans, Aeromonascaviae, Pseudomonas sp., Staphylococcus xylosus,* and *Streptomyces nouresei* (Malaviya and Singh, 2014; Arya and Liakopoulou-Kyriakides, 2015).

Based on the adverse impacts of these heavy metals, significant measures must be taken to successfully remove them from the environment.

20.5 ENZYMES USED IN BIOREMEDIATION

Recently, combined with the use of microorganisms, microbial enzymes isolated from their cells have been used for bioremediation (Thatoi *et al.*, 2014). Enzymes are intricate biological macromolecules that are catalysts for a range of biochemical processes involved in pollutant degradation (Kalogerakis *et al.*, 2017). By reducing the activation energy of molecules, enzymes might increase the rate of a reaction (Sharma *et al.*, 2018). Based on biodegradation ability, the enzymes involved are listed in Table 20.2.

20.5.1 OXIDOREDUCTASES

Different species of bacteria, fungi and higher plants generate and secrete oxidoreductases to detoxify compounds via oxidative coupling, which involves the oxidation of compounds by electron transfer from reductants to oxidants and results in the release of chloride ions, CO_2 and methanol (Sharma *et al.*, 2018). Through the breakdown of oxidoreductase contaminants, heat or energy is produced and microorganisms use this for their metabolic activities (Medina *et al.,* 2017). During the oxidation of natural and manmade pollutants, oxidoreductases are used. For example, *Bacillus safensis* CFA-06, a gram-positive bacterium, produces oxidoreductase to degrade petroleum compounds. In nature, lignin degradation produces a variety of phenolic chemicals that are polymerized and copolymerized by oxidoreductases into a new form (Husain, 2006). Color compounds formed in textile factories are introduced into the atmosphere and can be degraded by different oxidoreductases, such as peroxidases and laccase (Novotny *et al.*, 2014). White-rot fungi (*Panus tigrinus*) and their extracellular oxidoreductases, such as laccase, Mn-based peroxidase, and lignin peroxidase eliminate phenols, color, and the organic load from olive mill waste water (Annibale *et al.*, 2004). As a result of redox reactions, certain bacteria release oxidoreductase enzymes that help to remove radioactive metals (Sharma *et al.*, 2018).

20.5.1.1 Oxygenases

Oxygeneses are key enzymes in the aerobic degradation of aromatic compounds, and the step required for aromatic degradation is to catalyze the cleavage of the ring in them by adding one or two O_2 molecules. Oxygenases are divided into two subclasses based on the number of O_2 molecules involved: monooxygenase (catalyze the incorporation of a single O_2 molecule); and dioxygenase (catalyze the incorporation of two molecules of O_2). Isolated glyphosate oxidase (GOX) from *Pseudomonas sp.* LBr is involved in pesticide bioremediation. A broad variety of contaminants can be degraded by oxygenases, such as chlorinated compounds, which improves their reactivity and solubility by incorporating O_2 molecules. FAD/NADH/NADPH compounds are used as a cofactor for oxygenase activity (Sharma *et al.*, 2018). A significant number of halogens found in herbicides, fungicides, and pesticides are present in the environment, which is degraded by oxygenase by

TABLE 20.2
Enzymes Involved in Bioremediation and Their Functions

Classification of Enzyme	Examples	Functions	References
Oxidoreductase	Oxygenases	Catalyze oxidation by introducing one or two O_2 molecules into aromatic compounds, such as chlorinated biphenyls, aliphatic olefins and make them vulnerable to further transformation and mineralization	Chakraborty *et al.* (2014)
	Laccases	They cleave the rings found in aromatic compounds and eliminate one O_2 atom in water, creating free radicals	Shraddha *et al.* (2011)
	Peroxidases	In the presence of peroxides, such as H_2O_2, catalyze reduction reactions and produce reactive free radicals after organic compound oxidation	Bansal and Kanwar (2013)
Hydrolases	Lipases	Break down triglycerol into glycerol and fatty acids, commonly used for the treatment of wastewater and degradation of polyaromatic hydrocarbons	Mehta *et al.* (2017)
	Cellulases	Complex cellulosic materials are broken down into basic sugars and are widely used in the treatment of agricultural residues, such as cotton waste, *Khaya ivorensis,* sawdust and rice straw	Bhardwaj *et al.* (2017)
	Carboxylesterases	Catalyzes carboxyl ester bond hydrolysis found in synthetic pesticides, such as organophosphates with water addition	Singh (2014)
	Phosphotriesterases	Catalyze phosphotriesters hydrolysis, the major components of organophosphorus compounds found globally in pesticides and induces significant toxicity and death	Santillan *et al.* (2016)
	Haloalkane dehalogenases	Used in the biodegradation of halogenated aliphatic compounds, such as 1,2,3-trichloropropane	Nagata *et al.* (2015)

catalyzing their dehalogenation reactions. Some marine bacteria generate oxygenase for the degradation of organic contaminants (Sivaperumal *et al.*, 2017).

20.5.1.1.1 Monooxygenases

Monooxygenases catalyze the breakdown of aromatic compounds by incorporating one molecule of O_2, which increases their reactivity and solubility. Monooxygenases are involved in dehalogenation, desulphurization, denitrification, and hydroxylation of aromatic compounds (Arora *et al.*, 2010). Based on the cofactor used, monooxygenases are divided into two groups: flavin-dependent monooxygenases; and P450 monooxygenases. In flavin-dependent monooxygenases, which are reduced by NAD(P)H, a closely bound flavin cofactor is present. The two-component flavin diffusible monooxygenase (TC-FDM) family of endosulfan diol (Esd) and endosulfan ether (Ese) have been used for the degradation of chlorine-containing pesticides, such as endosulfan (Bajaj *et al.*, 2010). *Bacillus megaterium* BM3-isolated P450 monooxygenase could remove several substrates, such as fatty acids and aromatic compounds (Roccatano, 2015). Methane monooxygenase (MMO) cometabolizes halide containing aliphatic compounds, aromatic compounds, and heavy metals. There are two types of MMO, which are located in the cytoplasmic membrane or the cytoplasm (Sharma *et al.*, 2018). *Methylocella palustris* isolated soluble MMOs are effective at the removal of a wide range of pollutants, such as hydrocarbons, aliphatic, and aromatic compounds (Singh and Singh, 2017).

Cytochrome P450 is an essential class of monooxygenase that is used in many industries to oxidize the contaminant produced. Approximately 200 P450 oxidoreductase subfamilies are present in prokaryotes and eukaryotes. All the other members of the P450 oxidoreductase group include Fe-containing porphyrin and use a noncovalently bound cofactor to regenerate their redox center (NAD(P)H is most commonly used) (Sharma *et al.*, 2018). To catalyze the oxidative dealkylation of phenyl urea herbicides, such as linuron, cytochrome CYP76B1 isolated from *Helianthus tuberosus* (Jerusalem artichoke) was reproduced and expressed in tobacco and Arabidopsis (Didierjean *et al.*, 2002). Researchers created a whole-cell biocatalyst on the surface of *Escherichia coli* expressing rat NADPH and cytochrome P450 oxidoreductase using *Pseudomonas syringae* ice nucleation protein for different applications, including the specific synthesis of new chemicals and pharmaceuticals, bioconversion, bioremediation, and biochip development (Yim *et al.*, 2006).

20.5.1.1.2 Dioxygenases

Dioxygenases catalyze the oxidation of aromatic pollutants by incorporating two O_2 molecules into the ring. Aromatic dioxygenases can be categorized into: (1) aromatic ring hydroxylation dioxygenases (ARHDs); and (2) aromatic ring cleavage dioxygenases (ARCDs) based on the mechanism of action (Sharma *et al.*, 2018). By incorporating two O_2 molecules into the ring, ARHDs degrade the chemical compounds, and ARCDs dissolve the aromatic compound rings (Parales and Ju, 2011). The decomposition of toluene is catalyzed by toluene dioxygenase (TOD) produced by *Pseudomonas putida* F1. This multicomponent enzyme works as a dioxygenase or a monooxygenase. For a range of pollutants, such as aromatics and aliphatic compounds, it acts as a dioxygenase. TOD acts as a monooxygenase for aromatic monocyclic compounds, aliphatic chain olefins, and other miscellaneous compounds (Mukherjee and Roy, 2013). Catechol dioxygenases are present in soil bacteria that induce aromatic precursor biotransformation into aliphatic products (Muthukamalam *et al.*, 2017). During bioremediation of quinaldine and 1H-4-oxoquinoline, ring-opening 2,4-dioxygenases catalyze the reduction of two carbon–carbon (C) bonds to form carbon monoxide (Ali *et al.*, 2017).

20.5.1.2 Laccases

Laccases are composed of Cu-containing oxidases that catalyze the oxidation of a broad variety of phenolic and aromatic compounds that are found in water and soil. They are present in various isoforms produced by various bacterial, fungal, insects, and plant species (Sharma *et al.*, 2018). Laccases are often formed in the cell, but extracellular secretions might be able to degrade ortho and para diphenols, amino groups that compose phenols, lignin, and diamine-containing aryl groups (Mai *et al.*, 2000). Laccase decolorizes azo dyes and converted them into less toxic substances in the atmosphere by oxidizing their bonds (Legerska *et al.*, 2016). Laccase formed by *Rhizoctonia. practicolahave* can dissolve and biotransform phenolic compounds (Strong and Claus, 2011). Solid support immobilization of laccase enhances stability, half-life, and tolerance to protease enzymes (Dodor *et al.*, 2004). Laccase isolated from *Trametes versicolor* was immobilized on porous glass beads. This is an effective enzyme for the bioremediation of a wide variety of toxins, such as phenolic compounds, heterocyclic aromatic compounds, and aromatic compounds that contain amines. By eliminating electrons from organic substrates, laccase reduces the pollutant's dioxygen molecules in the water (Chakroun *et al.*, 2010). In docking experiments with two-dimensional (2D) pollutant structures downloaded from the National Center for Biotechnology Information (NCBI) database, the X-ray crystal structures of laccases stored in the Protein Data Bank (PDB) were used. CORINA, an online tool, was used to transform 2D pollutant structures into three-dimensional (3D) structures.

In addition, protein-ligand docking program GOLD was used for protein–ligand docking. The best average GOLD fitness score for bacterial and fungal laccase enzymes was seen in approximately 30% and 17% of the chosen datasets, which indicated that laccase might be able to oxidize these contaminants (Suresh et al., 2008).

20.5.1.3 Peroxidases

Peroxidases are produced by animals, plants, fungi, and bacteria and are abundant in nature. Through H_2O_2 and a mediator, peroxidases help to degrade lignin and other aromatic pollutants. Phenolic radicals are formed and are precipitated by the oxidation of phenolic compounds, aggregates, and become less soluble (Sharma et al., 2018). Heme may or may not be found in the enzyme (Bansal and Kanwar, 2013). There are two classes of heme-containing peroxidases: one group is only found in animals and the other group is found in fungi, bacteria, and plants. In addition, peroxidases present in bacteria, fungi, and plants are classified into: (1) Class 1 includes intracellular enzymes, such as yeast-producing cytochrome c peroxidase, ascorbate peroxidase (APX) formed by some plant species, and bacterial catalase peroxidases; (2) Class 2 includes secreted fungal enzymes, such as lignin peroxidase (LiP) and manganese peroxidase (MnP); and (3) Class 3 is composed of plant secreted peroxides from horseradish plants, such as horseradish peroxidase (HRP). There are five types of nonheme peroxidase: thiol peroxidase, alkyl hydro peroxidase, halo peroxidase, manganese catalase, and NADH peroxidase (Koua et al., 2009). LiP and MnP have a greater capacity for the removal of harmful compounds than most researched enzymes (Sharma et al., 2018). To degrade thiazole compounds, soybean peroxidase and chloroperoxidase were studied (Alneyadi and Ashraf, 2016).

20.5.1.3.1 LnP

The monomeric heme-containing protein and secondary metabolite produced by fungi, such as *Phanerochaete chrysosporium* and *T. versicolor* and bacteria are LiPs (Xu et al., 2014). Fe (III) in LiPs is penta-coordinated with a histidine residue and four heme tetrapyrrole nitrogens. It catalyzes toxic pollutant oxidation in the presence of H_2O_2 cosubstrate and Ike veratryl alcohol facilitator. First, LiPs are involved in the two-electron oxidation by H_2O_2 of the native ferric enzyme [Fe(III)] to create Compound I (LiP 1) that is reduced by a decreasing substrate and gains one electron to form Compound II (LiP 1 that is deficient in and electron). Compound II acquires a second electron from the reduced substrate in the final step, and the enzyme returns to its native ferric oxidation state (Abdel-Hamid et al., 2013). For the treatment of wastewater and in bioremediation, LiPs have several potential applications. Compared with fungal peroxidases, the degradation of lignin by bacterial peroxidases has been successful in specificity and thermostability (Behbahani et al., 2016).

20.5.1.3.2 MnP

MnP is a heme-containing extracellular enzyme that is generated by lignin-degrading fungi that can oxidize Mn^{2+} by multistep reactions to Mn^{3+} (Sharma et al., 2018). Many other acidic amino acid residues and one heme group that contains the Mn-binding site are found in MnPs. In addition, Mn^{2+} can provide a single electron to MnP Compound I and act as the strongest reducing substrate. This chelator has an indirect effect on the degradation of lignin and xenobiotic compounds. They catalyze the degradation of a variety of phenols, amine-containing aromatic compounds, and dyes (Have and Teunissen, 2001). MnPTra-48424 was identified and isolated from *Trametes sp.* 48424 white-rot fungi. This enzyme can decolorize various forms of dyes, such as indigo, anthraquinone, azo, and triphenyl methane. In addition, dyes, such as indigo carmine and methyl green combined with heavy metal ions and organic solvent can be breakdown by MnPtra-48424. Various polycyclic aromatic hydrocarbons can be broken down by purification with MnPtra-48424 (Zhanga et al., 2016). During anthracene degradation, the Mn-dependent peroxidase encoding gene (*pimp1*) was identified in *Peniophora incarnata* KUC8836. This gene was later expressed in *S. cerevisiae* to facilitate bioremediation (Lee *et al.*, 2016).

20.5.2 Hydrolases

Hydrolytic enzymes are widely used to bioremediate toxins and insecticides and reduce their toxicity. Major chemical bonds, such as esters, peptide bonds, and C–halide bonds are broken by various hydrolytic enzymes (Sharma *et al.*, 2018). Extracellular hydrolases secreted by microbes catalyze the bioremediation of organic polymers and hazardous molecules <600 Da molecular weight that can move through cell pores (Vasileva-Tonkova and Galabova, 2003). The bioremediation of oil spills, organophosphates, and carbamate insecticides using hydrolytic enzyme has been very successful. Extracellular hydrolytic enzymes that have been used in the food and chemical industries include proteases, lipase, xylanase, DNAses, and amylase. Hemicellulase, cellulase, and glycosidase are used for the degradation of biomass (Porro *et al.*, 2003). The carbendazim (*CBM*) hydrolyzing enzyme gene was cloned and heterologously expressed in *E. coli* BL21 (DE3), from *Microbacterium sp.* d jl-6F by Lei et al., 2017 to increase the rate of the enzyme. This enzyme could hydrolyze carbendazim to 2-aminobenzimidazole (Lei *et al.*, 2017). In another analysis, the organophosphorus hydrolase was added to the outer membrane vesicles of gram-negative bacteria, which indicated increased activity for parathion chemical compound degradation (Su *et al.*, 2017).

20.5.2.1 Lipases

Lipases are prevalent and catalyze the degradation of triacylglycerols into glycerol and free-fatty acids, which are the main components of hydrocarbons. Lipase is generated by several species of bacteria, plants, actinomycetes, and animal cells (Shukla and Gupta, 2007). Hydrolysis, interesterification, esterification, alcoholic, and aminolysis reactions are performed by lipases (Prasad and Manjunath, 2011). The concentration of hydrocarbons in polluted soil has reduced considerably due to lipase activity. These enzymes hydrolyze fatty acids into triglycerol, diacylglycerol, monoacylglycerol, and glycerol (Ghafil *et al.*, 2016). Various computational methods have been used to refine the media involved to increase the production of microbial lipases. Using a statistical techniques medium, the development of a unique crude oil-degrading lipase from the *Pseudomonas aeruginosa* SL-72 was designed for the bioremediation of crude oils (Verma *et al.*, 2012).

20.5.2.2 Cellulases

Cellulases are essential enzymes for the degradation of cellulose, which is the most prevalent biopolymer on earth. Microorganism-generated cellulase might be cell-bound, associated with the cell envelope, and extracellular (Yang *et al.*, 2016). Cellulases are utilized in the processing industries of detergents, where these enzymes extract cellulose microfibrils formed during the processes (Sharma *et al.*, 2018). *Bacillus sp.* contain some alkaline cellulases and *Trichoderma sp.* and *Humicola sp.* generates neutral and acidic cellulases (Ben Hmad and Gargouri, 2017). These cellulases have been used for the bioremediation of ink during paper processing in the paper and pulp industry (Karigar and Rao, 2011). Cellulase generated by *Humicola sp.* quickly adapts to harsh environmental conditions, such as high pH and temperature, and can be used to break down hydrogen bonds in the detergent and washing powder industry (Imran *et al.*, 2016). *Bacillus amyloliquefacience* ASK11 was derived from leather tanning industrial waste that could generate 20 U/mL cellulases at 10 ppm of Cr (VI) (Aslam *et al.*, 2017).

20.5.2.3 Carboxylesterases

The breakdown of synthetic substances and natural products, such as organophosphates, carbamate ester bonds, and organic chlorine compounds has been catalyzed by carboxylesterases

(Cummins *et al.*, 2007). Carboxylesterases have been used in the field to remove pesticides, insecticides, and fungicide sprays (Sharma *et al.,* 2018). Carboxylesterase E2 from *P. aeruginosa* PA1 was expressed on the outer membrane of *E. coli* for the absorption of mercury (Hg) at the infected site (Yin *et al.*, 2016). Synthetic pyrethroids, a large class of insecticides, have been used in the field for the last 40 years. Using typical mechanisms for the degradation of all forms of pyrethroid insecticides, their ester bonds have been hydrolyzed by carboxylesterases. The active site of the carboxylesterase derived from *Lucilia cuprina* and *Drosophila melanogaster* were adjusted for pyrethroid hydrolysis using in vitro mutagenesis (Heidari *et al.*, 2005). Isolated carboxylesterases E3 from *L. cuprina* is involved in organophosphorus insecticide degradation (Scott *et al.*, 2008).

20.5.2.4 Phosphotriesterases

Phosphotriesterases can eliminate chemical waste released from factories and pesticides that are used in crop fields, such as parathion (Romeh and Hendawi, 2014). Parathion is a compound that contains organophosphate that is used as an ingredient in herbicides and insecticides (Gao *et al.*, 2014). Organophosphate is a phosphoric acid ester degraded by phosphotriesterases that are known as aryl dialkyl phosphatase and organophosphorus hydrolase (Sharma *et al.*, 2018). SsoPox W263F and SsoPox C258L/I261F/W263A are recombinant thermostable phosphotriesterases whose genes were extracted from wild type *Sulfolobus solfataricus*, SacPox isolated from *Sulfolobus acidocaldari* has been produced and purified (Restaino *et al.,* 2016). A few strains of marine bacteria can degrade the phosphate triester present in coastal oceanic environments, such as *Phaeobacter sp., Ruegeria mobilis*, and *Thalassospira tepidiphila* (Yamaguchi *et al.*, 2016). A new enzyme homologous to phosphotriesterases was characterized from *Geobacillus stearothermophilus* (GsP), which can hydrolyze compounds that containing lactone and organophosphate. GsP-isolated phosphotriesterase-like lactonase is highly thermostable and can be active at even a temperature of 100°C (Hawwa *et al.*, 2009).

20.5.2.5 Haloalkane Dehalogenases

Due to natural processes and manmade activities, halogenated compounds are found in soil everywhere and can be harmful, poisonous, mutagenic, or carcinogenic (Koudelakova *et al.*, 2013). Haloalkane dehalogenases that are used for C–halogen bond hydrolysis are found in the halogens contained in contaminants and produce alcohol and halides (Kotik and Famerova,2012). The haloalkane dehalogenase active site is located between two domains. The main enzyme domain is composed of eight stranded β-sheets surrounded by α-helices (Pavlova *et al.,*2007). The first haloalkane dehalogenase degraded 1,2-dichloroethane in *Xanthobacter autotrophicus* GJ10 (Nagata *et al.,* 2015). Some dehalogenases from gram-positive and negative haloalkane degrading bacteria have now been cloned and defined. The most studied haloalkane dehalogenases used to degrade pesticides were dehalogenase DhaA from *Pseudomonas pavonaceae* 170 and LinB from *Sphingomonas paucimobilis* UT26 (Poelarends and Whitman, 2010). *dspA* was artificially synthesized in the California purple sea urchin *Strongylocentrotus purpuratus* in conjunction with the sequence stored in the NCBI PDB under accession number XP-794172. It was the first enzyme of dehalogenase generated from nonmicrobial origin (Fortova *et al.*, 2013).

20.6 MECHANISMS OF BIOREMEDIATION

In biological molecules, heavy metals are thought to displace essential components, which affects the molecules' activities and modify the composition or activity of the enzyme, protein, or membrane transporter, and therefore, are harmful to plants (Tak *et al.*, 2013). Using indigenous microorganisms with pathways that can destroy certain heavy metals or genetically modified microorganisms

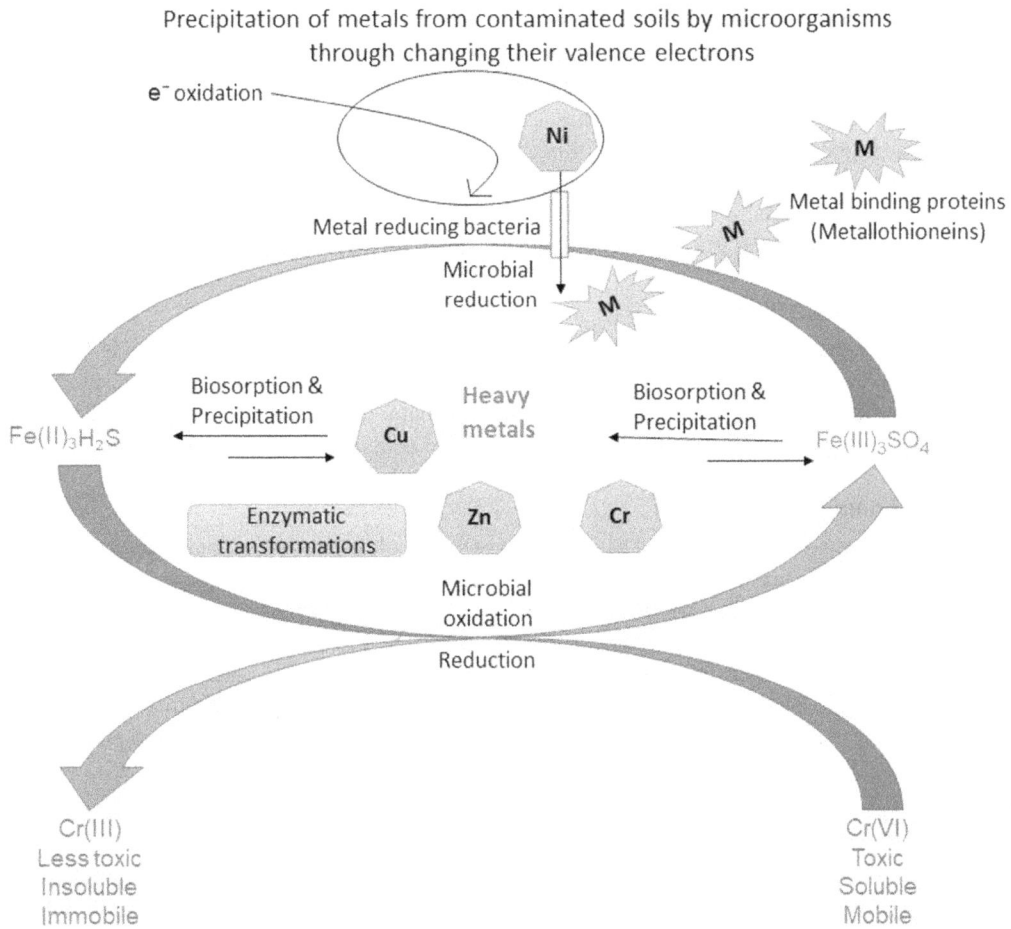

FIGURE 20.2 Mechanism for the removal of heavy metals from contaminated soils by microorganisms through precipitation, biosorption via sequestration of intracellular metal-binding proteins (i.e., metallothioneins), and conversion of metals into innocuous forms by enzymes (i.e., enzymatic transformation).

(GMOs) for the treatment of contaminated habitats by converting radioactive heavy metals into nonhazardous forms is an important method to eliminate toxic metals from the atmosphere and stabilize the ecosystem (Gupta *et al.,* 2016).

Using multiple methods, microorganisms convert heavy metals. They alter the metals ionic state, which impacts their solubility, bioavailability, and mobility (Ayangbenro and Babalola, 2017). When a consortium of microbial strains is used rather than a pure culture, the bioremediation of heavy metals could be effective (Ojuederie and Babalola, 2017). The mechanisms involved in the bioremediation of heavy metals from polluted soils are shown in Figure 20.2.

The metal processing pathways in various microorganisms are discussed in the following sections.

20.6.1 MOBILIZATION

The redox reaction system aims to remove radioactive metals and their radionuclides during mobilization and convert them into mineral, organic acids. In addition, it reduces the pH of the pollutant. Mobilization can be divided into four processes: enzymatic reduction, enzymatic oxidation, siderophores, and complexation (Pratush *et al.,* 2018).

20.6.1.1 Enzymatic Oxidation

A few inorganic compounds can occur in more than one oxidation state, and a higher oxidation state is less soluble than a lower ionic state. Under these conditions, enzymatic oxidation (i.e., catalyzed by enzymes released by microorganisms) has a vital role and increases the compound's solubility by oxidizing the higher state to a lower state. The elimination of inorganic species from the solution is an essential process. Heavy metals lose electrons in this step and are converted into a functional or less toxic state. The oxidation of uranium (U) from its ore by *Thiobacillus ferrooxidans* is the most common example of enzymatic oxidation (Cumberland *et al.*, 2016).

20.6.1.2 Enzymatic Reduction

This process is the reverse of oxidation. In their reduced state, inorganic compounds that have several oxidation states are insoluble. To extract specific elements from the solution, enzymatic reduction is useful. For *in situ* bioremediation, an enzymatic degradation process by optional and obligatory anaerobic microorganisms has been performed (Rabus *et al.*, 2016).

20.6.1.3 Complexation

Complexation involves the insertion of a ligand to create a complex with the inorganic metals. The toxic inorganic compounds are mobilized due to metal complex formation and can be quickly separated from solid waste (Ayangberno and Babalola, 2017). There are two types of microbial complexing agents: (1) low molecular weight organic acids (e.g., citric acid, alcohols, and tricarboxylic acids); and (2) high molecular weight ligands, toxic metal-binding compounds, and siderophores. Some amino acids that are synthesized by specific bacteria often serve as complex-forming agents in addition to this. The complexation of heavy metals and radionuclides in microorganisms depends on pH (Pratush *et al.*, 2018).

20.6.1.4 Siderophores

Some microbes generate Fe chelators, which are referred to as siderophores, during microbial growth in media that is Fe-deficient. Relevant binding groups, such as catecholate, hydroxamate, or phenolate have these siderophores. The complicated formation process for these siderophores is improved due to the presence of these unique groups and they form different complexes with toxic metal, and therefore, enhance their solubility (Khan *et al.*, 2017). A high proportion of siderophores or compounds that are similar to siderophores have been identified recently from various biological systems.

20.6.2 Immobilization

For the bioremediation of metal-contaminated soil, *in situ* and *ex situ* immobilization methods are used. In a heavily polluted area, *ex situ* methods are utilized. The soil in this region is removed and deposited in a new location where various microbial processes can be carried out to immobilize the metal ions that are present in it. For *in situ* process, the metal-polluted soil is processed at its original site. Immobilization is a biological process because numerous soil bacteria regulate it. Immobilization is carried out by active (i.e., energy-dependent) or passive (i.e., energy independent) processes. Immobilization of heavy metals is performed predominantly by precipitation, bioaccumulation, and biosorption. These methods are widely applied for the disposal of wastewater polluted with heavy metals (Ayangberno and Babalola, 2017).

The following sections offer a comprehensive explanation of the different phases of immobilization.

20.6.2.1 Precipitation or Solidification

Using various processes, metal ions can be solidified or precipitated in soil or solution. Sulfate reduction is the most typical example of precipitation. Sulfate-reducing bacteria (SRB) are used in engineering and natural environments, such as built wetlands for the treatment of metal pollutants. By producing metal sulfide precipitates, SRBs eliminates toxic metals from solutions. The solubility of most toxic metal sulfides is very poor, and their toxic effect on the atmosphere becomes almost negligible after precipitation. In solution, the SRBs produces conditions that favor the chemical removal of metals. In addition, metals can immobilize intracellular phosphates (Martinez *et al.*, 2014). *In situ* and *ex situ* precipitation can be carried out.

20.6.2.2 Biosorption

Biosorption is the physiochemical mechanism used by microorganisms, such as algae, bacteria, and fungi to remove metal ions. This is an energy-independent method, and the metal ions are removed by these species and their concentration in the solution is reduced (Ayangberno and Babalola, 2017). Biosorption is much more economical than other methods for the disposal of metal ions. The ability of biomass to remove metal ions is based on different physical, biological, chemical characteristics (Pratush *et al.,* 2018). For heavy metals, the common natural cellulosic materials are often utilized as possible sorbents (Malik *et al.*, 2017). Each biosorbent species has an affinity for a specific metal ion (Gupta and Singh, 2017). The biosorbent method has benefits over other techniques, for example, it is an inexpensive process, no sludge accumulates, and it is a regenerative process.

20.6.2.3 Bioaccumulation

Bioaccumulation is an energy-dependent mechanism in which the deposition of heavy metals is conducted by microorganisms. Bioaccumulation is a viable strategy, like biosorption. Metals, such as Pb, Hg, Cd, Ag, cesium, Ni, U, Cr, and Co are processed by microorganisms (Olaniran *et al.*, 2013). In general, the microorganism absorbs the metals via ion channels, ion pumps, lipid permeation, and endocytosis. Bioaccumulation is one of the most common and effective methods for recovering or eliminating heavy metals from the atmosphere. During bioaccumulation, heavy metal ions form an inactive complex with the other high-affinity ligands (Satyapal *et al.*, 2016).

20.7 TECHNOLOGY USED

Various problems are involved in the practical application of enzymes in bioremediation, such as lower stability, efficiency, and activity (Sharma *et al.*, 2018). Enzymes are macromolecules with complex structures. Any physical or chemical alteration results in the loss of enzyme function (Nigam and Shukla, 2015). Microbes will probably not produce adequate amounts of enzymes under normal environmental conditions (Sharma *et al.*, 2018).

Many environmental researchers are working to isolate novel microbial enzymes that can fully dissolve, or biotransform heavy metals and toxic pollutants into less toxic forms. Recombinant DNA technology could have a role in bioremediating heavy metal contaminants since it improves bioremediation. During the 1970s, due to the discovery of DNA ligases and restriction enzymes, the introduction of recombinant DNA technologies meant that the genome of living organisms could be modified. To date, the metabolic ability of some microorganisms has been investigated and some bacteria have been genetically engineered (Ojuederie and Babalola, 2017). Some of the newest approaches could solve the previously discussed disadvantages, such as genetic engineering and immobilization (Gupta and Singh, 2017).

20.7.1 Genetic Engineering

Genetically engineered microorganisms (GEMs) are microbial strains whose composition and genetic structure have been modified with the assistance of molecular biology techniques to increase biotransformation or biodegradation abilities (Zhang *et al.*, 2015). The development of

GEMs is predominantly accomplished by four key methods: (1) alteration of enzyme affinity and specificity; (2) regulation and construction of particular pathways; (3) bioprocess development for remediation and its control and monitoring; and (4) application and use of biosensors in chemical sensing, toxicity reduction, and end-point studies (Gupta and Singh, 2017). Based on these approaches, GEM's have been developed to mitigate heavy metal contamination. Due to the selective nature of GEMs, which cause relatively minimal health risks compared with other physicochemical approaches, the use of a genetically modified microbial system for heavy metal contamination bioremediation is a safe and low-cost option (Pratush *et al.*, 2018). Under natural conditions, enzyme generation by its native host is very limited. By isolating and moving the coding genes into another host for expression, the genetic engineering approach offers a method to increase the production of these enzymes (Gupta and Shukla, 2015). Enzyme overexpression using recombinant DNA technology is cost-effective and improves enzyme stability and function (Alcalde *et al.,* 2006). It is simpler to purify recombinant enzymes than the native strain. The enzyme produced by recombinant DNA technology is cost-effective to produce large quantities of enzymes, and therefore, improves enzyme stability and function. The self-life, substrate range, pH, enzyme temperature stability can be improved using genetic engineering techniques. Under contaminated environmental conditions, recombinant enzymes could have an increased capacity to degrade the contaminant (Sharma *et al.*, 2018). Some recent examples of the production of recombinant enzymes have listed in Table 20.3.

20.7.2 ENZYME ENGINEERING

Enzyme engineering involves a change or alteration in an enzyme's basic amino acid structure to enhance its properties, such as activity, resistance to stress, temperature, and pH (Rayu *et al.*, 2012). Enzyme engineering uses recombinant DNA technologies to make the desired modifications to an enzyme's amino acid sequences (Singh *et al.*, 2013). The fundamental structural and functional unit of an enzyme molecule is defined by its sequence of amino acids. In contrast, the enzyme's properties might be changed by a modification to the primary amino acid chain. However, the enzymatic properties are only altered if the amino acid in the protein region is distorted (Sharma *et al.*, 2018). In summary, various properties of an enzyme that can be altered by enzyme engineering are shown in Figure 20.3. Enzyme engineering for the selective and increased bioremediation of heavy metals and radionuclides has recently been studied.

20.7.3 IMMOBILIZED ENZYME TECHNOLOGY

Enzyme immobilization incorporates a free or soluble enzyme with a support, which reduces their mobility to improve activity and stability (Ahmad and Sardar, 2015). Enzymes might be immobilized using various methods: affinity-tag binding; adsorption on glass and alginate beads or a matrix; and covalently binding to an insoluble support, such as silica gel (Sirisha *et al.,* 2016). The method of immobilization has a significant effect on the enzyme properties and does not impair enzyme configuration and activity (Shaheen *et al.*, 2017). The support materials used for immobilization should be low-cost, have a wide surface area, and should avoid substrate and product diffusion limits for enzymatic reactions (Khan and Alzohairy, 2010).

The catalytic ability of enzymes has been improved by immobilization on solid support (Sharma *et al.*, 2018). Due to their stability and repeated use, the use of immobilized enzymes to degrade xenobiotic compounds could prove cost-effective (Mehta *et al.*, 2016). Immobilized enzymes have activity at a wide range of temperatures and pH, and improve enzyme thermal stability (Sharma *et al.*, 2018). In a cell-free environment, intracellular enzymes will often not be effective. Therefore, immobilization could increase the stability of certain enzymes (Skoronski *et al.*, 2017).

TABLE 20.3
Enzymes Produced by Recombinant DNA Technology Involved in Bioremediation

Recombinant Enzyme	Native Microorganism	Engineered Microorganism	Pollutants	References
Tetrahydrofuran monooxygenase (ThmD)	*Pseudonocardia sp.* KT	*E. coli*	Cyclic ethers 1,4-dioxane, Chlorinated solvents and tetrahydrofuran	Oppenheimer *et al.* (2010), Sales *et al.* (2013)
Laccase CueO	*E. Coli* K12	*P. pastoris* GS5115	Synthetic dye decolorization such as Congo red, malachite green	Ma *et al.* (2017)
Laccase	*B. vallismortis* fmb103	*E. coli* BL21	Synthetic dye decolorization such as Congo red, malachite green	Sun *et al.* (2017)
Laccase (lacIIIb), Versatile (vpl2) peroxidase, MnP, LiP	*T. Versicolor; P. eryngii, P. chrysosporium*	*P. chrysosporium*	synthetic dye, phenolic compounds	Coconi-Linaresa *et al.* (2015)
HRPs (C1A)	Horseradish plant *E. coli* BL21	*E. coli* BL21	Endocrine disruptive, phenolic compounds chemicals	Gundinger and Spadiut (2017), Bansal and Kanwar (2013)
Dye decolorizing peroxidase (DyP)	*G. candidum* Dec1	*A. Oryzae* RD005	Dye decolorization such as Congo red, malachite green	Sugano *et al.* (2000), Shakeri and Shoda (2010)
Flavodoxin-like protein (Pst2)	*S. cerevisiae*	*E. coli*	Phenolic component, quinine	Koch *et al.* (2017)
Carboxylesterase	Human liver	*E. coli* BL21	Carbamates, pesticides, chlorine compound	Boonyuen *et al.* (2015)
MnP	*I. lacteus* F17	*E. coli*	Phenols, dye, amine-containing aromatic compounds	Chen *et al.* (2015)
Mn-dependent peroxidase	*P. incarnate* KUC8836	*S. Cerevisiae* BY 4741	Phenols, dyes, anthracene	Lee *et al.* (2016)
Polyphenol oxidase (MnPPOA)	*M. medditerranea*	*E. coli*	Phenolic compounds	Tonin *et al.* (2016)

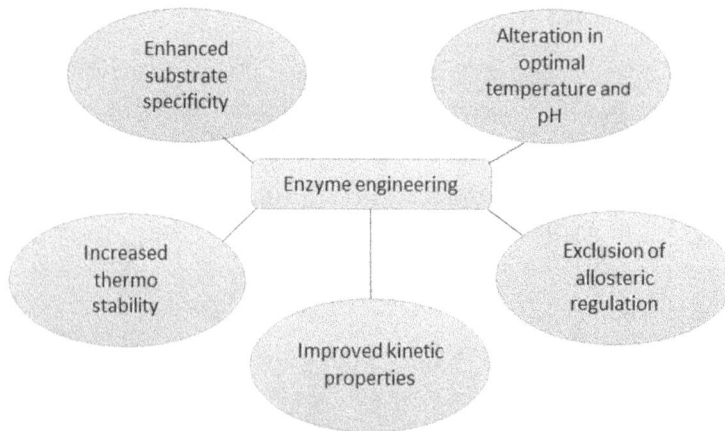

FIGURE 20.3   Enzyme engineering to improve enzymatic properties for bioremediation.

20.7.4 NANOZYMES

Nanozymes are nanoparticle-based enzyme mimics, which are known as artificial enzymes of the next generation with enzyme-like properties (Sharma *et al.*, 2018). They can catalyze substrate conversion and under physiological conditions, perform the same kinetics and natural enzyme's mechanisms (Gao and Yan, 2017). Several nanomaterials imitate different natural enzymes, such as oxidase, catalase, SOD, peroxidase, nuclease, esterase, phosphatase, ferroxidase, and protease (Sharma *et al.*, 2018). Nanozymes typically lack an active site; therefore, only a specific substrate can bind and perform chemical reactions (Wang *et al.*,2016).

In bioremediation, these nanozymes have different uses. They are used for the identification and degradation of pollutants, such as dyes, waste-containing lignin, and organic compounds (Liang *et al.*, 2017). Nanozymes based on nanoparticles are of interest to researchers because of their low cost and increased stability (Xie *et al.*, 2012).

20.8   OMICS: A SYSTEM BIOLOGY APPROACH TO BIOREMEDIATION

The effective use of metabolic engineering (ME) and system biology (SB) methods in various areas of life sciences are of interest to environmental researchers, because they could be used in bioremediation. In addition, this approach gives useful quantitative knowledge on biological processes. ME can be used to exploit this information to alter microbial metabolic pathways and neutralize one or multiple toxins simultaneously (Sharma *et al.*, 2018). The omics techniques (e.g., proteomics, genomics, metabolomics, and transcriptomics,) are commonly applied in SB analysis (Rayu *et al.*, 2012; Dangi *et al.*, 2017). The techniques that are used in SB are discussed in the following sections.

20.8.1 GENOMICS

By offering a global perspective on genetic materials, such as the RNA and DNA that are expressed in microorganisms via exposure to pollutants, genomics could help bioremediation This method includes bioinformatic analysis and gene sequencing that uses various algorithms and tools. Sequencing has been carried out for 270,567 genomes, according to a new study, and >46,000 genome sequencing programs are underway globally (Dangi *et al.*, 2019). The genomes of various microorganisms employed in bioremediation have been sequenced. The study of the genome sequencing (6.2 MB) of *Pseudomonas sp.* KT2440 proved the existence of genes that encode several proteins or enzymes, such as oxidoreductase, dehydrogenase, ferredoxin, oxygenase, glutathione-S transferase, cytochrome, efflux pumps, and sulfur-metabolizing proteins, which play an important role in the bioremediation of a variety of chemicals that are released from industrial effluents (Belda *et al.*, 2016). A metagenomic approach could help to examine the different degradation mechanisms found in poorly characterized and unknown microbial species. The information on such mechanisms could be used to construct more qualified customized consortia or strains for use in specific bioremediation (Dangi *et al.*, 2019).

20.8.2 TRANSCRIPTOMICS

Transcriptomics is a genomics-based approach that is applied to analyze a gene's differential expression, which in response to environmental contaminants are upregulated or downregulated. It could help to determine the function of previously unannotated genes. To evaluate a series of predetermined sequences, techniques, such as microarray and RNA sequencing could be implemented (Dangi *et al.*, 2019). The study of transcriptomes of the microbial community requires: (1) cDNA synthesis from total mRNA; (2) separation and improvement of total cellular mRNA; and (3) either complete sequencing of cDNA or the use of microarrays to hybridize cDNA (Zhu *et al.*, 2017).

20.8.3 PROTEOMICS

Combined with transcriptomics and genomics, proteomics is a valuable tool and is much more complex than genomics since the genome of an organism is stable; however, the proteome varies at different times and from cell to cell. Analysis of the sequence of proteins formed in environmental samples (i.e., metaproteomics) and by bacterial cultures (i.e., proteomics) could be used to detect variations in the structure and generation of proteins and to detect several proteins that are essential to a microbe's physiological response in the presence of contaminants (Dangi *et al.*, 2019).

20.8.4 METABOLOMICS AND METABOLIC FLUX ANALYSIS

A rapidly developing SB field at the intersection of chemistry and biological sciences is metabolomics, which analyses the complete collection of cellular metabolites that are produced in microbial cells (Klassen *et al.*, 2017). When optimizing bioremediation, the characterization of the primary and secondary metabolites that are produced by microbial cells in reaction to contaminant stress could be useful. In interspecies and intraspecies microbial interactions, microbial metabolites play a major role. By increasing bioavailability, stability, and biofilm formation, these interactions could improve bioremediation. Microbial communities secrete multiple metabolite forms under specific stresses; therefore, full knowledge of these metabolites could provide useful details on microbial interactions to degrade a specific pollutant (Tyc *et al.*, 2017). Using different techniques, such as target analysis, metabolite profiling for the detection and evaluation of a broad range of metabolic flux analysis, metabolic fingerprinting, and cellular metabolites, the metabolomics of microbes could be examined (Dangi *et al.*, 2019).

20.9 ADVANTAGES AND LIMITATIONS OF BIOREMEDIATION

Microorganisms play a vital role in daily life. They remove or biotransform pollutants or effluents by using them as a primary source of energy and C. No secondary contaminants are produced by microbes during biodegradation. The removal of target toxins is possible instead of moving chemicals from a specific environmental medium to medium, for example, from soil to air or water. In addition, it can be carried out on-site, often without allowing ongoing operations to be interrupted. Bioremediation technology is less expensive and highly efficient compared with conventional physical and chemical methods. However, it has some disadvantages, such as it is confined to biodegradable substances. Fast and total oxidation is not suitable for all substances. Biodegradation products can be more persistent or harmful than the parent compound. These processes are highly specific and rely on microbial communities, suitable conditions for environmental development, and sufficient levels of nutrients and pollutants. Extrapolation from the bench and pilot-scale experiments to large-scale field operations is challenging. In addition, bioremediation takes longer than most methods for recovery (Pratush *et al.*, 2018).

20.10 FUTURE PROSPECTS

Researchers are working to provide an efficient solution to the increasing contamination of the atmosphere. Therefore, the discovery of new enzymes and their role in bioremediation is required. A schematic diagram that describes the discovery of novel enzymes is shown in Figure 20.4.

Minimal cultivation technologies might have enormous bioremediation potential; however, a large number of microflora present in the ecosystem remain unknown. The microflora and the enzymes formed by them must be isolated and characterized. Metagenomics, metatranscriptomics, and metaproteomic approaches could solve this problem. These methods could be used without cultivating the microflora under laboratory conditions for the detection and characterization of novel microbes and enzymes present in nature. In addition, these technologies are important as effective methods to enhance the power of biodegradation. Under multiple environmental conditions, microbes exhibit complicated adaptive behavior. These are complex and involve communication networks between microbial communities, which in the presence of contaminants, play an important role in the adaptation of microbes. These microbial communities generate and secrete many enzymes and secondary metabolites. The dynamic behavior must be researched and the metabolites and their mechanisms of degradation established to identify the most appropriate solution to environmental

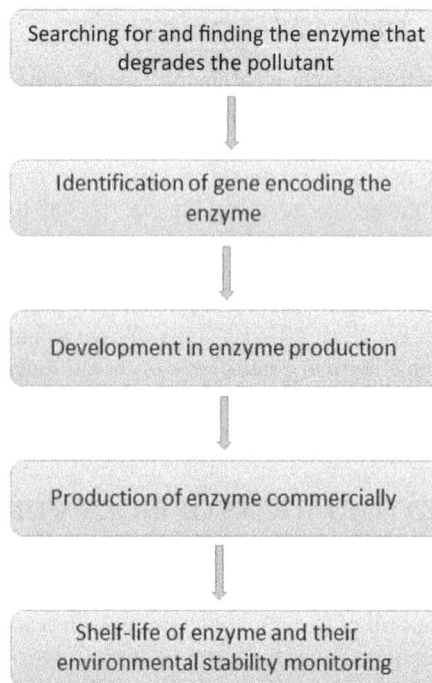

FIGURE 20.4   Schematic strategy for novel enzyme discovery and its validation.

contamination, which could be a significant driving force for its potential prospects (Sharma *et al.*, 2018).

The quantity of enzymes produced in the natural environment is very limited, and therefore, the processing of large quantities of enzymes produced by microorganisms is difficult for scientists that work on bioremediation. Recently, cost-effective techniques for the manufacture of nanoparticles and materials based on nanoparticles have received increased interest in diverse areas for their unique properties and enormous potential uses (Sharma *et al.*, 2018).

A new nanoparticle-related definition is a single enzyme nanoparticle, where each enzyme molecule is covered by a hybrid inorganic or organic polymer network. With a porous organic or inorganic structure (or armor), which is a few nanometers thick, each enzyme molecule is changed. This technique is a modern way of modifying and stabilizing enzymes, which produces a new kind of nanostructure (Hong *et al.*, 2017). These nanoparticles can bind to and entirely degrade xenobiotic compounds or convert them into less toxic derivatives that further help to clean the ecosystem.

However, the methods mentioned previously are fully developed and for effective bioremediation maximum technical intervention is required to discover reliable and environmentally sustainable strategies (Sharma *et al.*, 2018). Due to the significance of transgenic microbes that dramatically improve the degradation and detoxification of heavy metal and xenobiotics pollutants, further research is required to improve longevity when they are released for bioremediation into the ecosystem, because their survival is currently poor (Ojuederie and Babalola, 2017). In addition, further research is required to determine and understand the metabolic mechanisms of microbes that are used in bioremediation to ensure their efficacy and potential risks when used for bioremediation.

The potential of the microorganisms that are used in bioremediation to manage the indigenous microbial species is important for the effectiveness of bioremediation. (Ojuederie and Babalola, 2017). An assessment of bacteria that could function under a wide range of environmental conditions in the soil, such as pH, temperature, salinity, heavy metals, and supply of nutrients should be carried out.

20.11   CONCLUSIONS

Heavy metal pollution is a global problem. Many natural and industrial processes produce metal ions that enter our environment (e.g., soil and water) and interact with their natural composition. A significant number of experiments have been carried out in the last decade to reduce heavy metal emissions. Many research groups are actively focusing on developing alternative mechanisms and strategies to reduce these emissions. In addition, research groups are focused on discovering new potential microbial strains or species that could convert heavy metals more rapidly and effectively. In addition, novel methods for the bioremediation of heavy metal and the use of techniques, such as genetic engineering need to be developed to improve the potential of bioremediation. The cost of bioremediation remains high; therefore, a technique or technology that reduces the cost of bioremediation is required. To increase the biodegradation rate, researchers are experimenting with transgenic or GEMs. GEMs could successfully be used to remove heavy metals from the ecosystem; however, under controlled laboratory conditions, these transgenics did not demonstrate maximum potential. Transgenic *in situ* conditions need to be stabilized to increase the remediation rate of heavy metals from containment regions.

Advances in the use of microorganisms for the treatment of contaminated soils could be accomplished due to the application of modern analytical and molecular tools. Genetic engineering techniques allow targeted genes that encode enzymes in the metabolic pathways to be modified. Therefore, microorganisms with greater capacity for degradation or that can degrade other pesticides can be constructed. In addition, genomic and proteomic techniques that are applied to microorganisms of environmental significance could reveal main the key steps of the pathways in biodegradation and the ability of pesticide degraders to respond to varying environmental conditions.

Biodegradation is influenced by several abiotic factors, such as temperature, pH, nutrients, humidity, organic matter content, the initial concentration of pesticides, and additional C sources; however, more studies related to the interactions between these microorganisms and the soil environment are required needed before their implementation in field-scale bioremediation.

A variety of microbes from multiple natural sources have been explored and enzymes have been isolated that possess biodegradable capacity. A broad family of enzymes, such as oxidoreductase, laccases, and peroxidases, has been isolated and characterized that perform bioremediation.

Initially, enzyme-based bioremediation was not successful due to the very low amount of enzyme secreted by microorganisms under natural conditions. However, advances in recombinant DNA technology and supplying these microorganisms with optimal growth conditions could increase the production of enzymes. In addition, catalytic activity, shelf-life, and enzyme stress stabilization could be improved to higher levels by enzyme engineering and immobilization techniques. In addition, immobilized enzymes can be reused several times, which could remove contaminants at higher rates.

This chapter addressed the impact of heavy metal pollution on the environment and the potential health risks that have been caused by anthropogenic studies. In addition, different pathways and various enzymatic processes used by the microbes to efficiently remediate contaminated habitats have been discussed. This demonstrated the viability of various types of bioremediations as a better alternative for the removal and disposal of various heavy metals from polluted areas compared with the physicochemical approaches that are less effective and more expensive due to the amount of energy used.

Microorganisms have innate mechanisms that allow them to survive under heavy metal stress and degrade metals from the surrounding environment. However, the environmental conditions must be correct for effective bioremediation. Environmental factors play a vital role in the effectiveness of bioremediation because the microbes used would be impaired if the environmental conditions were not suitable. Transgenic microbes could efficiently remediate polluted areas that have heavy metals and organic contaminants; however, their use needs to be monitored and controlled due to strict biosafety protocols to ensure that there are no risks to the environment or humans.

It is important to determine more effective ways of utilizing transgenic microbes, which could allow the successful bioremediation of contaminated environments without the horizontal transition of recombinant plasmids to indigenous species, which remains a challenge.

Metagenomic techniques and metabolic analysis could be used to examine the functional structure of the microorganisms around toxic sites for metal tolerance genes that could be used to improve microbial strains for heavy metal depletion.

For their successful use, the public viewpoint on the use of gene technology for bioremediation would need to be changed; this will involve coordination between researchers and environmentalists.

Different omics-based tools, such as genomics, transcriptomics, proteomics, metabolomics and phenomics, and computational data analysis tools provide valuable knowledge to help to understand the complex behavior of the microbes that play a vital role in bioremediation. These methods must be analyzed and implemented from a theoretical level to a practical level to enable significant developments in the use of bioremediation for environmental remediation.

REFERENCES

Abdel-Hamid, A.M., Solbiati, J.O., and Cann, I.K., (2013) 'Insights into lignin degradation and its potential industrial applications', *Advances in Applied Microbiology*, 82(1), p. 28.

Ahmad, R. and Sardar, M. (2015) 'Enzyme immobilization: an overview on nano particles as immobilization matrix', *Biochemistry and Analytical Biochemistry*, 4(2), p. 1.

Aislabie, J., Saul, D.J. and Foght, J.M. (2006) 'Bioremediation of hydrocarbon contaminated polar soils', *Extremophiles*, 10 (3), p. 171-179.

Alcalde, M. *et al.* (2006) 'Environmental bio catalysis: from remediation with enzymes to novel green processes', *Trends in Biotechnology*, 24, pp. 281–287.

Ali, M.I., Khatoon, N. and Jamal, A. (2017) Polymeric pollutant biodegradation through microbial oxidoreductase; a better strategy to safe environment. *International Journal of Biological Macromolecules*. doi:10.1016/j.ijbiomac.2017.06.047.

Alneyadi, A.H. and Ashraf, S.S. (2016) 'Differential enzymatic degradation of thiazole pollutants by two different peroxidases e a comparative study,' *Chemical Engineering Journal*, 303, pp. 529-538. doi:10.1016/j.cej.2016.06.017.

Annibale, A.D. *et al.* (2004) '*Panus tigrinus* efficiently removes phenols, color and organic load from olive-mill wastewater', *Research Microbiology*, 155, 596-603. doi:10.1016/j.resmic.2004.04.009.

Arora, P.K., Srivastava, A. and Singh, V.P. (2010) 'Application of monooxygenases in dehalogenation, desulphurization, denitrification and hydroxylation of aromatic compounds', *Journal of Bioremediation and Biodegradation* 1, p. 112. doi:10.4172/2155-6199.1000112.

Aryal, M., and Liakopoulou-Kyriakides, M. (2015) 'Bioremoval of heavy metals by bacterial biomass', *Environmental Monitoring and Assessment*, 187(4173), pp. 1–26

Aslam, S., Hussain, A. and Qazi, J.I. (2017) 'Production of cellulase by *Bacillus amyloliquefaciens*-ASK11 under high chromium stress', *Waste and Biomass Valorization*, 10(1), pp. 53–61.

Ayangbenro, A.S,. and Babalola, O.O. (2017) 'A new strategy for heavy metal polluted environments: A review of microbial biosorbents', *International Journal of Environmental Research and Public Health*, 14(1), p. 94.

Azimi A. *et al.* (2017) 'Removal of heavy metals from industrial wastewaters: A review', *ChemBioEng Reviews*, 4, pp. 37–59.

Azimi, S., and Moghaddam, M.S. (2013) 'Effect of mercury pollution on the urban environment and human health', *Environmental Ecology Research*, 1(1), pp. 12–20.

Azubuike, C.C., Chikere, C.B., and Okpokwasili, G.C. (2016) 'Bioremediation techniques: Classification based on site of application: principles, advantages, limitations and prospects', *World Journal of Microbiology and Biotechnology*, 32(11), p. 180.

Bajaj, A. *et al.* (2010) 'Isolation and characterization of a *Pseudomonas sp.* strain IITR01 capable of degrading an endosulfan and endosulfan sulfate', *Journal of Applied Microbiology*, 109(6), 2135–2143.

Bansal, N., and Kanwar, S.S. (2013) 'Peroxidase(s) in environment protection', *Science World Journal*. doi:10.1155/2013/714639.

Behbahani, M., Mohabatkar, H. and Nosrati, M. (2016) 'Analysis and comparison of lignin peroxidases between fungi and bacteria using three different modes of Chou's general pseudo amino acid composition', *Journal of Theoretical Biology*, 411, pp. 1–5. doi:10.1016/j.jtbi.2016.09.001.

Belda, E. *et al.* (2016) 'The revisited genome of *Pseudomonas putida* KT2440 enlightens its value as a robust metabolic chassis', *Environmental Microbiology*, 18, pp. 3403–3424.

Ben Hmad, I., and Gargouri, A. (2017) 'Neutral and alkaline cellulases: production, engineering, and applications', *Journal of Basic Microbiology*, 57(8) pp. 653–658.

Bhalerao, S.A., and Sharma, A.S. (2015) 'Chromium: as an environmental pollutant', *International Journal of Current Microbiology & Applied Science*, 4(4), pp. 732–746.

Bhardwaj, V., Degrassi, G., and Bhardwaj, R.K. (2017) 'Bioconversion of cellulosic materials by the action of microbial cellulases', *International Journal of Engineering and Technology*, 4(8). pp. 2395–0056.

Boonyuen, U. *et al.* (2015) 'Efficient in vitro refolding and functional characterization of recombinant human liver carboxylesterase (CES1) expressed in *E. coli.' Protein Expression and Purification*, 107, pp. 68–75.

Brown, L.D. *et al.* (2017) 'Bioremediation of oil spills on land', in *Oil Spill Science and Technology*, 2nd edn., pp. 699–729. doi:/10.1016/B978-0-12-809413-6.00012-6

Chakraborty, J. *et al.* (2014) 'Ring-hydroxylating oxygenase database: a database of bacterial aromatic ring-hydroxylating oxygenases in the management of bioremediation and bio catalysis of aromatic compounds', *Environmental Microbiology Reports*, 6(5), pp. 519–523.

Chakroun, H. *et al.* (2010) 'Purification and characterization of a novel laccase from the ascomycete *Trichoderma atroviride*: application on bioremediation of phenolic compounds', *Process Biochemistry*, 45, pp. 507–513. doi.org/10.1016/j.procbio.2009.11.009.

Chen, H., and Wang, L. (2017) 'Microbial cell refining for biomass conversion', in *Technologies for Biochemical Conversion of Biomass*. Cambridge, MA. Academic Press. pp. 101–135.

Chen, W. *et al.* (2015) 'Cloning and expression of a new manganese peroxidase from *Irpex lacteus* F17 and its application in decolorization of reactive black 5', *Process Biochemistry*, 50 (11), 1748–1759.

Cheyns, K. *et al.* (2017) 'Arsenic release from food stuffs upon food preparation', *Journal of Agriculture and Food Chemistry*, 65(11), pp. 2443–2453.

Coconi-Linaresa, N. *et al.* (2015) 'Recombinant expression of four oxidoreductases in *Phanerochaete chrysosporium* improves degradation of phenolic and non-phenolic substrates', *Journal of Biotechnology*, 209, pp. 76–84. doi.org/10.1016/j.jbiotec.2015.06.401.

Coulon, F. *et al.* (2010) 'When is a soil remediated? Comparison of biopiled and windrowed soils contaminated with bunker-fuel in a full-scale trial', *Environmental Pollution*, 158 (10), pp. 3032–3040.

Cumberland, S.A. *et al.* (2016) 'Uranium mobility in organic matter-rich sediments: a review of geological and geochemical processes', *Earth Sciences Review*, 159, pp. 160–185.

Cummins, I. *et al.* (2007) 'Structure activity studies with xenobiotic substrates using carboxylesterases isolated from *Arabidopsis thaliana*', *Phytochemistry*, 68, pp. 811–818. doi.org/10.1016/j.phytochem.2006.12.014.

Dadzie, E. (2012) 'Assessment of Heavy Metal Contamination of the Densu River, Weija From Leachate'. Unpublished Master's Thesis, Kwame Nkrumah University of Science and Technology, Kumasi, Ghana.

Dangi, AK. *et al.* (2019) 'Bioremediation through microbes: systems biology and metabolic engineering approach', *Critical Reviews in Biotechnology*, 39(1), pp. 79–98. doi:10.1080/07388551.2018.1500997

Das, S. *et al.* (2014) '2-heavy metals and hydrocarbons: adverse effects and mechanism of toxicity', *Microbial Biodegradation and Bioremediation*, 2014, pp. 23–54.

Didierjean, L. *et al.* (2002) 'Engineering herbicide metabolism in tobacco and *Arabidopsis* with CYP76B1, a cytochrome P450 enzyme from Jerusalem artichoke', *Plant Physiology*, 130, 179–189. doi.org/10.1104/pp.005801.

Dodor, D.E., Hwang, H.M., and Ekunwe, S.I. (2004) 'Oxidation of anthracene and benzo[a]pyrene by immobilized laccase from *Trametes versicolors*', *Enzyme Microbial Technology*, 35, pp. 3210–3217. doi.org/10.1016/j.enzmictec.2004.04.007.

Driscoll, C.T. *et al.* (2013) 'Mercury as a global pollutant: sources, pathways, and effects', *Environmental Science & Technology*, 47, pp. 4967–4983.

Dua, M. *et al.* (2002) 'Biotechnology and bioremediation: successes and limitations', *Applied Microbiology & Biotechnology*, 59, pp. 143–152. doi.org/10.1007/s00253-002-1024-6.

Dzionek, A., Wojcieszynska, D., and Guzik, U. (2016) 'Natural carriers in bioremediation: A review', *Electronic Journal of Biotechnology*, 23, pp. 28–36. doi.org/10.1016/j.ejbt.2016.07.003.

Flora, G.J. (2012) 'Arsenic toxicity and possible treatment strategies: Some recent advancement', *Current Trends in Biotechnology & Pharmacology*, 6, pp. 280–289.

Fortova, A. *et al.* (2013) 'DspA from *Strongylocentrotus purpuratus*: the first biochemically characterized haloalkane dehalogenase of non-microbial origin', *Biochimie*, 95(11), pp. 2091–2096. doi.org/10.1016/j.biochi.2013.07.025

Frutos, F.J.G. *et al.* (2012) 'Remediation trials for hydrocarbon-contaminated sludge from a soil washing process: evaluation of bioremediation technologies,' *Journal of Hazardous Materials*, 199, pp. 262–271. doi:10.1016/j. jhazmat.2011.11.017

Gao, L., and Yan, X. (2017) 'Nanozymes: an emerging field bridging nanotechnology and biology', *Science China Life Science*, 59, 400e402. doi.org/10.1007/s11427-016-5044-3.

Gao, Y. *et al.* (2014) 'Bioremediation of pesticide contaminated water using an organophosphate degrading enzyme immobilized on nonwoven polyester textiles', *Enzyme and Microbial Technology*, 54, 38e44.

Ghafil, J.A., Hassan, S.S., and Zgair, A.K. (2016) 'Use of immobilized lipase in cleaning up soil contaminated with oil', *World Journal of Experimental Biosciences*, 4, 53e57.

Gomes, H.I., Ferreira, C.D., and Ribeiro, A.B. (2013) 'Overview of in situ and ex situ remediation technologies for PCB-contaminated soils and sediments and obstacles for full-scale application', *Science of the Total Environment*, 15(445–446), pp. 237–260. doi.org/10.1016/j.scitotenv.2012.11.098.

Govind, M. (2014) 'Heavy metals causing toxicity in animals and fishes', *Research Journal of Animal Veterinary and Fishery Science*, 2(2), pp. 17–23.

Gundinger, T., and Spadiut, O. (2017) 'A comparative approach to recombinantly produce the plant enzyme horseradish peroxidase in *Escherichia coli*', *Journal of Biotechnology*, 248, 15e24. doi.org/10.1016/j.jbiotec.2017.03.003.

Gupta, S., and Singh, D. (2017) 'Role of genetically modified microorganisms in heavy metal bioremediation', in Kumar R. (ed.) *Advances in Environmental Biotechnology*: Singapore: Springer Nature Pvt. doi.org/10.1007/978-981-10-4041-2_12

Gupta, A. *et al.* (2016) 'Microbes as potential tool for remediation of heavy metals: A review', *Journal of Microbial & Biochemical Technology,* 8, pp. 364–372.

Gupta, S.K., and Shukla, P. (2015) 'Advanced technologies for improved expression of recombinant proteins in bacteria: perspectives and applications', *Critical Reviews in Biotechnology,* 36, pp. 1089–1098. doi.org/10.3109/07388551.2015.1084264.

Have, R.T., and Teunissen, P.J. (2001) 'Oxidative mechanisms involved in lignin degradation by white-rot fungi', *Chemical Reviews,* 101, pp. 3397–3413.

Hawwa, R. *et al.* (2009) 'Structural basis for thermostability revealed through the identification and characterization of a highly thermostable phosphotriesterase-like lactonase from *Geobacillus stearothermophilus*', *Archives of Biochemistry and Biophysics,* 488, pp. 109–120. doi.org/10.1016/j.abb.2009.06.005.

Heidari, R. *et al.* (2005) 'Hydrolysis of organophosphorus insecticides by in vitro modified carboxylesterase E3 from *Lucilia cuprina*', *Insect Biochemistry and Molecular Biology*, 35, pp. 597–609. doi.org/10.1016/j.ibmb.2004.01.001.

Hong, S.G. *et al.* (2017) 'Single enzyme nanoparticles armored by a thin silicate network: single enzyme caged nanoparticles', *Chemical Engineering Journal,* 322, pp. 510–515. doi.org/10.1016/j.cej.2017.04.022F.

Husain, Q. (2006) 'Potential applications of the oxidoreductive enzymes in the decolorization and detoxification of textile', *Critical Reviews in Biotechnology*, 26, pp. 201–221. doi.org/10.1080/07388550600969936

Imran, M. *et al.* (2016) 'Cellulase production from species of fungi and bacteria from agricultural wastes and its utilization in industry: A review', *Advances in Enzyme Research*, 4 (02), p. 44.

Jebara, S.H. *et al.* (2015) 'Identification of effective Pb resistant bacteria isolated from *Lens culinaris* growing in lead-contaminated soils', *Journal of Basic Microbiology*, 55, pp. 346–353.

Joshi, PK. *et al.* (2011) 'Bioremediation of heavy metals in liquid media through fungi isolated from contaminated sources', *Indian Journal of Microbiology,* 51, pp. 482–487.

Kalogerakis, N., Fava, F., and Corvine, P.F. (2017) 'Bioremediation advances', *New Biotechnology*, 38, pp. 41–42. doi.org/10.1016/j.nbt.2017.07.004.

Karigar, C.S., and Rao, S.S. (2011) 'Role of microbial enzymes in the bioremediation of pollutants: A review', *Enzyme Research*, 7. doi.org/10.4061/2011/805187,805187.

Khan, A., Singh, P., and Srivastava, A. (2017) 'Synthesis, nature and utility of universal iron chelator–Siderophore: a review', *Microbiology Research*, 212–213, pp. 103–111.doi.org/10.1016/j.micres.2017.10.012

Khan, A.A., and Alzohairy, M.A. (2010) 'Recent advances and applications of immobilized enzyme technologies: A review', *Research Journal of Biological Sciences*, 5 (8), pp. 565–575.

Klassen, A. *et al.* (2017) 'Metabolomics: definitions and significance in systems biology', in Sussulini, A. (ed.) *Metabolomics: from fundamentals to clinical applications'*, Cham; Springer International Publishing, pp. 3–17.

Koch, K. *et al.* (2017) 'Structure, biochemical and kinetic properties of recombinant Pst2p from Saccharomyces cerevisiae, an FMN-dependent NAD(P)H: quinone oxidoreductase', *Biochimica et Biophsyica Acta – Protein and Proteonomics,* doi.org/ 10.1016/j.bbapap.2017.05.005.

Kotik, M., and Famerova, V. (2012) 'Sequence diversity in haloalkane dehalogenases, as revealed by PCR using family-specific primers', *Journal of Microbiological Methods*, 88, pp. 212–217.doi.org/10.1016/j.mimet.2011.11.013.

Koua, D. *et al.* (2009) 'Peroxi base: a database with new tools for peroxidase family classification', *Nucleic Acids Research,* 37, pp. D261–D266.

Koudelakova, T. *et al.* (2013) 'Haloalkane dehalogenases: biotechnological applications', *Biotechnology Journal,* 8 (1), pp. 32–45.

Kushwaha, A. *et al.* (2015) 'Heavy metal detoxification and tolerance mechanisms in plants: Implications for phytoremediation', *Environmental Review*, 24, pp. 39–51.

Layton, AC. *et al.* (2014) 'Metagenomes of microbial communities in arsenic- and pathogen-contaminated well and surface water from Bangladesh', *Genome Announcements*, 2(6): pp. e01170–e01114.

Lee, A.H. *et al.* (2016) 'Heterologous expression of a new manganese-dependent peroxidase gene from *Peniophora incarnata* KUC8836 and its ability to remove anthracene in *Saccharomyces cerevisiae'*, *Journal of Bioscience & Bioengineering,* 122, pp. 716–721. doi.org/10.1016/j.jbiosc.2016.06.006.

Legerska, B. *et al* (2016) 'Degradation of synthetic dyes by laccase- A mini review', *Nova Biotechnologica et Chimica*, p. 15. doi.org/10.1515/nbec-2016-0010.

Lei, J. *et al.* (2017) 'Hydrolysis mechanism of carbendazim hydrolase from the strain *Microbacterium sp.* djl-6F,' *Journal of Environmental Science,* 54, pp. 171–177. doi.org/10.1016/j.jes.2016.05.027.

Liang, H. *et al.* (2017) 'Multicopper laccase mimicking nanozymes with nucleotides as ligands', *ACS Applied Materials & Interfaces 9*, 2, pp. 1352–1360. doi.org/10.1021/acsami.6b15124.

Ma, X. *et al.* (2017) 'High-level expression of a bacterial laccase, CueO from *Escherichia coli* K12 in *Pichia pastoris* GS115 and its application on the decolorization of synthetic dyes', *Enzyme and Microbial Technology*, 103, pp. 34–41. doi.org/10.1016/j.enzmictec.2017.04.004.

Mai, C. *et al.* (2000) 'Enhanced stability of laccase in the presence of phenolic compounds', *Applied Microbiology and Biotechnology,* 54, pp. 510–514.

Malaviya, P., and Singh, A. (2014) 'Bioremediation of chromium solutions and chromium-containing wastewaters', *Critical Reviews in Microbiology*, 42(4), pp. 607–633.

Malik, D.S., Jain, C.K., and Yadav, A.K. (2017) 'Removal of heavy metals from emerging cellulosic low-cost adsorbents: A review', *Applied Water Science*, 7, pp. 2113–2136.

Martinez, R.J., Beazley, M.J,. And Sobecky, P.A. (2014) 'Phosphate-mediated remediation of metals and radionuclides,' *Advances in Ecology*, pp. 1–14. doi: 10.1155/2014/786929,

Medina, J.D.C. *et al.* (2017) 'Current developments in biotechnology and bioengineering production, isolation and purification of industrial products', *Peroxidases*, 217, pp. 217–232.

Mehta, A., Bodh, U., and Gupta, R. (2017) 'Fungal lipases: A review,' *Journal of Breath Research*, 8, pp. 58–77.

Mehta, J. *et al.* (2016) 'Recent advances in enzyme immobilization techniques: metal-organic frameworks as novel substrates', *Coordination Chemistry Reviews*, 322, pp. 30–40.

Mukherjee, P., and Roy, P. (2013) 'Cloning, sequencing and expression of novel trichloroethylene degradation genes from *Stenotrophomonas maltophilia* PM102: a Case of Gene Duplication', *Journal of Bioremediation & Biodegradation,* 4, p. 2. doi.org/10.4172/2155-6199.1000177.

Muthukamalam, S. *et al.* (2017) 'Characterization of dioxygenases and biosurfactants produced by crude oil degrading soil bacteria', *Brazilian Journal of Microbiology,* 48(4), pp. 637–647. doi.org/10.1016/j.bjm.2017.02.007.

Nagata, Y., Ohtsubo, Y., and Tsuda, M. (2015) 'Properties and biotechnological applications of natural and engineered haloalkane dehalogenases', *Applied Microbiology and Biotechnology*, 99(23), pp. 9865–9881.

Naik, M.M., and Dubey, S.K. (2013) 'Lead resistant bacteria: lead resistance mechanisms, their applications in lead bioremediation and biomonitoring', *Ecotoxicology & Environmental Safety*, 98, pp. 1–7.

Nigam, V.K., and Shukla, P. (2015) 'Enzyme based biosensors for detection of environmental pollutants: A review', *Journal of Microbiology and Biotechnology*, 25 (11), pp. 1773–1781.

Novotny, C. *et al.* (2004) 'Ligninolytic fungi in bioremediation: extracellular enzyme production and degradation rate', *Soil Biology and Biochemistry*, 36, pp. 1545–1551.

Ojuederie, O.B,. and Babalola, O.O. (2017) 'Microbial and plant-assisted bioremediation of heavy metal polluted environments: A Review', *International Journal of Environmental Research & Public Health,* 14(12), p. 1504. doi:10.3390/ijerph14121504. PMID: 29207531; PMCID: PMC5750922.

Olaniran, A.O., Balgobind, A., and Pillay, B. (2013) 'Bioavailability of heavy metals in soil: impact on microbial biodegradation of organic compounds and possible improvement strategies', *International Journal of Molecular Science,* 14(5), pp. 10197–10228.

Oppenheimer, M. *et al.* (2010) 'Recombinant expression, purification, and characterization of ThmD, the oxidoreductase component of tetrahydrofuran monooxygenase', *Archives of Biochemistry and Biophysics,* 496, 123–131. doi.org/10.1016/j.abb.2010.02.006.

Parales, R.E. and Ju, K.S. (2011) 'Degradation of Aromatic Hydrocarbons', in Moo-Young, M. (ed.) *Comprehensive Biotechnology*, 2nd edn. Amsterdam: Elsevier pp. 115–134.

Pavlova, M. *et al.* (2007) 'The identification of catalytic pentad in the haloalkane dehalogenase DhmA from *Mycobacterium avium* N85: reaction mechanism and molecular evolution', *Journal of Structural Biology,* 157, pp. 384–392. doi.org/10.1016/j.jsb.2006.09.004.

Philp, J.C., and Atlas, R.M. (2005) 'Bioremediation of contaminated soils and aquifers', in, Atlas, R.M. and Philp, J.C. (eds.) *Bioremediation: applied microbial solutions for real-world environmental cleanup.* Washington, DC: American Society for Microbiology (ASM) Press, pp. 139–236.

Poelarends, G.J., and Whitman, C.P. (2010) 'Mechanistic and structural studies of microbial dehalogenases: how nature cleaves a carbon halogen bond. Comprehensive natural products II, chemistry and biology', *Enzyme Mechanics*, 8, pp. 89–123.

Porro, C.S. *et al.* (2003) 'Diversity of moderately halophilic bacteria producing extracellular hydrolytic enzymes', *Journal of Applied Microbiology*, 94, pp. 295–300.

Prasad, M.P., and Manjunath, K. (2011) 'Comparative study on biodegradation of lipid-rich wastewater using lipase producing bacterial species', *Indian Journal of Biotechnology*, 10, pp. 121–124.

Pratush, A., Kumar, A., and Hu, Z. (2018) 'Adverse effect of heavy metals (As, Pb, Hg, and Cr) on health and their bioremediation strategies: a review', *International Microbiology*, 21(3), pp. 97–106. doi:10.1007/s10123-018-0012-3.

Rabus, R. *et al.* (2016) 'Anaerobic microbial degradation of hydrocarbons: from enzymatic reactions to the environment', *Journal of Molecular Microbiology and Biotechnology*, 26(1–3), pp. 5–28.

Rayu, S., Karpouzas, D.G., and Singh, B.K. (2012) 'Emerging technologies in bioremediation: constraints and opportunities', *Biodegradation*, 23(6), pp. 917–926.

Restaino, O.F. *et al.* (2016) 'Biotechnological process design for the production and purification of three recombinant thermophilic phosphotriesterases', *New Biotechnology*, 33, p. 15. doi.org/10.1016/j.nbt.2016.06.776.

Roccatano, D. (2015) 'Structure, dynamics, and function of the monooxygenase P450BM-3: insights from computer simulations studies', *Journal of Physics: Condensed Matter*, 27, 273–302. doi.org/10.1088/0953-8984/27/27/273102.

Rodriguez, E. *et al.* (2012) 'Chromium (VI) induces toxicity at different photosynthetic levels in pea', *Plant Physiology & Biochemistry*, 53, pp. 94–100.

Romeh, A.A., and Hendawi, M.Y. (2014) 'Bioremediation of certain organophosphorus pesticides by two biofertilizers, *Paenibacillus (Bacillus) polymyxa (Prazmowski)* and *Azospirillum lipoferum (Beijerinck)*. *Journal of Agricultural Science and Technology*, 16 (2), pp. 265–276.

Sales, C.M. *et al.* (2013) 'Oxidation of the cyclic ethers 1, 4-dioxane and tetrahydrofuran by a monooxygenase in two *Pseudonocardia* Species', *Applied Environmental Microbiology*, 79, pp. 247702–247708.

Santillan, J.Y., Dettorre, L.A., Lewkowicz, E.S., and Iribarren, A.M. (2016) 'New and highly active microbial phosphotriesterase sources', *FEMS Microbiology Letters*, 363(24). doi: 10.1093/femsle/fnw276.

Sarubbo L.A., *et al* (2015) 'Some aspects of heavy metals contamination remediation and role of biosurfactants', *Chemistry and Ecology*, 31(8), pp. 707–723. doi: 10.1080/02757540.2015.1095293.

Satyapal, G.K. *et al.* (2016) 'Potential role of arsenic resistant bacteria in bioremediation: Current status and future prospects', *Journal of Microbial & Biochemical Technology*, 8(3), pp. 256–258.

Scott, C. *et al.* (2008) 'The enzymatic basis for pesticide bioremediation', *Indian Journal of Microbiology*, 48(1), p. 65.

Shaheen, R. *et al.* (2017) 'Immobilized lignin peroxidase from *Ganoderma lucidum* IBL-05 with improved dye decolorization and cytotoxicity reduction properties', *International Journal of Biological Macromolecules*, 103, pp. 57–64. doi.org/10.1016/j.ijbiomac.2017.04.040.

Shakeri, M., and Shoda, M. (2010) 'Efficient decolorization of an anthraquinone dye by recombinant dye-decolorizing peroxidase (rDyP) immobilized in silica-based mesocellular foam', *Journal of Molecular Catalysis B: Enzymatic*, 62, pp. 277–281. doi.org/ 10.1016/j.molcatb.2009.11.007.

Sharma, B., Dangi, A.K., and Shukla, P. (2018) 'Contemporary enzyme based technologies for bio-remediation: A review', *Journal of Environmental Management*, 15(210), pp. 10–22. doi: 10.1016/j.jenvman.2017.12.075.

Shraddha, R.S. *et al.* (2011) 'Laccase: microbial sources, production, purification and potential biotechno-logical applications', *Enzyme Research*, 2011. doi: 10.4061/2011/217861.

Shukla, P., and Gupta, K. (2007) 'Ecological screening for lipolytic molds and process optimization for lipase production from *Rhizopus oryzae* KG-5. *Journal of Applied Sciences in Environmental Sanitation*, 2, pp. 35–42.

Singh, B. (2014) 'Review on microbial carboxylesterase: general properties and role in organophosphate pesticides degradation', *Biochemistry and Molecular Biology*, 2, pp. 1–6.

Singh, J.S., and Singh, D.P. (2017) 'Methanotrophs: an emerging bioremediation tool with unique broad spec-trum methanotrophs', in Sing, J.S and Seneviratne, G. (eds.) *Agro-Environmental Sustainability*: Vol II', New York: Springer. doi.org/10.1007/978-3-319-49727-3_1.

Singh, R.K. *et al.* (2013) 'From protein engineering to immobilization: promising strategies for the upgrade of industrial enzymes', *International Journal of Molecular Science*, 14 (1), pp. 1232–1277.

Sinha, R.K. *et al.* (2009) 'Bioremediation of contaminated sites: a low-cost nature's biotechnology for environmental cleanup by versatile microbes, plants & earthworms', *Solid Waste Management and Environmental Remediation*, New York: Nova Science Publishers Inc. pp. 1–72

Sirisha, V.L., Jain, A., and Jain, A. (2016) 'Enzyme immobilization: an overview on methods, support material, and applications of immobilized enzymes', *Advances in Food Nutrition and Research*, 79, pp. 179–211. doi.org/10.1016/bs.afnr.2016.07.004.

Sivaperumal, P., Kamala, K., and Rajaram, R. (2017) 'Bioremediation of industrial waste through enzyme producing marine microorganisms', *Advances in Food Nutrition and Research*, 80, pp. 165–179. doi.org/10.1016/bs.afnr.2016.10.006.

Skoronski, E., et al. (2017) 'Immobilization of laccase from *Aspergillus oryzae* on graphene nanosheets', *International Journal of Biological Macromolecules*, 99, pp. 121–127. doi: 10.1016/j.ijbiomac.2017.02.076.

Smith, E. *et al.* (2015) 'Remediation trials for hydrocarbon-contaminated soils in arid environments: evaluation of bio slurry and biopiling techniques', *International Biodeterioration & Biodegradation*, 101, pp. 56–65. doi:10.1016/j.ibiod.2015.03.029

Strong, P.J., and Claus, H. (2011) 'Laccase: a review of its past and its future in bioremediation', *Critical Reviews in Environmental Science and Technology*, 41, pp. 373–434. doi.org/10.1080/10643380902945706.

Su, F.H. *et al.* (2017) 'Decorating outer membrane vesicles with organophosphorus hydrolase and cellulose binding domain for organophosphate pesticide degradation', *Chemical Engineering Journal*, 308, pp. 1–7.

Sugano, Y. *et al.* (2000) 'Efficient heterologous expression in *Aspergillus oryzae* of a unique dye-decolorizing peroxidase, dyp, of *Geotrichum candidum* dec 1', *Applied and Environmental Microbiology*, 66, pp. 1754–1758.

Sun, J. *et al.* (2017) 'Heterologous production of a temperature and pH-stable laccase from *Bacillus vallismortis* fmb-103 in *Escherichia coli* and its application. *Process Biochemistry*, 55, pp. 77–84. doi.org/10.1016/j.procbio.2017.01.030.

Suresh, P.S. *et al.* (2008) 'An insilco approach to bioremediation: laccase as a case study', *Journal of Molecular Graphics and Modelling*, 26, pp. 845–849. doi.org/10.1016/j.jmgm.2007.05.005.

Tak, H.I., Ahmad, F. and Babalola, O.O. (2013) 'Advances in the application of plant growth-promoting rhizobacteria in phytoremediation of heavy metals', in Whitacre, D.M. (ed.), *Reviews of Environmental Contamination and Toxicology*, New York: Springer:, pp. 33–52.

Tang, W. *et al.* (2014) 'Heavy metal contamination in the surface sediments of representative limnetic ecosystems in eastern China', *Science Reports*, 4, p. 7152.

Thatoi, H. *et al.* (2014) 'Bacterial chromate reductase, a potential enzyme for bioremediation of hexavalent chromium: A review', *Journal of Environmental Management*, 146, pp. 383–399.

Tonin, F. *et al.* (2016) 'Different recombinant forms of polyphenol oxidase A, a laccase from *Marinomonas mediterranea*', *Protein Expression and Purification*, 123, pp. 60–69. doi.org/10.1016/j.pep.2016.03.011.

Tyc, O. *et al.* (2017) 'The ecological roleof volatile and soluble secondary metabolites produced by soil bacteria', *Trends in Microbiology*, 25, pp. 280–292.

Uqab, B., Mudasir, S., and Nazir, R. (2016) 'Review on bioremediation of pesticides', *Journal of Bioremediation & Biodegradation*, 7(3), pp. 1–5.

Vasileva-Tonkova, E., and Galabova, D. (2003) 'Hydrolytic enzymes and surfactants of bacterial isolates from lubricant-contaminated wastewater', *Zeitschrift fur Naturforschung C: A Journal of Biosciences*, 58(1–2), pp. 87–92.

Verma, S. *et al.* (2012) 'Medium optimization for a novel crude-oil degrading lipase from *Pseudomonas aeruginosa* SL-72using statistical approaches for bioremediation of crude-oil', *Biocatalysis and Agricultural Biotechnology*, 1, pp. 321–329.

Wang, X., Hu, Y., and Wei, H. (2016) 'Nanozymes in bio nanotechnology: from sensing to therapeutics and beyond', *Inorganic Chemistry Frontiers*, 3, pp. 41–60. doi.org/10.1039/C5QI00240K.

Xie, J. *et al.* (2012) 'Analytical and environmental applications of nanoparticles as enzyme mimetics', *Trends in Analytical Chemistry*, 39, pp. 114–129.

Xu, P. *et al.* (2014) 'Heavy metal-induced glutathione accumulation and its role in heavy metal detoxification in *Phanerochaete chrysosporium*. *Applied Microbiology and Biotechnology*, 98 (14), pp. 6409–6418.

Xu, R. *et al.* (2017) 'A new method for extraction and heavy metals removal of abalone visceral polysaccharide', *Journal of Food Processing and Preservation,* 41, p. 13023.

Yamaguchi, H. *et al.* (2016) 'Phosphotriesterase activity in marine bacteria of the genera *Phaeobacter, Ruegeria,* and *Thalassospira,*' *International Biodeterioration and Biodegradation,* 115, pp. 186–191. doi.org/10.1016/j.ibiod.2016.08.019.

Yang, C. *et al.* (2016) 'Discovery of new cellulases from the metagenome by a metagenomics-guided strategy', *Biotechnology for Biofuels,* 9 (1), p. 138.

Yim, S.K. *et al.* (2006) 'Functional expression of mammalian NADPH ecytochrome P450 oxidoreductase on the cell surface of *Escherichia coli*', *Protein Expression and Purification,* 49, pp. 292–298.

Yin, K. *et al.* (2016) 'Simultaneous bioremediation and bio detection of mercury ion through surface display of carboxylesterase E2 from *Pseudomonas aeruginosa* PA1', *Water Research,* 103, pp. 383–390. doi.org/10.1016/j.watres.2016.07.053.

Zhang, R. *et al.* (2015) 'Source of lead pollution, its influence on public health and the countermeasures', *International Journal of Health, Animal Science and Food Safety,* 2, pp. 18–31.

Zhanga, H. *et al.* (2016) 'Characterization of a manganese peroxidase from white-rot fungus *Trametes sp.* 48424 with strong ability of degrading different types of dyes and polycyclic aromatic hydrocarbons', *Journal of Hazardous Materials,* 320, pp. 265–277. doi.org/10.1016/j.jhazmat.2016.07.065.

Zhu, Y. *et al.* (2017) 'Genomic and transcriptomic insights into calcium carbonate biomineralization by marine *Actinobacterium Brevibacterium linens* BS258', *Frontiers in Microbiology,* 8, p. 602.

Index

For Product Safety Concerns and Information please contact our EU
representative GPSR@taylorandfrancis.com
Taylor & Francis Verlag GmbH, Kaufingerstraße 24, 80331 München, Germany